Diagnosis of Heritage Buildings by Non-Destructive Techniques

Woodhead Publishing Series in Civil and Structural Engineering

Diagnosis of Heritage Buildings by Non-Destructive Techniques

Edited by

Blanca Tejedor Herrán

Dept. of Project and Construction Engineering Polytechnic University of Catalonia Barcelona Spain

David Bienvenido-Huertas

Dept. of Building Construction University of Granada Granada Spain

WOODHEAD PUBLISHING

ELSEVIER An imprint of Elsevier

ISBN: 978-0-443-16001-1 (print)
ISBN: 978-0-443-16002-8 (online)

For information on all Woodhead Publishing publications
visit our website at https://www.elsevier.com/books-and-journals

Publisher: Matthew Deans
Acquisitions Editor: Chiara Giglio
Editorial Project Manager: Emily Thomson
Production Project Manager: Erragounta Saibabu Rao
Cover Designer: Mark Rogers

Typeset by MPS Limited, Chennai, India

Working together
to grow libraries in
developing countries

www.elsevier.com • www.bookaid.org

Contents

Section III   Laser Scanning and Photogrammetry

Section IV   Heritage Building Information Modeling (BIM) and Digital Twins (DTs)

Section V   Other Techniques for Heritage Building Diagnosis

List of contributors

Kalare Agrasar-Santiso TICBE Research Group, Department of Architecture, Faculty of Engineering of Gipuzkoa, University of the Basque Country UPV/EHU, Donostia-San Sebastián, Spain

Joaquín Aguilar-Camacho Department of Graphic Engineering, University of Seville, Seville, Spain

Amin Al-Habaibeh Product Innovation Centre, School of Architecture, Design and the Built Environment, Nottingham Trent University, Nottingham, United Kingdom

José-Lázaro Amaro-Mellado Departamento de Ingeniería Gráfica, Universidad de Sevilla, Seville, Spain

D. Ambrosini Department of Industrial and Information Engineering and Economics, University of L'Aquila, L'Aquila, Italy

Daniel Antón Departamento de Expresión Gráfica e Ingeniería en la Edificación, Escuela Técnica Superior de Ingeniería de Edificación, Seville, Spain; The Creative and Virtual Technologies Research Lab & Product Innovation Centre, School of Architecture, Design and the Built Environment, Nottingham Trent University, Nottingham, United Kingdom

F. Asdrubali Department of International Human and Social Sciences, Perugia Foreigners' University, Perugia, Italy

María José Ayora-Cañada Department of Physical and Analytical Chemistry, Universidad de Jaén, Jaén, Spain

Abderrahmane Baïri Laboratoire Thermique Interfaces Environnement (LTIE), Université de Paris, EA 4415, 50 Rue de Sèvres, F-92410 Ville d'Avray, France

Jesús Balado CINTECX, University of Vigo, Vigo, Spain

David Bienvenido-Huertas Department of Building Construction, University of Granada, Spain

Elena Cabrera-Revuelta Department of Mechanical Engineering and Industrial Design, University of Cadiz, Cadiz, Spain

Sara Calandra Department of Earth Sciences, University of Florence, Florence, Italy; Department of Chemistry, University of Florence, Sesto Fiorentino, Italy

Vieri Cardinali Department of Architecture, University of Florence, Florence, Italy

Amaia Casado-Rezola TICBE Research Group, Department of Architecture, School of Architecture, University of the Basque Country UPV/EHU, Donostia-San Sebastián, Spain

Irene Centauro Department of Earth Sciences, University of Florence, Florence, Italy

Anna Livia Ciuffreda Department of Earth Sciences, University of Florence, Florence, Italy

Alessio Cordisco Institute for Construction Technologies - Italian National Research Council, L'Aquila, Italy

Lloyd A. Courtenay CNRS, PACEA UMR 5199, Université de Bordeaux, Bât B2, Allée Geoffroy Saint Hilaire, CS50023, 7 Pessac, France

R. De Lieto Vollaro Department of Industrial, Electronic and Mechanical Engineering, Roma Tre University, Rome, Italy

T. de Rubeis Department of Civil, Construction-Architectural and Environmental Engineering, University of L'Aquila, L'Aquila, Italy

Laurent Debailleux Architectural and Urban Engineering Unit, Faculty of Engineering, University of Mons, Mons, Belgium

Susana del Pozo Department of Cartographic and Land Engineering, Universidad de Salamanca, Ávila, Spain

Pablo Díaz-Cañete Departamento de Expresión Gráfica e Ingeniería en la Edificación, Escuela Técnica Superior de Ingeniería de Edificación, Seville, Spain

Ana Domínguez-Vidal Department of Physical and Analytical Chemistry, Universidad de Jaén, Jaén, Spain

Tessa Donigaglia Department of Earth Sciences, University of Florence, Florence, Italy

L. Evangelisti Department of Industrial, Electronic and Mechanical Engineering, Roma Tre University, Rome, Italy

Giovanni Fabbrocino Institute for Construction Technologies - Italian National Research Council, L'Aquila, Italy; Department of Biosciences and Territory, StreGa Lab, University of Molise, Campobasso, Italy

Morgane Palma Fanfone Architectural and Urban Engineering Unit, Faculty of Engineering, University of Mons, Mons, Belgium

María Fernández-Alconchel Department of Graphical Expression and Building Engineering, ETSIE, University of Seville, Seville, Spain

Simona Fontul Transportation Department, LNEC - National Laboratory for Civil Engineering, Lisbon, Portugal

Iván Garrido Defense University Center, Spanish Naval Academy, Marín, Spain

Russell Gentry Georgia Institute of Technology, Atlanta, GA, United States

Di Benedetto Giacomo Enginlife Engineering Solutions, Rome, Italy

Marco Giallonardo Institute for Construction Technologies - Italian National Research Council, L'Aquila, Italy

Diego González-Aguilera Department of Cartographic and Terrain Engineering, Universidad de Salamanca, Ávila, Spain; Department of Cartographic and Land Engineering, Universidad de Salamanca, Ávila, Spain

Enrique González-González Department of Cartographic and Terrain Engineering, Universidad de Salamanca, Ávila, Spain

C. Guattari Department of Philosophy, Communication and Performing Arts, Roma TRE University, Rome, Italy

David Hernández-López Institute for Regional Development (IDR), University of Castilla La Mancha, Albacete, Spain

Javier Irizarry Georgia Institute of Technology, Atlanta, GA, United States

Castro-Gomes João Civil Engineering and Architecture Department, University of Beira Interior, Covilha, Portugal

Durán-Suárez Jorge A. Sculpture Department, University of Granada, Granada, Spain

Rodríguez-Gordillo José Mineralogy and Petrology Department, University of Granada, Granada, Spain

A.K. Kasthurba National Institute of Technology, Calicut, Kerala, India

Manogna Kavuru Architrave p.c., Washington, DC, United States

Susana Lagüela Department of Cartographic and Terrain Engineering, Universidad de Salamanca, Ávila, Spain

Iñigo Leon TICBE Research Group, Department of Architecture, Faculty of Engineering of Gipuzkoa, University of the Basque Country UPV/EHU, Donostia-San Sebastián, Spain

Botao Li Georgia Institute of Technology, Atlanta, GA, United States

Junshan Liu Georgia Institute of Technology, Atlanta, GA, United States; Auburn University, Auburn, AL, United States

Jorge López-Rebollo Department of Cartographic and Terrain Engineering, Universidad de Salamanca, Ávila, Spain

Elena Lucchi Department of Architecture, Built Environment and ConstructionEngineering (DABC), Politecnico di Milano, Milan, Italy

Sáez-Pérez Maria Paz Building Constructions Department, University of Granada, Granada, Spain

David Marín-García Department of Graphical Expression and Building Engineering, ETSIE, University of Seville, Seville, Spain

Adriana Marra Institute for Construction Technologies - Italian National Research Council, L'Aquila, Italy

Alexander Martín-Garín TICBE Research Group, Department of Architecture, Faculty of Engineering of Gipuzkoa, University of the Basque Country UPV/EHU, Donostia-San Sebastián, Spain

Miguel Ángel Maté-González Department of Cartographic and Terrain Engineering, Universidad de Salamanca, Ávila, Spain

Alberto Meiss GIR Arquitectura & Energía, Dpto. Construcciones Arquitectónicas, Ingeniería del Terreno y Mecánica de los Medios Continuos y Teoría de Estructuras, Universidad de Valladolid, Valladolid, Spain

Jose Antonio Millan-Garcia ENEDI Research Group, Department of Energy Engineering, Faculty of Engineering of Gipuzkoa, University of the Basque Country UPV/EHU, San Sebastián, Spain

Alberto Morcillo Department of Cartographic and Land Engineering, Universidad de Salamanca, Ávila, Spain

Juan Moyano Department of Graphical Expression and Building Engineering, ETSIE, University of Seville, Seville, Spain

Iole Nardi ENEA Casaccia Research Center, I 00123 S.M. di Galeria, Rome, Italy

Alejandra Ospina-Bohórquez Department of Cartographic and Terrain Engineering, Universidad de Salamanca, Ávila, Spain

Juan Pedro Otaduy-Zubizarreta TICBE Research Group, Department of Architecture, Faculty of Engineering of Gipuzkoa, University of the Basque Country UPV/EHU, Donostia-San Sebastián, Spain

Miguel Ángel Padilla-Marcos GIR Arquitectura & Energía, Dpto. Construcciones Arquitectónicas, Ingeniería del Terreno y Mecánica de los Medios Continuos y Teoría de Estructuras, Universidad de Valladolid, Valladolid, Spain

D. Paoletti Department of Industrial and Information Engineering and Economics, University of L'Aquila, L'Aquila, Italy

Erica Isabella Parisi Department of Environmental and Civil Engineering di Ingegneria Civile e Ambientale, State Univeristy of Florence (IT), Italy

G. Pasqualoni Department of Industrial and Information Engineering and Economics, University of L'Aquila, L'Aquila, Italy

Mahesh Patil Sardar Vallabhbhai National Institute of Technology, Surat, Gujarat, India

Supriya Patil School of Architecture, D.Y. Patil College of Engineering and Technology, Kolhapur, Maharashtra, India

Vega Pérez-Gracia Dpt Resistencia de Materials i Estructures a l'Enginyeria, RMEE, Universitat Politècnica de Catalunya UPC-Barcelona Tech, Barcelona, Spain

José Javier Pérez-Martínez TICBE Research Group, Department of Architecture, School of Architecture, University of the Basque Country UPV/EHU, Donostia-San Sebastián, Spain

Javier Pisonero Department of Cartographic and Terrain Engineering, Universidad de Salamanca, Ávila, Spain

Irene Poza-Casado GIR Arquitectura & Energía, Dpto. Construcciones Arquitectónicas, Ingeniería del Terreno y Mecánica de los Medios Continuos y Teoría de Estructuras, Universidad de Valladolid, Valladolid, Spain

Tarek Rakha Georgia Institute of Technology, Atlanta, GA, United States

Fernando Rico-Delgado Departamento de Expresión Gráfica e Ingeniería en la Edificación, Escuela Técnica Superior de Ingeniería de Edificación, Seville, Spain

Pablo Rodríguez-Gonzálvez Department of Mining Technology, Topography and Structures, Universidad de León, Av. Astorga s/n, Ponferrada, Spain

Elisabetta Rosina Department of Architecture, Built Environment and Construction Engineering, Polytechnic of Milan, Milan, Italy

Teresa Salvatici Department of Earth Sciences, University of Florence, Florence, Italy

Luis Javier Sánchez-Aparicio Construction and Building Technology, Universidad Politécnica de Madrid, Madrid, Spain

Francesca Savini Institute for Construction Technologies - Italian National Research Council, L'Aquila, Italy

María Senderos Laka TICBE Research Group, Department of Architecture, Faculty of Engineering of Gipuzkoa, University of the Basque Country UPV/EHU, Donostia-San Sebastián, Spain

Mario Soilán CINTECX, GeoTECH Research Group, Universidade de Vigo, Vigo, Spain

Mercedes Solla CINTECX, GeoTECH Research Group, Universidade de Vigo, Vigo, Spain

Diego Tamayo-Alonso GIR Arquitectura & Energía, Dpto. Construcciones Arquitectónicas, Ingeniería del Terreno y Mecánica de los Medios Continuos y Teoría de Estructuras, Universidad de Valladolid, Valladolid, Spain

Marco Tanganelli Department of Architecture, University of Florence, Florence, Italy

Blanca Tejedor Herrán Universitat Politècnica de Catalunya (UPC), Department of Project and Construction Engineering, Group of Construction Research and Innovation (GRIC), 08222 Terrassa (Barcelona), Spain

Marta Torres Gonzalez Department of Architectural Constructions II, University of Seville, Seville, Spain

Ilaria Trizio Institute for Construction Technologies - Italian National Research Council, L'Aquila, Italy

Danielle S. Willkens Georgia Institute of Technology, Atlanta, GA, United States

Section I

Basic Foundations of Non-Destructive Testing (NDT) Techniques

A comprehensive overview of NDT: From theoretical principles to implementation

Blanca Tejedor Herrán[1], David Bienvenido-Huertas[2], Elena Lucchi[3] and Iole Nardi[4]

[1]Universitat Politècnica de Catalunya (UPC), Department of Project and Construction Engineering, Group of Construction Research and Innovation (GRIC), 08222 Terrassa (Barcelona), Spain, [2]Department of Building Construction, University of Granada, Spain, [3]Department of Architecture, Built Environment and ConstructionEngineering (DABC), Politecnico di Milano, Milan, Italy, [4]ENEA Casaccia Research Center, I 00123 S.M. di Galeria, Rome, Italy

1.1 Introduction

The conservation of patrimonial assets has become an important concern for the AEC (architecture, engineering, and construction) sector, especially with the reduction of new building workmanships. Cultural heritage (CH) buildings are characterized by being traditional constructions that followed specific 50- to 70-year-old techniques and used local raw materials, and apparently without artistic significance. Within EU context, CH construction represents to be 35% of the existing building stock (Caro & Sendra, 2021; Eurostat CensusHub HC53, 2021), and most of them are located on Southern Europe regions (i.e., Italy, Spain, France, Portugal, and Greece) (Tejedor et al., 2022). Nevertheless, there is not a systematic and integrated approach for the establishment of preventive practices based on regular inspections with non-destructive testing (NDT) techniques (Tejedor et al., 2022). NDT technique allows acquiring geometric data and architectural information through the on-site monitoring, for the subsequent characterization of materials, detection of pathologies, or structure modeling that contribute to the decision-making on the protection of CH assets. It should be highlighted that refurbishment strategies and adaptive reuse of heritage buildings could help to minimize the energy consumption (Angrisano et al., 2021) and accomplish with the decarbonization targets (Caro & Sendra, 2021). Along this line, several funded EU projects have been proposed in recent years. By way of example, some of them are Open Heritage, STORM, CLIC, ROCK, RIBUILD, HERICOAST, HERITAGECARE, and so on (European Union, 2020; Ramos et al., 2018; Tejedor et al., 2022).

Diagnosis of Heritage Buildings by Non-Destructive Techniques. DOI: https://doi.org/10.1016/B978-0-443-16001-1.00001-2

1.2   Foundations of non-destructive testing techniques

1.2.1   Qualitative and quantitative infrared thermography

Infrared thermography (IRT) can be defined as a noninvasive technique that measures remotely the radiant thermal energy distribution that is emitted from an element's surface. Subsequently, the thermal image of the entire element is provided as a matrix of temperature values (Tejedor et al., 2017). Usually, IRT has been applied to detect pathologies in building elements, following international regulations and guidelines (i.e., RESNET (RESNET, 2012)). The evolution of this NDT technique over the years is shown in Fig. 1.1.

The qualitative approach involves the identification of thermal irregularities (cold and hot spots) compared to the surroundings (Fox et al., 2016; Kylili et al., 2014). In the case of heritage buildings, IRT is coupled with other NDT technique to complement the diagnosis or to facilitate the installation of sensors. The anomalies detected in the building structure can be moisture regions, thermal bridges, air or water leakages, detachments, cracks, lack of insulation, and so on (Barbosa et al., 2021; Bisegna et al., 2014; Kavuru & Rosina, 2021; Kordatos et al., 2013; Lerma et al., 2014; Odgaard et al., 2017; Paoletti et al., 2013; Valluzzi et al., 2019).

The quantitative approach is focused on the characterization of thermal properties of the building element, such as the thermal transmittance under the hypothesis of steady-state heat transfer conditions. In fact, it is considered an alternative technique to the heat-flux meter (HFM) method (Tejedor et al., 2019). Normally, the equipment consists of an IRT camera, a reflector, and a blackbody. The IRT camera, connected to a computer, allows monitoring the building element over a test period that can vary in function of the sample (2−3 hours for heterogeneous specimens and 30 minutes for homogenous specimens). For the environmental

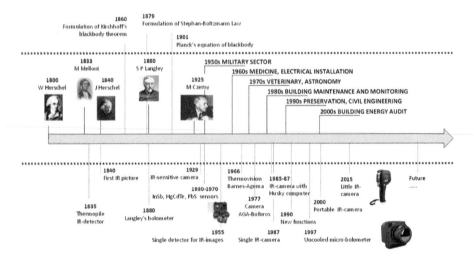

Figure 1.1 History of the IRT measurement. *IRT*, Infrared thermography.

parameters, two thermocouples with a data logger are needed to determine the temperatures inside and outside the building. During the data postprocessing, the numerical model of the U-value is computed, taking the instantaneous readings as inputs of the algorithm. The accuracy ranges between 1% and 3% respect to the theoretical value (Tejedor et al., 2017, 2019). Most of the studies in this topic were developed in residential buildings, following the initial recommendations established by Fokaides and Kalogirou (2011), Albatici and Tonelli (2010), Albatici et al. (2015), and Nardi et al. (2015). Indeed, Tejedor et al. (2022) highlighted that there is a lack of researches on CH buildings and a protocol for diagnosis is required. Grinzato (2010) analyzed the decay of heritage building surfaces covered by frescoes. Tavukçuoğlu (2018) decided to use two NDT techniques, ultrasonic pulse velocity and quantitative infrared thermography (QIRT), for the assessment of historical structures with pathologies. Tejedor et al. (2020, 2021) proposed a 2D U-value map to obtain the thermal transmittance in any point of the wall area with partial defects or nonuniform thermal pattern, replicating the most representative European construction technologies. Despite implementing the method in a laboratory with a climatic chamber, the proposal was interesting in terms of quantifying the energy expenditure without destructive surveys, especially for masonries of brick and stone that can be found in CH buildings.

1.2.2 Photogrammetry and laser scanning

Usually, the technical documentation of heritage buildings is not available and consequently, the current state of the construction is unknown (Moyano et al., 2020, 2021). To solve this problem, the technicians may represent virtually complex architectural restorations by means of laser scanning (TLS) or structure from motion (SfM) (Gómez-Zurdo et al., 2021; Remondino et al., 2014). Along this line, and according to Tejedor et al. (2022), SfM and TLS could be adopted as consolidated techniques for CH preservation and retrofit strategies. The main differences between the aforementioned methods are shown in Fig. 1.2.

Photogrammetry is defined as a science whose aim is to obtain geometrical details of a structure from 2D or 3D data, taking automatically images from different positions that are called "control points" by means of unmanned aerial vehicles or cameras located on the ground (Gómez-Zurdo et al., 2021; Moyano et al., 2020; Tumeliene et al., 2017). The few SfM studies were focused on evaluating wall deformation from outside the building with an accuracy of ± 2 mm in countries such as Peru (Pierdicca et al., 2016), Turkey (Erenoglu et al., 2017), or Spain (Antón et al., 2018). Only one case was found to be executed inside a building, and this was developed to generate 3D models of paintings in Lithuania (Tumeliene et al., 2017).

Regarding laser scanning, the 3D mesh is created by the control point measurements of the elements located between 100 and 300 m from the scanner. This implies that the technical characteristics of the equipment and the sample can influence to the precision of the final textured model (i.e., angle of incidence, environmental conditions, and surface object properties). In fact, Remondino et al. (2014)

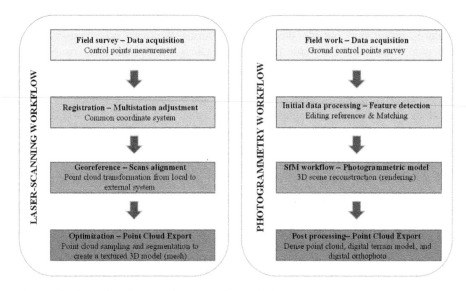

Figure 1.2 Comparison between laser scanning and photogrammetry.

reviewed studies for the analysis of structures from inside and outside the building with SfM. However, this technology for CH buildings is still ongoing. Antón et al. (2018) and Moyano et al. (2020) implemented TLS in Spain, such as obtaining an indoor 3D model that may be used with building information modelling (BIM) tools for archaeology applications. Ramos et al. (2015) and Balado et al. (2021) evaluated the outdoor structures of Southern European countries with automatic modeling and accuracies that ranged between 2.6 and 10.6 mm.

It should be noted that some authors combined TLS and SfM to assess walls from outside with the purpose of carrying out a comparative study between these techniques (Bolognesi et al., 2015; Grussenmeyer et al., 2008; Moyano et al., 2021), and to create 3D models of structures from both sides of the building (Brumana et al., 2014). Most of them extrapolated that laser scanning tends to present greater accuracy than photogrammetry and a low equipment cost. Nevertheless, TLS could not be suitable for short-range elements where the pictorial information is important.

1.2.3   Heat-flux meter

HFM is the most common procedure to quantitatively evaluate the thermal transmittance (also called U-value) of opaque building components, such as facades or ceilings, when these ones have a thermal gradient between 10°C and 15°C (Gaspar et al., 2018). The measuring apparatus is composed by a data logger for collecting and storing the data, two or more thermocouples for temperature measurements, and one heat-flux plate for heat flow measurements (Fig. 1.3). Its history starts in the 14th century with the development of the thermocouples for measuring air

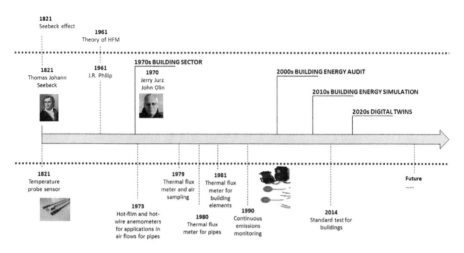

Figure 1.3 HFM apparatus during an on-site monitoring. *HFM*, Heat-flux meter.

temperature (T_a). Otherwise, the production of the HFM apparatus started from 1970s in United States with the development of the heat flow plate (Fig. 1.4).

This method allows to minimize the inaccuracy due to the definition of the stratigraphy, the presence of damage, and the amount of moisture content that characterizes other QIRT and analytical calculations (Lucchi et al., 2019; Lucchi, 2017). The procedure for the HFM measurement is described by the international standard ISO 9869-1 (International Organization for Standardization, 2014) that details the HFM device as well as the monitoring period, installation, data processing, and calibration procedures. For this procedure, the standard monitoring period must be at least consecutive 72 hours according to component features and T_a variations. Otherwise, tests should reach the thermal stability. Thus the monitoring period for thick walls must be higher than 72 hours with integer multiple of 24 hours, to contemplate thermal mass effects (Lucchi, 2017). This aspect is particularly important for historic walls that usually have a thickness that ranges from 0.40 to 1.20 m.

It should be pointed out that HFM measurement is affected by several parameters, such as (1) apparatus location, (2) homogeneity of the building element, (3) heat-flux perturbations due to the device, and (4) boundary conditions. HFM location requires uniform boundary conditions to collect reliable data. Thus, actions like protection from weather perturbation and absence of indoor (e.g., thermal and electrical devices) or outdoor (e.g., sun, wind, snow) sources should be guaranteed in Northern Façades (Baker, 2011; Ficco et al., 2015; Lucchi, 2016, 2018). To this purpose, the qualitative IRT supports the proper installation of the sensors visualizing the presence of thermal anomalies related to heat-flux transmission or building heterogeneities (Lucchi et al., 2018). Also, the influence of vertical temperature stratification, both T_a and contact temperature, should be minimized inserting the sensors about half-way between thermal breaks (e.g., floors-ceiling, windows-corner, windows-doors, and corners)

Figure 1.4 History of the HFM measurement. *HFM*, Heat-flux meter.

(Baker, 2011). Finally, an optimum thermal contact for all the HFM sensors is required. To this purpose, the use of glues, pastes, or adhesive tapes with high thermal resistance is suggested for fixing the probes without improving the thermal conduction between the building component and the HFM apparatus (Gaspar et al., 2021).

HFM is appropriate for heritage buildings thanks to its NDT approach. Despite this, the fixing of the probes requires a low intrusion in historic masonries. In valuable heritage components, such as frescoes, wall paintings, stuccos, or original plasters, this test is forbidden due to the presence of outstanding heritage values (and constrains). In heritage buildings, HFM is commonly used for creating shared databases of thermal performances of historic components made by bricks (Evangelisti et al., 2015, 2020), stones (Lucchi, 2017), or mixed materials (Baker, 2011; Williamson et al., 2014). These applications demonstrated that historic masonries have better measured thermal performances of a 20%−30% than calculated thanks to the high quality of old materials, construction systems, and application techniques. Besides, the impact of external environments, data refinement, and calibration are analyzed (Cortellessa & Iacomini, 2019; Gumbarević et al., 2020, 2021).

1.2.4 Airtightness measurements

Major sources of air infiltration in heritage buildings are related to wall joints, windows (e.g., frames, and rolling shutters), cracks, thermal and electrical ducts, lighting, chimneys, and other cavities (Martín-Garín et al., 2020). Their rate depends on several building characteristics (i.e., age, shape, dimension, typology, use, conservation level, construction quality, materials, and technologies used), environmental conditions (i.e., wind speed and direction, difference indoor/outdoor air temperature), and surroundings (i.e., presence of shadows or air blockers) (Dimitroulopoulou, 2012; Salehi et al., 2017). Airtightness and ventilation rates have a direct impact on energy efficiency and human comfort, influencing positively in the energy performance of a building but negatively with the presence of allergic diseases. Their measurements provide the air permeability value of the building, and the estimate of natural ventilation performances, as well as indoor air changes, and construction quality for planning correct retrofits following ISO 9972:2015 (International Organization for Standardization, 2015). It should be highlighted that airtightness measurements are divided into qualitative and quantitative. Qualitative approach refers to the visualization of airflows with qualitative IRT thanks to the increasing of the pressure difference (ΔP) between indoor and outdoor environments (Ferdyn-Grygierek et al., 2019). Qualitative tests are difficult to achieve for the variability of the air flow and the thermal buoyancy generated by different air densities, as seen in Fig. 1.5.

Quantitative measurements are classified into instantaneous and continuous. In the first case, anemometers and barometers are used, respectively, to measure the air velocity and the air flow in rooms and openings (Ferdyn-Grygierek et al., 2019). In the second case, tracer gas and fan pressurization methods directly estimate the air flow rates of the building. Tracer gas allows estimating the air infiltration rates through direct and reiterating assessments of gas concentration provided by constant injection, constant concentration, or concentration decay of the gas inside the building (Almeida et al., 2020; Roque et al., 2020). The first two methods are not commonly applied to heritage building for the cost due to long time and expensive equipment. Also, they can be applied only when building components (i.e., windows or ducts) can be closed for long time (Pasos et al., 2020). Concentration decay tests are the most practiced on heritage buildings thanks to its easiness, low costs, and times. In this case, the gas tracer is introduced inside the building and uniformly distributed with a fan. Then, the gas is naturally diluted with outdoor air infiltrations. The air change rate (ACR) value is calculated by a mathematical regression according to the concentration decay monitored by calibrated sensors (Pasos et al., 2020). This test is difficult to achieve in naturally ventilated rooms, due to the difficulties in obtaining a perfect air mix (Van Buggenhout et al., 2009) and controlling user behaviors (e.g., window or door openings) (Lee et al., 2016), the impact of boundary conditions (Hamid et al., 2020), and huge building dimensions on the accuracy of the results (Abolfazl, 2017). To overcome these limitations, the fan pressurization method, also known also as blower door test (BDT), measures the air flow rates across the building envelope considering a range of

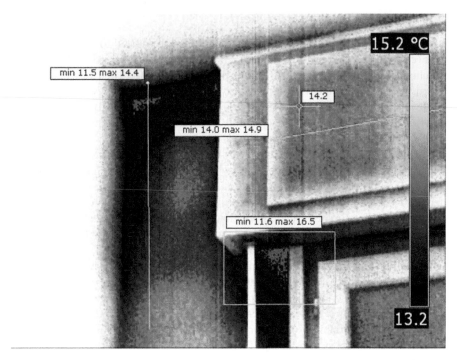

Figure 1.5 Visualization of airflows with qualitative IRT. *IRT*, Infrared thermography.

induced ΔP between indoor and outdoor. In this way, it calculates q and air leakage rate (n) across the building envelope (ISO 9972:2015). Concerning its history, this is related to development of window and door mounted tests with the support of a fan (Fig. 1.6).

The procedure of BDT is also defined by the international standard ISO 9972 for measuring q at 50 Pa (q50) and ACR at 50 Pa (n50), respectively, for single- or multizone buildings. It introduced three different test methods: (1) "building in use" where natural ventilation is closed, and mechanical ventilation or air conditioning are sealed; (2) "building envelope" where windows, doors, and manholes are closed, and all the intentional opening are sealed; and (3) "specific purposes" where the treatment of intentional opening is adapted to the local legislation. In the first two cases, the procedure requires at least a ΔP of 50 Pa between indoor and outdoor. In the third case, the conditions are specified at local level by municipal or regional regulations. The test equipment is composed by: (1) a fan to guarantee constant air flow at ΔP; (2) mobile telescopic door to close and seal the opening; (3) pressure device to gauge the ΔP with the accuracy of ± 2 Pa in the range 0–60 Pa, and (4) eventually T_a devices to measure the indoor/outdoor ΔT (Fig. 1.7). To obtain correct results, the test should be conducted with low ΔT and wind speeds (<6 m/s or < level 3 on Beaufort scale).

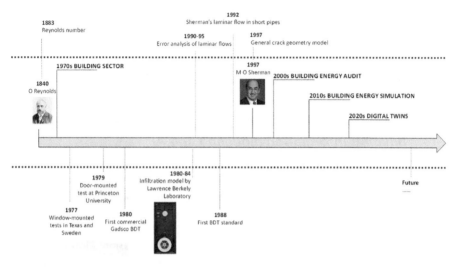

Figure 1.6 History of the BDT. *BDT*, Blower door test.

Figure 1.7 BDT apparatus. *BDT*, Blower door test.

BDT on heritage buildings follows this procedure, without defining specific techniques for considering the building features that may affect the results, such as building age, construction types, huge volumes, complex shapes, and high ventilation rates. Old buildings (pre-1950) have higher leakages (Sinnott & Dyer, 2012), due to the presence of cracks and decay, while recent homes (post-1980) reduce their rates for the presence of air barriers, airtightness, and weather-stripped windows (Sherman & Dickerhoff, 1994). Otherwise, older and smaller buildings have higher leakage areas than the larger ones (Chan et al., 2005). The unique study on heritage buildings demonstrated the importance of defining specific tests because their performance are in a large range both of n50 ($68-37.12\ \mathrm{h}^{-1}$) and q50 value

$(0.50-20.46 \text{ m}^3/\text{m}^2\text{h})$. This is due to different shapes, constructive systems, conservation levels, and building repairs (Martín-Garín et al., 2020). Despite this, there is a general agreement to use high permeability ranges ($q50 = 5-10 \text{ m}^3/\text{m}^2\text{h}$) for permeable (i.e., wooden construction), or damaged materials, large openings (Akkurt et al., 2020), or nonretrofitted buildings (Salehi et al., 2017).

1.2.5 Interoperability of traditional approaches with heritage building information modeling and AI

The development of a model of a cultural asset may be limited, since it requires from technical information and historical information (Andriasyan et al., 2020; Bruno et al., 2018; Delegou et al., 2019; Nieto-Julián et al., 2021; Yang et al., 2020). Along this line, some authors combined a specific NDT technique and IoT sensor networks to compute 3D HBIM (heritage building information modeling) (Bruno et al., 2018; Delegou et al., 2019). This type of integration allows characterizing building materials, preserving the state of the construction, and generating a planning for maintenance interventions (Delegou et al., 2019). Nevertheless, several drawbacks were detected. The implementation is often a challenging task, since there is a lack of interoperability between BIM and diagnostic tools (Acierno et al., 2017; Ilter et al., 2015; Nieto-Julián et al., 2021; Volk et al., 2014). In fact, the optical properties of irregular structures difficult the identification of the elements, and this can lead to high computation time of data postprocessing and the application of expensive and different tools to create workflows from point clouds to BIM (Chacón et al., 2021; Mineo et al., 2019; Nieto-Julián et al., 2021; Zhong et al., 2019). Based on this, some researchers proposed to adopt multidisciplinary approaches: (1) BIM-GIS, (2) BIM-IRT, and (3) BIM-photogrammetry.

Nowadays, the use of machine learning (ML) and deep learning (DL) also increases in heritage buildings, since both are flexible mathematical methods that make possible to compute as task without human intervention (Fiorucci et al., 2020; Garrido et al., 2021; Hatir et al., 2020). However, some limitations were highlighted, such as the quality and limited access of input datasets as well as the short range of applicability with IRT or HFM (i.e., material classification, defect detection, and segmentation of artistic objects) (Bang et al., 2020; Duan et al., 2019; Luo et al., 2019; Yousefi et al., 2018). In the case of photogrammetry or laser scanning, ML or DL are still ongoing. A few studies developed a routine inspection based on on-site monitoring, to estimate structure deformation in heritage buildings (Wojtkowska et al., 2021; Zou et al., 2019).

1.3 Conclusion

This book chapter summarizes the foundations of the most common NDT techniques and how these ones are being integrated with the novelties of the current scenario. Nowadays, HBIM and AI methods are based on the modeling of architectural restorations, avoiding the quantitative diagnosis that could help to determine the impact of pathologies on the thermal performance of CH assets or to propose a deep energy retrofit for a greater resilience of the building. IRT, HFM, or airtightness measurements are not implemented to enhance HBIM energy models. Several technical barriers could be the following: (1) standards and protocols of NDT technique are required, to establish a measurement pattern for the diagnosis of heritage buildings, because the majority of existing studies were carried out in residential buildings; (2) HFM needs a large sensor network, since historic masonries present a higher degree of nonhomogeneity; (3) a lack of interoperability between platforms exists; (4) a decentralization of the information was observed, because a common data environment between the potential stakeholders is not created; (5) high air flow rates across the building facades are detected in CH buildings, which increase the complexity of the inspection; and (6) data privacy, network security, and quality of datasets do not allow easily generating cloud points of the segmentation areas of building materials for the application of ML and DL. Hence, a holistic approach should be adopted in future research steps, especially to accomplish the Sustainable Development Goals.

Declaration of competing interest

The authors declare that they have no known competing financial interests or personal relationships that could have appeared to influence the work reported in this chapter.

References

Abolfazl, H. (2017). *Natural ventilation and air infiltration in large single-zone buildings: measurements and modelling with reference to historical churches*. Gävle University Press Unpublished content.

Acierno, M., Cursi, S., Simeone, D., & Fiorani, D. (2017). Architectural heritage knowledge modelling: An ontology-based framework for conservation process. *Journal of Cultural Heritage*, 24, 124–133. Available from https://doi.org/10.1016/j.culher.2016.09.010.

Akkurt, G. G., Aste, N., Borderon, J., Buda, A., Calzolari, M., Chung, D., Costanzo, V., Del Pero, C., Evola, G., Huerto-Cardenas, H. E., Leonforte, F., Lo Faro, A., Lucchi, E., Marletta, L., Nocera, F., Pracchi, V., & Turhan, C. (2020). Dynamic thermal and hygrometric simulation of historical buildings: Critical factors and possible solutions. *Renewable and Sustainable Energy Reviews*, 118(July 2019), 109509. Available from https://doi.org/10.1016/j.rser.2019.109509.

Albatici, R., & Tonelli, A. M. (2010). Infrared thermovision technique for the assessment of thermal transmittance value of opaque building elements on site. *Energy and Buildings*, 42(11), 2177–2183. Available from https://doi.org/10.1016/j.enbuild.2010.07.010.

Albatici, R., Tonelli, A. M., & Chiogna, M. (2015). A comprehensive experimental approach for the validation of quantitative infrared thermography in the evaluation of building thermal transmittance. *Applied Energy*, *141*, 218−228. Available from https://doi.org/10.1016/j.apenergy.2014.12.035, https://doi.org/10.1016/j.apenergy.2014.12.035.

Almeida, R. M. S. F., Barreira, E., & Moreira, P. (2020). A discussion regarding the measurement of ventilation rates using tracer gas and decay technique. *Infrastructures*, *5* (10), 1−13. Available from https://doi.org/10.3390/infrastructures5100085.

Andriasyan, M., Moyano, J., Nieto-Julián, J. E., & Antón, D. (2020). From point cloud data to Building Information Modelling: An automatic parametric workflow for heritage. *Remote Sensing*, *12*(7). Available from https://doi.org/10.3390/rs12071094.

Angrisano, M., Fabbrocino, F., Iodice, P., & Girard, L. F. (2021). The evaluation of historic building energy retrofit projects through the life cycle assessment. *Applied Sciences (Switzerland)*, *11*(15). Available from https://doi.org/10.3390/app11157145.

Antón, D., Medjdoub, B., Shrahily, R., & Moyano, J. (2018). Accuracy evaluation of the semi-automatic 3D modeling for historical building information models. *International Journal of Architectural Heritage*, *12*(5), 790−805. Available from https://doi.org/10.1080/15583058.2017.1415391.

Baker, P. (2011). *Technical Paper 10: U-values and traditional buildings-In situ measurements and their comparisons to calculated values Historic Scotland*, Conservation Group. Unpublished content.

Balado, J., Díaz-Vilariño, L., Azenha, M., & Lourenço, P. B. (2021). Automatic detection of surface damage in round brick chimneys by finite plane modelling from terrestrial laser scanning point clouds. Case study of Bragança Dukes' Palace, Guimarães, Portugal. *International Journal of Architectural Heritage*, 1−15. Available from https://doi.org/10.1080/15583058.2021.1925779.

Bang, H. T., Park, S., & Jeon, H. (2020). Defect identification in composite materials via thermography and deep learning techniques. *Composite Structures*, *246*(January), 112405. Available from https://doi.org/10.1016/j.compstruct.2020.112405.

Barbosa, M. T. G., Rosse, V. J., & Laurindo, N. G. (2021). Thermography evaluation strategy proposal due moisture damage on building facades. *Journal of Building Engineering*, *43* (March), 102555. Available from https://doi.org/10.1016/j.jobe.2021.102555.

Bisegna, F., Ambrosini, D., Paoletti, D., Sfarra, S., & Gugliermetti, F. (2014). A qualitative method for combining thermal imprints to emerging weak points of ancient wall structures by passive infrared thermography − A case study. *Journal of Cultural Heritage*, *15*(2), 199−202. Available from https://doi.org/10.1016/j.culher.2013.03.006.

Bolognesi, M., Furini, A., Russo, V., Pellegrinelli, A., & Russo, P. (2015). Testing the low-cost rpas potential in 3D cultural heritage reconstruction. *International Archives of the Photogrammetry, Remote Sensing and Spatial Information Sciences - ISPRS Archives*, *40*, 229−235. Available from https://doi.org/10.5194/isprsarchives-XL-5-W4-229-2015.

Brumana, R., Oreni, D., Cuca, B., Binda, L., Condoleo, P., & Triggiani, M. (2014). Strategy for integrated surveying techniques finalized to interpretive models in a byzantine church, Mesopotam, Albania. *International Journal of Architectural Heritage*, *8*(6), 886−924. Available from https://doi.org/10.1080/15583058.2012.756077.

Bruno, S., De Fino, M., & Fatiguso, F. (2018). Historic Building Information Modelling: Performance assessment for diagnosis-aided information modelling and management. *Automation in Construction*, *86*(November 2017), 256−276. Available from https://doi.org/10.1016/j.autcon.2017.11.009.

Caro, R., & Sendra, J. J. (2021). Are the dwellings of historic Mediterranean cities cold in winter? A field assessment on their indoor environment and energy performance. *Energy and Buildings*, *230*, 110567. Available from https://doi.org/10.1016/j.enbuild.2020.110567.

Chacón, R., Puig-Polo, C., & Real, E. (2021). TLS measurements of initial imperfections of steel frames for structural analysis within BIM-enabled platforms. *Automation in Construction, 125.* Available from https://doi.org/10.1016/j.autcon.2021.103618.

Chan, W. R., Nazaroff, W. W., Price, P. N., Sohn, M. D., & Gadgil, A. J. (2005). Analyzing a database of residential air leakage in the United States. *Atmospheric Environment, 39* (19), 3445−3455. Available from https://doi.org/10.1016/j.atmosenv.2005.01.062.

Cortellessa, G., & Iacomini, L. (2019). A novel calibration system for heat flow meters: Experimental and numerical analysis. *Measurement: Journal of the International Measurement Confederation, 144,* 105−117. Available from https://doi.org/10.1016/j. measurement.2019.05.053.

Delegou, E. T., Mourgi, G., Tsilimantou, E., Ioannidis, C., & Moropoulou, A. (2019). A multidisciplinary approach for historic buildings diagnosis: The Case Study of the Kaisariani Monastery. *Heritage, 2*(2), 1211−1232. Available from https://doi.org/ 10.3390/heritage2020079.

Dimitroulopoulou, C. (2012). Ventilation in European dwellings: A review. *Building and Environment, 47*(1), 109−125. Available from https://doi.org/10.1016/j.buildenv.2011. 07.016.

Duan, Y., Liu, S., Hu, C., Hu, J., Zhang, H., Yan, Y., Tao, N., Zhang, C., Maldague, X., Fang, Q., Ibarra-Castanedo, C., Chen, D., Li, X., & Meng, J. (2019). Automated defect classification in infrared thermography based on a neural network. *NDT and E International, 107*(July), 102147. Available from https://doi.org/10.1016/j.ndteint.2019. 102147.

Erenoglu, R. C., Akcay, O., & Erenoglu, O. (2017). An UAS-assisted multi-sensor approach for 3D modeling and reconstruction of cultural heritage site. *Journal of Cultural Heritage, 26,* 79−90. Available from https://doi.org/10.1016/j.culher.2017.02.007.

European Union. (2020). *Built cultural heritage.* Unpublished content.

Eurostat CensusHub HC53 (2021). https://ec.europa.eu/CensusHub2/selectHyperCube? clearSession = true.

Evangelisti, L., Guattari, C., De Lieto Vollaro, R., & Asdrubali, F. (2020). A methodological approach for heat-flow meter data post-processing under different climatic conditions and wall orientations. *Energy and Buildings, 223,* 110216. Available from https://doi. org/10.1016/j.enbuild.2020.110216.

Evangelisti, L., Guattari, C., Gori, P., & De Lieto Vollaro, R. (2015). In situ thermal transmittance measurements for investigating differences between wall models and actual building performance. *Sustainability (Switzerland), 7*(8), 10388−10398. Available from https://doi.org/10.3390/su70810388.

Ferdyn-Grygierek, J., Baranowski, A., Blaszczok, M., & Kaczmarczyk, J. (2019). Thermal diagnostics of natural ventilation in buildings: An integrated approach. *Energies, 12*(23). Available from https://doi.org/10.3390/en12234556.

Ficco, G., Iannetta, F., Ianniello, E., D'Ambrosio Alfano, F. R., & Dell'Isola, M. (2015). U-value in situ measurement for energy diagnosis of existing buildings. *Energy and Buildings, 104,* 108−121. Available from https://doi.org/10.1016/j.enbuild.2015.06.071.

Fiorucci, M., Khoroshiltseva, M., Pontil, M., Traviglia, A., Del Bue, A., & James, S. (2020). Machine learning for cultural heritage: A survey. *Pattern Recognition Letters, 133,* 102−108. Available from https://doi.org/10.1016/j.patrec.2020.02.017.

Fokaides, P. A., & Kalogirou, S. A. (2011). Application of infrared thermography for the determination of the overall heat transfer coefficient (U-Value) in building envelopes. *Applied Energy, 88*(12), 4358−4365. Available from https://doi.org/10.1016/j.apenergy. 2011.05.014.

Fox, M., Goodhew, S., & De Wilde, P. (2016). Building defect detection: External versus internal thermography. *Building and Environment, 105*, 317–331. Available from https://doi.org/10.1016/j.buildenv.2016.06.011.

Garrido, I., Erazo-Aux, J., Lagüela, S., Sfarra, S., Ibarra-Castanedo, C., Pivarčiová, E., Gargiulo, G., Maldague, X., & Arias, P. (2021). Introduction of deep learning in thermographic monitoring of cultural heritage and improvement by automatic thermogram preprocessing algorithms. *Sensors (Switzerland), 21*(3), 1–44. Available from https://doi.org/10.3390/s21030750.

Gaspar, K., Casals, M., & Gangolells, M. (2021). Influence of HFM thermal contact on the accuracy of in situ measurements of façades' U-value in operational stage. *Applied Sciences (Switzerland), 11*(3), 1–14. Available from https://doi.org/10.3390/app11030979.

Gaspar, K., Casals, M., & Gangolells, M. (2018). In situ measurement of façades with a low U-value: Avoiding deviations. *Energy and Buildings, 170*, 61–73. Available from https://doi.org/10.1016/j.enbuild.2018.04.012.

Grinzato, E. (2010). Humidity and air temperature measurement by quantitative infrared thermography. *Quantitative InfraRed Thermography Journal, 7*(1), 55–72. Available from https://doi.org/10.3166/qirt.7.55-72.

Grussenmeyer, P., Landes, T., Voegtle, T., & Ringle, K. (2008). Comparison methods of terrestrial laser scanning, photogrammetry and tacheometry data for recording of cultural heritage buildings. *2008 21st ISPRS International Congress for Photogrammetry and Remote Sensing, 37*, 213–218.

Gumbarević, S., Milovanović, B., Gaši, M., Bagarić, M. (2021). *Thermal transmittance prediction based on the application of artificial neural networks on heat flux method results.* arXiv preprint arXiv.

Gumbarević, S., Milovanović, B., Gaši, M., & Bagarić, M. (2020). Application of multilayer perceptron method on heat flow method results for reducing the in-situ measurement time. *Engineering Proceedings* (November), 8272. Available from https://doi.org/10.3390/ecsa-7-08272.

Gómez-Zurdo, R. S., Martín, D. G., González-Rodrigo, B., Sacristán, M. M., & Marín, R. M. (2021). Aplicación de la fotogrametría con drones al control deformacional de estructuras y terreno. *Informes de la Construcción, 73*(561), e379. Available from https://doi.org/10.3989/ic.77867.

Hamid, A. A., Johansson, D., & Bagge, H. (2020). Ventilation measures for heritage office buildings in temperate climate for improvement of energy performance and IEQ. *Energy and Buildings, 211*, 109822. Available from https://doi.org/10.1016/j.enbuild.2020.109822.

Hatir, M. E., Barstuğan, M., & İnce, İ. (2020). Deep learning-based weathering type recognition in historical stone monuments. *Journal of Cultural Heritage, 45*, 193–203. Available from https://doi.org/10.1016/j.culher.2020.04.008.

Ilter, D., Ergen, E., Ilter, D., Ergen, E., Kassem, M., Kelly G., Dawood, N., Serginson, M., Lockley, S. (2015). Environment Project, Asset Management, John Rogers, Heap-yih Chong, Christopher Preece, Architectural Management, *Article information: BIM for building refurbishment and maintenance: Current status and research directions.*

International Organization for Standardization. (2015). *ISO 9972:2015 - Thermal performance of buildings—Determination of air permeability of buildings—Fan pressurization method.* Unpublished content.

International Organization for Standardization. (2014). *ISO 9869-1:2014 Thermal insulation—Building elements—In-situ measurement of thermal resistance and thermal transmittance—* Part 1: Heat flow meter method Unpublished content.

International Organization for Standardization, ISO 9972:2015 - Thermal performance of buildings — Determination of air permeability of buildings — Fan pressurization method, 2015.

Kavuru, M., & Rosina, E. (2021). IR thermography for the restoration of Colonial Architecture in India—Case study of the British Residency in Hyderabad, Telangana. *Journal of Cultural Heritage*, *48*, 24−28. Available from https://doi.org/10.1016/j.culher.2021.01.009.

Kordatos, E. Z., Exarchos, D. A., Stavrakos, C., Moropoulou, A., & Matikas, T. E. (2013). Infrared thermographic inspection of murals and characterization of degradation in historic monuments. *Construction and Building Materials*, *48*, 1261−1265. Available from https://doi.org/10.1016/j.conbuildmat.2012.06.062.

Kylili, A., Fokaides, P. A., Christou, P., & Kalogirou, S. A. (2014). Infrared thermography (IRT) applications for building diagnostics: A review. *Applied Energy*, *134*, 531−549. Available from https://doi.org/10.1016/j.apenergy.2014.08.005.

Lee, S., Park, B., & Kurabuchi, T. (2016). Numerical evaluation of influence of door opening on interzonal air exchange. *Building and Environment*, *102*, 230−242. Available from https://doi.org/10.1016/j.buildenv.2016.03.017.

Lerma, C., Mas, A., Gil, E., Vercher, J., & Peñalver, M. J. (2014). Pathology of building materials in historic buildings. Relationship between laboratory testing and infrared thermography. *Materiales de Construccion*, *64*(313). Available from https://doi.org/10.3989/mc.2013.06612.

Lucchi, E. (2018). Review of preventive conservation in museum buildings. *Journal of Cultural Heritage*, *29*, 180−193. Available from https://doi.org/10.1016/j.culher.2017.09.003.

Lucchi, E. (2017). Thermal transmittance of historical stone masonries: A comparison among standard, calculated and measured data. *Energy and Buildings*, *151*, 393−405. Available from https://doi.org/10.1016/j.enbuild.2017.07.002.

Lucchi, E. (2016). Multidisciplinary risk-based analysis for supporting the decision making process on conservation, energy efficiency, and human comfort in museum buildings. *Journal of Cultural Heritage*, *22*, 1079−1089. Available from https://doi.org/10.1016/j.culher.2016.06.001.

Lucchi, E., Pereira, L. D., Andreotti, M., Malaguti, R., Cennamo, D., Calzolari, M., & Frighi, V. (2019). Development of a compatible, low cost and high accurate conservation remote sensing technology for the hygrothermal assessment of historic walls. *Electronics (Switzerland)*, *8*(6). Available from https://doi.org/10.3390/electronics8060643.

Lucchi, E., Roberti, F., & Alexandra, T. (2018). Definition of an experimental procedure with the hot box method for the thermal performance evaluation of inhomogeneous walls. *Energy and Buildings*, *179*, 99−111. Available from https://doi.org/10.1016/j.enbuild.2018.08.049.

Luo, Q., Gao, B., Woo, W. L., & Yang, Y. (2019). Temporal and spatial deep learning network for infrared thermal defect detection. *NDT and E International*, *108*(March), 102164. Available from https://doi.org/10.1016/j.ndteint.2019.102164.

Martín-Garín, A., Millán-García, J. A., Hidalgo-Betanzos, J. M., Hernández-Minguillón, R. J., & Baïri, A. (2020). Airtightness analysis of the built heritage ǁ field measurements of nineteenth century buildings through blower door tests. *Energies*, *13*(24). Available from https://doi.org/10.3390/en13246727.

Mineo, C., Pierce, S. G., & Summan, R. (2019). Novel algorithms for 3D surface point cloud boundary detection and edge reconstruction. *Journal of Computational Design and Engineering*, *6*(1), 81−91. Available from https://doi.org/10.1016/j.jcde.2018.02.001.

Moyano, J., Nieto-Julián, J. E., Bienvenido-Huertas, D., & Marín-García, D. (2020). Validation of close-range photogrammetry for architectural and archaeological heritage: Analysis of point density and 3d mesh geometry. *Remote Sensing, 12*(21). Available from https://doi.org/10.3390/rs12213571.

Moyano, J., Nieto-julián, J. E., Lenin, L. M., Bruno, S., Moyano, J., Nieto-julián, J. E., Lenin, L. M., Bruno, S., & Moyano, J. (2021). Operability of Point Cloud Data in an Architectural Heritage Information Model. *International Journal of Architectural Heritage, 16*(10), 1−20. Available from https://doi.org/10.1080/15583058.2021. 1900951.

Nardi, I., Ambrosini, D., De Rubeis, T., Sfarra, S., Perilli, S., & Pasqualoni, G. (2015). A comparison between thermographic and flow-meter methods for the evaluation of thermal transmittance of different wall constructions. *Journal of Physics: Conference Series, 655*(1). Available from https://doi.org/10.1088/1742-6596/655/1/012007.

Nieto-Julián, J. E., Lara, L., & Moyano, J. (2021). Implementation of a teamwork-hbim for the management and sustainability of architectural heritage. *Sustainability (Switzerland), 13*(4), 1−26. Available from https://doi.org/10.3390/su13042161.

Odgaard, T., Bjarløv, S. P., & Rode, C. (2017). Interior insulation − Experimental investigation of hygrothermal conditions and damage evaluation of solid masonry façades in a listed building. *Building and Environment*. Available from https://doi.org/10.1016/j. buildenv.2017.11.015.

Paoletti, D., Ambrosini, D., Sfarra, S., & Bisegna, F. (2013). Preventive thermographic diagnosis of historical buildings for consolidation. *Journal of Cultural Heritage, 14*(2), 116−121. Available from https://doi.org/10.1016/j.culher.2012.05.005.

Pasos, A. V., Zheng, X., Smith, L., & Wood, C. (2020). Estimation of the infiltration rate of UK homes with the divide-by-20 rule and its comparison with site measurements. *Building and Environment, 185*(August), 107275. Available from https://doi.org/ 10.1016/j.buildenv.2020.107275.

Pierdicca, R., Frontoni, E., Malinverni, E. S., Colosi, F., & Orazi, R. (2016). Virtual reconstruction of archaeological heritage using a combination of photogrammetric techniques: Huaca Arco Iris, Chan Chan, Peru. *Digital Applications in Archaeology and Cultural Heritage, 3*(3), 80−90. Available from https://doi.org/10.1016/j.daach.2016.06.002.

Ramos, L., Marchamalo, M., Rejas, J. G., & Martínez, R. (2015). Aplicación del Láser Escáner Terrestre (TLS) a la modelización de estructuras: Precisión, exactitud y diseño de la adquisición de datos en casos reales. *Informes de la Construccion, 67*(538). Available from https://doi.org/10.3989/ic.13.103.

Ramos, L.F., Masciotta M.G., Morais, M.J., Azenha, M., Ferreira, T., Pereira, E.B., Lourenço, P.B. (September, 2018). *HeritageCARE: Preventive conservation of built cultural heritage in the south-west Europe. Innovative Built Heritage Models - Edited contributions to the International Conference on Innovative Built Heritage Models and Preventive Systems, CHANGES 2017*, 135−142. Available from https://doi.org/10.1201/ 9781351014793-16.

Remondino, F., Spera, M. G., Nocerino, E., Menna, F., & Nex, F. (2014). State of the art in high density image matching. *Photogrammetric Record, 29*(146), 144−166. Available from https://doi.org/10.1111/phor.12063.

RESNET. (2012). *Interim guidelines for thermographic inspections of buildings.* Unpublished content.

Roque, E., Vicente, R., Almeida, R. M. S. F., Mendes da Silva, J., & Vaz Ferreira, A. (2020). Thermal characterisation of traditional wall solution of built heritage using the simple hot box-heat flow meter method: In situ measurements and numerical simulation.

Applied Thermal Engineering, 169(January), 114935. Available from https://doi.org/10.1016/j.applthermaleng.2020.114935.

Salehi, A., Torres, I., & Ramos, A. (2017). Experimental analysis of building airtightness in traditional residential Portuguese buildings. *Energy and Buildings, 151,* 198−205. Available from https://doi.org/10.1016/j.enbuild.2017.06.037.

Sherman, M. H., & Dickerhoff, D. J. (1994). Proceedings, *15th Air Infiltration and Ventilation Centre Conference Air Tightness of U.S. Dwellings Unpublished content Air Tightness of U.S. Dwellings.*

Sinnott, D., & Dyer, M. (2012). Air-tightness field data for dwellings in Ireland. *Building and Environment, 51,* 269−275. Available from https://doi.org/10.1016/j.buildenv.2011.11.016.

Tavukçuoğlu, A. (2018). Non-destructive testing for building diagnostics and monitoring: Experience achieved with case studies. *MATEC Web of Conferences, 149.* Available from https://doi.org/10.1051/matecconf/201714901015.

Tejedor, B., Barreira, E., Almeida, R. M. S. F., & Casals, M. (2020). Thermographic 2D U-value map for quantifying thermal bridges in building façades. *Energy and Buildings, 224,* 110176. Available from https://doi.org/10.1016/j.enbuild.2020.110176.

Tejedor, B., Barreira, E., Almeida, R. M. S. F., & Casals, M. (2021). Automated data-processing technique: 2D Map for identifying the distribution of the U-value in building elements by quantitative internal thermography. *Automation in Construction, 122* (November 2020), 103478. Available from https://doi.org/10.1016/j.autcon.2020.103478.

Tejedor, B., Casals, M., Gangolells, M., & Roca, X. (2017). Quantitative internal infra-red thermography for determining in-situ thermal behaviour of façades. *Energy and Buildings, 151,* 187−197. Available from https://doi.org/10.1016/j.enbuild.2017.06.040.

Tejedor, B., Casals, M., Macarulla, M., & Giretti, A. (2019). U-value time series analyses: Evaluating the feasibility of in-situ short-lasting IRT tests for heavy multi-leaf walls. *Building and Environment, 159,* 106123. Available from https://doi.org/10.1016/j.buildenv.2019.05.001.

Tejedor, B., Lucchi, E., Bienvenido-Huertas, D., & Nardi, I. (2022). Non-destructive techni-ques (NDT) for the diagnosis of heritage buildings: Traditional procedures and futures perspectives. *Energy and Buildings, 263,* 112029. Available from https://doi.org/10.1016/j.enbuild.2022.112029.

Tumeliene, E., Nareiko, V., & Suziedelyte Visockiene, J. (2017). Photogrammetric measure-ments of heritage objects. *ISPRS Annals of the Photogrammetry, Remote Sensing and Spatial Information Sciences, 4,* 71−76. Available from https://doi.org/10.5194/isprs-annals-IV-5-W1-71-2017.

Valluzzi, M. R., Lorenzoni, F., Deiana, R., Taffarel, S., & Modena, C. (2019). Non-destructive investigations for structural qualification of the Sarno Baths, Pompeii. *Journal of Cultural Heritage, 40,* 280−287. Available from https://doi.org/10.1016/j.culher.2019.04.015.

Van Buggenhout, S., Van Brecht, A., Eren Özcan, S., Vranken, E., Van Malcot, W., & Berckmans, D. (2009). Influence of sampling positions on accuracy of tracer gas mea-surements in ventilated spaces. *Biosystems Engineering, 104*(2), 216−223. Available from https://doi.org/10.1016/j.biosystemseng.2009.04.018.

Volk, R., Stengel, J., & Schultmann, F. (2014). Building Information Modeling (BIM) for existing buildings - Literature review and future needs. *Automation in Construction, 38,* 109−127. Available from https://doi.org/10.1016/j.autcon.2013.10.023.

Williamson, J. B., Stinson, J., Garnier, C., & Currie, J. (2014). In-situ monitoring of thermal refurbishment on pre-1919 properties in Scotland. *REHAB 2014 - International Conference on Preservation, Maintenance and Rehabilitation of Historical Buildings and Structures* 2(1), 1037−1046. Available from https://doi.org/10.14575/gl/rehab2014/105.

Wojtkowska, M., Kedzierski, M., & Delis, P. (2021). Validation of terrestrial laser scanning and artificial intelligence for measuring deformations of cultural heritage structures. *Measurement: Journal of the International Measurement Confederation, 167*(July 2020), 108291. Available from https://doi.org/10.1016/j.measurement.2020.108291.

Yang, X., Grussenmeyer, P., Koehl, M., Macher, H., Murtiyoso, A., & Landes, T. (2020). Review of built heritage modelling: Integration of HBIM and other information techniques. *Journal of Cultural Heritage, 46*, 350−360. Available from https://doi.org/10.1016/j.culher.2020.05.008.

Yousefi, B., Kalhor, D., Usamentiaga, R., Lei, L., Ibarra-Castanedo, C., Maldague, X. (2018) Application of deep learning in infrared non-destructive testing. *14th Quantitative InfraRed Thermography Conference* (pp. 1−9). Available from https://doi.org/10.21611/qirt.2018.p27.

Zhong, S., Zhong, Z., & Hua, J. (2019). Surface reconstruction by parallel and unified particle-based resampling from point clouds. *Computer Aided Geometric Design, 71*, 43−62. Available from https://doi.org/10.1016/j.cagd.2019.04.011.

Zou, Z., Zhao, X., Zhao, P., Qi, F., & Wang, N. (2019). CNN-based statistics and location estimation of missing components in routine inspection of historic buildings. *Journal of Cultural Heritage, 38*, 221−230. Available from https://doi.org/10.1016/j.culher.2019.02.002.

Section II

Infrared Thermography (IRT)

Advancement of infrared thermography for built heritage

2

Elisabetta Rosina[1], Manogna Kavuru[2], and Erica Isabella Parisi[3]
[1]Department of Architecture, Built Environment and Construction Engineering, Polytechnic of Milan, Milan, Italy, [2]Architrave p.c., Washington, DC, United States, [3]Department of Environmental and Civil Engineering di Ingegneria Civile e Ambientale, State Univeristy of Florence (IT), Italy

2.1 Introduction

Investigation on buildings and traditional system for dry stone retain walls by IR thermal scanning is useful to collect information regarding building technology and elements, their shape, their material characteristics, and their state of conservation/ damage. Different kinds of discontinuities affecting building structures are detected by thermal analysis of the surface temperature, taken at particular boundary conditions. The application of IR thermography to historic buildings is a specific use case of the thermal testing for buildings that are under protection of the government or municipality for their value as cultural heritage.

Cultural heritage is defined according to the tradition, cultural significance, and identity of the community that reside in the building region. Cultural heritage includes artifacts, monuments, a group of buildings and sites, and museums that have a diversity of values, including symbolic, historic, artistic, esthetic, ethnological or anthropological, scientific and social significance (http://uis.unesco.org/en/glossary-term/cultural-heritage).

In November 2018, UNESCO's Representative List of the Intangible Cultural Heritage of Humanity added "the art of dry stone walling, knowledge, and techniques" to their list (UNESCO Intergovernmental Committee, 2018). This recognition highlights the continued use of dry-stone walls through ancient times as important components of infrastructure in mountainous and hilly regions worldwide. These walls are traditional elements of the environment and landscape and serve as bases for roads, buildings, and agricultural terraces. Dry-stone walls are anthropic elements that strongly shape the landscapes, creating a diffuse heritage, since they represent an ancient long-standing footprint of the relationship between humans and nature. They are often strictly connected to terracing as agricultural systems functioning as retaining slope systems. The combined presence of dry-stone walls and terraces, if properly maintained, plays a significant role in preventing landslides, floods, erosion, and hydrogeological risk while also enhancing biodiversity and soil fertility and creating suitable microclimatic conditions for agriculture. Moreover, the use of steep

Diagnosis of Heritage Buildings by Non-Destructive Techniques. DOI: https://doi.org/10.1016/B978-0-443-16001-1.00002-4

slopes at higher altitudes for agriculture is becoming more common to face temperature variations induced by the effects of climate change (Parisi & Tyc, 2021).

Different components of construction often include different bonding components, which could result from different walling phases. In fact, due to many causes (historic stratification, change of function and consequent refurbishment, collapse, and reconstruction of a part, and so on), several types of structures could have been put together: in some cases, their different bonding is apparent on sight; sometimes the bonding is masked by a homogeneous coating such as a surface layer of paint. Many detailed studies have been carried out about all these technologies and materials. It is helpful to bear in mind their results during investigations (Castillo et al., 2012). Indeed, a basic knowledge of the building's structural pattern is required for a correct evaluation of thermograms.

Infrared thermography is applied with similar procedures for both modern and ancient buildings; nevertheless, the following differences exist for historic buildings (Avdelidis & Moropoulou, 2004; Danese et al., 2010; Ludwig & Rosina, 2006; Meola, 2007; Rosina et al., 2008).

The objective of this investigation is the detection and evaluation of thermal anomalies corresponding to discontinuities due to damage and building elements that are not visible. The first step for obtaining a successful result of infrared thermography (IRT) investigation is the examination of all the available documents regarding the building and the components of the structure.

Nevertheless, in some cases, a little preliminary information is available. In other cases, the archival documentation could take a long time, while the assessment of the building condition could require short time, especially in the case of structural failures and risk of collapse.

Because IR thermography is mostly used for preliminary investigation, surveys and lab tests on materials are usually still in progress during the planning phase. Therefore a direct survey of the materials and their damage is required to assess the real state of the test surface. The preliminary assessment and planning phase includes identifying areas of interest and specifying test techniques and even the integration with other testing methods. Finally, the planning of the scanning has a range of approximations that must be considered prior to starting this assessment. For instance, the heating time may vary depending on unexpected changes in the structure; additionally, in case of natural irradiation, an accurate plan of the time, weather, and ambient condition is mandatory. In fact, the planning phase also includes the evaluation of natural/artificial sources of heating that affect the IR recapture and the choice of their best exploitation or reduction.

Moreover, IRT has a major application as a predictive technique in the maintenance of building stock, both before and after restoration.

2.2   Methods and purpose of the inspection

The use of IRT for buildings' investigation is quite common for identifying evident and incoming issues, detachments, and critical points of large portions of the building envelope, thanks to the advantages of being a contactless technique (Di Maio et al., 2015; Kirimtat & Krejcar, 2018; Tejedor et al., 2022). In fact, papers and works that employ IRT in heritage preservation are devoted to the diagnosis of the opaque surface envelope, including identification of cracks; health and structural state; moisture, humidity, or rising damp; and air leakages (Ascione et al., 2015; Avdelidis & Moropoulou, 2004; Barbosa et al., 2021; Bisegna et al., 2014; Di Maio et al., 2015; Falchi et al., 2018; Georgescu et al., 2017; Grinzato et al., 2002; Kilic, 2015; Kordatos et al., 2013; Lerma et al., 2014; Meola, 2013; Paoletti et al., 2013; Rosina, 2018; Tejedor et al., 2022).

Very often walls of ancient buildings are not regular; their thickness, structure, and number of layers may change unpredictably. For this reason, procedures have to be flexible to meet the requirement of the primary investigation. The examples presented in the following paragraphs will help in planning the procedure. Because of the large thickness of walls, some thermographic non-destructive testing (NDT) techniques developed for industrial purposes have to be carefully evaluated and adopted, considering the necessary changes to be made.

Usually for ancient buildings only NDT can be applied. Consequently, the integration with destructive methods is allowed under exceptional circumstances and strict limitations are also imposed on the maximum value of temperature that the building can be subjected to in case of artificial heat stimulation.

Moreover, thermal imaging on dry-stone wall can serve as dual-purpose NDT by enabling the assessment of structural stability for built structures such as dry-stone walls, while also evaluating the microclimatic impact of their thermal inertia in traditional agro-hydraulic systems.

As reported by Warren et al. (2016), thermal imaging can be used to determine the structural stability of such built elements. NDT, like thermography, helps identify possible critical construction features for risk assessment remotely, without interfering with structural stability. The particular constructive characteristics of dry-stone walls entail that the heat capacity depends on the connections between the backside (in contact with the soil) and the front side, which surface is exposed to atmospheric agents and solar radiation. Different temperatures will thus be present, between the two sides, detectable by thermal imaging, which has high sensitivity and can reveal (1) aspects of the hidden construction of the wall, (2) voiding behind the wall, (3) depth of backfills, and (4) areas of high moisture and water buildup. However, since thermal imaging does not measure temperature directly but relies on an assumption of high emissivity in the object being examined, misleading may be produced for dry-stone walls built with different kind of stones and may show different thermal behavior, for example, due to the different emissivity values of the constituting materials.

Currently, there are specific software that can help in the optimization of the temperature span of thermograms with the aim of detecting the researched anomalies, as well as extracting the temperature data on some specific areas of the thermograms. These tools are very helpful, because old surfaces often have heterogeneous colors; materials and their state of conservation are variable. As a consequence, many false alarms may be raised which need further scrutiny. In these cases, the tandem processing of raw thermograms and use of software filters based on visual analysis are effective tools to reduce undesired information. In addition, it is helpful for having information on the inner layer of the structure, because of the temperature of the surface is a function of heat flow crossing the wall and local boundary conditions.

Moreover, differences in surface temperature because of the different thermal properties of construction elements such as timber, bricks, stone, and mortar can be visualized at proper time as a *footprint* of their shapes projected on the overlapping plaster surface above them. This aids in identifying the different segments of construction beneath the opaque surface layers.

Any thin delamination of the coating strongly reduces the heat transfer and adds its signal to that given by the structure on the thermograms. The postprocessing phase is very helpful for analyzing different sets of thermograms, shot at different times for locating the anomalies in different layers of the structure.

In the following subsections, the main applications are described, from the methods to the results because each application requires specific methods that cannot be generally described only in one subsection.

2.2.1   Inspection of the building elements

In many historic buildings, an external finishing covers the masonry. Depending on the materials, conservation, and location in the building (interior, exterior, and orientation), some application of IRT can be helpful in the assessment and documentation phase. The more common usages of IRT on these finishing materials are described in the following sections.

2.2.1.1   Vertical structures

Finishing detachment

The detachments or lack of adhesion between the finishing surface, including ceramic tiles, and the substrate are very common discontinuities. The discontinuities are due to bad setting of the finishing or the crystallization cycles of soluble salts. The parget/plaster layers may embody an air gap inside or between the wall and the plaster. The detection of the discontinuity is more reliable in transient conditions (Ludwig & Rosina, 2006).

Or more generally speaking, it is detectable during and after a thermal stimulation of the surface. Such stimulation could be due to insolation (solar radiation) dependent on the time of day. In case of artificial heating, the power of the heat source depends on the material, location of the defect, air temperature. Moreover, the availability of energy supply for the artificial heating should be taken into consideration as heritage buildings do not have high-voltage power lines.

Natural sources also include air streams which provide passive heating, and finally solar irradiation may be also properly shielded to enhance the transient behavior (Ludwig & Rosina, 2005).

This kind of thermal discontinuity appears as a warmer area when the net heat flux enters the building—that is, usually during the first phase of the heating or cooling, following the thermal excitation of the surface. The heat remains in the area insulated by the air layer instead of flowing inside the structure. Villa Mirabello in Monza (Figs. 2.1 and 2.2) was captured an IR thermograph that presents examples of these discontinuities.

In fact, thermal discontinuities may also appear during the heating phase. A discontinuity can last for a particular time depending on the depth and thickness of the discontinuity.

The localization of discontinuities can be obtained by comparing the response of the anomalous areas with a reference at an optimal time.

Cracking

In most cases, IRT investigates the issues of surface anomalies, due to the continuity of coating. However, the most important information to collect is about the location of anomalies in depth, especially cracks—whether the cracking is only across the coating, across the thickness of the brick, or across the whole thickness of the masonry. Under transient condition, it is possible to survey the main occurrences of cracking by IRT (Ludwig & Rosina, 2006; Ludwig et al., 2005).

Figure 2.1 Villa Mirabello, Monza. Image of the interior.

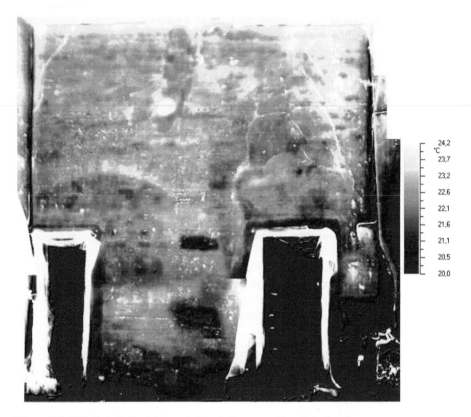

Figure 2.2 Villa Mirabello, Monza. infrared thermography of the interior.

An example of this is illustrated in the infrared image of the ceiling in Villa Carrera, Genova (Figs. 2.3 and 2.4). The IR thermograms were captured on a summer day where surface temperature of the cracking reached around 34°C, allowing them to be detected better.

A crack represents a higher thermal resistance perpendicular to the surface, detectable when a heat flux parallel to the surface exists. Therefore proper thermal stimulation is crucial to obtain the best result.

2.2.1.2 Hidden structures

In brick and stone buildings, the masonry is massive, but it could include voids and frames. In addition, also the pattern of the masonry gives much information to the restorers, but often the surface outside is plastered.

Usually, the thickness of the wall depends on its components and their interlocking. The thinnest wall consists of a single leaf of half brick, which measures about 60 mm (2.4 in.). Greater thickness depends on the bond (texture of the masonry), where two or more stretcher and header elements are laid across the wall.

Figure 2.3 Villa Carreca, Genova 2010. Image of the ceiling.

Figure 2.4 Villa Carreca, Genova. IRT of the ceiling.

The most common types of structural masonry are composed of bricks or stone ashlars and joined by lime mortar. The bond type used is the characteristic of the age of the building and the place and depends on the size of the elements, as well as the thickness. A widespread typology consists of rubble walls: stones of different shapes, dimensions, and provenience are embedded in lime mortar, in regular courses. Stretched and herringbone bonding are the typical courses. Thickness is not less than 0.4 m (16 in.). Often relieving arches are used in the masonry. Their pattern is slightly different from the bond of the remaining wall.

The survey of the structural pattern of the building provides important design information for its historic restoration. Also, in these cases, infrared thermography can be usefully applied to detect such structural elements, thanks to the different thermal properties of materials constituting ashlars, bricks, and mortar (Casapulla et al., 2018; Rosina et al., 2017; Quagliarini et al., 2012).

Another spread typology of structural wall bonding is cavity walls: they are composed of two leaves of external bricks, with stretched bond, mortar, and rubble (or pieces of bricks, tiles, and other materials) filling the cavity inside, without a regular disposition. The thickness of the whole wall is more than 0.5−0.6 m (20−24 in.); the usual thickness is 0.80−1.00 m (31−39 in.). Infrared thermography can detect a lack of adhesion between the leaves of bricks and the filling if the thickness is not prohibitive for obtaining imaging and effective heating of the layer under investigation is possible.

Example of discontinuity inside solid masonry

The infrared images of Faini factory, in Lecco (Figs. 2.5−2.7), reveal the different types of masonry underneath the plaster and an infilled opening in the western facade. The recapture was shot at sunset, after 4 hours of solar irradiation.

The IR thermograms show both the texture and the damage of the plaster at the same time: for the evaluation of the results (interpretation of the thermographic images), it is important to compare these with the photos, for distinguishing between the contributions due to the surface defects and visible damage.

Thermal imaging is mainly used for investigation of shallow discontinuities, from 0 to 30 mm (0 to 1.2 in.) long. Nevertheless, it can also help to detect voids

Figure 2.5 Faini factory, Lecco. Image of the exterior.

Figure 2.6 Faini factory, Lecco. IRT1.

Figure 2.7 Faini factory, in Lecco. IRT2.

such as cavities, weak pockets, chimney stacks, smoke tubes, dead spaces, or thickness variations. In these cases, active techniques are preferable, and the thermal excitation of the wall must be properly designed (Daffra et al., 2008; Rosina et al., 2017).

Mathematical models are very useful for this purpose. The models must take into account the thermophysical properties of the investigated materials and the environmental conditions during the tests. Unfortunately, such data are often not available.

An earlier appearance of a historic building is a very important topic for NDT and evaluation. The presence of closed openings provides context with respect to the usage of rooms and infrared thermography is well suited to this task. The fenestration is often the result of refurbishment during the life of the building. Therefore the position of the openings may depend on the changes of the internal arrangement of rooms and their furniture.

According to the new plan of the elevations, the ancient doors and windows may be walled up while new openings are made. Moreover, often a new coating covers the facades and hides the irregularity of the masonry caused by those alterations. Infrared thermography makes it possible to locate filled openings that are no longer visible. In fact, the structural elements of the opening (lintels, edges, abutments, thresholds, doorsteps, and others) can be recognized by their thermal properties. The new openings constitute weaknesses in the masonry because their presence changes the distribution of the vertical loads. Therefore cracking of the masonry may occur.

In the following cases, infrared thermography makes it possible to localize the cracking as described earlier.

What is more important in the long run is that the additional documentation and assessment analysis of the building during these thermal surveys makes it possible to ascertain the many steps of the historic evolution up to the final asset of the masonry, increasing the knowledge about the building.

In case the plaster covers the structure, it is not advisable to scrap the parget to dig out the traces of alterations occurring over centuries. Infrared and thermal testing are very effective tools for obtaining the required information about the state of the components of the wall.

Infrared scanning has to be applied after an adequate heating of the surface. Best results are achieved where materials with different thermal characteristics are juxtaposed, for example, stone-to-lime mortar masonry. The cases presented were obtained using both active and passive heating. In the latter case, a direct irradiation or a natural warm draught licking the surfaces for few hours was exploited. Hence, a high daily thermal excursion (temperature increasing more than $+10K$ [$+10°C = +18°F$]), high average air temperature [higher than $288K = 15 °C = 59°F$], and low relative humidity [relative humidity below 60%] are required.

In case of surface discontinuities (lack of finishing, damage to the finishing or colored decorations, and so on) convective heating can be more homogeneous. It can be also more effective to reduce the volume of the air near the surface under the investigation.

Example of the thermal behavior of dry-stone retaining walls as diffuse heritage

The effect of thermal inertia of these retaining systems can have an impact on the close surroundings, coupled with the effect of the overall terracing system (including the soil). The analysis of microclimate modifications induced by alterations of hillslope and by dry-stone walls is of particular interest for the valuation of benefits and drawbacks of terrace cultivation, a global land management technique (Tucci et al., 2019).

The thermal behavior of terraced vineyards with dry-stone walls in the Chianti area (Tuscany, Italy) has been studied in the last few years (Parisi & Tyc, 2021; Parisi et al., 2019; Tucci et al., 2019; Tyc et al., 2021, 2022) to assess the impact of microclimate-induced variation on the temperature patterns of the crop and thus on its growth, phenology, ripening, and consequently on quality. Ground-based and aerial-based, qualitative and quantitative, thermal imaging tests have been carried out to study the daily thermal behavior of all the constituting elements of the terraced context, that is, the dry-stone walls, the vine rows, and the soil.

For what concerns the built structures, has to be kept in mind that they will warm up slowly during the day and they will passively release heat during the night, because of the poor heat conductivity of the stones. The exact evaluation of thermal characteristic of stone, depends on its porosity, mineral composition, and specific heat conductivity. Such characteristic usually entails low heat transfer from the irradiated surface to the inner part of the rock (Fig. 2.8).

The dissipation of the heat cumulated during the day by the stones reduced temperature excursion at night, positively affecting the vines' adaptation. Also, the local microclimate may be affected by high-temperature values resulting from solar heating during the day. Furthermore, the thermal inertia of the stones may greatly enhance dewfall occurrence, both on stones and on leaves, suggesting a thermal gradient between the stones and the nearby environment (Fig. 2.9). Finally, in this kind of microclimate evaluation, the shadowing effect related to the orientation, exposition, and timing of measurements must be considered.

Remotely sensed methods and NDT, like thermal imaging, are proved to be effective both in the structural assessment of built elements, such as dry-stone walls, as well as in the monitoring of their own induced microclimatic effects on the proximities.

2.2.1.3 Horizontal structures

In construction, horizontal structures distribute the loads to the vertical elements. Briefly, they can be classed as follows.

Floors (girders, shingles, laths, and paving)

The main supporting structures, with larger cross-section, are the beams or girders. Common construction materials used are timber, iron, and cast iron. They can be hidden in the thickness of the floor or can be external, sometimes with parget or plaster facing or just a painted surface.

The timber and iron shingles, smaller in cross-section, contribute to distribute the loads and to release them on to the main supporting structures. Another typology, widespread in structures of the last decades of the 19th century and the beginning of the 20th century, is that of small vaults made of hollow terracotta wares and supported by iron beams with average spacing of 0.8−1.2 m (2.6−3.9 ft).

Timber lath ceilings are often embedded in mortar or sand, and the paving finishes the floors. Paving materials are usually made of timber, stone, bricks, or

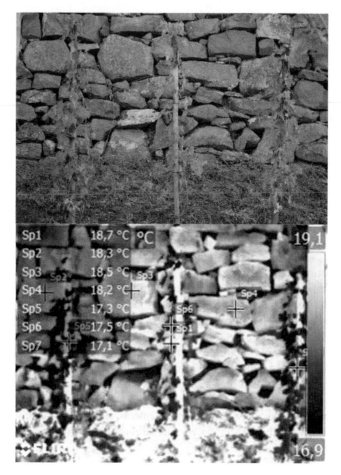

Figure 2.8 The terraced vineyards with dry-stone walls in the Chianti area, (Tuscany, Italy), IRT1.
Source: From Parisi, E. I. (2022).

tile and arranged with bonding that is stretched, herringbone, or other configurations. The development of floor tiling produced different kinds of patterned tiles with different thermal properties.

Vaults and ceilings

Structural masonry (mortar and bricks or stone quoins) is usually used as supporting structures for vaults, dome vaults, and domes.

Shuttering bricks/stones are used to flank the bases of the vaults because in case of the different thickness of the structure: thermograms show a varying temperature from the basis to the apex, at least in near steady-state conditions.

Figure 2.9 The terraced vineyards with dry-stone walls in the Chianti area, (Tuscany, Italy), IRT2.
Source: From Parisi, E. I. (2022).

Ceilings can be structural elements or decorative layer, hung or suspended in the interior of the rooms. In both the cases, thermal analysis at transient condition can help identify the structure and its condition.

Roofs
In most cases, a suitable roof covers the structure. Timber beams and rafters are the simplest supporting structure with double-pitch and hipped roofs. Trusses and a third layer of shingles are used in larger roofs. Additionally, the outermost coating lies on a lath layer. The roof tiles are the most external layers that are used to withstand the weather. Other coverings are made of wooden shingles, arranged in a multilayer arrangement or slabs of stones (slate, gneiss, or other). Metallic plates or

occasionally vegetable materials such as straw can be also found as roofing in historic buildings. The inspection of roofs is mostly finalized to assess the structure condition and the presence of water/air leakage.

2.2.1.4 Moisture diffusion and water content

Moisture diffusion in porous materials, their exchanges with air, is one of the causes of the most common causes of damage. When rising damp evaporates from a wet surface due to temperature changes, both seasonal and occasional, it can cause significant harm, such as salt crystallization and condensation (Grinzato et al., 1998; Ludwig et al., 2004; Ludwig et al., 2018; Rosina et al., 2003). Understanding the distribution and content of water within walls is fundamental to the analysis of decay. In temperate climates during winter, the thermal inertia due to the thickness of the ancient walls (more than 0.50 m) prevents frost buildup inside the wall. Damage is limited to the surface, 0.03 or 0.04 m (1 or 2 in.), because of continual cycles of freezing and thawing. In addition, in cold climactic conditions, it is usually a key factor only when temperatures stay below zero for several months. In those cases, the water volume grows in the form of frost and generates strength within porous wall materials, therefore generating cracks in the structure especially during long and cold winters.

In the recent past, the effects of climate change, the alteration of dry seasons (with high temperature and no rain for months), and almost monsoon type seasons (at mild temperature but heavy and prolonged rains) dramatically affect the distribution of rising damp in porous materials of masonry, as well as the water content over time. This phenomenon has been studied in the tropical climate of India to understand optimal and suboptimal conditions for thermal imaging in the different climatic conditions the country presents along with the different ways in which suboptimal conditions can be managed for better thermal imaging (Olmi et al., 2016).

An example of this is presented in Oratorio Mantegazza, located near Milan (Figs. 2.10 and 2.11). The IR thermography, taken on April 20, 2021, presents a lower temperature on the base of the northeast and southeast sides of the nave, due to high water content, confirmed by gravimetric tests.

A commonly used approach in applying thermography is to qualitatively detect the moisture distribution caused by the cooling effect of water evaporation (Della Torre & Rosina, 2008). Another approach is based on water's high thermal capacity that is revealed during transient thermal tests, although the feasibility of conducting this in the field is remote.

A second level is the quantitative estimation of water content using the surface temperature distribution in space and time or additional measurements with alternative techniques on selected points. Note that measuring the water content of a wall by thermography alone appears practically impossible and an estimation of the moisture in a shallow layer is the maximum achievable result by investigating with surface detection methods, such as IRT.

The relative approach is used much more than the absolute quantification of the water content. In this case, the investigation of damp areas is based on the

Figure 2.10 Oratorio Mantegazza, Sesto Milanese. Image of the interior.

Figure 2.11 Oratorio Mantegazza, Sesto Milanese. IRT of the interior wall.

comparison of the thermal behavior of dry areas with damp ones. Therefore, in the same thermogram, there can be both situations, assuming the same heating and boundary conditions. Of course, by using a passive procedure only, the environment has affected the surface.

Finally, the following problems may be anticipated:

1. A preliminary scanning of the surface of the wall makes it possible to set camera parameters and to identify the most evident anomalies, a set of further measurements closer to the surface yields the actual analysis.
2. Spot heating due to active elements such as hot pipelines or electric cables may disturb the test.
3. During an active procedure, the presence of plaster detachment and stains or colored parts may affect results.
4. The winter heating active inside the building could prevent a correct moisture detection of the wall.
5. The heating tends to dry the wall surface, masking the moisture map in case of low to medium water content.
6. Thermal bridges and thinner walls, the imperfect connection between walls, and the presence of windows and doors close to the investigated area—all these may cause false alarms.

The higher the variability of the environment change, the higher is the time needed for monitoring. The total time needed to record data representative of the recurrent changes of the parameters under analysis may be for one or more years.

By monitoring the presence and distribution of the water, it is possible to obtain indication of the cause of the water and support the choice of the most appropriate intervention, reducing the risk to apply not effective and expensive products and preventing an oversize intervention. The objective of moisture analysis is to find out the sources of water and to plan the correct intervention to dry the walls and save their surfaces. The main paths of water coming into building materials are (1) from the ground (capillary rising), (2) by condensation from the air, and (3) from infiltration because of broken piping or guttering or leakage from the roof.

Fast scan working techniques on wide surfaces have a considerable advantage over techniques that give prompt results, after a long processing data time. Collection of documentation about the building, damage location, its evolution in time, and the use across years is mandatory to work out an effective diagnostic plan, especially for a preliminary test.

Therefore the triage setting of multispectral analysis (IR, visual—including sliding light photos and macro photos, UV) is an important component of surface monitoring, for early detection, as well as for prolonged monitoring, especially if coupled with some less destructive, quantitative methods (Bison et al., 2011; Kylili et al., 2014).

Capillary rising

Most ancient buildings are affected by the presence of moisture due to the absorption of water from the ground. The presence of water in the ground could be due to the following: shallow groundwater; impermeable ground layers (such as clay) that prevent rain drainage; proximity of rivers, sea, lakes, and channels; proximity of sewage and drainage system failures; slope of the ground; and inadequate (highly porous, lacking waterproof layers or broken) pavement near the building.

The bases of the walls are the typical areas affected by rising damp. The water content and its distribution inside the wall may vary depending on the particular source supplying the underground moisture volume.

Figs. 2.12 and 2.13 illustrate the historical cemetery of San Fiorano. The IR thermography presents a thermal anomaly on the base of the masonry wall with lower

Figure 2.12 Cemetery of San Fiorano.

Figure 2.13 Cemetery of San Fiorano, IRT.

temperature. After further gravimetric analysis, the hypothesis that the IR thermography showed was confirmed, with WC reaching values up to 16%.

The main characteristic is an almost horizontal frontier, bending higher from the ground at the corners or where the masonry is thicker. The height of the damp areas may vary during the year. Salt efflorescence is frequently generated at the border of the damp areas.

Water condensation

Where relative humidity is high and temperature drops, vapor condenses on the colder surfaces. Risk of condensation increases in cases of high relative humidity and the presence of cold surfaces. It is possible on surfaces subject to evaporation, which can see this in the presence, for instance, due to a crowd in the rooms. Additionally, inadequate ventilation, for instance, lack of windows, can cause this.

Any part of the wall may be affected by dew, depending on the temperature of the wall. The spots where thermal imbalance occurs are the preferred areas of condensation. Condensation may disappear within a short space of time and occurs periodically (in the same season).

The condensation areas often correspond to stains or mold. The dew point may be monitored to measure the wall temperature and relative humidity of the air.

Leakage and infiltration

Infiltration of rain or snow may occur where the roof is not well maintained. Typical stains may be found on the ceilings (or vaults) on the upper floor. In case of wide infiltration, even the walls nearby may absorb much more water than the floor itself (depending on the thickness and the absorption characteristics of the walls).

Figs. 2.14−2.16 illustrate the infiltration of rain on the vault of Santa Maria in Strada church, in Monza. It is interesting to observe the correlation between water staining on the surface and the actual water content of each point. Such a visual

Figure 2.14 Santa Maria in Strada church, Monza. Image of the nave area—counter-facade infiltration.

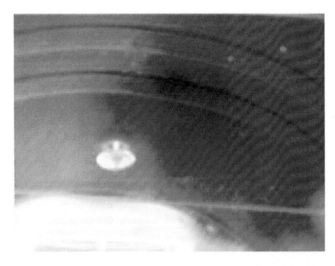

Figure 2.15 Santa Maria in Strada church, Monza. Thermogram of the vault, infiltration 1.

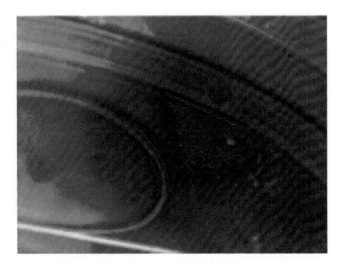

Figure 2.16 Santa Maria in Strada church, Monza. Thermogram of the vault, infiltration 2.

indication is important but sometimes misleading. In fact, it appears suddenly when moisture concentrates but remains after the surface dries out, although the water content decreases.

Broken pipes cause localized high water concentration: the wall materials soon get saturated. Although in many cases the extent of the infiltration will not appear visually and not close to the failure, infrared images show the real extent of its spread. Careful analysis is needed for accurate diagnosis.

2.2.1.5   Testing procedure of the passive approach

A preliminary scanning from a point as far from the masonry as possible makes it possible to set the camera parameters and to detect the anomalies in a few images, covering the whole surface. Further measurements closer to the surface permit more precise analysis for small discontinuities.

The classification of the surface in homogeneous areas is achieved in almost steady thermal state because of natural boundary conditions, based on the following steps:

1. Collect the available description and plans of the building; the building technology; information about the masonry, finishing, and all materials used; and the survey of the present state of damage due to moisture.
2. A suitable geometrical drawing of the interior and exterior of the building at adequate scale represents the template on which the visual state, thermograms, and others ancillary data will be superimposed.
3. The actual boundary conditions and in particular the temperature, relative humidity, and solar irradiation have been recorded for at least 24 hours before the thermogram is recorded.

Boundary conditions are controlled, which are as follows:

1. The inspected surface must be kept out of direct heating for about 12 hours before the scanning, because different absorption coefficients of the surface cause effects contrasting the cooling due to evaporation.
2. Environmental conditions should have a medium to high transpiration (relative humidity lower than 80% and air temperature not below 279K [6°C = 43°F] in the air surface boundary layer). Note that adverse microclimatic conditions may prevent a reliable testing. Most environmental circumstances require increasing the level of transpiration. During outdoor surveys, the weather can change quickly, so the operator has to measure temperature and relative humidity periodically. The indoor microclimate may vary too but in a longer time scale than that needed for the scanning. Hence, relative humidity and air temperature can be measured only at the beginning of the test, after it, and at any significant change of the setting as needed.
3. Normally, the infrared camera is set at the maximum sensitivity. In fact, the cooling effect of evaporation gives small differences of temperature. The manual setting of the readout range is preferable, even if some detail of the infrared image is out of scale. Additionally, the average function is usually applied in case of noisy images to create smooth temperature variations.
4. The choice of the field of view depends on the lens and on the distance. Normally, a doubling of the scale is recommended.
5. The identification of the damp areas is achieved by comparing the temperatures of the dry surfaces to those of the moist surfaces. In the case of a wide surface study, the investigation may require the shooting of numerous thermograms, which will be composed into a composite image.

2.2.1.6   Results and discussion

The anticipated methods of each application have been presented together with some example of the obtained results, for the better understanding of the achievable information and assessment that can be obtained. The aim of this subsection is to further discuss some aspects of the application that has to be taken in account to achieve reliable results and to present the results of the advanced application of the IRT.

2.2.2 Constraints on the in-field applications of IRT

After a period of laboratory experimentation and theoretical modeling, the application of infrared thermography has been optimized in the field for detecting the abovedescribed thermal anomalies. The effectiveness of the analyzed techniques and influences of some conditions in real tests are discussed in the following (Fokaides & Kalogirou, 2011).

On-site applications require the procedure to be as fast as possible, to keep the condition of the scanning constant and reduce costs. Passive techniques have the advantage of being much more productive and easily applied to surfaces larger than 10 m^2 (108 ft^2).

In contrast, using artificial active heating sources means that only a small area can be homogeneously heated. Moreover, effective radiating requires a powerful power supply (more than 3 kW). The sources must be supported close to the surface with expensive scaffoldings, needed to survey roofs or such areas out of reach. As a consequence, it is necessary to scan the whole surface in smaller areas of $1-2 \text{ m}^2$ ($10-22 \text{ ft}^2$). Additional time is needed for the transient analysis. In fact, a sequence of thermograms is recorded and each pixel processed in its turn. The only advantage of the active procedure is the reduced dependence on environmental conditions. Broadly, the active approach can best be applied in addition to the passive approach in case of the presence of other (not removable) sources of heating, because it is based on temperature change. On the contrary, the passive approach works at steady state.

Surface condition. Often historical buildings lack homogeneous surfaces, either in color or material. Differences of heating absorption affect the thermal analysis, causing false alarms and altering the thermal response of the structure. Particularly in the case of frescoes and paintings, the artificial heating (for instance by lighting) highlights such a difference. Furthermore, salt deposits modify reflectance and emissivity. Soluble salts are ever present in a surface damaged by damp. Generally, deposit of salts decreases the emissivity, resulting in an apparent temperature on the thermogram which is lower than the actual.

Heating. Artificial heating must be applied as perpendicularly to the surface and as diffusely as possible. In cases of heating from long distances or from the bottom of elevations, artificial heating could be very uneven. An autonomous power supply could be necessary. In many cases, the temperature increase must be limited to only a few degrees, because of the preciousness of the surfaces.

The heating of all the layers of the structure is hardly achievable: the low thermal diffusivity and the wideness of a building require lengthy and powerful heating. The direction of the heat flux should be unidirectional but the discontinuities of the surface modify the propagation of the heat inside the structures.

Visual analysis. The visible damage of the materials does not correspond to the actual state of the masonry. For example, cracks in the coating, in the plasters, and in the external layers sometimes can be smaller than inside damage and traces of damp. Stains, salt deposits, increases of porosity, delaminating of plasters, voids, and other kinds of damage may occur even if water has dried. Furthermore, the damage of the surface often modifies locally the imposed heat flux, complicating the diagnosis. Nevertheless, the availability of the picture of the surface is extremely useful.

2.3 Conclusion

In conclusion, the utilization of IRT techniques in historical buildings has demonstrated great feasibility in evaluating, assessing, and monitoring cultural heritage across a wide variety of buildings and purposes. Although there are certain constraints and limitations of the application, primarily dependence on the climatic and environmental condition, it is possible to obtain reliable results in many cases regardless of their geographic location (Northern Europe and America, tropical areas, and so on) even if they are not present in the temperate zone.

In addition, the use of IRT requires the collaboration with many scientific disciplines (history of architecture, the technology of ancient building materials, the technology of historical buildings, and so on) both in the planning phase of the test and the evaluation of the results. The collaborative evaluation of the results often results in a better understanding of the building itself and an improved assessment, far beyond the initial aim of testing the building.

Acknowledgments

The authors thank the collaboration on image editing of Megi Zala and the coauthors quoted in the references.

References

Ascione, F., Bianco, N., De Masi, R. F., De'Rossi, F., & Vanoli, G. P. (2015). Energy retrofit of an educational building in the ancient center of Benevento. Feasibility study of energy savings and respect of the historical value. *Energy and Buildings*, *95*, 172−183. Available from https://doi.org/10.1016/j.enbuild.2014.10.072, https://www.journals.elsevier.com/energy-and-buildings.

Avdelidis, N. P., & Moropoulou, A. (2004). Applications of infrared thermography for the investigation of historic structures. *Journal of Cultural Heritage*, *5*(1), 119−127. Available from https://doi.org/10.1016/j.culher.2003.07.002, http://www.elsevier.com.

Barbosa, M. T. G., Rosse, V. J., & Laurindo, N. G. (2021). Thermography evaluation strategy proposal due moisture damage on building facades. *Journal of Building Engineering*, *43*. Available from https://doi.org/10.1016/j.jobe.2021.102555, http://www.journals.elsevier.com/journal-of-building-engineering/.

Bisegna, F., Ambrosini, D., Paoletti, D., Sfarra, S., & Gugliermetti, F. (2014). A qualitative method for combining thermal imprints to emerging weak points of ancient wall structures by passive infrared thermography - A case study. *Journal of Cultural Heritage, 15* (2), 199−202. Available from https://doi.org/10.1016/j.culher.2013.03.006, http://www. elsevier.com.

Bison, P., Cadelano, G., Capineri, L., Capitani, D., Casellato, U., Faroldi, P., Grinzato, E., Ludwig, N., Olmi, R., Priori, S., Proietti, N., Rosina, E., Ruggeri, R., Sansonetti, A., Soroldoni, L., & Valentini, M. (2011). Limits and advantages of different techniques for testing moisture content in masonry. *Materials Evaluation, 69*(1), 111−116. Available from https://asnt.org/MajorSiteSections/Publications/Periodicals/ME.aspx.

Casapulla, C., Maione, A., & Argiento, L. U. (2018). Infrared thermography for the characterization of painted vaults of historic masonry buildings. *International Journal of Structural Glass and Advanced Materials Research, 2*(1), 46−54. Available from https://doi.org/ 10.3844/sgamrsp.2018.46.54.

Castillo, V., Pérez-Lara, M. A., Rivera-Muñoz, E., Arjona, J. L., Rodríguez-García, M. E., Acosta-Osorio, A., & Galván-Ruiz, M. (2012). Thermal imaging as a non-destructive testing implemented in heritage conservation. *Journal of Geography and Geology, 4*, 1−12.

Daffra, E., Grazioli, R., & Rosina, E., Contessi, L. (2008). *Convegno Scienza e Beni Culturali XXIV 941−950 Un ciclo del Cinquecento o un ambiente dell'Ottocento? Indagini propedeutiche al restauro del Salone di Ulisse nel Palazzo della Provincia di Bergamo.*

Danese, M., Demšar, U., Masini, N., & Charlton, M. (2010). Investigating material decay of historic buildings using visual analytics with multi-temporal infrared thermographic data. *Archaeometry, 52*(3), 482−501. Available from https://doi.org/10.1111/j.1475-4754. 2009.00485.x.

Della Torre, S., & Rosina, E. (2008). Rapid techniques for monitoring historic fabric in preservation plan. *International Workshop SMW08, In situ monitoring of monumental surface* (pp. 421−426).

Falchi, L., Slanzi, D., Balliana, E., Driussi, G., & Zendri, E. (2018). Rising damp in historical buildings: A Venetian perspective. *Building and Environment, 131*, 117−127. Available from https://doi.org/10.1016/j.buildenv.2018.01.004, http://www.elsevier.com/inca/publications/store/2/9/6/index.htt.

Fokaides, P. A., & Kalogirou, S. A. (2011). Application of infrared thermography for the determination of the overall heat transfer coefficient (U-Value) in building envelopes. *Applied Energy, 88*(12), 4358−4365. Available from https://doi.org/10.1016/j.apenergy. 2011.05.014, http://www.elsevier.com/inca/publications/store/4/0/5/8/9/1/index.htt.

Georgescu, M. S., Ochinciuc, C. V., Georgescu, E. S., & Colda, I. (2017). Heritage and climate changes in Romania: The St. Nicholas Church of Densus, from degradation to restoration. *Energy Procedia, 133*, 76−85. Available from https://doi.org/10.1016/j. egypro.2017.09.374, http://www.sciencedirect.com/science/journal/18766102, 18766102 Elsevier Ltd Romania.

Grinzato, E., Bison, P. G., & Marinetti, S. (2002). Monitoring of ancient buildings by the thermal method. *Journal of Cultural Heritage, 3*(1), 21−29. Available from https://doi. org/10.1016/S1296-2074(02)01159-7.

Grinzato, E., Vavilov, V., & Kauppinen, T. (1998). Quantitative infrared thermography in buildings. *Energy and Buildings, 29*(1), 1−9. Available from https://doi.org/10.1016/ s0378-7788(97)00039-x, https://www.journals.elsevier.com/energy-and-buildings.

Kilic, G. (2015). Using advanced NDT for historic buildings: Towards an integrated multidisciplinary health assessment strategy. *Journal of Cultural Heritage, 16*(4), 526−535. Available from https://doi.org/10.1016/j.culher.2014.09.010, http://www.elsevier.com.

Kirimtat, A., & Krejcar, O. (2018). A review of infrared thermography for the investigation of building envelopes: Advances and prospects. *Energy and Buildings*, *176*, 390−406. Available from https://doi.org/10.1016/j.enbuild.2018.07.052.

Kordatos, E. Z., Exarchos, D. A., Stavrakos, C., Moropoulou, A., & Matikas, T. E. (2013). Infrared thermographic inspection of murals and characterization of degradation in historic monuments. *Construction and Building Materials*, *48*, 1261−1265. Available from https://doi.org/10.1016/j.conbuildmat.2012.06.062.

Kylili, A., Fokaides, P. A., Christou, P., & Kalogirou, S. A. (2014). Infrared thermography (IRT) applications for building diagnostics: A review. *Applied Energy*, *134*, 531−549. Available from https://doi.org/10.1016/j.apenergy.2014.08.005, http://www.elsevier.com/inca/publications/store/4/0/5/8/9/1/index.htt.

Lerma, C., Mas, Á., Gil, E., Vercher, J., Peñalver, M. J., Odgaard, T., Bjarløv, S. P., & Rode, C. (2014). Interior insulation − Experimental investigation of hygrothermal conditions and damage evaluation of solid masonry façades in a listed building. *Relationship between laboratory testing and infrared thermography, 64*.

Ludwig, N., Redaelli, V., Rosina, E., & Augelli, F. (2004). Moisture detection in wood and plaster by IR thermography. *Infrared Physics & Technology*, *46*(1-2), 161−166. Available from https://doi.org/10.1016/j.infrared.2004.03.020.

Ludwig, N., Redaelli, V., & Rosina, E. (2005). *IRT for mapping restoration plasters by convective heating*. ART2005. AIPnD: Lecce., 1−16.

Ludwig, N., & Rosina, E. (2005). Dynamic IRT for the frescoes assessment, the study case of Danza Macabra in Clusone (Italy). *Proceedings of SPIE - The International Society for Optical Engineering*, *5782*, 272−279. Available from https://doi.org/10.1117/12.604648, 0277786X Italy.

Ludwig, N., & Rosina, E. (2006). International conference on quantitative infrared thermography VIII, *Proceeding CNR-ITC Restoration mortars at IRT: optical and Hygroscopic properties of surfaces*.

Ludwig, N., Rosina, E., & Sansonetti, A. (2018). Evaluation and monitoring of water diffusion into stone porous materials by means of innovative IR thermography techniques. *Measurement*, *118*, 348−353. Available from https://doi.org/10.1016/j.measurement.2017.09.002.

Di Maio, R., Piegari, E., Mancini, C., & Chiapparino, A. (2015). Quantitative analysis of pulse thermography data for degradation assessment of historical buildings. *European Physical Journal Plus.*, *130*(6). Available from https://doi.org/10.1140/epjp/i2015-15105-6, https://www.springer.com/journal/13360.

Meola, C. (2007). Infrared thermography of masonry structures. *Infrared Physics and Technology*, *49*(3), 228−233. Available from https://doi.org/10.1016/j.infrared.2006.06.010.

Meola, C. (2013). Infrared thermography in the architectural field. *The Scientific World Journal*, *2013*, 1−8. Available from https://doi.org/10.1155/2013/323948.

Olmi, R., Riminesi, C., Rosina, E. (2016). Integration of EFD, MRM and IRT for moisture mapping on historic masonry: Study cases in Northern Italy emerging technologies in non-destructive testing VI. *Proceedings of the 6th International Conference on Emerging Technologies in Nondestructive Testing*, ETNDT 2016 9781138028845 471−476 CRC Press/Balkema Italy. Available from https://doi.org/10.1201/b19381-78.

Paoletti, D., Ambrosini, D., Sfarra, S., & Bisegna, F. (2013). Preventive thermographic diagnosis of historical buildings for consolidation. *Journal of Cultural Heritage*, *14*(2), 116−121. Available from https://doi.org/10.1016/j.culher.2012.05.005, http://www.elsevier.com.

Parisi, E. I., Suma, M., Güleç Korumaz, A., Rosina, E., & Tucci, G. (2019). Aerial platforms (uav) surveys in the VIS and TIR range. Applications on archaeology and agriculture. *ISPRS Annals of the Photogrammetry, Remote Sensing and Spatial Information*

Sciences, 42(2), 945−952. Available from https://doi.org/10.5194/isprs-Archives-XLII-2-W11-945-2019, 21949050 Copernicus GmbH Italy, http://www.isprs.org/publications/annals.aspx.

Parisi, E.I., & Tyc, J. (2021). Multi-scale and multi-domain approaches for cultural terraced landscapes. *ARQUEOLÓGICA 2.0 − 9th International Congress & 3rd GEORES − GEOmatics and PRESservation* (pp. 317−324).

Quagliarini, E., Lenci, S., & Seri, E. (2012). On the damage of frescoes and stuccoes on the lower surface of historical flat suspended light vaults. *Journal of Cultural Heritage, 13*(3), 293−303. Available from https://doi.org/10.1016/j.culher.2011.11.008, http://www.elsevier.com.

Rosina, E., Cadelano, G., Ferrarini, G., Bortolin, A., & Bison, P. (2017). The instrumental investigation at Padri Serviti Convent: traditional and innovative active approach of IRT in historic solid masonry. *Poliscript, Milan*, 67−90, http://ISBN 978-88-6493-040-4.

Rosina, E., Ludwig, N., Della Torre, S., Ascola, S., Sotgia, C., & Cornale. (2008). Thermal and hygroscopic characteristic of restored plasters with different surface textures. *Materials Evaluation, 66*, 1271−1276.

Rosina, E., Ludwig, N., Redaelli, V., & Robison, E. (2003). IRT techniques for the detection of timber moisture. *Proceedings of SPIE - The International Society for Optical Engineering, 5073*, 100−108. Available from https://doi.org/10.1117/12.486890, Italy.

Rosina, E. (2018). When and how reducing moisture content for the conservation of historic building. A problem solving view or monitoring approach? *Journal of Cultural Heritage, 31*, S82−S88. Available from https://doi.org/10.1016/j.culher.2018.03.023, http://www.elsevier.com.

Rosina, E., Palazzo, M., & Tasso, F. (2017). *The preliminary thermo-hygrometric research and environmental monitoring of Sala delle Asse*. 0 130-143.

Tejedor, B., Lucchi, E., Bienvenido-Huertas, D., & Nardi, I. (2022). Non-destructive techniques (NDT) for the diagnosis of heritage buildings: Traditional procedures and futures perspectives. *Energy and Buildings, 263*112029. Available from https://doi.org/10.1016/j.enbuild.2022.112029.

Tucci, G., Parisi, E. I., Castelli, G., Errico, A., Corongiu, M., Sona, G., Viviani, E., Bresci, E., & Preti, F. (2019). Multi-sensor UAV application for thermal analysis on a dry-stone terraced vineyard in rural Tuscany landscape. *ISPRS International Journal of Geo-Information, 8*(2). Available from https://doi.org/10.3390/ijgi8020087, https://www.mdpi.com/2220-9964/8/2/87/pdf.

Tyc, J., Parisi, E. I., Tucci, G., Hensel, D. S., & Hensel, M. U. (2022). a data-integrated and performance-oriented parametric design process for terraced vineyards. *Journal of Digital Landscape Architecture, 2022*(7), 504−521. Available from https://doi.org/10.14627/537724049, https://gispoint.de/fileadmin/user_upload/paper_gis_open/DLA_2022/537724049.pdf.

Tyc, J., Sunguroğlu Hensel, D., Parisi, E. I., Tucci, G., & Hensel, M. U. (2021). Integration of remote sensing data into a composite voxel model for environmental performance analysis of terraced vineyards in tuscany, italy. *Remote Sensing, 13*(17). Available from https://doi.org/10.3390/rs13173483, http://www.mdpi.com/journal/remotesensing.

UNESCO Intergovernmental Committee. (2018). 2023 1 10 *Decision of the Intergovernmental Committee: 13.COM 10.B.10*. https://ich.unesco.org/en/decisions/13.COM/10.B.10.

Warren, L. A., Briggs, K. M., & McCombie, P. F. (2016). Thermal imaging assessment of drystone retaining walls. *Proceedings of the Institution of Civil Engineers: Forensic Engineering, 169*(3), 111−120. Available from https://doi.org/10.1680/jfoen.16.00012, http://www.icevirtuallibrary.com/content/serial/feng.

Evaluation of the envelope airtightness by means of combined infrared thermography and pressurization tests in heritage buildings: A case study

3

Irene Poza-Casado, Diego Tamayo-Alonso, Miguel Ángel Padilla-Marcos, and Alberto Meiss
GIR Arquitectura & Energía, Dpto. Construcciones Arquitectónicas, Ingeniería del Terreno y Mecánica de los Medios Continuos y Teoría de Estructuras, Universidad de Valladolid, Valladolid, Spain

3.1 Introduction

3.1.1 Background

Nowadays focus has been put on the reduction of the energy use of buildings. It is estimated that around 35% of the EU's buildings are over 50 years old and almost 75% of the building stock is energy-inefficient (Filippidou & Navarro, 2019). Hence, improving the energy performance not only of new buildings but also of the existing built stock is crucial to reach the EU's objectives (Directive 2018/844 amending Directive 2010/31/EU on the energy performance of buildings and Directive 2012/27/EU on energy efficiency, 18/844 Amending Directive 2010/31/EU on the Energy Performance of Buildings and Directive 2012/27/EU on Energy Efficiency, 2018). Strong long-term renovation strategies must be encouraged fostering deep energy retrofitting of existing buildings.

Therefore establishing solutions for the renovation of national building stocks and promoting their transformation into nearly zero-energy buildings seems necessary. The problem of energy loss caused by conduction through the building envelope has been significantly reduced by the development of advanced insulating materials, but it is frequently forgotten that the presence of air infiltration has a significant impact on the overall energy performance of buildings. Previous research estimated that 10%−30% of the heating demand is caused by air infiltration, which is mostly determined by the permeability of the building (Domínguez-Amarillo et al., 2019; Huang et al., 1999; Jokisalo et al., 2009; Jones et al., 2015; Kalamees, 2007; Meiss & Feijó-Muñoz, 2015; Simson et al., 2020; Zheng et al., 2020).

Diagnosis of Heritage Buildings by Non-Destructive Techniques. DOI: https://doi.org/10.1016/B978-0-443-16001-1.00003-6

Consequently, actions toward better energy efficiency will necessarily involve the improvement of the thermal performance of the exterior envelope including airtight solutions. Because leakages are hidden in walls and other cavities, it is difficult to increase the airtightness of existing buildings. Instead, leakage paths must be individually identified and sealed (Bohac et al., 2016), so special attention needs to be paid to the diagnosis stage.

In this context, historic buildings play a key role. Age in buildings hardly affects the presence of envelope leakages, but historic building techniques could not avoid them (Lerma et al., 2014). Nowadays technology allows for achieving almost perfect airtight envelopes by using special tapes and new building systems, which are effective for airtight joints between materials and mechanical parts and giving continuity to the built surfaces.

To assess historic buildings, non-destructive techniques (NDTs) need to be implemented. In the field of airtightness, several methods have been developed over the years (Charlesworth, 1988; Kronvall, 1980; McWilliams, 2003; Priestner & Steel, 1991; Zheng et al., 2020). Fan pressurization tests prevail due to their simplicity, time requirements, availability of commercial equipment, and mitigation of weather conditions by the pressure differential (Ashrae, 2021). Thus this NDT is considered reasonably accurate and reproducible (Sherman & Chan, 2004).

Pressurization tests provide quantitative information on the airtightness of a building, but this method does not address leakage location. Thus several methods have been reported in the literature to qualitatively assess leakages: smoke detection, anemometers, the soap bubble method, hand inspection, or helium-filled balloons (Pickering et al., 1987), but IR thermography (IRT), often combined with depressurization, seems to be the most widespread method (Kronvall, 1980). IRT technology is applied in old and new buildings to identify leakages in the building without affecting the finishing work (Lucchi, 2018).

The aim of this study consists of the development of a non-destructive methodology to evaluate the airtightness performance of the envelope and its application in a historic building. IRT was used while depressurizing the volume to locate the leakages along the envelope of the conditioned space. Knowing that the space is a residential retrofitted unit in a historical building, the evaluation of leakages can improve the global performance of the envelope to establish a strategy for energy retrofitting. This kind of building has been normally refurbished covering those existing historical structures, walls, floors, or roofs, which promotes indirect air leakage through unidentified air chambers in the building.

3.2 Materials and methods

3.2.1 Case study

3.2.1.1 History and context of the building

The NDT methodology developed within this research was applied to a room of a historic building of Universidad de Valladolid (Spain), that is, the residence hall

Figure 3.1 Palacio Santa Cruz. View from Cardenal Mendoza Street.
Source: From Fundación Joaquín Díaz.

"Colegio Mayor Santa Cruz" (Fig. 3.1). This building was founded in 1483 and finished in 1491 as a protohumanist work whose purpose was to provide greater access to culturally inclined individuals accessing the university. The sponsor of this great institution was Don Pedro González de Mendoza, Cardinal Mendoza, whose purpose was to build a symbolic building for the institution. Today, this building is known as "Palacio Santa Cruz" and is the headquarter of the Rector's Office of Universidad de Valladolid (Ordax, 2005).

Later on, as a consequence of the time limit rule for the residents, the so-called Hospedería del Colegio Mayor Santa Cruz, a guesthouse dependent on the residence hall, was created. It seems that this category of "guest" residents was generalized at the end of the 16th century. In the guesthouse, although they were not proper residents, former residents continued to enjoy almost all the privileges—except the right to vote. Outsiders could also stay when they purchased a letter of brotherhood.

Historic sources make reference to the establishment of the guesthouse in the 16th century, in 1588. In addition, the need to expand its capacity made the residence hall buy several annex buildings between 1641 and 1649 (Fernández et al., 2010). This explains why when a guesthouse is built in the second half of the 17th century, it is referred to as a "new guesthouse" and the remains of an "old guesthouse" are mentioned in Fig. 3.2.

In 1675, Master Francisco de la Torre was commissioned to build a new guesthouse, whose common facilities and cells were organized around a cloister with four arches. This building has two floors and a square layout, and it was smaller in size than the residence hall. The exterior walls are made of a stone plinth and a

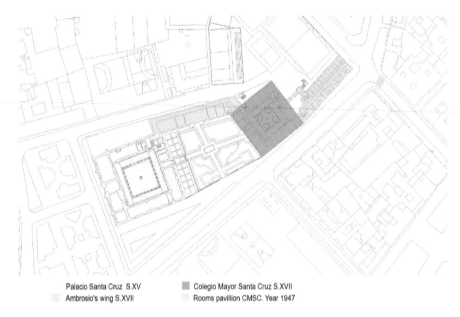

Palacio Santa Cruz S.XV   Colegio Mayor Santa Cruz S.XVII
Ambrosio's wing S.XVII   Rooms pavillion CMSC. Year 1947

Figure 3.2 Santa Cruz complex. Historical evolution.

combination of masonry and brick, according to 17th-century preferences Fig. 3.3. The masonry is quite correct, with stones of great size, while the brickwork is done with a relatively thick tendril in which cracks are marked achieving a play of light and shade.

At the end of the 18th century, after many years of improper practices in the residence halls, guesthouses were eliminated, and the privileges of the residents definitively disappeared.

During the 19th and early 20th centuries, the Santa Cruz complex had different uses, such as the Episcopal Palace, the residence of General Wellington after the liberation of Valladolid from the French occupation, the prison, the School of Arts and Crafts, the Art Museum, Archaeological Museum, and headquarter of the Real Academy of the Purísima Concepción. Later on, in 1935, it became the Rector's Office of the University of Valladolid and the Archaeological Museum until 1968. Since 1929, the building has recovered its use as a residence hall (Ordax, 2005).

So many changes in its use led to continuous renovations over time, which were hardly ever documented. Only the renovations made by the end of the 20th century as a residence hall have been registered: a glass closure was installed in the first-floor corridor around the courtyard, and rooms with bathrooms and the director's house, which later became facilities for the students, were designed.

Currently, the guesthouse has six rooms with bathrooms, which are rented for short stays to university visitors. The building also comprises several facilities for the students who stay in the adjacent building, such as a study room, video library, dining room, living room, and auditorium.

Figure 3.3 Guesthouse (current Santa Cruz Residence Hall). Main facade.

3.2.1.2 Building characterization

The building under study, "Hospedería del Colegio Mayor Santa Cruz," is located parallel to Cardenal Mendoza Street in the city center of Valladolid, near other historic buildings of the University. Its main facade faces southwest and can be accessed from the garden located between the guesthouse and Palacio Santa Cruz. The square-shaped building, 34 m on each side, stands free on all sides except for two corners, where it connects to the San Ambrosio wing and the rooms pavilion of the residence hall, as shown in Fig. 3.4.

The building is configured with a double concentric gallery with respect to the central courtyard. The exterior gallery is supported on 110 cm load-bearing walls made of brick and masonry, and the inner gallery used as a corridor is supported on the intermediate load-bearing wall and the arches on stone columns of the courtyard. Between these supports, wooden beams are placed, and on top of them, a wooden platform. Internal partitions are made of simple-hollow brick with plaster on both sides. In the bathrooms of the rooms, a marble floor is installed on this platform, as well as the walls' finishing. The rooms have a false ceiling with perimeter molding. The bathrooms also have a false ceiling, but in this case with an open perimeter groove. The windows are relatively recent, made of wood with double glazing with a chamber.

A suite room on the southeast facade, facing Cardenal Mendoza Street, was chosen as the area to be studied, according to the floor plan in Figs. 3.5−3.7. The name of the room honors a former resident, Juan de Marquina. Tests were carried out considering the whole suite but also excluding the volume of the bathroom.

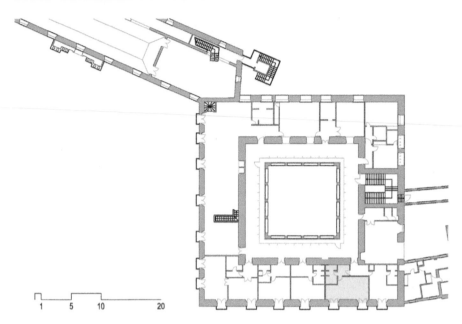

Figure 3.4 Guesthouse. First floor.

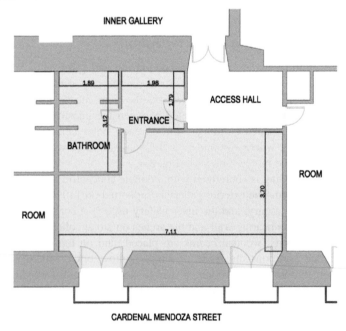

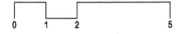

Figure 3.5 Room of the guesthouse under study. Testing area floor plan (dimensions in meters).

Figure 3.6 Room of the guesthouse under study. Elevation.

Figure 3.7 Room of the guesthouse under study. Interior layout.

The room is adjacent to two other rooms on the same floor, the exterior wall, and the nonconditioned corridor and access hall. Below, there are conditioned common facilities, whereas there is a crawl space above the room. This is a ventilated space, and it is partially insulated.

Table 3.1 shows the main dimensions of the room.

The room presents hardly any pathology. Only some cracks in the false ceiling were found, probably due to some movement of the building, after an intervention (Fig. 3.8).

The building's conditioning system, installed in the second half of the 20th century, consists of a diesel boiler, currently replaced by a gas boiler, and water radiators in the rooms. Regarding ventilation, there is only mechanical exhaust ventilation in the bathroom commanded by the bathroom light (Fig. 3.9). Fresh air enters the room through leakages in the envelope as a consequence of the depressurization caused by the exhaust.

3.2.2 Pressurization of the building envelope

The air permeability of the room envelope was measured by means of the fan pressurization method, also known as the blower door test, which has become the most widely method used to evaluate the airtightness of a building (Priestner & Steel, 1991). The equipment used was Minneapolis Blower Door Model 4. The test was performed according to ISO 9972,72 (2015). This test is meant to assess the airtightness of a whole building or a part of it, in this case, a room of the guesthouse. Therefore the room is treated as a single-zone building by opening interior doors in the adjacent rooms. In this sense, it is important to highlight that the envelope includes the boundary with the outdoor air but also with other conditioned rooms and unconditioned spaces, and this test does not provide information regarding the location of the leaks. Guarded-zone airtightness tests were not performed.

For the test, a fan temporarily replaces the main door of the space to be measured and generates a pressure difference between the inside and the outside of the room (Fig. 3.10). Depending on the position of the fan, the depressurization or pressurization of the room forces the entrance/exhaust of the outdoor/indoor air, respectively. In this way, the airtightness of the envelope is measured.

High wind speeds and indoor/outdoor temperature difference compromise the accuracy of the results. However, under favorable conditions, the generated pressure difference overcomes weather impact on the measurement.

The assessed room was prepared according to Method 2 described in ISO 9972. That is the measurement of the building envelope, with all intentional openings sealed. In this way, the bathroom exhaust was sealed, and the windows were closed.

Automated tests were carried out and monitored by the software provided by the blower door manufacturer (TECTITE Express). Before inducing any pressure by means of the fan, and also at the end of the test, the zero-flow pressure difference is evaluated over a period of 30 seconds to verify that the test has been performed according to the standard. Then, two sets of measurements, both pressurization and depressurization, are induced taking measurements of the airflow rate over a range of pressure differences between ± 16 and ± 70 Pa in increments of 6 Pa. These measurements determine the leakage function (log-log plot), which establishes a relationship between the airflow rate and the pressure difference (Ashrae, 2021) through the so-called power law:

$$Q_{pr} = c_L \cdot (\Delta p_r)^n$$

Table 3.1 Main dimensions of the room under study.

Room	Floor area	Height	Volume	Outdoor envelope area	Heated envelope area	Unconditioned envelope area	Total envelope area	Windows
Units	(m²)	(m)	(m³)	(m²)	(m²)	(m²)	(m²)	(m)
Whole room	34.54	2.98	102.77	24.82	61.19	57.67	143.68	2 (1.4 × 2.4)
Room excluding the bathroom	28.75	2.98	85.68	134.56	28.75	20.56	134.56	2 (1.4 × 2.4)

Figure 3.8 Room of the guesthouse under study. False ceiling pathology.

Figure 3.9 Room of the guesthouse under study. Bathroom extraction.

where Q_{pr} is the airflow rate of the opening at a reference pressure difference $[m^3/h]$; C_L is the air leakage coefficient $[m^3/(hPa^n)]$; Δp_r is the reference pressure difference $[Pa]$; n is the airflow exponent $[-]$.

An ordinary least square regression was used for the calculation of the airflow characteristics (ISO 9972,72, 2015). The coefficient n value must be in the range 0.5 and 1, and it provides information concerning the relative size of the leaks. Values near 0.5 are related to large leakages and short paths, whereas n values close to 1 are usually found in airtight envelopes due to a steady laminar flow through small leakages (Allen, 1985).

The correct calibration of the equipment was ensured to maintain accuracy specifications of 1% of sampling, or 0.15 Pa to reduce uncertainties. The overall uncertainty in the parameters obtained was considered to remain below 10% under the calm conditions reported (ISO 9972,72, 2015).

Results are referred at a reference pressure of 50 Pa and normalized by both the inner volume (V) and envelope area (A_E), obtaining the air change rate (n_{50}) and the specific air leakage rate (q_{50}), as shown in the following equations:

$$n_{50} = Q_{50}/V$$

Figure 3.10 Airtightness test layout. Blower door system.

where n_{50} is the air change rate at $50\,\mathrm{Pa}\left[\mathrm{h}^{-1}\right]$; Q_{50} is the air leakage rate at $50\,\mathrm{Pa}\left[\mathrm{m}^3/\mathrm{h}\right]$; and V is the internal volume $\left[\mathrm{m}^3\right]$.

$$q_{50} = Q_{50}/A_E$$

where q_{50} is the specific leakage rate per the building envelope area at $50\,\mathrm{Pa}\left[\mathrm{m}^3/\mathrm{hm}^2\right]$; Q_{50} is the air leakage rate at $50\,\mathrm{Pa}\left[\mathrm{m}^3/\mathrm{h}\right]$; and A_E is the envelope area $\left[\mathrm{m}^2\right]$.

3.2.3 Infrared thermography

IRT images consist of a series of plain thermal data. This data is analyzed using IRT processing software, which interpolates temperature values captured by the camera. For this purpose, the obtention of materials' parameters, especially IR emissivity (ε), is needed. IR emissivity is the property that materials have for emitting thermal radiant energy to the media. This depends on multiple characteristics of the material such as density, texture, and composition, among others, but not color or lighting.

Perfect emissivity ($\varepsilon = 1$) is only theoretically defined by the "black body." Other materials have an emissivity value below 1. To set up the emissivity in the IRT camera, black thermal insulation tape is used. This tape glued to the surface under study acquires the temperature of that surface in a couple of minutes. The emissivity is set by screening the temperature registered on the tape. Also, the picture can be adjusted using the image treatment software.

Thermal radiant energy comes from everywhere in space, so it is necessary to discard the energy reflected by the analyzed surface, other objects, or people. The most common process to do so is configuring the ambient temperature in the thermal IR imaging camera and discarding the thermal radiant energy reflected in the space. This is possible using Lambert's radiator, which provides the value of the emissivity part of the energy captured by the IR sensor within the camera. Lambert's radiator creates a diffuse reflection whose radiant temperature is a mix of the radiant temperatures coming from all the surfaces in the space. To set the ambient temperature of the reflected thermal radiation, the radiant temperature of Lambert's radiator is measured, adjusting its emissivity as a perfect black body ($\varepsilon = 1$). This is commonly known as radiant temperature correlation (RTC).

Every IRT equipment captures the IR thermal radiation emitted by the surface under study, the reflection of the IR thermal radiant energy of the ambient and the transmission of the IR thermal radiant energy along the object or surface, which is usually negligible.

This non-destructive evaluation is normally used to evaluate energy losses, for detecting a lack of insulation in the envelope of conditioned spaces (Fox et al., 2014), but also to locate moisture in building components and as a diagnosis tool of the building envelope (Barreira et al., 2016). A more advanced method is the detection of leakages through the envelope (Kirimtat & Krejcar, 2018). This is possible since air infiltration through building elements involves temperature differences on its surfaces that can be measured by IRT. The application of IRT to assess the airtightness and thermal insulation of the building envelope was thoroughly detailed by Pettersson and Axen (1980).

Leakages allow the air in and out through the building envelope. If the space is normally conditioned, it impacts the global energy efficiency of the building. Using IRT, air leakages can be located by analyzing the temperature difference between the air leakage affection area and a reference temperature in a close surface. The reference temperature is commonly obtained from the close nonaffected surface.

However, IRT also involves some limitations. It is demonstrated that sometimes the air is exchanged through different paths, such as building systems, crawl spaces, or ducts. In multiple-layer walls, the air may enter the air chamber between fabric or plaster layers from cracks in the external layer and get inside through leaks in the inner layer. This promotes a phenomenon of indirect air infiltration, which is not suitable for its full evaluation using IRT because of the preheating process in the air chamber (Janssens, 2003). In this case, alternative evaluation processes are needed.

Although IRT is usually applied to locate leakages, some authors have also developed a method for the quantitative assessment of air leaks: to evaluate the contribution of airtightness of building components (Barreira et al., 2017), to take an approximation to the surface area of the gap (Dufour et al., 2009), and to calculate the airflow through the gap (Baker et al., 1987), which all together define the air infiltration rate in buildings (Liu et al., 2018). This kind of analysis requires the evaluation of the characteristics of the building materials in the affection area, specially referred to as energy heat transfer. Its use in the detection of pathologies in historical and heritage buildings is useful (Tavukçuoğlu et al., 2018). This knowledge can give technicians a great amount of information for the evaluation of retrofitting strategies.

3.2.4 Infrared thermography combined with pressurization

The use of combined IRT and pressurization of the building envelope has become a common practice (Eskola et al., 2015; Feijó-Muñoz et al., 2018; Gillott et al., 2016; Tanyer et al., 2018). Pressurization tests are used to quantify the airtightness but no information regarding leakage location is given. This combined method allows for the identification of leakages provided that there is enough temperature difference between the indoor and outdoor environments (ISO 9972,72, 2015).

Outdoor air enters the room forced by the depressurization created by the fan through the leaks, causing a quick temperature difference in the surface surrounding the gap of the envelope of the conditioned space. In this sense, it is important to avoid confusing leakage paths and thermal bridges, which often are concentrated around the same areas (Charlesworth, 1988). To avoid it, image subtraction or image comparison between natural and pressurized conditions is often performed (Kalamees, 2007; Vollmer & Möllmann, 2017). In such cases, the reference temperature used is that registered before the air exchange, identifying the air leakage impact by comparing temperature data along the time (Fox et al., 2015).

The first step in the study is to prepare the space to take good-quality IR thermal images. A good way is to cover all the lamps and bulbs which use incandescent technologies with rugs, clothes, or rigid panels because they emit thermal IR radiation (longwave radiation). This can badly affect the identification of air leakages using IR thermal imaging.

After that, several IRT images are taken inside, focusing especially on the usual leakage paths (Fig. 3.11). A Flir e75 IR thermal camera is used to picture leakages. These are typically found in joints such as window frames, materials joints, wall cracks, electrical and sanitary grids, wall-mounted and ceiling lighting and other security facilities such as smoke detectors or emergency lighting in ceilings, and so on. The IR thermal camera has a 320×240 (76,800 pixels) resolution with an object temperature range of $-20°C$ to $120°C$ in an 18 mm lens.

For that, it is necessary to identify each material's emissivity. A black thermal insulation tape is positioned close to the area to be IR thermally captured. After 5 minutes, the surface of the tape acquires the temperature of the surface object of the IR thermal picture. The IRT camera can be set up by adjusting the emissivity as previously defined or evaluated with software. These IR thermal images are used as reference temperature indicators in the assumed areas of affection of the air leakages.

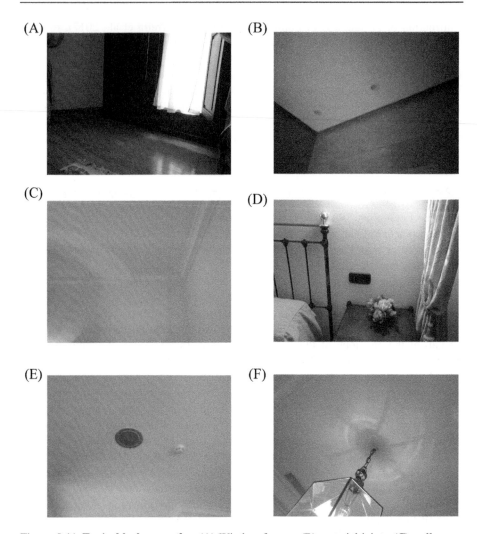

Figure 3.11 Typical leakage paths. (A) Window frames, (B) material joints, (C) wall cracks, (D) electrical grids, (E) ceiling holes for facilities, and (F) plafonds.

The emissivity values reached for the different construction materials in the room are shown in Table 3.2.

Before turning on the blowing fan, further thermal data must be registered by a term hygrometer. The outdoor temperature (2.8°C) is lesser than indoors (22.4°C), with a thermal difference of 19.6°C. Humidity is constant inside and outside. RTC is adjusted by placing the Lamberts' radiator close to the surface to be photographed. RTC value is the same (22.4°C) so no adjustment is needed.

Next, the blowing fan is set up in the position of depressurization, forcing the air coming from outdoors to enter the space through the leakage paths of the envelope. Depressurization is set up to a 50 Pa pressure difference (between inside and outside)

Table 3.2 Materials'emissivity in the room.

Position	Surface material	Emissivity (ε)
Ceiling	Painted plain plaster	0.90
Walls	Painted rough plaster	0.89
Walls (skirting board)	Vanished plain wood	0.87
Walls (bathroom)	Polished tiles	0.92
Windows	Painted plain wood	0.94
Windows	Glass	0.95
Windows	Painted metal	0.97
Floor	Vanished specular wood	0.89

in "cruise configuration." This pressure difference guarantees a quick transfer of thermal energy by the air entering the space and the air leakage affection areas. In winter conditions, the incoming air makes the temperature of the envelope surface decrease from that registered as a reference. After 5−10 minutes, the thermal IR images are taken. The surface differential temperature will focus the attention on the defying infiltration gap.

After the images are taken, every image is processed using specialized software by adjusting those parameters that are difficult to set during the test.

3.3 Results

3.3.1 Air permeability

The pressurization test to measure the air permeability of the guesthouse room was carried out between 10:00 a.m. and 12:00 p.m. The average outdoor temperature was 2°C with 75% relative humidity. The wind speed was on average 1.39 m/s from the northwest. Therefore environmental conditions were favorable and within the limits stated in ISO 9972. Indoor/outdoor temperature difference, expressed in Kelvin, multiplied by the height, expressed in meters, was lesser than 250 mK. Regarding wind speed, the limits were not reached and kept always below 3 on the Beaufort scale.

Initially, the test was performed considering the whole volume of the room (Test 1). Extremely high permeability results and the hypothesis that most of the airflow could come from the bathroom motivated the performance of a second pressurization test excluding its volume (Test 2). The configuration of both tests is shown in Fig. 3.12.

Table 3.3 and Fig. 3.13 show the results obtained regarding the airtightness of the envelope.

The obtained coefficient of determination r^2 was greater than 0.98, thus the tests can be considered valid.

The air change rate at 50 Pa obtained in Test 1 is extremely high. Results improve significantly when excluding the bathroom volume (Test 2), so it can be

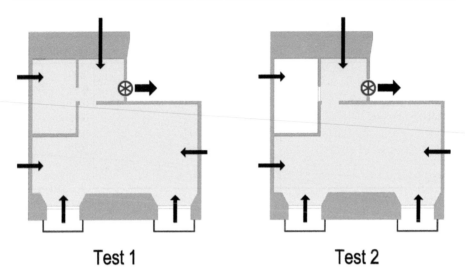

Figure 3.12 Airtightness test configuration. Test 1 (whole room) and Test 2 (excluding the bathroom).

stated that important leakages are located in this area. Although results are not comparable due to the different typologies and construction technologies, the airtightness of the room is close to the average n_{50} values reported for multifamily buildings in this area (Feijó-Muñoz et al., 2019).

The coefficient n values obtained close to 0.5 are a symptom of the leaky envelope. It must be noted that especially for Test 1, the airflow is almost fully turbulent, and it is clear the presence of large leakage paths. Regarding Test 2, a slight improvement can be observed, by adding a laminar component. Nevertheless, the values obtained for Test 2 are in agreement with other test results reported in this area (Feijó-Muñoz et al., 2019).

Finally, it can be observed in the graphs that the airflow rates during the pressurization stage are larger in both cases than the ones during depressurization. This can be explained by the fact that the building is subject to unnatural conditions, which can alter the leakage characteristics, especially false ceilings, traps, or some kinds of windows (Charlesworth, 1988). Leakage curves do not overlap as a result of a possible valving effect or asymmetric geometry of some leakages in relation to the flow direction (Allen, 1985; Baker et al., 1986).

3.3.2 Evaluation of leakages

The analysis of the permeability values shows the lack of airtightness of the envelope. However, the blower door test did not provide information regarding leakage location and characterization. IRT is then useful for its location and identification.

Fig. 3.14 shows a comparison of the IR thermal images taken before and after depressurization. The air leakage affection area is in this way shown. All the

Table 3.3 Pressurization tests results.

	Depressurization					Pressurization					Average values		
	$[m^3/h]$	$n_{50}[h^{-1}]$	$q_{50}[m^3/m^2h]$	$[-]$	$r^2[-]$	$Q_{50}[m^3/h]$	$n_{50}[h^{-1}]$	$q_{50}[m^3/m^2h]$	$[-]$	$r^2[-]$	$Q_{50}[m^3/h]$	$n_{50}[h^{-1}]$	$q_{50}[m^3/m^2h]$
Test 1	5296	51.53	36.86	0.51	0.996	6859	66.74	47.74	0.57	0.989	6078	59.14	42.30
Test 2	628	7.33	4.67	0.61	0.999	692	8.07	5.14	0.61	0.995	660	7.70	4.90

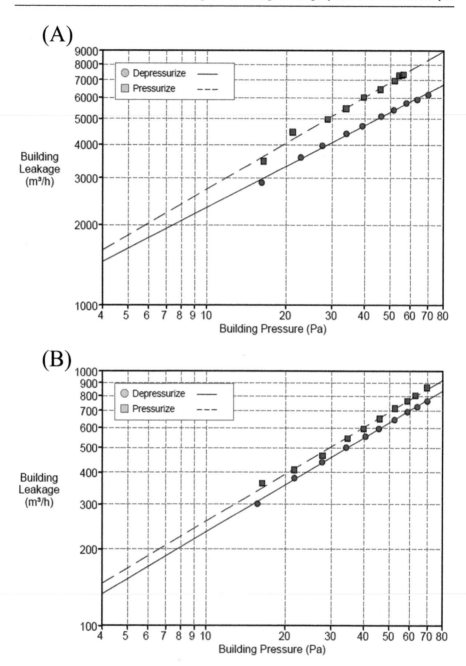

Figure 3.13 Pressurization tests graphs. (A) Test 1 graph (whole volume of the room) and (B) Test 2 graph (volume of the room excluding the bathroom).

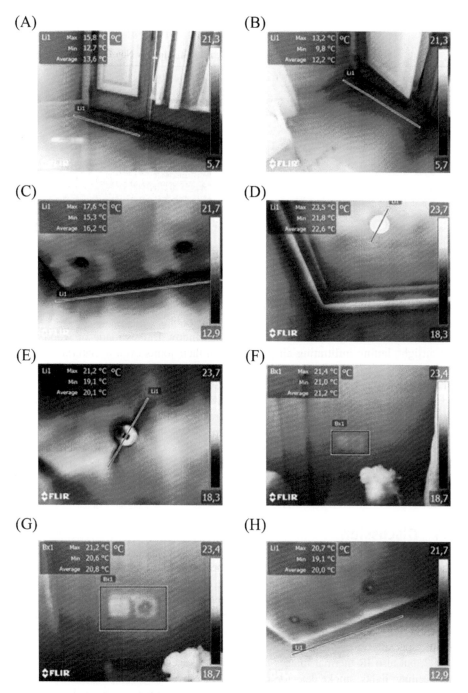

Figure 3.14 Comparison of the IR thermal images taken before and after depressurization. (A) Representative wooden door-window before depressurization test (floor emissivity ε = 0.89), (B) Representative wooden door-window after depressurization test (floor emissivity ε = 0.89), (C) Materials joints in bathroom before depressurization test (wall tiles emissivity ε = 0.92), (D) Materials joints in bathroom after depressurization test (wall tiles emissivity ε = 0.92), (E) Ceiling holes for smoke detector before depressurization test (ceiling emissivity ε = 0,90), (F) Ceiling holes for smoke detector after depressurization test (ceiling emissivity ε = 0,90), (G) Electrical grids before depressurization test (wall plaster emissivity ε = 0,89), (H) Electrical grids after depressurization test (wall plaster emissivity ε = 0,89). *IR*, Infrared.

(A)   (B)

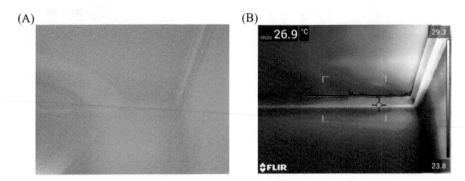

Figure 3.15 Corner cracks. (A) Picture and (B) IR thermal image. *IR*, Infrared.

coupled IR thermal images were adjusted to the same temperature range and a reference temperature line is defined for their comparison.

From these results, it can be assumed that window frames made of wood were not airtight, letting infiltrating air pass through their joints even at natural building pressure difference. Wall-mounted electrical grids did not involve any important air leakage. However, most of the air infiltration has its origin in the ceiling. It is assumed that there is a highly ventilated crawl space above the room between the room and the roof made of tiles. This was presumed from the observation of three different facts: (1) most of the air enters through the bathroom ceiling perimeter, which lacks joints to the walls; (2) the air infiltration affection area temperature decayed close to the holes made on the ceiling; and (3) corner cracks between ceiling and walls were identified (Fig. 3.15).

3.4 Discussion

The application of pressurization tests combined with IRT allowed both the quantification of the airtightness of the room envelope and the location of the main leakages.

Results show that the room is extremely leaky when including the bathroom volume. The evaluation of the envelope led to the hypothesis that the false ceiling is somehow connected to the crawl space above, letting unconditioned air enter the room, especially through the bathroom. This is supported by the results obtained in Test 2, but also IR thermal images show clear leakages in every discontinuity of the false ceiling: lights, smoke detector, or cracks. To prove this assumption, the ventilated roof crawl space was accessed, and the construction configuration was evaluated.

The hypothesis of a nonfinished false ceiling and insulation was then proved. As can be observed in Fig. 3.16, the wooden platform of the ceiling and the insulation show important discontinuities and even holes of great dimensions. Furthermore, the ventilation exhaust of the bathroom is open and not connected to the exhaust duct. This clearly explains the airtightness test results, which constituted a key element in the diagnosis of the energy performance of the room.

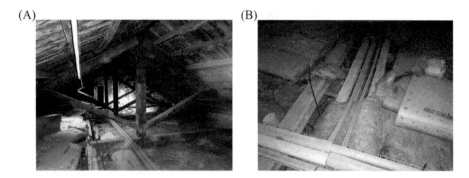

Figure 3.16 Upper crawl space. (A) General view and (B) detail of the construction elements.

The perimeter of windows also concentrates important leakage paths, especially in the joint with the pavement. In contrast, electrical grids and other discontinuities in the walls did not involve important temperature difference. This can be explained due to the masonry construction system of the dividing walls. This construction practice is now uncommon since new buildings usually integrate light plaster walls, which promote air exchange through the inner air chamber. When this chamber connects different spaces or even the outdoor air, the air can be freely exchanged through electrical grids, pipes, or any other open device.

3.5 Conclusion

The evaluation of the envelope of historic buildings requires the application of NDT, which does not alter its characteristics. In the case of energy audits and energy retrofitting solutions, a good diagnosis is key in the decision-making process. In this sense, the airtightness of the envelope is the main characteristic that involves air infiltration. Uncontrolled airflows across the building envelope led to a phenomenon that involves air mass exchange between the indoors and outdoors, and, thus, energy transfer and unnecessary energy use.

This chapter presents a methodology based on the combination of pressurization tests and IRT to quantify the airtightness of the envelope and to locate the main leakages. The method was then applied to a room in the Hospedería del Colegio Santa Cruz. Although in plain sight no pathologies were found and no inadequate performance had been reported, the assessment of the envelope airtightness showed hidden construction issues which seriously compromise the energy performance of the building.

The study demonstrated that air infiltration detected using IRT technology is due to several circumstances. Leaks were identified around the window frames. This can be related to material compatibility. The wood in the window frames can change its dimensions along its life, and, when no elastic joints are included, discontinuities and cracks allow the air to infiltrate.

On the other hand, the main leakages are related to the lack of airtightness and insulation of the room ceiling and the roof crawl space above, especially in the bathroom area. The inclusion of a light plaster false ceiling hides this and makes it almost impossible to identify this construction problem. This false ceiling contains multiple holes and leakages around several devices (speakers), lighting or security (smoke sensors or emergency lighting), and also some cracks near one of the corners of the room. This was further evaluated by observation from the crawl space. Clear gaps and unfinished construction elements explained the lack of airtightness. That creates a path for air transfer between spaces with different thermal conditions, which impacts the surface temperature close to those leakages. Therefore solving this construction problem constitutes a priority when implementing energy measures and retrofitting actions in the building.

The application of IRT while depressurizing the volume under study was found to be a useful method to identify weak points of the envelope without degrading the construction. This NDT proved to be crucial to evaluate the performance of the building envelope before proposing energy retrofitting solutions.

References

Allen, C. (1985). *Technical Note AIC 16. Leakage distribution in buildings.*

ASHRAE. (2021). *ASHRAE Handbook - Fundamentals.*

Baker, P. H., Sharples, S., & Ward, I. C. (1987). Air flow through cracks. *Building and Environment, 22*(4), 293–304. Available from https://doi.org/10.1016/0360-1323(87)90022-9.

Baker, P. H., Sharples, S., & Ward, I. C. (1986). Air flow through asymmetric building cracks. *Building Services Engineering Research & Technology, 7*(3), 107–108. Available from https://doi.org/10.1177/014362448600700302.

Barreira, E., Almeida, R. M. S. F., & Delgado, J. M. P. Q. (2016). Infrared thermography for assessing moisture related phenomena in building components. *Construction and Building Materials, 110,* 251–269. Available from https://doi.org/10.1016/j.conbuildmat.2016.02.026.

Barreira, E., Almeida, R. M. S. F., & Moreira, M. (2017). An infrared thermography passive approach to assess the effect of leakage points in buildings. *Energy and Buildings, 140,* 224–235. Available from https://doi.org/10.1016/j.enbuild.2017.02.009.

Bohac, D., Schoenbauer, B., Fitzgerald, J. (2016). Using an aerosol sealant to reduce multi-family envelope leakage. *ACEEE Summer Study on Energy Efficiency in Buildings,* pp. 1–16.

Charlesworth, P.S. (1988). *Air exchange rate and airtightness measurement techniques - An applications guide.* Air Infiltration and Ventilation Centre.

Directive 2018/844 amending Directive 2010/31/EU on the energy performance of buildings and Directive 2012/27/EU on energy efficiency. 156 (2018).

Domínguez-Amarillo, S., Fernández-Agüera, J., Campano, M. Á., & Acosta, I. (2019). Effect of airtightness on thermal loads in legacy low-income housing. *Energies, 12* (9). Available from https://doi.org/10.3390/en12091677, https://www.mdpi.com/1996-1073/12/9.

Dufour, M. B., Derome, D., & Zmeureanu, R. (2009). Analysis of thermograms for the estimation of dimensions of cracks in building envelope. *Infrared Physics and Technology*, 52(2−3), 70−78. Available from https://doi.org/10.1016/j.infrared.2009.01.004.

Eskola, L., Alev, Û., Arumägi, E., Jokisalo, J., Donarelli, A., Sirén, K., & Kalamees, T. (2015). Airtightness, air exchange and energy performance in historic residential buildings with different structures. *International Journal of Ventilation*, 14(1), 11−26. Available from https://doi.org/10.1080/14733315.2015.11684066, http://ijovent.org/doi/pdf/10.5555/2044-4044-14.1.11.

Feijó-Muñoz, J., González-Lezcano, R. A., Poza-Casado, I., Padilla-Marcos, M. Á., & Meiss, A. (2019). Airtightness of residential buildings in the Continental area of Spain. *Building and Environment*, 148, 299−308. Available from https://doi.org/10.1016/j.buildenv.2018.11.010, http://www.elsevier.com/inca/publications/store/2/9/6/index.htt.

Feijó-Muñoz, J., Poza-Casado, I., González-Lezcano, R. A., Pardal, C., Echarri, V., De Larriva, R. A., Fernández-Agüera, J., Dios-Viéitez, M. J., Campo-Díaz, V. J. D., Calderín, M. M., Padilla-Marcos, M. Á., & Meiss, A. (2018). Methodology for the study of the envelope airtightness of residential buildings in Spain: A case study. *Energies*, 11 (4), 704. Available from https://doi.org/10.3390/en11040704.

Fernández, Cantera, García. (2010). Universidad de, Valladolid, Locus sapientiae: la Universidad de Valladolid en sus edificios. Consejo Social,.

Filippidou, F., Navarro, J.P. (2019). *Achieving the cost-effective energy transformation of Europe's buildings*. Publications Office of the European Union. 29906, Available from https://doi.org/10.2760/278207.

Fox, M., Coley, D., Goodhew, S., & De Wilde, P. (2015). Time-lapse thermography for building defect detection. *Energy and Buildings*, 92, 95−106. Available from https://doi.org/10.1016/j.enbuild.2015.01.021, https://www.journals.elsevier.com/energy-and-buildings.

Fox, M., Coley, D., Goodhew, S., & De Wilde, P. (2014). Thermography methodologies for detecting energy related building defects. *Renewable and Sustainable Energy Reviews*, 40, 296−310. Available from https://doi.org/10.1016/j.rser.2014.07.188.

Gillott, M. C., Loveday, D. L., White, J., Wood, C. J., Chmutina, K., & Vadodaria, K. (2016). Improving the airtightness in an existing UK dwelling: The challenges, the measures and their effectiveness. *Building and Environment*, 95, 227−239. Available from https://doi.org/10.1016/j.buildenv.2015.08.017, http://www.elsevier.com/inca/publications/store/2/9/6/index.htt.

Huang, J., Hanford, J., Yang, F. (1999). *Residential heating and cooling loads component analysis*.

ISO 9972. (2015). *Thermal performance of buildings. Determination of air permeability of buildings. Fan pressurization method*. 9972.

Janssens, A. (2003). Methodology for measuring infiltration heat recovery for concentrated air leakage. In Research in Building Physics: *Proceedings of the Second International Conference on Building Physics* (pp. 14−18), CRC Press.

Jokisalo, J., Kurnitski, J., Korpi, M., Kalamees, T., & Vinha, J. (2009). Building leakage, infiltration, and energy performance analyses for Finnish detached houses. *Building and Environment*, 44(2), 377−387. Available from https://doi.org/10.1016/j.buildenv.2008.03.014.

Jones, B., Das, P., Chalabi, Z., Davies, M., Hamilton, I., Lowe, R., Mavrogianni, A., Robinson, D., & Taylor, J. (2015). Assessing uncertainty in housing stock infiltration rates andassociated heat loss: English and UK case studies. *Building and Environment*, 92, 644−656. Available from https://doi.org/10.1016/j.buildenv.2015.05.033, http://www.elsevier.com/inca/publications/store/2/9/6/index.htt.

Kalamees, T. (2007). Air tightness and air leakages of new lightweight single-family detached houses in Estonia. *Building and Environment*, *42*(6), 2369−2377. Available from https://doi.org/10.1016/j.buildenv.2006.06.001.

Kirimtat, A., & Krejcar, O. (2018). A review of infrared thermography for the investigation of building envelopes: Advances and prospects. *Energy and Buildings*, *176*, 390−406. Available from https://doi.org/10.1016/j.enbuild.2018.07.052.

Kronvall, J. (1980). *Airtightness measurements and measurement methods*. Swedish Council for Building Research, Stockholm, Sweden. https://www.aivc.org/resource/airtightness-measurement-and-measurement-methods-matningar-och-matmetoder-lufttathet.

Lerma, C., Mas, Á., Gil, E., Vercher, J., & Peñalver, M. J. (2014). Pathology of building materials in historic buildings. Relationship between laboratory testing and infrared thermography. *Materiales de Construcción*, *64*(313), e009. Available from https://doi.org/10.3989/mc.2013.06612.

Liu, W., Zhao, X., & Chen, Q. (2018). A novel method for measuring air infiltration rate in buildings. *Energy and Buildings*, *168*, 309−318. Available from https://doi.org/10.1016/j.enbuild.2018.03.035, https://www.journals.elsevier.com/energy-and-buildings.

Lucchi, E. (2018). Applications of the infrared thermography in the energy audit of buildings: A review. *Renewable and Sustainable Energy Reviews*, *82*, 3077−3090. Available from https://doi.org/10.1016/j.rser.2017.10.031, https://www.journals.elsevier.com/renewable-and-sustainable-energy-reviews.

McWilliams, J. (2003). *AIVC Annotated Bibliography 12. Review of air flow measurement techniques*. Air Infiltration and Ventilation Centre.

Meiss, A., & Feijó-Muñoz, J. (2015). The energy impact of infiltration: a study on buildings located in north central Spain. *Energy Efficiency*, *8*(1), 51−64. Available from https://doi.org/10.1007/s12053-014-9270-x, http://www.springer.com/environment/journal/12053.

Ordax, A. (2005). Santa Cruz, arte e iconografía: el Cardenal Mendoza, El Colegio y los colegiales. *Ediciones Institucionales*, *124*.

Pettersson, B., Axen, B., (1980). *Thermography: Testing of the thermal insulation and air-tightness of buildings*. Swedish Council for Building Research, Stockholm.

Pickering, P. L., Cucchiara, A. L., Gonzales, M., & McAtee, J. L. (1987). Test ventilation with smoke, bubbles, and balloons. *ASHRAE Transactions*, *2*.

Priestner, R., Steel, A.C., (1991). *Air flow patterns within buildings measurement techniques*. Technical Note AIVC. Air Infiltration and Ventilation Centre, Coventry, Great Britain, https://www.aivc.org/resource/tn-34-air-flow-patterns-within-buildings-measurement-techniques.

Sherman, M.H., & Chan, R., (2004). *Building airtightness: Research and practice*.

Simson, R., Rebane, T., Kiil, M., Thalfeldt, M., Kurnitski, J. (2020). The impact of infiltration on heating systems dimensioning in Estonian climate. *E3S Web of Conferences*. 10.1051/e3sconf/202017205004 22671242 EDP Sciences Estonia. http://www.e3s-conferences.org/ 172.

Tanyer, A. M., Tavukcuoglu, A., & Bekboliev, M. (2018). Assessing the airtightness performance of container houses in relation to its effect on energy efficiency. *Building and Environment*, *134*, 59−73. Available from https://doi.org/10.1016/j.buildenv.2018.02.026, http://www.elsevier.com/inca/publications/store/2/9/6/index.htt.

Tavukçuoğlu, A., Diouri, A., Boukhari, A., Ait Brahim, L., Bahi, L., Khachani, N., Saadi, M., Aride, J., & Nounah, A. (2018). Non-destructive testing for building diagnostics and monitoring: Experience Achieved with case studies. *MATEC Web of Conferences, 149,* 01015. Available from https://doi.org/10.1051/matecconf/201814901015.

Vollmer, Michael, & Möllmann, Klaus-Peter (2017). *Fundamentals of infrared thermal imaging* (pp. 1—106). Wiley. Available from 10.1002/9783527693306.ch1.

Zheng, Xiaofeng, Cooper, Edward, Gillott, Mark, & Wood, Christopher (2020). A practical review of alternatives to the steady pressurisation method for determining building airtightness. *Renewable and Sustainable Energy Reviews, 132*110049. Available from https://doi.org/10.1016/j.rser.2020.110049.

Integration of historical studies and ND techniques for the structural characterization of the masonry walls in Palazzo Vecchio, Florence

Sara Calandra[1,2], Vieri Cardinali[3], Irene Centauro[1],
Anna Livia Ciuffreda[1], Tessa Donigaglia[1], Teresa Salvatici[1], and
Marco Tanganelli[3]

[1]Department of Earth Sciences, University of Florence, Florence, Italy, [2]Department of Chemistry, University of Florence, Sesto Fiorentino, Italy, [3]Department of Architecture, University of Florence, Florence, Italy

4.1 Introduction

Monumental masonry buildings constitute a heritage that needs to be preserved in the centuries for the future generations (ICOMOS, 2003). In fact, they represent the highest testimony of the development of societies in the past in the architectural and cultural field. For these reasons, monumental structures must be protected and the classical principles for restoration of compatibility, reversibility, and sustainability are applied (Acampa et al., 2022; Carbonara, 2012; Coli et al., 2022). Nevertheless, the safeguard of the existing buildings is not sufficient by guaranteeing an effective maintenance. Different risk sources threaten the heritage goods, both of anthropic derivation as from natural hazards (Arrighi et al., 2022). To perform risk analysis, the three risk components, hazard, exposure, and vulnerability, need to be identified (Fuchs et al., 2012). The hazard is generally assumed as the risk source; in performance-based assessment, it is generally identified by means of hazard curves and probabilistic return periods. The exposure, representing the metric of the exposed asset, can vary depending on the considered hazard sources (e.g., it can be the ground level for floods and the whole building for earthquakes). In any case, analyzing a specific risk, it is generally constant inside the case studies, therefore it is sometimes neglected. Finally, vulnerability expresses the propension of a structure to suffer damage. In this context, building characteristics can lead to different vulnerabilities, driven by the structural specificities of each construction. As mentioned, if the hazard computation depends on external factors independent of the investigated monumental structure, the vulnerability represents an intrinsic

Diagnosis of Heritage Buildings by Non-Destructive Techniques. DOI: https://doi.org/10.1016/B978-0-443-16001-1.00004-8

property of buildings. That is why the risk reduction of existing structures is generally addressed at mitigation strategies aimed at decreasing their vulnerability (Chieffo et al., 2019). In this area of interest, concerning the existing buildings, the comprehension of the involved structural organism becomes crucial. The so-called knowledge path represents the methodological procedure targeted at characterizing the structural system of a given building to successively execute structural analyses (MIT, 2019). When applied in every point, the knowledge path can be divided according to the following phases:

- study of the historical evolution of the building, comprehensive of archive research and literature review, targeted at acquiring all the information available on the construction and its phases;
- geometrical architectural survey correlated by degradation maps and survey of the crack pattern, finalized at the spatial understanding of the building;
- geometrical structural survey obtained by analyzing the connections between parts, their structural details, as characterizing the load-bearing system in plan and height;
- mechanical characterization of the adopted materials by executing different in situ and laboratory tests targeted at identifying the mechanical values for the numerical analysis.

Dealing with monumental masonry structures, some of the previous points become more complicated to be fulfilled (MIBACT, 2011). Historical buildings can have a traced history rich of documentation and archive information, but in most of the cases the undergone alterations are lost in the centuries (Caprili et al., 2017; Caranti et al., 2022). This aspect goes in line with the acquisition of the structural surveys of structure. The more accurate the documentation is, the less destructive site surveys are needed to understand the load-bearing system and its dimensioning. In the absence of detailed historical information, the comprehension of the structural system requires the removal of the finishing layers to check the structural consistency of the constructions. With regards to the last point of the knowledge path, the definition of the mechanical properties of materials, this issue is still really demanding. In fact, the direct determination of mechanical values passes only by the execution of destructive tests (Maio et al., 2018; Pelà et al., 2017). Since the monumental structures represent important heritage buildings, these applications are generally discouraged. Alternatively, in the last decades, the application of non-destructive (ND) and minor destructive (MD) techniques took place in the field of cultural heritage buildings. Although they are not able to directly estimate the mechanical properties of the investigated structures, these applications can provide important qualitative information on the structural survey. Moreover, it is possible to correlate the obtained results to international databases to have some clue on the mechanical values. These techniques are particularly effective when combined together (Cardinali et al., 2023; Coli et al., 2022). In this way, it is possible to double-check each outcome by the evidence of the other adopted techniques (Biscarini et al., 2020). Due to the invasiveness of the destructive techniques and the good results that can be obtained through ND and MD approaches, except in special cases, these investigations have become the only type of application accepted for cultural heritage structures. The knowledge path is at the basis of the further steps concerning the structural assessment of constructions. Whether the structural assessment will regard advanced numerical strategies or simplified mechanical approaches, the characterization of the case study is crucial; it is worth noting that in case of high uncertainties given by nonrobust cognitive procedures, the risk is to obtain results analytically correct for a certain condition which is not the one of the real structures (Cardinali et al., 2022).

In this chapter, the integration of historical studies and ND techniques is applied to characterize the masonry walls of Palazzo Vecchio, in Florence. The building represents the political headquarter of the city since the Medieval times. It has grown along the centuries through a series of additions and renovations that has led to a big significant aggregate of six main structural units. In the following sections, the methodological framework of the research will be presented, together with the adopted ND techniques and the structural characterization obtained as outcome of the work.

4.2 Methodological framework

The structural and mechanical characterization of masonry buildings should be carried out on the basis of historical data, visual surveys, and experimental investigations (Marghella et al., 2016), such as ND techniques which can be performed on-site. In Fig. 4.1, the flowchart of the research is presented. The work took the

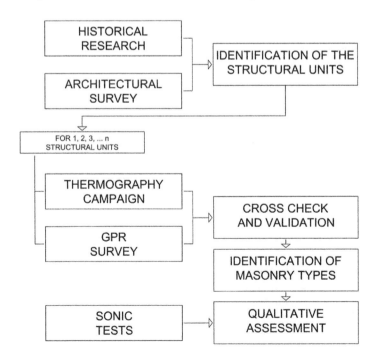

Figure 4.1 Methodological flowchart of the work.

advantage of different phases developed consequently over the investigated building. To characterize the bearing masonry walls of the cultural heritage building, the following procedure has been adopted. Initially, historical information has been combined with the evidence of the architectural survey to identify the main structural units composing the monumental aggregate. This phase is important to define the masonry typologies inside each structure, yet limiting the heterogeneity of the outcomes, making possible proper classifications of the bearing system. Once the structural units have been identified, for each building two ND campaigns have been executed, that is, thermography campaign and ground-penetrating radar (GPR) survey. After the identification of masonry types, a selected number of walls for each typology has been chosen to execute a qualitative assessment through sonic tests.

Thermography inspections are based on remote sensing acquisition by photographical instruments able to highlight the intensity given by the infrared variations of the electromagnetic spectrum. Thanks to this principle, it is possible to obtain color maps accounting on emissions as a function of their temperature. The intensity of the radiation and temperature are proportionally related as described by the Stefan−Boltzmann law. In the field of architecture, thermography acquisitions are used to identify the structural consistency of goods (Carosena, 2013). In practice, their adoption is mostly predominant in finding electric and hydraulic systems under the plaster layer of constructions. Thermography investigations are used to determine the structural characteristics of constructions, their variations in terms of adopted materials, and consequently emanated infrared (Caranti et al., 2022). In the field of historical buildings, its pros are connected to the noninvasive and noncontact requirement of the investigation. This acquisition system is used to determine structural peculiarities of the constructions hidden by plaster layers for both vertical and horizontal elements (Casapulla et al., 2018; Gusella et al., 2021).

The GPR is a well-known ND technique. It had its development for subsoil objects (Fuliagar & Livleybrooks, 1994) and for defining lithological contacts (Basson et al., 1994; Van Heteren et al., 1994), faults (Deng et al., 1994), and fracture bands in rock mass (Bjelm, 1980; Moffat & Puskar, 1976), such as for investigating the competence of concrete in bridges and tunnels; in some instance also the water table can be detected (Beres & Haeni, 1991; Overmeeren, 1994). In the last few years, GPR has been used for investigating masonry (Coli et al., 2019; Coli, Donigaglia, et al., 2018; Coli, Papeschi, et al., 2018; Pieraccini et al., 2004) and cultural heritage objects too (Coli & Micheloni, 2019; Coli et al., n.d.; Coli et al., 2019).

The GPR measures the two-way travel time of an electromagnetic impulse in the radar range (100 MHz−100 GHz) between the input and its return (Fig. 4.2). The radar impulse travels in the material at a velocity mainly related to the magnetic conductivity and permeability of the medium (GSSI, 1987).

The boundary between materials with different electromagnetic properties partially backscatter the signal to the antenna. The entity of penetration of the signal into the material is a function of the material properties and the signal frequency: lower is the frequency and higher is the penetration, but less are the details of the investigation because they are functions of the wavelength.

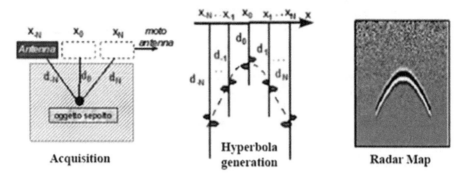

Figure 4.2 GPR functioning principles: an antenna inputs a radar signal in the medium, the boundary between materials with different electromagnetic properties backscatters the signal to the receiver determining the two-way travel time. *GPR*, Ground-penetrating radar.

The knowledge of the wave velocity in the medium or that of a distance of a sure signal are compulsory for fixing the scale of the investigation and therefore for correctly placing the material interfaces in the surveyed body, voids included. In investigating a manufacture, a good practice is measuring the thickness of the wall and placing a still/iron plate as reference to the opposite side of investigation for having a clear end-signal of the masonry, to fix the scale of resulting radargrams.

These techniques are absolutely noninvasive and they do not interfere with the standard use of constructions. Although the parsimonious strategy, the combination of thermography and GPR allows to structurally identify the different parts of an historical construction providing information on the composition of the main parts. Both ND results are then compared and critically examined to cross-check the outcome of each investigation, aimed at synthesizing the products identifying the main structural types of each structure. The validation of the cross-checking is also validated executing MD experiments, that is, drilling perforation of masonry walls investigating their structural consistency through endoscopy recordings. This approach does not require the removal of the external plaster layer; the destructive area of the sample has a circular dimension of around 20 mm diameter, which can be considered compatible with the significant value of cultural heritage buildings. Although only locally, in this way, it is possible to verify the hypothesis made through the ND investigations, especially with the outcomes of GPR. Moreover, the dimension of the external leaves of the panels, highlighting the quality of the inner parts, and the eventual presence of voids without mortar or incoherent materials can be observed and their dimensions can be measured.

As ND techniques, thermography, and GPR do not allow characterizations in terms of mechanical properties of the materials, although they permit reconducting the investigated consistencies to available databases. In case of existing masonry buildings, the Italian reference is given by MIT (2019), specifically with Tab. C8.5. I. The latter identifies a list of eight different masonry typologies. For each one of them, a range of mechanical values for the different properties characterizing bearing walls are listed. The code establishes also the parameters to be adopted depending on the reached knowledge level; for the lowest cognitive level in the absence of destructive tests, the mean values for the elastic moduli (E and G) and the minimum values for the resistances (compressive strength, shear strength) are suggested. In this work, the classification of the masonry walls has critically followed two distinct lists, from one side adopting Tab. C8.5.I from the Italian code and on the other side considering the codification given by the Tuscany region (Ferrini et al., 2003). The latter is in fact more detailed in terms of structural identification of the masonry walls, allowing to highlight more the variabilities within the walls.

The combination of thermography and GPR is then integrated with another ND technique, the sonic test, to complete the evaluation of the quality and homogeneity of the walls.

Sonic test is a widely used ND technique for masonry quality assessment and is part of acoustic investigation methods (Binda & Saisi, 2001; Cantini, 2016; Grazzini, 2019; Miranda et al., 2013; Valluzzi et al., 2018; Van Eldere et al., 2019). This method is used to evaluate the homogeneity of masonry by measuring the propagation time of a compression wave through the wall element.

Sonic investigation has multiple uses for the diagnostics of masonry; indeed it can be useful to:

- identify voids, inhomogeneity, and internal defects and identify crack patterns and detachments in coatings;
- assess the state of conservation and the mechanical characteristics of masonry materials; and
- verify and monitor the effectiveness of restoration interventions.

This technique is based on the generation of elastic waves (with frequencies generally between 20 Hz and 20 kHz) from a mechanical impulse, by means of an instrumented hammer, and on the detection of wave propagation times, detected by an accelerometer. The main measured parameter is the *time of flight* (expressed in μs) that the pulse takes to cover the distance between the transmitter and the receiver: knowing the distance between them, it is therefore possible to obtain the sonic velocity (m/s). The velocity of an elastic wave passing through a solid material such as masonry is theoretically proportional to the density, dynamic modulus, and Poisson ratio of the material. A decrease in velocity values generally indicates a worse state of conservation or the presence of internal inhomogeneities. Each individual measure is performed at a specific point in the structure, around the material region where the transmitter and receiver stations are located. Depending on the position of hammer and receiver meter, the sonic test can be performed according to three modes: direct transmission (hammer and accelerometer are placed on opposite faces), indirect transmission (hammer and accelerometer are placed on the same face), and semidirect transmission (hammer and accelerometer are placed on adjacent faces). Generally, measurements are performed according to a grid scheme of equidistant points (RILEM Recommendation TC 127-Ms, 1996), which allows a complete investigation of the wall element. The data acquired and expressed in tabular format can be shown also in graphical ways to improve the readability of the results, such as false color maps of sonic velocity distribution. The instrument calibration and a careful design of the measurement scheme, according to the type of artifact to be investigated, have a very important role in the testing campaign. The execution of the sonic tests within the framework of Fig. 4.1 indicated a hierarchical procedure where, after the masonry typologies are identified, a qualitative discriminant is researched. This information is still important to check the evidence of the endoscopy tests inside the masonry panels, to understand if the continuity (or the nonhomogeneity) is representative of a wider structural area rather than a single point.

4.3 Structural characterization of the masonry walls of Palazzo Vecchio

4.3.1 Localization and historical evolution

Palazzo Vecchio is one of the most iconic buildings in the historical center of Florence, the city worldwide known as the Capital of Renaissance. The building represents the political power of the city since its construction, which is dated back in the early 14th century. Palazzo Vecchio has an articulated historical evolution that lies on the archaeological remains of the Roman age of the city. The Palace is located on the northern side of the Arno river, close to the southern border of the old castrum Roman, the Roman quadrilateral built in the first century BCE (Arrighi et al., 2022). At the time, the ancient theater of the city was located in the building position, as demonstrated by the archaeological excavations that have been carried out (Bruttini, 2013). The city has expanded during the Medieval period and the construction of Palazzo Vecchio has started with the erection of the oldest and most iconic part facing Piazza della Signoria, Palazzo dei Priori. In Fig. 4.3, a satellite image of the historical center of Florence at the current days with the identification of Palazzo Vecchio is shown. In Fig. 4.4, an image of the main facade of the building facing Piazza della Signoria is presented.

As the main historical Palace of the city since the Medieval time, the history of Palazzo Vecchio covers almost a millennium of years, with many interventions not documented and never clarified. However, due to the same reason, a lot of researchers have separately studied several aspects concerning the Palace. A brief overview concerning the historical evolution from a structural perspective of the building can be found in the work of Paoletti et al. (2020). The complex urban aggregate of Palazzo Vecchio is the result of an evolution starting from the Medieval time. The first historical nucleus, the so-called Palazzo dei Priori, was designed by Arnolfo Di Cambio, which incorporated an older preexisting tower, Torre della Vacca, inside the masonry walls of the new Palace. On the remains of this Medieval tower, the predominant vertical element of the building has been then erected, that is, the Tower of Arnolfo (the main tower shown in Fig. 4.4). During the centuries, the

Figure 4.3 View of the historical center of Palazzo Vecchio. In red, the position of Palazzo Vecchio is highlighted; source Google maps.

Palace grew along the east direction, slowly saturating an entire aggregate that, in the early middle age, was divided by streets and properties of different families. Several interventions have occurred during the Athen's Duke period, in the 14th century, as under the management of Medici's family. During the period of Cosimo de' Medici, architect Michelozzo promoted different renovations of the structure. Successively, other important architects such as Giuliano da Maiano and Vasari worked inside Palazzo Vecchio. Vasari realized the stairwells inside the Dogana Courtyard, carrying out important renovations inside the Salone dei Cinquecento. Further works were conducted under architect Bartolomeo Ammannati, completing part of the renovations facing the Third Courtyard and the eastern part of the complex. Starting from the 14th century, Palazzo Vecchio stopped being the court palace for the regency family, as it moved to Palazzo Pitti, maintaining only the political representative power. Further interventions were made during the 19th century and the years when Florence became the Capital of Italy. Finally, since 1872, Palazzo Vecchio hosts the city council of the municipality. Other interventions have been carried out during the last century. Besides the standard renovations due to architectural adaptations of historical buildings to contemporary uses, the older part of the building, Palagio dei Priori was strengthened by Sisto Mastrodicasa through the installation of steel tie rods and bolt connections at the merlon level of the Palace.

Nowadays, Palazzo Vecchio represents a complex building hosting many different activities. The offices of the city hall are inside the structure, so that the citizens can access the structure to get married or for variations to their Register Office. The council of Palazzo Vecchio is still in the oldest part of the Palace and the political commissions are developed in hall rooms inside the structure. In addition, a public museum with a significant number of visitors explores the museal path of Palazzo

Figure 4.4 View of Palazzo Vecchio and its main facade fronting Piazza della Signoria.

Vecchio every day with over 200,000 tickets per year. As the Palace constitutes an important monumental heritage, it has been the objective of different studies along the years. However, they mostly concern the historical evolution of the building and the main interventions occurred. In terms of structural investigations, the contributions are less numerous. This can be justified by different reasons. From one side, the development of ND techniques in the architectural field is quite recent and on the other side, both the huge dimensions of the complex as its complexity did not encourage systematic application finalized at executing structural assessment of the building. In 1990, Chiarugi et al. (1990) have presented the results of a dynamic identification of the Tower of Arnolfo. New ND acquisitions have been carried out, together with contributions describing the structural features of the building. Giorgi (2017) presented a study on the constructive features of slabs and ceiling of the 15th century in Palazzo Vecchio, showing interesting structural apparati. Regarding the application of ND techniques, other contributions have been found. Pieraccini et al. (2006) have used GPR surveys to investigate the masonry consistency of the walls of the Salone dei Cinquecento, particularly behind the frescos decorating the hall. Other studies on the Sala Grande by ND tests are shown in the work of Coli et al. (2019). Napolitano et al. (2019) have recently integrated ND campaigns, laser scanner survey, and FE modeling to interpret the damage assessment of the Room of the Elements.

4.3.2 Investigation procedure

Palazzo Vecchio has been recently the objective of an interdisciplinary investigation targeted at defining the structural performances of the urban complex under static and seismic loads. To this aim, a wide knowledge path has been developed. The historical evolution has been integrated to an extent architectural survey of the whole palace is pursued by remote sensing laser scanner acquisitions (Verdiani, 2019). The total volume of the Palace is over 190,000 m^3, with a planar extension bigger than 6000 m^2.

The acquisition of the geometrical and historical information has allowed identifying the main structural units of the building. Six different structural units have been recognized (Fig. 4.5). The older part of Palazzo Vecchio, that is, Palagio dei Priori identified as capital letter A, has been divided within three main structures,

Figure 4.5 Identification of the main units of the building: the Arnolfo's Tower in gray; the units around the Michelozzo's Courtyard in light orange; the Sala d'Armi hall in red; Cortile della Dogana space in blue, the Salone dei Cinquecento hall, in green; structural units D and E in dark orange.

the Tower of Arnolfo (gray color—A0), the units around the Michelozzo's Courtyard (light orange—A1), and the Sala d'Armi hall (red color—A2). Structural unit B has been identified as the Cortile della Dogana space (blue color), while structural unit C is recognized as the Salone dei Cinquecento hall (green color). Finally, structural units D and E are identified in Fig. 4.5 by dark orange. The two final units have been mapped by a unified color due to their historical evolution. After the decline of the Roman period, the area was occupied by different Medieval urban aggregates separated by civic streets. The contemporary structural units incorporate the remains of several structures still recognizable from the facade and the archaeological underground of the Palace. Nonetheless, they have been significantly redesigned under the power of Ferdinando I Grand Duke. For this reason, besides the structural evolutions, they can be identified as two main units circumscribing the Third Courtyard of the building.

Once the structural units are identified, it is possible to proceed with the following steps of the research. To this aim, the ND procedures have been executed according to the methodological flowchart shown in Fig. 4.1. The tests have regarded a significant and extensive number of ND tests. In Fig. 4.6, a plan of the intermediate floor of the Third Courtyard of Palazzo Vecchio with the different tests executed is shown.

Initially, thermography acquisition and GPR surveys have been adopted to identify the structural characteristics of Palazzo Vecchio. The two ND techniques have

Figure 4.6 Intermediate plan of the Third Courtyard of Palazzo Vecchio. ND tests. *ND*, Non-destructive.

been used to investigate the structural consistency of the bearing walls, as to identify the technologies adopted for vaults and slabs. Concerning the thermography, an FLIR T460 instrument with a thermal sensitivity lower than 0.030°C and a spectral range of 7.5−13 μm has been adopted. The quality of the obtained information is not always the same, as it varies as a function of the conditions of the environmental temperatures, thermal exchange, and thickness of the plaster layers. The external walls between the inner parts of the system and the outside are the most sensitive, nonetheless, regarding the facades of the building, these are mostly covered by ashlar sandstones. In any case, the thermography campaign has allowed obtaining qualitative information. In Fig. 4.7, a thermal orthophoto pointing out the variability of the results within the northern facade of the third courtyard is shown. The composition of the masonry walls hidden under the plaster layer is exhibited in the figure; in particular, it is worth noting that the remaining of bolt steel element is present inside the finish layer, which demonstrate the presence of steel tie rods preventing out-of-plane mechanisms.

In Fig. 4.8, some examples pointing out the qualitative information provided by the thermography campaign are shown. Fig. 4.8A points out the presence of an irregular masonry wall with the use of stone elements with different dimensions. On the contrary, in Fig. 4.8B, thermography exhibits the presence of horizontal courses, as the adoption of regular structural elements with coherent dimensions between them. It is worth mentioning that most of the walls of Palazzo Vecchio are covered by plaster, so without investigation it would not be possible to understand their structural features.

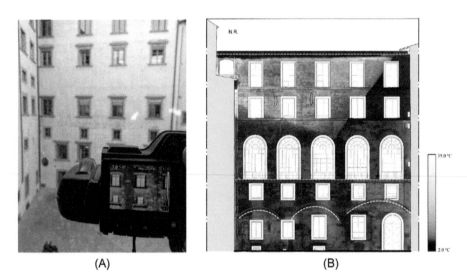

(A)   (B)

Figure 4.7 (A) Thermography orthophoto of the northern side of the Third Courtyard of Palazzo Vecchio and (B) in situ execution of the ND test in the southern side. *ND*, Non-destructive.

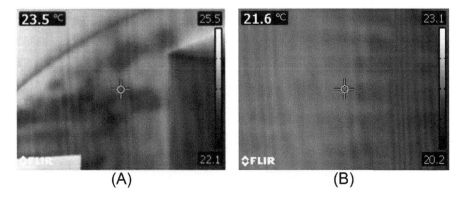

Figure 4.8 Thermal photos of inner walls of Palazzo Vecchio, showing (A) the presence of walls made by irregualr stone elements, (B) a regular wall with listed horizontal layers.

The GPR survey on the Palazzo Vecchio walls was carried out by using the IDS Georadar C-thrue System, which is a 2 GHz system and for this reason this is defined as high frequency. The identification made by thermographical and GPR acquisition has led to support the identification of the masonry panels of the building. The usage of GPR along the masonry walls of the Palace has allowed obtaining structural information on the masonry consistency, as to their quality. In fact, GPR permits the detection of the variations inside the thickness of the panels. In the research work, the dimension of the external masonry leaves could be measured, providing by subtraction of the thickness of the inner sacco filling. In addition, qualitative information on the quality of the inner parts have been obtained. In Fig. 4.9, an abacus listing the masonry radargrams according to the presence of cavities is shown.

The GPR survey has been carried out both on the walls as horizontally and on the intradoxes of slabs and vaults. Regarding the masonry structures, it allowed pointing out the presence of steel elements inside the external layers of the masonry, which could be ascribed to strengthening interventions that occurred during the centuries, such as steel dressing of piers or columns. In Fig. 4.10, two examples are shown.

The combination of these information has allowed the first definition of the structural typologies of masonry walls. The ND tests have been conducted widely all over the different structural units of the building, as they do not alter in any way the buildings. Then, the hypotheses have been verified through a more limited number of endoscopy tests. Drilling perforations along the walls have been realized. For the endoscopy, a Novatest—Model RA350M has been used. Due to the architectural quality of the Palace, which is not always the same, endoscopy validation has not been realized along all the structural units. In the structural unit A, the tests were conducted in the A1 unit, thanks to the presence of older holes or obsolete electric plants. Then, few tests have been conducted in structural unit C, especially

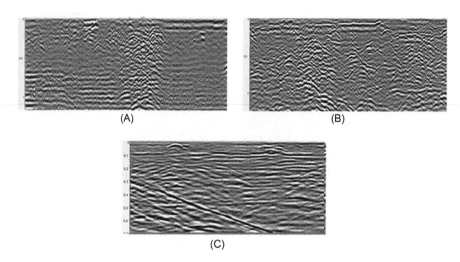

Figure 4.9 Abacus showing the results in terms of radargrams for different masonry typologies; (A) homogeneous masonry wall with few cavities; (B) homogeneous masonry wall with many cavities; and (C) homogeneous masonry wall.

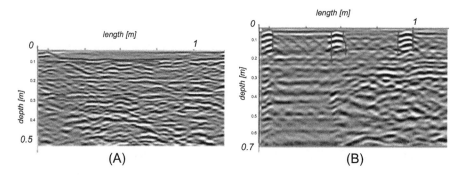

Figure 4.10 GPR surveys; (A) radargram showing the thickness of the external leaf and (B) presence of steel elements spaced around 50 cm between them, probably undocumented strengthening intervention on a corner portion of the Palace. *GPR*, Ground-penetrating radar.

at the ground level. A more numerous number of endoscopies have been conducted in structural units D and E, although limiting their execution at the underground and ground level for logistic reasons. In Figs. 4.11 and 4.12, two sheets presenting the endoscopy tests executed are shown. In many cases, the endoscopy tests have allowed validating the response of the GPR signal. Concerning the case presented in Fig. 4.12, the research took advantage of restoration works conducted by Fabbrica delle Belle Arti of Palazzo Vecchio, permitting documenting the masonry walls of the oldest part of the Palace.

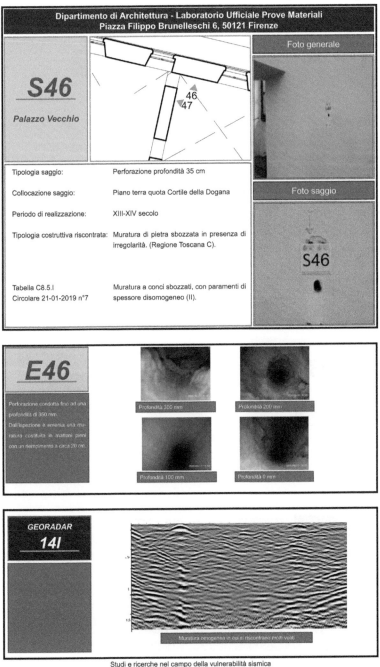

Figure 4.11 MD sheet for Palazzo Vecchio. In the document, the results of the endoscopy validates the outcome of the GPR survey. *GPR*, Ground-penetrating radar; *MD*, minor destructive.

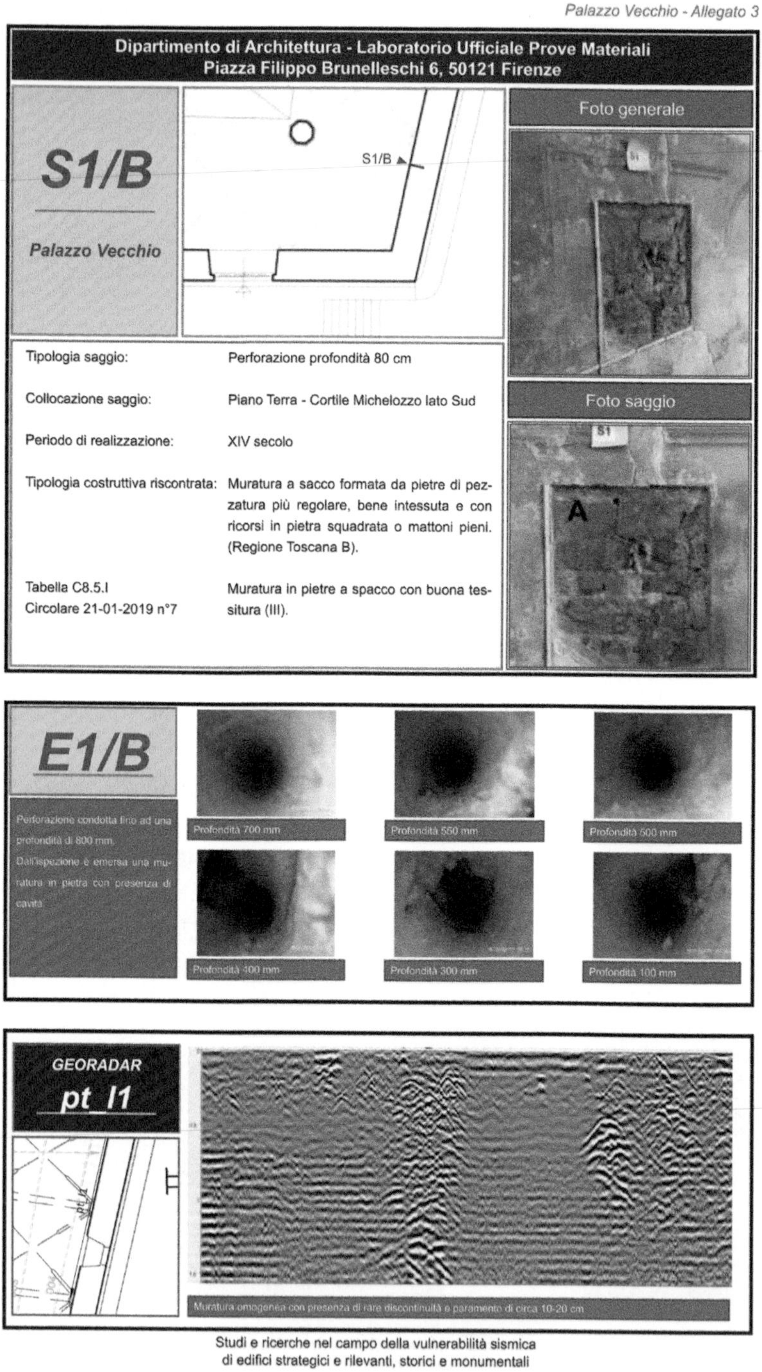

Figure 4.12 Sheet for the tests conducted in the Michelozzo Courtyard during the restoration of the frescoes of the hall.

According to the flowchart in Fig. 4.1, sonic tests have been realized to support the quality assessment of the masonry, detecting the variations inside the wall thickness. The sonic test campaign was conducted on 120 portions of different masonry, located on all floors of the palace. The investigations have been carried out mainly by direct transmission on rectangular or square areas: a grid of points has been marked on the two opposite faces of the investigated wall portion, providing for the acquisition of the values according to horizontal scan lines that run along the entire length of the area. The measuring points were aligned and positioned at regular distances, so that the hammer and the receiver were always on the same plane (Fig. 4.13). The dimensions of the investigated areas were different depending on the type of masonry and operating conditions.

The instrumentation used is Novasonic U5200 CSD of IMG Ultrasuoni Srl; data has been recorded through a detection system directly connected to the hammer and implemented in a totally customized management system (Centauro et al., 2022). The tests have been performed for each area measuring the wall thickness, which is the distance between the hammer and the accelerometer; then, 3 times of flight have been recorded for each measuring point of the area, finally obtaining the average velocity for each point and each investigated portion of the wall.

To improve the readability of the results, false color maps of sonic velocity distribution have been processed through GIS software. Indeed, for the qualitative assessment of the wall, it is extremely important to analyze how the velocity values are distributed along the investigated surface. This analysis allows to obtain

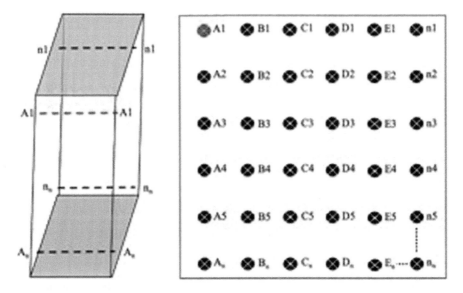

Figure 4.13 Representation of measurement method for sonic velocity test.

information on the homogeneity of a masonry, on its regularity and on the presence of defects or detachments.

In addition, reading the average values allows to define the different states of masonry. Three different classes delimited by the ranges of average velocity change could be identified for investigated masonry:

- $V < 800$ m/s identifies heavily damaged masonry, characterized by large internal voids or detachments and fractures of plaster.
- 800 m/s $< V < 1600$ m/s represents the majority of existing masonry; values of V less than 1000 m/s may indicate the presence of voids, defects, and irregularities.
- $V > 1600$ m/s indicates carefully constructed and preserved masonry with high strength.

Fig. 4.14 shows the distribution of 3 classes for the investigated masonry and a few representative examples of sonic velocity distribution maps obtained. All these masonries have a plaster coating.

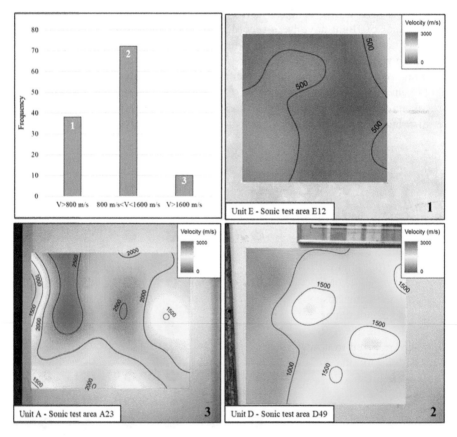

Figure 4.14 Results of sonic velocity tests—frequency distribution of different masonry classes: damaged masonry (1), inhomogeneous (2), and compact masonry (3).

The sonic area E12 represents one of the oldest walls in the building, one side of masonry consists of frescoed plaster. The measured velocity distribution is uniform, indicating a homogeneous masonry with low velocities due to the plaster detachment. This area falls into class 1. In the sonic area, D49 was recorded an inhomogeneous velocity distribution with large discontinuities or internal unworked stones mixed with filler mortar. In fact, some points reach a high velocity of about 1700 m/s, besides points showing much lower velocities of about 800 m/s. This area falls in class 2 and is representative of most historical walls of the Palazzo Vecchio. The sonic area A23 belongs to class 3 and is a rare example of modern brickwork. It is homogeneous and characterized by good compactness, as well as the high values of sonic velocity.

4.3.3 Identification of masonry typologies and determination of the mechanical values

In this work, the masonry walls of Palazzo Vecchio have been listed according to two distinct classifications. The first one is specifically referred to the Tuscany region, and it is divided within 14 different masonry typologies (Ferrini et al., 2003). The second classification is provided by the current Italian code for constructions, as it deals also with the existing structures (MIT, 2019). The list realized by the Tuscany region takes into account masonry typologies specific for the territory where Florence is located. It considers 13 distinct masonry types and, in general, it is more detailed than the classification shown in (MIT, 2019) (Table 4.1). In fact, it also considers reinforced masonry typologies or modern ones. On the other hand, the pros of the (MIT, 2019)classification are that, on the basis of each masonry category, it specifies a series of mechanical parameters that can be adopted in the absence of more detailed analysis (Table 4.2). Hence, they constitute the main reference for assessing the existing masonry structures in Italy. As specified in the MIT recommendations, the range of mechanical values provided by the code is referred to masonry walls characterized by the use of a standard mortar (resistance between 0.7 and 1.5 MPa). In case of irregular textures, the list is referred to structures with the absence of coursed layers and external leaves with a bad disposition or not connected. For regular masonry disposition, it involves structures following the good workmanlike; in both cases, these values shown in Tab. 2 are referred to nonconsolidated structures. In the case of positive or negative features of the masonry typologies of a structure, some detrimental or improvement coefficients can be used. They are referred to the presence of horizontal coursed layers, binding stones within the thickness depth connecting the two leaves of a panel, and usage of a good quality mortar (please see Tab. C8.5.II of MIT, 2019). In the current work, the masonry typologies have been connected to both classification, the Regional one and the Italian one, developing interdependencies between the two systems.

Table 4.1 Identification of the masonry typologies according to the classification provided by the Tuscany Region.

Tuscany Region classification (Ferrini et al., 2003)	
A	Masonry realized by inner nucleus and two external leaves realized with variable stones, bad disposed and without a connection between the two sides of the panel
B	Masonry realized by an inner nucleus and two external leaves realized with stones of regular dimensions, well-disposed and with connections between the two sides of the panel, horizontal layers of chiseled stones or clay bricks
C	Rough stone masonry with the presence of irregularity
D	Rough stone masonry with horizontal layers of chiseled stones or clay bricks
E	Rubble stone masonry of variable dimensions without horizontal dispositions of chiseled stones or clay bricks
F	Rubble stone masonry of variable dimensions with horizontal dispositions of chiseled stones or clay bricks
G	One-leaf masonry made by blocks of tuff of chiseled stones with constant dimensions
H—I	Prefabricated concrete blocks with ordinary or light inerts homogenous along the panel dimension
L	Full or semifull clay masonry
M	Clay block masonry with the hole's dimension major than 45%
T	Mixed structure, considered as the combination of one of more of the previous typologies
U	Confined masonry
V	Reinforced masonry

Concerning Palazzo Vecchio, the following relationships have been provided.

- A1—masonry realized by inner nucleus and two external leaves realized with variable stones, bad disposed and without a connection between the two sides of the panel/rubble stone masonry;
- C2—rough stone masonry with the presence of irregularity/rough-block masonry with nonhomogeneous thickness of the external faces;
- B3—masonry realized by an inner nucleus and two external leaves realized with stones of regular dimensions, well-disposed and with connections between the two sides of the panel, horizontal layers of chiseled stones or clay bricks/split stone masonry with good texture disposition;
- L-7—full or semifull clay masonry/clay bricks and lime mortar masonry; and
- G5—one-leaf masonry made by blocks of tuff of chiseled stones with constant dimensions/regular block masonry made of soft stone (tuff, calcarenite, and so on).

In Fig. 4.15, the results of the work are shown. The knowledge path developed has allowed obtaining thematic maps presenting the different plans of Palazzo Vecchio indicating the different units and their structural features. The same work has been done for the slabs, defining all the different dead loads (structural and nonstructural) as the live loads involved in the building.

Table 4.2 Mechanical parameters from the Italian code for the different masonry typologies listed in Tab. C8.5.I of MIT2019.

	f_m (Mpa)	t_0 (Mpa)	f_{vo} (Mpa)	E (Mpa)	G (Mpa)	r (kN/m³)
Rubble stone masonry	1	0.018	–	690	230	19
	2	0.032	–	1050	350	
Rough-block masonry with nonhomogeneous thickness of the external faces	2.0	0.035	–	1020	340	20
		0.051	–	1440	480	
Split stones masonry with good texture disposition	2.6	0.056	–	1500	500	21
	3.8	0.074	–	1980	660	
Irregular soft stone masonry (tuff, calcarenite, and so on)	1.4	0.028	–	900	300	13 + 16
	2.2	0.042	–	1260	420	
Regular block masonry made by soft stone (tuff, calcarenite, and so on)	2.0	0.04	0.10	1200	400	
	3.2	0.08	0.19	1620	500	
Stone squared blocks masonry	5.8	0.09	0.18	2400	800	22
	8.2	0.12	0.28	3300	1100	
Clay bricks and lime mortar masonry	2.6	0.05	0.13	1200	400	18
	4.3	0.13	0.27	1800	600	
Semifull bricks with cement mortar masonry (Double UNI with hollow part ≤ 40%)	5.0	0.08	0.20	3500	875	15

Regarding the bearing structures, the thematic maps shown in Fig. 4.15 permit to conduct any masonry panels to a list of mechanical values to be adopted for the structural analyses. Concerning the quality of the masonry panels, this information has not been plotted in the maps, although the research collected qualitative insights by the use of GPR and sonic investigations. In fact, due to the noninvasiveness of these operations, the outcomes should be carefully pondered based on the level of the investigation executed for the structural assessment. In case of simplified analysis (e.g., the LV1 level for cultural heritage building provided by MIBACT, 2011), the adoption of the standard values coming from Tab. 2 is in line with the raw structural approach. On the other side, these data could be further implemented to help calibrating high-detailed finite-element models on the basis of the evidence of the dynamic identification of the different structural units via ambient vibration tests.

CARATTERIZZAZIONE DELLE MURATURE
Palazzo Vecchio

PIANO PRIMO

Figure 4.15 Structural characterization of the masonry walls of Palazzo Vecchio: first floor.

4.4 Conclusion

In this chapter, the results of an integrated procedure targeted at characterizing the bearing walls of Palazzo Vecchio in Florence has been presented. The methodology defines a holistic approach where ND and MD tests are arranged according to a hierarchical order. Although there is intrinsic limits of using such types of tests, the combination of unrelated outcomes coming from different investigations allows a more effective structural characterization. The cross-correlation of the diagnostic campaigns conducted in the chapter permits to define: (1) the masonry typologies of the investigated structures and (2) a qualitative distinction within the bearing system. Therefore the proposed methodological path is particularly reliable toward monumental structures where extensive destructive campaigns cannot be executed. Nevertheless, the proposed hierarchical scheme can still be contextualized into more detailed campaigns where DTs can be executed. In this case, the ND and MD investigations permit an extension of the results to wider masonry typologies inside a structure, still providing useful information for the localization of the areas where to conduct the destructive examinations. This research is at the basis for the following steps of structural analysis of the building. To the authors' perspective, in the structural performance assessment of masonry construction, the clear identification of the structural system and its characterization are crucial aspects influencing the results of the work. Hence, the proposed methodology is addressed at guaranteeing reliable levels of knowledge with a focus on the use of noninvasive tests. It is worth noting that, beyond the issues related to the invasivity of the investigations, ND campaigns generally require lower budgets compared to the destructive ones.

In this chapter, the conceived path has been applied to the monumental aggregate of Palazzo Vecchio, in Florence. The building represents the administrative headquarter of the city since the Middle Age, and it has grown along the centuries through several interventions. The conducted investigations have taken advantage of an extensive knowledge path. Historical information and architectural surveys have been integrated to identify the structural units composing the building complex. Then, different ND tests have been combined to characterize the structural consistency of walls, vaults, and slabs of each structure. Initially, thermography campaigns and GPR surveys have identified the plausible constructive techniques of the masonry walls; hence, they have been later validated executing MD techniques investigating the inner parts of the walls through endoscopy tests. Finally, sonic tests provided qualitative information of the masonry walls. The identification of the structural features of the mechanical walls of Palazzo Vecchio has been reconducted to available classification for the Tuscan and Italian territory, allowing to define a range of likely mechanical parameters. The methodology has allowed to determine important structural information at the basis of following steps of structural assessment. The research is still going on. Further analysis will be conducted on the building, through the application of simplified or more sophisticated procedures to evaluate its performance toward static and seismic actions.

The chapter shows how the implementation of different ND techniques inside a coherent methodological framework allows to determine the structural features of cultural heritage buildings where invasive tests cannot be carried out. The adoption of mechanical properties coming from regional and national databases extends the possibilities of application of this procedure to other monumental buildings in different contexts. In this regard, the methodology can be enriched by implementing further ND and MD investigations available due to the technological advances, to determine more comprehensive research still limiting the effects on the cultural heritage.

Acknowledgments

This research belongs to the protocol signed between the University of Florence by Profs. Mario De Stefano and Massimo Coli and the Municipality of Florence for the seismic vulnerability assessment of monumental historical buildings. The authors acknowledge Giorgio Verdiani and his group for the in situ surveys, the work of scholars Barbara Paoletti and Anna Caranti, and the studies done by Elisa Mammoli, Beatrice Fossatelli, Francesca Meli, and Agnese Gasparotti during their Master's thesis.

References

Acampa, G., Francini, C., & Grasso, M. (2022). Study of the cloisters of the Historical Center of Florence: Methodological approach for the definition of restoration intervention priorities. In F. Calabrò, L. Della Spina, & M. J. Piñeira Mantiñán (Eds.), *New metropolitan perspectives. NMP 2022. Lecture notes in networks and systems* (482). Cham: Springer. Available from https://doi.org/10.1007/978-3-031-06825-6_247.

Arrighi, C., Tanganelli, M., Cristofaro, M. T., Cardinali, V., Marra, A. M., Castelli, F., & De Stefano, N. (2022). Multi-risk assessment in a historical city. *Nat Hazards, 119*. Available from https://doi.org/10.1007/s11069-021-05125-6.

Basson, U., Enzel, Y., Amit, R., & Ben-Avraham, Z. (1994). Detecting and mapping recent faults with a ground-penetrating radar in the alluvial fans of the Arava Valley. *Proceedings of the Fifth International Conference on Ground Radar*, Waterloo centre for Groundwater research, Waterloo, Canada, pp. 777–788.

Beres, L., & Haeni, H. (1991). Application of ground-penetrating radar methods in hydrogeologic studies. *Groundwater, 29*(375-86), 1991.

Binda, L., & Saisi, A. (2001). Application of NDTs to the diagnosis of historic structures. Non-destructive Testing in Civil Engineering. *NDT & E International, 34*(2), 123–138.

Biscarini, C., Catapano, I., Cavalagli, N., Ludeno, G., Pepe, F., & Ubertini, F. (2020). UAV photogrammetry, infrared thermography and GPR for enhancing structural and material degradation evaluation of the Roman masonry bridge of Ponte Lucano in Italy. *NDT E Int., 115*102287.

Bjelm, L. (1980). *Geologic interpretation of SIR data from peat deposit in Northern Sweden.* Unpublished manuscript, Department of Engineering Geology, Lund Institute Technology, Lund, Sweden, 1980.

Bruttini, J. (2013). *Archeologia urbana a Firenze. Lo scavo della terza corte di Palazzo Vecchio (indagini 1997−2006)*, Firenze, All'Insegna del Giglio.

Cantini, L. (2016). The diagnostic investigations plan for historic masonry buildings: The role of sonic tests and other minor destructive techniques. In Modena, C., Da Porto, F., Valluzzi M.R. (eds.), *IB2MAC2016, 16th International Brick and Block Masonry Conference*, 1469−1476. CRC Press, London.

Caprili, S., Mangini, F., Paci, S., Salvatore, W., Bevilacqua, M. G., Karwacka, E., Squeglia, N., Barsotti, R., Bennati, S., Scarpelli, G., et al. (2017). A knowledge-based approach for the structural assessment of cultural heritage, a case study: La Sapienza Palace in Pisa. *Bull. Earthq. Eng.*, *15*, 4851−4886.

Caranti, A., Cardinali, V., Ciuffreda, A. L., Coli, M., De Stefano, M., Le Pera, E., & Tanganelli, M. (2022). Seismic vulnerability assessment of a medieval urban cluster identified as a complex historical palace: Palagio di Parte Guelfa in Florence. *Heritage*, *5*, 4204−4227. Available from https://doi.org/10.3390/heritage5040217.

Carbonara, G. (2012). An Italian contribution to architectural restoration. *Frontiers of Architectural Research*, *1*(1). Available from 10.1016/j.foar.2012.02.007.

Cardinali, V., Cristofaro, M.T., De Stefano M., Tanganelli, M. (2022). Cultural heritage buildings and strategic uses: seismic vulnerability assessment in Florence, *REHABEND* 2022, 1148−1155. https://www.scopus.com/record/display.uri?eid = 2-s2.0-85100380213&origin = resultslist. ISSN; 23868198, ISBN; 978-840917871-1.

Cardinali, V., Castellini, M., Cristofaro, M. T., Lacanna, G., Coli, M., De Stefano, M., & Tanganelli, M. (2023). Integrated techniques for the structural assessment of cultural heritage masonry buildings: Application to Palazzo Cocchi-Serristori in Florence. *J. Cult. Heritage Manag. Sustain. Dev.*, *13 (1)*, 123−145.

Carosena, M. (2013). Infrared Thermography in the Architectural Field. *The Scientific World Journal, 2013*.

Casapulla, C., Maione, A., & Argiento, L. U. (2018). Infrared thermography for the characterization of painted vaults of historic masonry buildings. *International Journal of Structural Glass and Advanced Materials Research*, *2*(1), 46−54.

Centauro, I., Calandra, S., Salvatici, T., Garzonio, C.A. (2022). System integration for masonry quality assessment: A complete solution applied to sonic velocity test on historic buildings. In R. Furferi, R. Giorgi, K. Seymour, A. Pelagotti (eds.) *The Future of Heritage Science and Technologies*. Florence Heri-Tech 2022. Advanced Structured Materials, vol 179.

Chieffo, N., Formisano, A., & Ferreira, T. M. (2019). Damage scenario-based approach and retrofitting strategies for seismic risk mitigation: an application to the historical Centre of Sant'Antimo (Italy). *European Journal of Environmental and Civil Engineering*, *25:11*, 1929−1948.

Coli, M., Ciuffreda, A. L., Caciagli, S., & Agostini, B. (2022). Principles and practices for conservation of historical buildings: The case history of the Saint John Baptistery at Florence, Italy. In R. Lancellotta, C. Viggiani, A. Flora, F. de Silva, & L. Mele (Eds.), Geotechnical *engineering for the preservation of monuments and historic sites* III. CRC Press.

Coli, M., Ciuffreda, A. L., & Donigaglia, T. (2022). Technical analysis of the masonry of the Bargello' Palace, Florence (Italy). *Appl. Sci.*, *12*, 2615.

Chiarugi, A., Foraboschi, P., Nistri, P., *The Arnolfo Tower: structural identification*, Software for Engineering Workstation, 1990, Vol. 6, April.

Coli, M., Donigaglia, T., Papeschi, P., & Boscagli, F. (2018a). GPR investigation for historical masonry: Case histories from Florence Italy cultural heritage monumental buildings. *3rd International Conference on Techniques, Measurements & Materials in Art & Archaeology*, Jerusalem, Israel, 2018.

Coli, M., & Micheloni, M. (2019). Structural analysis of the walls supporting the resurrection of Christ by Piero della Francesca mural painting at Sansepolcro, Italy. COMPDYN 2019 *7th ECCOMAS Thematic Conference on Computational Methods in Structural Dynamics and Earthquake Engineering*. M. Papadrakakis, & M. Fragiadakis (eds.), Crete, Greece, 24–26 June 2019.

Coli, M., Papeschi, P., Boscagli, F., Innocenti, L., & Agostini, B. (2018b). GPR investigation on the masnory of the Brunelleschi's Cupola, Florence Italy. *Giornale delle prove non distruttive, Monitoraggio, Diagnostica AIPnD, 1*(52-7), 2018.

Coli, M., Vecchio, P., & Grande, S. (2019). *Indagini non invasive sulle murature, estratto da La Sala Grande di Palazzo Vecchio e la Battaglia di Anghiari di Leonardo Da Vinci*, Editore Leo S. Olschki, Firenze.

Deng, S., Zhengrong, Z., & Wang, H. (1994). The application of ground penetrating radar to detectionof shallow faults and caves. *Proceedings of the Fifth International Conference on Ground Radar*, Waterloo centre for Groundwater research, Waterloo, Canada, 1115–1133.

Ferrini, M., Melozzi, A., Pagliazzi, A., & Scarparo, S. (2003). *Rilevamento della Vulnerabilità Sismica degli Edifici in Muratura*. Toscana, Italy: Manuale per la Compilazione Della SCHEDA GNDT/CNR di II Livello-Versione Modificata Dalla Regione Toscana; Regione Toscana.

Fuchs, S., Birkmann, J., & Glade, T. (2012). Vulnerability assessment in natural hazard and risk analysis: current approaches and future challenges. *Nat Hazards, 64*, 1969–1975. Available from https://doi.org/10.1007/s11069-012-0352-9.

Fuliagar, P. K., & Livleybrooks, D. (1994). *Trial of tunnel radar for cavity and one detection in the Sudbury Mining Camp, Ontario*. Proceedings of the Fifth International Conference on Ground Radar (pp. 883–894). Waterloo, Canada: entre for Groundwater Research.

Giorgi, L. (2017), *I solai e soffitti quattrocenteschi di Palazzo Vecchio a Firenze, I Convegno della Società Italiana per il Restauro dell'Architettura (SIRA)*, Roma, 26–27 settembre 2016.

Grazzini, A. (2019). Sonic and impact test for structural assessment of historical masonry. *Appl. Sci., 9*, 5148.

GSSI Geophysical Survey Systems, Inc. (1987). *Operations Manual for Subsurface Interface Radar System-3*. Manual #MN83–728. Geophysical System, North Salem, New Hampshire.

Gusella, V., Cluni, F., & Liberotti, R. (2021). Feasibility of a thermography nondestructive technique for determining the quality of historical Frescoed Masonries: Applications on the Templar Church of San Bevignate. *Appl. Sci., 11*, 281. Available from https://doi.org/10.3390/app11010281.

ICOMOS. (27–31 October 2003). Principles for the analysis, conservation and structural restoration of architectural heritage. In *Proceedings of the ICOMOS 14th General Assembly*, Victoria Falls, Zimbabwe.

Maio, R., Santos, C., Ferreira, T. M., & Vicente, R. (2018). Investigation techniques for the seismic response assessment of buildings located in historical centers. *International Journal of Architectural Heritage*, 1–14.

Marghella, G., Marzo, A., Carpani, B., Indirli, M., Formisano, A.: Comparison between in situ experimental data and Italian code standard values. In: Brick and Block Masonry: Trends, Innovations and Challenges - Proceedings of the 16th International Brick and Block Masonry Conference (IBMAC 2016), Padova, Italy, pp. 1707–14 (2016).

Miranda, L., Cantini, L., Guedes, J., Binda, L., & Costa, A. (2013). Applications of sonic tests to masonry elements: Influence of joints on the propagation velocity of elastic waves. *Journal of Materials in Civil Engineering, 25*(6), 667–682.

MIBACT (2011). Linee Guida per la Valutazione e Riduzione del Rischio Sismico del Patrimonio Culturale Allineate alle nuove Norme (in Italian)

MIT (2019). Circolare 21 gennaio 2019, n. 7 Istruzioni per l'applicazione dell'«Aggiornamento delle "Norme tecniche per le costruzioni"» di cui al decreto ministeriale 17 gennaio 2018 (in Italian)

Moffat, D., & Puskar, R. (1976). A subsurface electromagnetic pulse radar. *Geophysics, 41* (506-18), 1976.

Napolitano, R., Hess, M., & Glisic, B. (2019). Integrating non-destructive testing, laser scanning, and numerical modeling for damage assessment: The room of the elements. *Heritage, 2*(1), 151−168. Available from https://doi.org/10.3390/heritage2010012.

Overmeeren, R.A. (1994). High speed georadar data acquisition for groundwater exploration in the Netherlands. *Proceedings of the Fifth International Conference on Ground Radar*, Waterloo Centre for Groundwater Research, Waterloo, Canada, 1057−1073.

Paoletti, B., Coli, M., Ferretti, E., & Tanganelli, M. (2020). Multidisciplinary approach to the study of the structural evolution of Palazzo Vecchio, Florence (Italy). In: *Construction Pathology, Rehabilitation Technology and Heritage Management*, Spagna, Granada, Rehabend congress, pp. 867−874, ISBN:978-84-09-17873-5.

Pelà, L., Roca, P., & Aprile, A. (2017). Combined in-situ and laboratory minor destructive testing of historical mortars. *International Journal of Architectural Heritage*. Available from https://doi.org/10.1080/15583058.2017.1323247.

Pieraccini, M., Fratini, M., Parrini, F., Macaluso, G., & Atzeni, C. (2004). High-speed CW step-frequency coherent radar for dynamic monitoring of civil engineering structures. *Electronics Letters, 40*(14), 907−908.

Pieraccini, M., Noferini, L., Mecatti, D., Luzi, G., Atzeni, C., Persico, R., & Soldovieri, F. (2006). Advanced processing techniques for step-frequency continuous-wave penetrating radar: The case study of "Palazzo Vecchio" Walls (Firenze, Italy). *Research in Nondestructive Evaluation, 17*(2), 71−83. Available from https://doi.org/10.1080/09349840600689475.

RILEM Recommendation TC 127-MS. (1996). MS.D.1 Measurement of mechanical pulse velocity for masonry. *Materials and Structures., 29*, 463−466, [CrossRef].

Valluzzi, M., Cescatti, E., Cardani, G., Cantini, L., Zanzi, L., Colla, C., & Casarin, F. (2018). Calibration of sonic pulse velocity tests for detection of variable conditions in masonry walls. *Construction and Building Materials, 192*, 272−286.

Van Eldere, H., Ramos, L. F., Verstrynge, E., Shetty, N., Van Balen, K., Barroso, C. E., & Oliveira, D. V. (2019). The application of sonic testing on double-leaf historical Portuguese masonry to obtain morphology and mechanical properties. In R. Aguilar, et al. (Eds.), *Structural Analysis of Historical Constructions, RILEM Bookseries* (18). Heidelberg: Springer.

Van Heteren, S., Fitzgerald, D.M., & Mckinlay, P.S. (1994). Application of ground-penetrating radar in the coastal stratigraphic studies. *Proceedings of the Fifth International Conference on Ground Radar*, Waterloo Centre for Groundwater Research, Waterloo, Canada, 869−881.

Verdiani, G. (2019). Digital survey: from new technology to everyday use, a knowledge path and challenge for scholars. *EGE Revista de Expresion Grafica en la Edificacion, 11*, 94−105.

Applications of deep learning to infrared thermography for the automatic classification of thermal pathologies: Review and case study

Susana Lagüela[1], Iván Garrido[2], Jesús Balado[3], Jorge López-Rebollo[1], and Javier Pisonero[1]

[1]Department of Cartographic and Terrain Engineering, Universidad de Salamanca, Ávila, Spain, [2]Defense University Center, Spanish Naval Academy, Marín, Spain, [3]CINTECX, University of Vigo, Vigo, Spain

5.1 Introduction

Infrared thermography (IRT) is an ideal tool for the detection of thermal pathologies. The anomalous thermal behavior of pathological areas versus the temperature distribution of the unaltered surroundings makes it possible to predict pathologies from thermal images. Applications of this technique range from new materials such as composites (Alhammad et al., 2022) to energy installations such as photovoltaic panels (de Oliveira et al., 2020; Dunderdale et al., 2020) and transformers (dos Santos et al., 2018). There is also a wide range of applications for industrial maintenance (Li et al., 2020; Venegas et al., 2022), while the widest set of applications can be found in the field of medicine, where IRT can be utilized for different uses such as tumor detection (Mačianskytė & Adaškevičius, 2022; Mashekova et al., 2022), the measurement of infectious diseases (Svantner et al., 2021), and sports medicine (Fernández-Cuevas et al., 2016). Another important application of IRT for the detection of thermal pathologies is in the field of building diagnosis, since this technique can be applied to the inspection of the building envelope regarding the detection of energy losses, as well as to the mechanical and electrical inspection toward the clarification of the operating conditions of the installations and the identification of problems under operation (Balaras & Argiriou, 2002).

In the case of cultural heritage, the application of IRT allows the detection of thermal pathologies, and of other types of pathologies that imply a change in the thermal footprint. For the case of heritage buildings and bridges, and the inspection of facades and walls, thermal bridges are among the first type (Garrido et al., 2018), especially in the case of buildings under use, while structural pathologies

Diagnosis of Heritage Buildings by Non-Destructive Techniques. DOI: https://doi.org/10.1016/B978-0-443-16001-1.00005-X

such as delamination, fractures, cracks, and detachments can be included in the second type (Solla et al., 2019). Additionally, water-related problems can also be detected using IRT based on the different thermal conductivity of water with respect to the thermal conductivity of construction materials (Solla et al., 2013). These are pathologies common to heritage buildings and new constructions, where the structural diagnosis is also of interest following an earthquake or any other extreme event, regardless the previous level of deterioration of the building structure (Kirimtat & Krejcar, 2018).

IRT can also be applied to other cultural heritage elements such as statues, frescoes, and mosaics. For those cases, inspections toward the detection of defects such as the adhesion between different layers and moisture in paintings and frescoes, or stratigraphy and water content in mosaics are also ideal studies in which to apply IRT (Rasulo et al., 2022). Provided the non-destructive, noncontact, and continuous nature of the IRT technique, its application for the detection of pathologies in cultural heritage implies several advantages such as the fact of providing continuous information of the state of the surface and subsurface of the cultural heritage element under study, from a remote position and without performing any disturbance on the element. The fact of having continuous temperature data of the elements allows for the integral interpretation of the pathology, enabling a more precise identification of its nature and its severity, in comparison to individual or punctual measurements.

In this respect, and toward maximizing the precision of the thermographic inspection, different strategies have been developed toward its automation, both for the data acquisition and data processing. In the case of data acquisition, automation has been aimed through the integration of the thermographic camera in mobile platforms, and the sequential acquisition of images during displacement. In this line, López-Fernández et al. (2016) have developed a methodology for IRT data acquisition from a mobile cart, pushed by an operator, for the thermographic inspection of interior building scenes, while Daffara et al. (2020) have acquired thermographic data from building envelopes by using UAVs (unmanned aerial vehicles) and Susana et al. (2014) have used a terrestrial vehicle, van-type, for the automatic inspection of building facades at neighborhood or city scale. Regarding the latest two, aerial platforms have the advantage of enabling data acquisition in difficult or no accessible areas. However, other approaches have focused on the automation of the temporal monitoring of objects with a thermographic camera, thus automating the control of the camera toward the independent acquisition of images in certain time intervals depending on the processing methodology to be applied (Bagavathiappan et al., 2013), or when a specific event happens. The latter is the case of Cai et al. (2023), who have developed a robotic system based on computer vision for the acquisition of thermographic data when a leak is detected in a HVAC (heating, ventilation and air conditioning) installation.

With respect to the automation of the thermographic image processing, most efforts have focused on the automation of the processing of the data from active thermography tests, with the development and evaluation of techniques such as pulse thermography (PT), principal component thermography (PCT), thermal signal reconstruction (TSR), and differential absolute contrast (DAC) (Panella et al., 2021).

These techniques are usually applied for the detection of subsurface defects. In contrast, regarding the processing of images from pasive thermography inspections and the identification of pathologies from single images instead of sequential image series, and based on their surface print, algorithms have been developed based on geometric and thermal criteria (Garrido et al., 2019), but their scalability to be used on all cases is a complex task. To cover this issue, artificial intelligence (AI) models appear as the solution for the prediction and classification of thermal pathologies from thermal images.

AI has shown in the recent years a wide range of uses regarding images, including image generation (Elasri et al., 2022), image recognition (Zhang et al., 2021), image classification (Jena et al., 2021), and image processing (Zhang & Dahu, 2019), which can focus on different applications from medicine (England & Cheng, 2018) to energy systems (Ahmad et al., 2021) and agriculture (Lee et al., 2020). Most of these cases involve the use of images in the visible band of the spectrum, but AI models have shown to be useful also when trained or developed for hyperspectral images (Huang et al., 2023), or for their application in specific bands such as X-rays (Schalekamp et al., 2022), near infrared, or thermal infrared (He et al., 2021).

Among the different strategies of AI applied to images, machine learning (ML) and deep learning (DL) algorithms appear as the most applied solutions (Aggarwal et al., 2022). While AI is defined as "the study and design of intelligent systems that include their environment and take measures that increase their chances of success," ML is a type of AI, where statistical methods are used to enable machines to improve their performance with experience. Provided this, DL is a type of ML, in which artificial neural networks are developed to mimic the human brain performance.

DL has improved the predictive capacity of computers, thanks to the superior learning capacity of the developed algorithms (Ahmad et al., 2019). For this reason, DL has been considered as a solution for multiple applications from image recognition and processing, and object detection in medicine (Chen et al., 2022), remote sensing (Ma et al., 2019), materials science (Choudhary et al., 2022), and other fields, to self-driving cars (Gupta et al., 2021). The capacity of DL to reuse the knowledge from one application to another, known as deep transfer learning, is also key for the spread in the utilization of this technology, since it solves one of the main drawbacks of the application of DL which is the scarceness of large-sized and well-annotated datasets to train the models (Iman et al., 2023), which is even more intense when applying DL to IRT imagery.

Deep transfer learning models have been applied in the field of cultural heritage, for different purposes and different types of datasets. In the work of Banar et al. (2023), transfer learning is applied for the identification of icon classes and their assignment of codes in a multilinguality and multimodality approach. In this case of study of icon classes, deep transfer learning approaches are required because the amount of training data is usually limited comparing to the number of labels to be assigned. Deep transfer learning approaches have also been applied to the data formats used in this chapter for the study of cultural heritage: images and point clouds.

Regarding the first, Janković Babić (2023) evaluated four ML algorithms for the classification of images of cultural heritage such as archaeological sites, frescoes, and monasteries available in Google Images and Flickr. These algorithms were pre-trained using the ImageNet dataset. With respect to point clouds, AI is applied for the segmentation of points in features toward the posterior generation of the 3D model of the element scanned. However, the application of DL models to point clouds from cultural heritage assets is still scarce, so that the data available for the training of the models is also limited. Thus, in the work of Matrone and Martini (2021), a deep transfer learning approach is applied for the semantic segmentation of the point cloud. In this work, a neural network is trained on a point cloud dataset specifically created for the evaluation of cultural heritage point clouds and then applied to the point cloud of the new scene. In this case, the complexity comes from the lack of standardization in the cultural heritage field; as a result, the geometric and radiometric parameters from one scene can be completely different from the same parameters of another scene, even if they come from the same type of asset, because the age of construction and its location also play a significant role in the design.

Examples of the application of deep transfer learning models' thermal infrared imagery are easily found in the field of medicine. According to Dey et al. (2022), trained models are applied for the screening of breast cancer from thermographic images, with the aim at providing an effective methodology for cancer detection in developing countries. In an approach open to wider diseases, Ornek and Ceylan (2022) presented the application of deep transfer learning models and methods to classify neonates in the NICU (neonatal intensive care unit) as healthy and unhealthy, obtaining results of 100% specificity, sensitivity, and accuracy for the detection of health status of these patients. These studies show the increase in accuracy of the results compared to the application of DL models specifically trained with the thermal infrared images available per study, showing the higher importance of having a high number of images versus having specific images.

Provided the issues discussed earlier, this chapter starts with a section on related work, continuing with a case study to demonstrate the potential of using DL models in IRT, focusing on the field of cultural heritage. In this case, a deep transfer learning approach was applied, to compensate for the scarcity of thermal images for the training of the model. This allows to evaluate the replicability of results from thermographic inspections in heritage buildings with respect to the most common pathologies.

5.2 Related work

Advances in AI learning methods, and specifically in DL models, have allowed their combination with IRT inspection techniques that have been reflected in numerous works in different fields. Applications can be found ranging from the field of medicine (Cruz-Vega et al., 2020; Magalhaes et al., 2021) to engineering

(Marani et al., 2016; Saeed, Abdulrahman, et al., 2019). In the latter, there are studies mainly dedicated to new materials such as composites or additive manufacturing (Luo et al., 2019), although others can also be found focused on more traditional materials related to construction and that can be associated with heritage such as concrete or masonry (Ichi & Dorafshan, 2022).

IRT is a widely used inspection technique in materials engineering for both analysis and monitoring of materials. In this sense, new materials, such as carbon fiber composites, fiber-reinforced polymers such as carbon (CFRPs) or glass, or those coming from new additive manufacturing techniques such as nylon or polylactic acid, need to be characterized. Moreover, IRT is an ideal technique to study the heterogeneity and possible internal defects associated with the manufacturing processes of this type of materials (Marani et al., 2016; Rodríguez-Martín et al., 2020). The combination of this technique with AI learning methods allows for the automation of these processes by eliminating the subjectivity of the operator and has been successfully used for the detection of interior defects in materials such as nylon (Rodríguez-Martín et al., 2022). In addition, other materials such as CFRPs have been analyzed by implementing DL models for automatic defect detection (Saeed, King, et al., 2019) or water detection in aircraft materials (D'Orazio et al., 2005), and a PVC (PolyVinyl Chloride)-infrared dataset has been created for the evaluation of the performance of different popular deep learning−based instance segmentation models for characterization tests with PT (Wei et al., 2023).

Although some of these materials are currently used for the rehabilitation and restoration of heritage elements (Yumnam et al., 2021), their applications are mainly focused on industry. In the field of infrastructures, other materials traditionally used for construction have been the object of study and application of IRT techniques in combination with AI. One of the topics of interest is the automatic detection of cracks in both steel (Yang et al., 2019) and concrete (Jang et al., 2019) elements by applying DL models together with thermal images, mainly in cases of bridges (Ali & Cha, 2019) and delamination of their decks (Cheng et al., 2020). The detection of internal defects has also been subject of analysis toward its resolution with DL algorithms: Qiang Fang et al. (2020) have performed an analysis of different DL algorithms for automatic defect detection and precise localization in the case of subsurface defects. Sometimes, the introduction of hybrid imaging (visible and IR) has been used to study large structures such as bridges or dams (Alexander et al., 2022; Yun-Kyu et al., 2018), while Yun-Kyu et al. (2018) have also applied hybrid images including vision camera, IR camera, and continuous-wave line laser for the identification of cracks in concrete based on a convolutional neural network. Other studies, such as Peng et al. (2021), have exploited the automation both regarding image acquisition from UAVs and defect detection applying DL methods, for the identification of debonding areas in building envelopes.

All these applications present the same drawback, that is the scarceness of images for the training of the DL models. For this reason, deep transfer learning models have also been applied to thermographic images of different materials. Saeed, King, et al. (2019) showed the application of deep transfer learning for the detection of defects in CFRP thermographic images. In this study, pretrained neural networks are selected for the identification of defects, and their performance is fine-tuned by performing their training with a small dataset of thermal infrared images consisting of 200 images of subsurface defects in a CFRP sample. Following this approach, the authors manage to detect subsurface defects using neural networks that had been previously trained to detect common features such as straight lines, curves, and shapes. On a similar approach toward defect detection, Liu and Wang (2022) analyzed the performance of different convolutional neural networks and the effect of training these models from scratch or benefiting from transfer learning for the detection and classification of cracks in asphalt pavement. Visible images, thermal infrared images, and hybrid visible and thermal images were used for the study. Their results showed that each DL model has a different behavior regarding training from scratch and being pretrained, and that the severity level of the crack to be detected and classified also had an influence in the performance of the DL model and the training process. In a double automation approach toward automating the image acquisition and the image processing, Zhou et al. (2022) presented the application of deep transfer learning for the detection of embankment leakage from thermal infrared images acquired from a UAV. In this case, the DL model used, Alex-Net based, is pretrained with a dataset of 10,000 thermal infrared images acquired from simulations of slope leakage in an open-air platform and then applied to the detection and classification of leakage in field tests. With this approach, 94.90% classification accuracy is reached.

In the case of cultural heritage, the application of DL techniques has been done for different uses, mostly aiming at protecting the sites and increasing their value through a greater interaction with the visitors. In this line of expert—user interaction, Sperlí (2021) has presented the application of DL to the creation of a chatbot to support tourist vist, while Kumar et al. (2020) have developed a method for the detection of damage in cultural heritage related to disasters through the recognition and classification of images uploaded by visitors on social media. Focusing on the use of imagery by cultural heritage experts and conservators, DL techniques have been applied to the detection of unknown cultural heritage sites (Monna et al., 2021) and archeological mounds (Orengo et al., 2020) from satellite imagery. In addition, images and 3D scans have been subjected to DL techniques for the generation of 3D models of cultural sites (Pierdicca et al., 2020), as well as for the interpretation of the images toward the extraction of semantic terms that could contribute to the digital documentation in the form of Heritage-BIM (Building Information Modeling) (HBIM) and decision-making of the site (Llamas et al., 2016). In this line, Díaz-Rodríguez et al. (2022) have developed an EXplainable Neural-Symbolic learning methodology to increase the understanding of cultural heritage sites by fusing DL interpretations with expert knowledge graphs.

Regarding specific applications of DL for the monitoring of cultural heritage and detection of pathologies, which are the main uses of IRT, several studies can also be found in the literature. As an example, a query performed in the study of Tejedor et al. (2022) using the terms "historic building," "machine learning," "deep learning," and their synonyms resulted in a total of 299 papers published between 2001 and 2021. An analysis of these papers shows that DL is mostly considered as necessary for the generation of HBIM, through its application to the recognition, interpretation, and processing of large datasets from photogrammetry and laser scanning techniques. The analysis performed in this paper also shows that there is a difference between the number of papers regarding advanced modeling techniques and the number of papers about non-destructive testing, thus showing a lack of integration of the results of historic building diagnosis, using IRT, for the update of the HBIM, where usually the applications of DL lie.

Adamopoulos (2021) developed an automatic method of classifying multispectral images including IRT bands that uses AI to detect the deteriorated areas of a weathered historic fortification. Such images had been used before to analyze historical facades and automatically detect different materials using supervised classification algorithms (Lerma, 2001; Lerma, 2005). Nevertheless, these works use multispectral cameras, which, in contrast to thermal cameras, are much more expensive and complex. The use of thermal imaging and its combination with DL models for specific applications in heritage studies is a field yet to be explored, where the introductory fundamentals are recently being laid for works focused on automatic detection and segmentation of defects (Garrido et al., 2021). In another study, Moradi et al. (2022) have evaluated the performance of a spatiotemporal deep neural network for defect identification, contrasting the results with those of conventional algorithms. Regarding artworks, Liu et al. (2023) have developed a strategy based on the incorporation of DL approaches to the traditional principal component analysis methodology toward the detection of defects, which lead to important improvements in the signal-to-noise ratio.

5.3   Case study

The case study selected was the church of Santiago de Cangas located in the town of Cangas, NW Spain (42°15′ 51.9″ N, 8°47′06.0″ W). The origin of the church of Santiago de Cangas dates back to 1493. It is in this year when we have evidence of the construction of the current temple with almost cathedral-like capacity, composed of three naves separated by airy columns with starred stone vaults and pointed arches in its nave. It emphasizes its plateresque facade of the 16th century with bodies of double Ionic and Corinthian starred columns. The four evangelists are sculpted in low relief, with busts and figures, all crowned by the Eternal Father in the center of a pediment. In 1542, it was erected as a collegiate church with a prior and six racioneros by order of the bishop of Compostela, Gaspar de Avalos. Subsequently, the bell tower was erected, in the Baroque style known as "de placas" (of plates). Its plan is square and consists of two sills crowned by a small dome externally adorned and with four pinnacles in its corners. Its height up to the end of the dome is about 37 m (Igrexa de Santiago de Cangas). Figs. 5.1 and 5.2 show global views of the facade of the church Santiago de Cangas from different points of view.

Figure 5.1 View 1: Global view of the facade of the church Santiago de Cangas.

Figure 5.2 View 2: Global view of the facade of the church Santiago de Cangas.

5.3.1   Acquisition process

The main objective of the inspection was to perform an integral characterization of the church, toward not only identifying the pathologies present but also to reach their complete description in terms of location, geometry, and severity.

To reach this objective, two non-destructive technologies are employed: IRT for the automatic detection and segmentation of thermal bridges and water-damaged areas on the front facade of the church and terrestrial laser scanning (TLS) for the subsequent geometric characterization of the areas where pathologies are detected.

Both technologies were used in the same morning, in a rainy season during the month of November. It should be noted that for the day of study, the acquisition conditions were optimal, that is, dry and cloudy day for a correct measurement of both equipments. In addition, rain had disappeared for the 24 hours prior the inspection, in such a way that the effect of the recent wetness in the material was avoided for the inspection.

5.3.1.1 Point cloud data

The survey of the front facade was performed with the TLS Faro Focus3D X330, which technical specifications are shown in Table 5.1. Due to the 26-m width of the facade and the existence of buttresses that produced occlusions in the portico, three scan positions were selected along the facade (Fig. 5.3): one scan position perpendicular to the portico and two other scan positions at oblique directions to the buttresses. The registration of the three point clouds acquired from the three different positions into a unique point cloud was performed manually, using as reference the central scan position (no. 2). In this procedure, target points in the overlapping areas between point clouds were identified and manually marked, in such a way that the rotation and translation matrix from point cloud 1 to the reference system of point cloud 2, and from point cloud 3 to the reference system of point cloud 2, were calculated. The registration resulted with a root mean square (RMS) error of 1.5 cm. The complete point cloud contains 44 million points. The scanning at the ground level showed the presence of occlusions in elevated areas above the cornices that could be avoided by the point cloud data acquisition from an aerial platform such as UAVs.

In parallel, to obtain an occlusion-free scan of the portico, a scan was performed with the iPad Pro 2021. The generated point cloud (Fig. 5.4) obtained more completeness at the base of the columns and the spacing between columns and buttresses. The point cloud from the iPad contains 1.2 million points and was registered to scan position no. 2 of Faro point cloud with an RMS error of 7 cm. The range of the iPad Pro 2021 sensor is only 5 m, so the upper area of the facade was left unscanned. Both surveys, with Faro Focus3D X330 and iPad Pro 2021 were performed in real RGB color.

Table 5.1 Technical specifications of the FARO Focus3D X330.

Model	Faro Focus3D X330
Measurement range	From 0.6−330 m
Ranging error (25 m, one sigma)	± 2 mm
Step size	0.009 degree
Field of view (vertical/horizontal)	300/360 degrees
Beam divergence	0.011 degree
Measurement rate (points per second)	122,000−976,000
Laser wavelength	1550 nm

(A)   (B)

Figure 5.3 Point cloud generated with Faro Focus3D X330: (A) Top view with scan positions and (B) perspective view of the facade.

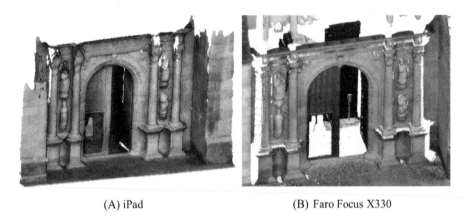

(A) iPad   (B) Faro Focus X330

Figure 5.4 Point cloud of the portico generated with (A) iPad Pro and (B) Faro Focus3D X330.

5.3.1.2 Thermal infrared data

The thermographic camera NEC TH9260 was used for the thermal inspection of the front facade. The technical specifications of the camera are shown in Table 5.2.

Table 5.2 Technical specifications of the thermographic camera NEC TH9260.

Model	NEC TH9260
Sensor type	Uncooled focal plane array (μbolometer)
Thermal image/pixels	640 (H) × 480 (V)
Resolution (°C)	0.1
Accuracy	± 2°C or ± 2% of reading, whichever is greater
Spectral range (μm)	8−14

As an important characteristic of the thermographic cameras to take into account when planning a thermographic inspection, the field of view of these cameras is usually smaller than that of photographic cameras. For this reason, the area that can be inspected in an image is usually limited, and the acquisition of several images is required for the inspection of the complete area under study. Thus, in the case of the church of Santiago de Cangas, the thermal inspection of the facades required the acquisition of more than one image per facade, in such a way that mosaics of thermal images had to be created to have a global thermal map. Another important issue to take into account when performing a thermographic inspection is the angle of acquisition of the thermal infrared images: images acquired from a frontal position to the element under study may be affected by reflections of the temperature of the camera operator on the surface under inspection, while images acquired from a position with a high angle regarding the surface under inspection may not record the real temperature of this surface due to the presence of angular reflections of the thermal radiation. For this reason, the thermographic inspection of any element should be from a nearly frontal position to the element under study, with an angle of 10 degrees from the line perpendicular to the surface. With this angle, the reflections of the operator on the surface are avoided, and the radiation artifacts due to a very inclined view are not present. This condition is especially difficult to fulfill in the inspection of the upper parts of the facade if the inspection is performed from a position on the ground. However, the angle of view can be approximated by performing the acquisition of those images from a further distance, reaching a compromise distance so that the angle is not maximum and the pixels are not too big that the temperature measured is not accurate.

The generation of mosaics requires the registration of the thermographic images to be united into the same reference system, in addition to the correction of artifacts introduced by the lens in the images in their acquisition. For both procedures, the knowledge of the geometric parameters of the camera are required, in such a way that the geometric calibration of the camera should be performed to allow any processing procedures. This type of calibration has been developed in the last years and is not included in the sale of the thermographic cameras, since the most common calibration for these sensors is the radiometric or temperature calibration (Usamentiaga et al., 2014). The radiometric or temperature calibration is the one that determines the level of accuracy of thermographic cameras in the measurement of temperatures.

The geometric calibration of cameras implies the determination of the inner geometric characteristics of the camera-lens system. This can be done mainly in two ways: during a geometric calibration performed at the same time as the image registration, in a procedure called vicarious calibration, or in a geometric calibration performed previously, using a geometric calibration field, specifically designed for the task. The design of geometric calibration fields was the first approach developed for the geometric calibration of thermographic cameras, based on the principles applied for the calibration of cameras in photogrammetric processes. Thus geometric calibration fields have been designed based on temperature differences (Lagüela, González-Jorge, et al., 2011) and in emissivity differences (Lagüela, Armesto, et al., 2012; Lagüela, González-Jorge, et al., 2012) between the targets points and the board, both obtaining adequate results. The vicarious calibration consists of the geometric calibration of the thermographic camera during measurement, using for calibration those images acquired for the inspection instead of images acquired specifically for calibration and from specific calibration boards (Usamentiaga et al., 2018). These types of calibrations enlighten the process, since the performance of an additional calibration step is avoided, but the results usually present higher error and are only applicable for the inspection for which the calibration was performed. As a contrast, the geometric calibration of thermographic cameras based on calibration boards is valid for all the inspections performed without changing the position and focus of the lens.

Regarding the generation of thermographic mosaics, after the geometric correction of the thermographic images through the elimination of the geometric drift and artifacts introduced by the camera-lens system, the following step is the registration of the thermal images into the same reference system. As stated by Lagüela et al. (2018), this step, denominated as image registration, requires the images that constitute the same mosaic are acquired with overlapping areas, where common points between images can be identified. Thus the image registration starts with the identification of points of interest in the images and the identification of those points among the points of interest that are common between images, which are named point pairs or corresponding points. Different studies have focused on the identification of points of interest, applying the different point descriptors that have been designed for visible images. However, provided the low spatial resolution of thermographic images in comparison with that of visible images, the thermographic image registration requires the application of advanced point descriptors and computer vision algorithms. As examples, Lagüela, Armesto, et al. (2012) and Lagüela, González-Jorge, et al. (2012) perform the image registration using an operator such as FAST (Features from Accelerated Segment Test) that identifies characteristic points, such as corners, by analyzing the difference between each pixel and its neighbors. The correspondences between point features in the different images are evaluated through the computation of the correlation parameter. On the other hand, González-Aguilera et al. (2013) uses a feature-based algorithm instead a technique based on areas such as the previous one. This algorithm is ASIFT (Affine Scale Invariant Feature Transform), which is selected due to its robustness to geometric and radiometric variations. Lastly, González-Aguilera et al. (2012) present a combination of both techniques, by applying an area-based method such as pyramidal images, followed by a feature-based method with the Harris detector, in such a way that the procedure could be applied for the registration of images from the same sensor, as in the generation of thermographic mosaics, as well as for multisensor registration.

In this work, the thermographic mosaics were created by applying the feature-based approach, through the identification of corresponding points in overlapping areas between thermal images, and using those points for the computation of the orientation and translation matrix of every image into the same reference system, as shown by Lagüela et al. (2018). The following figures show different thermal image mosaics created for the integral thermal representation of the church under study. Since the objective was to perform a qualitative analysis, that is, detection and geometric characterization of pathological areas, the calibration of the emissivity value (set to 1), the reflected temperature, and the ambient temperature and humidity (ambient conditions) were not taken into account. In this way, the study is performed based on apparent temperatures, instead of absolute temperatures (Marinetti & Cesaratto, 2012).

To perform an organized inspection of the facade, its surface was divided in the different forms it is composed of. In this way, the first thermographic images were acquired from the tower, in sequential rows from top to bottom, as shown in Figs. 5.5−5.7.

Then, the main facade was divided in vertical rows, so the mosaics represent different parts of the facades, from left to right, as shown in Figs. 5.8−5.10, where the main entrance is shown.

The generation of mosaics allow the experts to perform the interpretation of the state of the surfaces with higher accuracy, being able to detect pathologies that could have passed unnoticed if the surfaces were evaluated through their individual images. In addition, the type of pathology is also easier to identify when analyzed from a mosaic image, since this image combination allows the visualization of the complete form of the thermal print of the pathology.

Figure 5.5 Mosaic of thermal images from the top of the tower of the church.

(A)   (B)

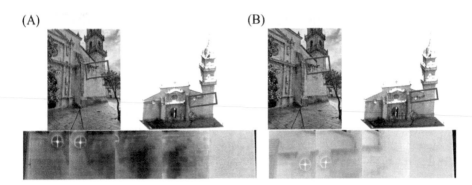

Figure 5.6 (Left and right) Mosaics of thermal images from the middle of the tower of the church.

(A)   (B)

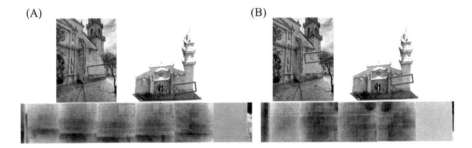

Figure 5.7 Mosaic of thermal images from the bottom of the tower of the church.

5.3.2 Deep learning application to infrared data

The Mask Region-based Convolution Neural Network (Mask R-CNN), which was developed by the Facebook AI Research (He et al., 2020), is the DL model used for the automatic interpretation of the mosaics of the thermal images acquired. This DL model is a simple, flexible, and general framework for object detection, segmentation, and classification tasks in images. It was selected for the application of pathology detection and classification from thermal infrared images because Mask R-CNN has outperformed all existing DL models in instance segmentation, object detection via bounding box, and person key point detection using the Microsoft COCO dataset (Lin et al., 2014) as input dataset. Proof of this is the modifications made on the basis of this DL model in the latest published papers (Bonhage et al., 2021; Hou et al., 2021; Wu et al., 2021; Zhang et al., 2020). Mask R-CNN is built on the top of Faster R-CNN, which is another DL model that returns the bounding box coordinates and the class label for each object detected in the input image. However, Mask R-CNN, apart from returning the same outputs of Faster R-CNN, also returns the object mask. In this case, the trained Mask R-CNN from the study of Garrido et al. (2022) is directly employed to predict the pathological areas, given that such a trained model was based on a wide variety of different infrastructures

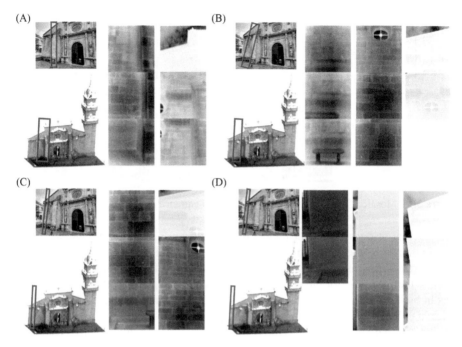

Figure 5.8 Mosaic of thermal images from the left side of the main facade of the church.

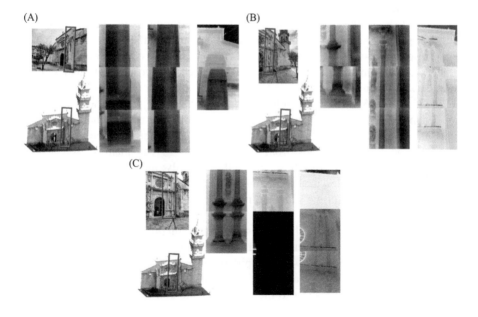

Figure 5.9 Mosaic of thermal images from the right side of the main facade of the church.

(A)   (B)

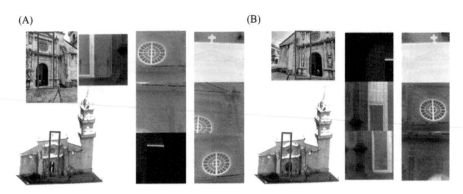

Figure 5.10 Mosaic of thermal images from the main entrance of the church.

with thermal bridges and water-related problems, from residential, public, and heritage buildings to the pillars of various bridges and regions inside tunnels. In this way, a transfer learning approach, specifically deep transfer learning, is also applied to show the coherence of the thermal inspection regarding interpretation of pathologies in buildings and avoid the scarceness of thermal infrared imagery from pathologies in cultural heritage sites.

First, the DL model is applied to the thermal images individually. Fig. 5.11 shows the results of the predictions made by the model in images from the bottom part of the main facade of the church. For all cases (Fig. 5.11 and following figures), a solid line represents the boundary of the pathology, and the broken line indicates its bounding box.

Fig. 5.12 shows the results of the prediction of pathologies by the DL model in the upper part of the main facade, including the rosette and the tower. The presence of the rosette shows a weakness of the DL model, since it detects the rosette as a pathology. The same happens in the balustrade, since some of the balusters and the space between them are also detected as pathologies.

Figs. 5.13–5.15 show the results of the predictions of pathologies of the DL model applied to the thermal mosaics. Fig. 5.13 shows the results obtained for the right side of the facade, Fig. 5.14 shows the results from the main entrance, while Fig. 5.15 includes the results from the right side of the main facade. These figures show similar results to those obtained previously: most pathologies detected correspond to pathologies in the masonry, and false positives are detected in the rosette, in the sculpture of the saint and in the bench located in front of the left side of the main facade.

In addition, the analysis of the figures and the predictions obtained with the DL model shows that correct detections and segmentations of areas related to thermal bridges (window frames) and water (capillary rise, water seepage from roofs) are obtained up to the top of the main nave. Due to the excessive angle of incidence of the thermal camera with respect to the surface of the bell tower, the predictions in that zone are not valid. Also, the DL model is not able to predict on the surfaces of the different decorative figures (saint sculptures, rosette) of the facade due to the

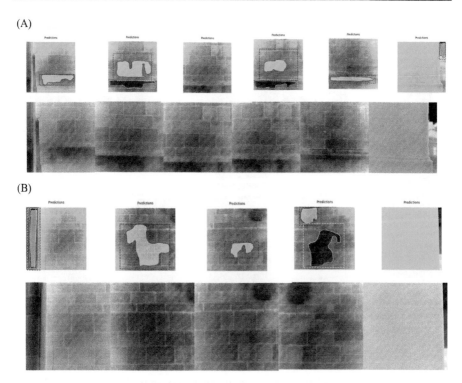

Figure 5.11 Pathology prediction results from the application of the DL to individual thermal images from the bottom of the main facade of the church. *DL*, Deep learning.

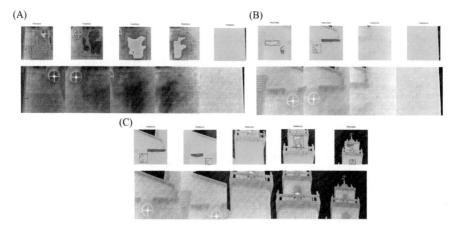

Figure 5.12 Pathology prediction results from the application of the DL to individual thermal images from the top of the main facade of the church, where the rosette and the balustrade are located. *DL*, Deep learning.

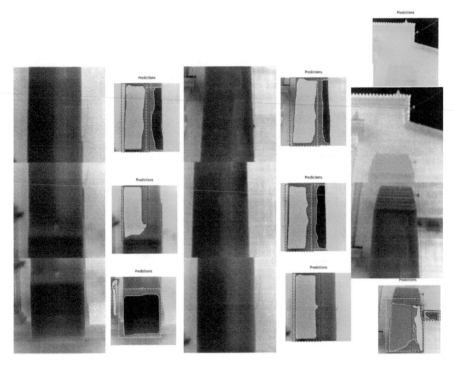

Figure 5.13 Pathology prediction results from the application of the DL to mosaic thermal images from the right side of the facade. *DL*, Deep learning.

lack of a high resolution in the thermal camera, a limit common to all thermal cameras.

5.3.2.1   Point cloud−based geometric characterization

With the pathological areas detected and segmented, it is possible to perform their geometric characterization based on the acquired point clouds. That is, the one- and two-dimensional dimensions of the segmented areas of both thermal bridges and water-related problems can be obtained in metric units. Figs. 5.16 and 5.17 show the geometric characterization of a thermal bridge and a water-problem area, respectively, as examples of the possibilities of the geometric characterization.

The geometric characterization of thermal pathologies can be done thanks to the registration of the thermal images (individual or mosaics) to the point clouds, in such a way that the position of each thermal images within the point cloud is known. Several methodologies can be applied for the performance of this procedure: Lagüela, Martínez, et al. (2011) presented the first methodology for thermal image-point cloud registration, based on the recognition of common points between the thermal images and the point clouds by an expert and the manual determination of the point pairs. Advances toward the automation of this methodology were performed by

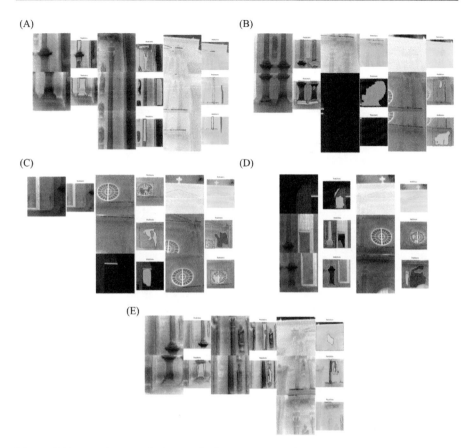

Figure 5.14 Pathology prediction results from the application of the DL to mosaic thermal images from the main entrance, at the center of the facade. *DL*, Deep learning.

Lagüela et al. (2013) and González-Aguilera et al. (2012). The main complexity of the registration process is that it is based on the resolution of a 2D−3D problem, since (thermal) images have information in 2D, while point clouds include information in 3D. Thus both papers mentioned approach the resolution of the problem in a different way. Lagüela et al. (2013) presented the identification of lines in 2D in the thermal images and of the same lines in 3D in the point cloud; based on the sets of lines in the images, the position of the images with respect to the point cloud can be calculated by resolving the collinearity equations (Liu & Stamos, 2005). Differently, González-Aguilera et al. (2012) approached the resolution of the 2D−3D problem by converting it in a 2D−2D problem, and then transforming the resolution to 3D. For this, the point cloud is transformed into range images, and the registration is performed through the identification of point pairs between thermal images and range images. In this case, the Harris operator is selected among existing operators due to the invariance to rotations and scale factors.

(A)

(B)

(C)

(D)

(E)

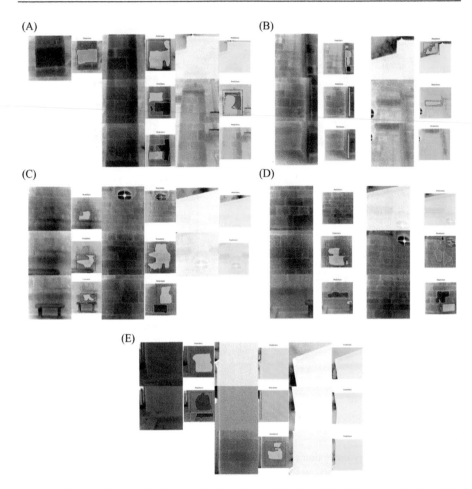

Figure 5.15 Pathology prediction results from the application of the DL to mosaic thermal images from the left side of the facade. *DL*, Deep learning.

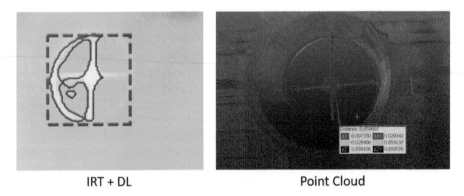

IRT + DL   Point Cloud

Figure 5.16 Longitudinal measurement of the detected and segmented vertical thermal bridge. Units in m.

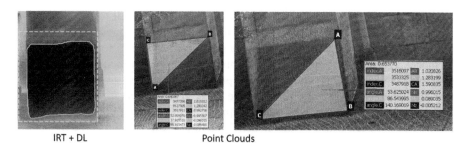

IRT + DL   Point Clouds

Figure 5.17 Area measurement of the detected and segmented water area by capillary rise. Units in m².

5.4 Conclusions and future challenges

This chapter introduced a review of the applications of AI algorithms to the detection and characterization of thermal pathologies in heritage buildings from thermal images, with special emphasis on ML algorithms, specially DL. With the purpose of demonstrating the performance of DL models in this application, the front facade of a heritage church was thermographically inspected and analyzed. Specifically, the deep transfer learning alternative was applied, since the DL model selected, the Mask R-CNN, had been trained with a different dataset, composed of thermal infrared images from a wide variety of different infrastructures with the pathologies most commonly present in infrastructures and cultural heritage and most commonly detected with IRT, such as thermal bridges and water-related problems, from residential, public, and heritage buildings to the pillars of various bridges and regions inside tunnels. The results of the application of this model with the training done using images from different infrastructures to the main facade of the church of Santiago de Cangas show that deep transfer learning can be applied to thermal images of heritage buildings, since the thermal prints of the most common pathologies maintain a similar shape and temperature difference in all infrastructures. This fact can also be seen through the application of the DL model to mosaic thermal images, and individual thermal images, since the detection of pathologies is similar in both cases. The advantages of the use of mosaic thermal images instead of individual thermal images can be seen by the fact that more pathologies are detected in the mosaics, due to the fact that the pathologies appear in their totality instead of partially.

The usefulness of the combination of non-destructive techniques in the thermographic inspection of heritage buildings is shown through the combination of the results of the thermographic inspections with the data acquired in a laser scanning inspection. As a results of the combination of thermal data with geometric data in the form of point clouds results in the possibility of performing the geometric characterization of the pathologies detected.

However, some drawbacks have been detected for the application of DL models to thermal imagery, such as the detection of false positives and the lack of capability to predict pathologies on the surfaces of the different decorative figures of the facade. These problems are mainly due to the low spatial resolution in thermal infrared cameras, and to the number of images used for the training of the model. It should be noted that the training dataset did not include cultural heritage assets, specifically no churches were present. For this reason, most false positives and false negatives were found in the areas of the balustrade and the rosette, since these decorative elements did not appear in the training dataset and consequently had not been learnt by the model. Thus future research should be in the line of the creation of large datasets of properly labeled thermal infrared images, as well as in the strengthening of the analysis of the applicability of deep transfer learning approaches to compensate for the lack of specific thermal images. The use of DL models previously trained in a standard dataset and fine-tuned in the dataset under study or a similar one should also be evaluated. In addition, future works should deepen in the adaptation of existing DL models, or the creation of new ones, with applicability to low resolution imagery.

References

Adamopoulos, E. (2021). Learning-based classification of multispectral images for deterioration mapping of historic structures. *Journal of Building Pathology and Rehabilitation*, 6 (1), 41. Available from https://doi.org/10.1007/s41024-021-00136-z, https://doi.org/10.1007/s41024-021-00136-z.

Aggarwal, K., Mijwil, M. M., Sonia., Al-Mistarehi, A.-H., Alomari, S., Gök, M., Anas, M., Zein, A., & Abdulrhman, S. H. (2022). Has the future started? The current growth of artificial intelligence, machine learning, and deep learning. *Iraqi Journal For Computer Science and Mathematics*, 3(1), 115−123. Available from https://doi.org/10.52866/ijcsm.2022.01.010.013, https://journal.esj.edu.iq/index.php/IJCM/article/view/100.

Ahmad, J., Farman, H., Jan, Z., Khan, M., Jan, B., & Farman, H. (2019). *Deep learning methods and applications* (pp. 31−42). Singapore: Springer Singapore. Available from https://doi.org/10.1007/978-981-13-3459-7_3, 10.1007/978-981-13-3459-7_3.

Ahmad, T., Zhang, D., Huang, C., Zhang, H., Dai, N., Song, Y., & Chen, H. (2021). Artificial intelligence in sustainable energy industry: Status quo, challenges and opportunities. *Journal of Cleaner Production*, 289125834. Available from https://doi.org/10.1016/j.jclepro.2021.125834, https://www.sciencedirect.com/science/article/pii/S0959652621000548.

Alexander, Q. G., Hoskere, V., Narazaki, Y., Maxwell, A., & Spencer, B. F. (2022). Fusion of thermal and RGB images for automated deep learning based crack detection in civil infrastructure. *AI in Civil Engineering*, 1(1), 3. Available from https://doi.org/10.1007/s43503-022-00002-y, https://doi.org/10.1007/s43503-022-00002-y.

Alhammad, M., Avdelidis, N. P., Ibarra Castanedo, C., Maldague, X., Zolotas, A., Torbali, E., & Genest, M. (2022). Multi-label classification algorithms for composite materials under infrared thermography testing. *Quantitative InfraRed Thermography Journal*, 1−27. Available from https://doi.org/10.1080/17686733.2022.2126638, https://doi.org/10.1080/17686733.2022.2126638.

Ali, R., & Cha, Y.-J. (2019). Subsurface damage detection of a steel bridge using deep learning and uncooled micro-bolometer. *Construction and Building Materials*, 226,

376−387. Available from https://doi.org/10.1016/j.conbuildmat.2019.070.293, https://www.sciencedirect.com/science/article/pii/S0950061819319671.

Bagavathiappan, S., Lahiri, B. B., Saravanan, T., Philip, John, & Jayakumar, T. (2013). Infrared thermography for condition monitoring − A review. *Infrared Physics & Technology*, *60*, 35−55. Available from https://doi.org/10.1016/j.infrared.2013.030.006, https://www.sciencedirect.com/science/article/pii/S1350449513000327.

Balaras, C. A., & Argiriou, A. A. (2002). Infrared thermography for building diagnostics. *TOBUS - a European method and software for office building refurbishment.*, *34*(2), 171−183. Available from https://doi.org/10.1016/S0378-7788(01)00105-0, https://www.sciencedirect.com/science/article/pii/S0378778801001050.

Banar, N., Daelemans, W., & Kestemon, M. (2023). Transfer learning for the visual arts: The multi-modal retrieval of Iconclass Codes. *Journal on Computing and Cultural Heritage*, *16*(2), 1−16. Available from https://doi.org/10.1145/3575865.

Bonhage, A., Eltaher, M., Raab, T., Breuß, M., Raab, A., & Schneider, A. (2021). A modified Mask region-based convolutional neural network approach for the automated detection of archaeological sites on high-resolution light detection and ranging-derived digital elevation models in the North German Lowland. *Archaeological Prospection*, *28*(2), 177−186. Available from https://doi.org/10.1002/arp.1806, https://doi.org/10.1002/arp.1806.

Cai, W., Huang, L., & Zou, Z. (2023). Actively-exploring thermography-enabled autonomous robotic system for detecting and registering HVAC thermal leaks. *Automation in Construction*, *152*104901. Available from https://doi.org/10.1016/j.autcon.2023.104901, https://www.sciencedirect.com/science/article/pii/S0926580523001619.

Cheng, C., Shang, Z., & Shen, Z. (2020). Automatic delamination segmentation for bridge deck based on encoder-decoder deep learning through UAV-based thermography. *NDT & E International*, *116*102341. Available from https://doi.org/10.1016/j.ndteint.2020.102341, https://www.sciencedirect.com/science/article/pii/S0963869520303224.

Chen, X., Wang, X., Zhang, K., Fung, K., Thai, T. C., Moore, K., Mannel, R. S., Liu, H., Zheng, B., & Qiu, Y. (2022). Recent advances and clinical applications of deep learning in medical image analysis. *Medical Image Analysis*, *79*102444. Available from https://doi.org/10.1016/j.media.2022.102444, https://www.sciencedirect.com/science/article/pii/S1361841522000913.

Choudhary, K., DeCost, B., Chen, C., Jain, A., Tavazza, F., Cohn, R., Park, C. W., Choudhary, A., Agrawal, A., Billinge, S. J. L., Holm, E., Ong, S. P., & Wolverton, C. (2022). Recent advances and applications of deep learning methods in materials science. *npj Computational Materials*, *8*(1), 59. Available from https://doi.org/10.1038/s41524-022-00734-6, https://doi.org/10.1038/s41524-022-00734-6.

Cruz-Vega, I., Hernandez-Contreras, D., Peregrina-Barreto, H., Rangel-Magdaleno, J. D., & Ramirez-Cortes, J. M. (2020). Deep learning classification for diabetic foot thermograms. *Sensors*, *20*(6). Available from https://doi.org/10.3390/s20061762.

D'Orazio, T., Guaragnella, C., Leo, M., & Spagnolo, P. (2005). Defect detection in aircraft composites by using a neural approach in the analysis of thermographic images. *NDT & E International*, *38*(8), 665−673. Available from https://doi.org/10.1016/j.ndteint.2005.04.005, c.

Daffara, C., Muradore, R., Piccinelli, N., Gaburro, N., de Rubeis, T., & Ambrosini, D. (2020). A Cost-effective system for aerial 3D thermography of buildings. *Journal of Imaging*, *6*. Available from https://doi.org/10.3390/jimaging6080076.

de Oliveira, A. K. V., Aghaei, M., & Rüther, R. (2020). Aerial infrared thermography for low-cost and fast fault detection in utility-scale PV power plants. *Solar Energy*, *211*, 712−724. Available from https://doi.org/10.1016/j.solener.2020.090.066, https://www.sciencedirect.com/science/article/pii/S0038092X20310197.

Dey, S., Roychoudhury, R., Malakar, S., & Sarkar, R. (2022). Screening of breast cancer from thermogram images by edge detection aided deep transfer learning model. *Multimedia Tools and Applications*, *81*(7), 9331–9349. Available from https://doi.org/ 10.1007/s11042-021-11477-9.

Dunderdale, C., Brettenny, W., Clohessy, C., & Ernest van Dyk, E. (2020). Photovoltaic defect classification through thermal infrared imaging using a machine learning approach. *Progress in Photovoltaics: Research and Applications*, *28*(3), 177–188. Available from https://doi.org/10.1002/pip.3191, https://doi.org/10.1002/pip.3191.

Díaz-Rodríguez, N., Lamas, A., Sanchez, J., Franchi, G., Donadello, I., Tabik, S., Filliat, D., Cruz, P., Montes, R., & Herrera, F. (2022). EXplainable Neural-Symbolic Learning (X-NeSyL) methodology to fuse deep learning representations with expert knowledge graphs: The MonuMAI cultural heritage use case. *Information Fusion*, *79*, 58–83. Available from https://doi.org/10.1016/j.inffus.2021.090.022, https://www.sciencedirect. com/science/article/pii/S1566253521001986.

dos Santos, G. M., de Aquino, R. R. B., & Lira, M. M. S. (2018). Thermography and artificial intelligence in transformer fault detection. *Electrical Engineering*, *100*(3), 1317–1325. Available from https://doi.org/10.1007/s00202-017-0595-2, https://doi.org/ 10.1007/s00202-017-0595-2.

Elasri, M., Elharrouss, O., Al-Maadeed, S., & Tairi, H. (2022). Image generation: A review. *Neural Processing Letters*, *54*(5), 4609–4646. Available from https://doi.org/10.1007/ s11063-022-10777-x, https://doi.org/10.1007/s11063-022-10777-x.

England, J. R., & Cheng, P. M. (2018). Artificial intelligence for medical image analysis: A guide for authors and reviewers. *American Journal of Roentgenology*, *212*(3), 513–519. Available from https://doi.org/10.2214/AJR.18.20490, https://doi.org/10.2214/ AJR.18.20490.

Fernández-Cuevas, I., Arnáiz Lastras, J., Escamilla Galindo, V., & Gómez Carmona, P. (2016). Application of Infrared Thermography in Sports Science. Available from 10.1007/978-3-319-47410-6_4In J. Priego Quesada (Ed.), *Biological and medical physics, biomedical engineering infrared thermography for the detection of injury in sports medicine*. Cham: Springer.

Garrido, I., Erazo-Aux, J., Lagüela, S., Sfarra, S., Ibarra-Castanedo, C., Pivarčiová, E., Gargiulo, G., Maldague, X., & Arias, P. (2021). Introduction of deep learning in thermographic monitoring of cultural heritage and improvement by automatic thermogram pre-processing algorithms. *Sensors*, *21*(3). Available from https://doi.org/10.3390/s21030750.

Garrido, I., Lagüela, S., Arias, P., & Balado, J. (2018). Thermal-based analysis for the automatic detection and characterization of thermal bridges in buildings. *Energy and Buildings*, *158*, 1358. Available from https://doi.org/10.1016/j.enbuild.2017.110.031.

Garrido, I., Lagüela, S., Fang, Q., & Arias, P. (2022). Introduction of the combination of thermal fundamentals and Deep Learning for the automatic thermographic inspection of thermal bridges and water-related problems in infrastructures. *Quantitative InfraRed Thermography Journal*, 1–25. Available from https://doi.org/10.1080/17686733.2022.2060545, https://doi. org/10.1080/17686733.2022.2060545.

Garrido, I., Lagüela, S., Sfarra, S., Madruga, F., & Arias, P. (2019). Automatic detection of moistures in different construction materials from thermographic images. *Journal of Thermal Analysis and Calorimetry*, *138*, 1649. Available from https://doi.org/10.1007/ s10973-019-08264-y.

González-Aguilera, D., Lagüela, S., Rodríguez-Gonzálvez, P., & Hernández-López, D. (2013). Image-based thermographic modeling for assessing energy efficiency of buildings façades.

Energy and Buildings, 65, 29–36. Available from https://doi.org/10.1016/j. enbuild.2013.050.040, https://www.sciencedirect.com/science/article/pii/S0378778813003289.

González-Aguilera, D., Rodriguez-Gonzalvez, P., Armesto, J., & Lagüela, S. (2012). Novel approach to 3D thermography and energy efficiency evaluation. *Energy and Buildings*, *54*, 436–443. Available from https://doi.org/10.1016/j.enbuild.2012.070.023, https:// www.sciencedirect.com/science/article/pii/S0378778812003684.

Gupta, A., Anpalagan, A., Guan, L., & Khwaja, A. S. (2021). Deep learning for object detection and scene perception in self-driving cars: Survey, challenges, and open issues. *Array.*, *10*100057. Available from https://doi.org/10.1016/j.array.2021.100057, https:// www.sciencedirect.com/science/article/pii/S2590005621000059.

He, Y., Deng, B., Wang, H., Cheng, L., Zhou, K., Cai, S., & Ciampa, F. (2021). Infrared machine vision and infrared thermography with deep learning: A review. *Infrared Physics & Technology*, *116*103754. Available from https://doi.org/10.1016/j.infrared.2021.103754, https://www.sciencedirect.com/science/article/pii/S1350449521001262.

He, K., Gkioxari, G., Dollár, P., & Girshick, R. (2020). Mask R-CNN. *IEEE Transactions on Pattern Analysis and Machine Intelligence*, *42*(2), 386–397. Available from https://doi. org/10.1109/TPAMI.2018.2844175.

Hou, F., Lei, W., Li, S., Xi, J., Xu, M., & Luo, J. (2021). Improved Mask R-CNN with distance guided intersection over union for GPR signature detection and segmentation. *Automation in Construction*, *121*103414. Available from https://doi.org/10.1016/j.autcon.2020.103414, https://www.sciencedirect.com/science/article/pii/S0926580520309948.

Huang, H., Liu, Z., Philip Chen, C. L., & Zhang, Yun (2023). Hyperspectral image classification via active learning and broad learning system. *Applied Intelligence*, *53*(12), 15683–15694. Available from https://doi.org/10.1007/s10489-021-02805-5, https://doi. org/10.1007/s10489-021-02805-5.

Ichi, E., & Dorafshan, S. (2022). Effectiveness of infrared thermography for delamination detection in reinforced concrete bridge decks. *Automation in Construction*, *142*104523. Available from https://doi.org/10.1016/j.autcon.2022.104523.

Iman, M., Arabnia, H. R., & Rasheed, K. (2023). A review of deep transfer learning and recent advancements. *Technologies*, *11*(2). Available from https://doi.org/10.3390/ technologies11020040.

Jang, K., Kim, N., & An, Y.-K. (2019). Deep learning–based autonomous concrete crack evaluation through hybrid image scanning. *Structural Health Monitoring*, *18*(5–6), 1722–1737. Available from https://doi.org/10.1177/1475921718821719, https://doi.org/ 10.1177/1475921718821719.

Janković Babić, R. (2023). A comparison of methods for image classification of cultural heritage using transfer learning for feature extraction. *Neural Computing and Applications*. Available from https://doi.org/10.1007/s00521-023-08764-x, https://doi.org/10.1007/ s00521-023-08764-x.

Jena, B., Saxena, S., Nayak, G. K., Saba, L., Sharma, N., & Suri, J. S. (2021). Artificial intelligence-based hybrid deep learning models for image classification: The first narrative review. *Computers in Biology and Medicine*, *137*104803. Available from https:// doi.org/10.1016/j.compbiomed.2021.104803, https://www.sciencedirect.com/science/ article/pii/S0010482521005977.

Kirimtat, A., & Krejcar, O. (2018). A review of infrared thermography for the investigation of building envelopes: Advances and prospects. *Energy and Buildings*, *176*, 390–406. Available from https://doi.org/10.1016/j.enbuild.2018.070.052, https://www.sciencedirect.com/science/article/pii/S0378778818312398.

Kumar, P., Ofli, F., Imran, M., & Castillo, C. (2020). Detection of disaster-affected cultural heritage sites from social media images using deep learning techniques. *Journal on Computing and Cultural Heritage*, *13*(3), 1−31. Available from https://doi.org/10.1145/3383314.

Lagüela, S., Armesto, J., Arias, P., & Herráez, J. (2012). Automation of thermographic 3D modelling through image fusion and image matching techniques. *Automation in Construction*, *27*, 24−31. Available from https://doi.org/10.1016/j.autcon.2012.050.011, https://www.sciencedirect.com/science/article/pii/S0926580512000842.

Lagüela, S., Díaz-Vilariño, L., Martínez, J., & Armesto, J. (2013). Automatic thermographic and RGB texture of as-built BIM for energy rehabilitation purposes. *Automation in Construction*, *31*, 230−240. Available from https://doi.org/10.1016/j.autcon.2012.120.013, https://www.sciencedirect.com/science/article/pii/S092658051200252X.

Lagüela, S., González-Jorge, H., Armesto, J., & Arias, P. (2011). Calibration and verification of thermographic cameras for geometric measurements. *Infrared Physics & Technology*, *54*(2), 92−99. Available from https://doi.org/10.1016/j.infrared.2011.010.002, https://www.sciencedirect.com/science/article/pii/S135044951100003X.

Lagüela, S., González-Jorge, H., Armesto, J., & Herráez, J. (2012). High performance grid for the metric calibration of thermographic cameras. *Measurement Science and Technology*, *23*(1)015402. Available from https://doi.org/10.1088/0957-0233/23/1/015402, https://doi.org/10.1088/0957-0233/23/1/015402.

Lagüela, S., Martínez, J., Armesto, J., & Arias, P. (2011). Energy efficiency studies through 3D laser scanning and thermographic technologies. *Energy and Buildings*, *43*(6), 1216−1221. Available from https://doi.org/10.1016/j.enbuild.2010.120.031, https://www.sciencedirect.com/science/article/pii/S0378778811000041.

Lagüela, S., Solla, M., Puente, I., & Prego, F. J. (2018). Joint use of GPR, IRT and TLS techniques for the integral damage detection in paving. *Construction and Building Materials*, *174*, 749−760. Available from https://doi.org/10.1016/j.conbuildmat.2018.040.159, https://www.sciencedirect.com/science/article/pii/S0950061818309723.

Lee, J., Nazki, H., Baek, J., Hong, Y., & Lee, M. (2020). Artificial intelligence approach for tomato detection and mass estimation in precision agriculture. *Sustainability*, *12*. Available from https://doi.org/10.3390/su12219138.

Lerma, J. L. (2001). Multiband versus multispectral supervised classification of architectural images. *The Photogrammetric Record*, *17*(97), 89−101. Available from https://doi.org/10.1111/0031-868X.00169, https://doi.org/10.1111/0031-868X.00169.

Lerma, J. L. (2005). Automatic plotting of architectural facades with multispectral images. *Journal of Surveying Engineering.*, *131*(3), 73−77. Available from https://doi.org/10.1061/(ASCE)0733-9453(2005)131:3(73).

Lin, T.-Y., Maire, M., Belongie, S., Hays, J., Perona, P., Ramanan, D., Dollár, P., & Zitnick, C. L. (2014). *Microsoft COCO: Common Objects in Context* (pp. 740−755). Cham: Springer International Publishing.

Liu, L., & Stamos, I. (2005) Automatic 3D to 2D registration for the photorealistic rendering of urban scenes. *IEEE Computer Society Conference on Computer Vision and Pattern Recognition (CVPR'05)*, 137−143 2 doi: 10.1109/CVPR.2005.80 1063−6919.

Liu, F., Liu, J., & Wang, L. (2022). Deep learning and infrared thermography for asphalt pavement crack severity classification. *Automation in Construction*, *140*104383. Available from https://doi.org/10.1016/j.autcon.2022.104383, https://www.sciencedirect.com/science/article/pii/S0926580522002564.

Liu, Y., Wang, F., Jiang, Z., Sfarra, S., Liu, K., & Yao, Y. (2023). Generative deep learning-based thermographic inspection of artwork. *Sensors, 23*. Available from https://doi.org/10.3390/s23146362.

Li, Y., Du, X., Wan, F., Wang, X., & Yu, H. (2020). Rotating machinery fault diagnosis based on convolutional neural network and infrared thermal imaging. *Chinese Journal of Aeronautics, 33*(2), 427−438. Available from https://doi.org/10.1016/j.cja.2019.080.014, https://www.sciencedirect.com/science/article/pii/S1000936119303127.

Llamas, J., Lerones, P. M., Zalama, E., & Gómez-García-Bermejo, J. (2016). *Applying deep learning techniques to cultural heritage images within the INCEPTION Project* (pp. 25−32). Cham: Springer International Publishing.

Luo, Q., Gao, B., Woo, W. L., & Yang, Y. (2019). Temporal and spatial deep learning network for infrared thermal defect detection. *NDT & E International, 108*102164. Available from https://doi.org/10.1016/j.ndteint.2019.102164, https://www.sciencedirect.com/science/article/pii/S0963869519301355.

López-Fernández, L., Lagüela, S., González-Aguilera, D., & Lorenzo, H. (2016). Thermographic and mobile indoor mapping for the computation of energy losses in buildings. *Indoor and Built Environment, 26*(6), 771−784. Available from https://doi.org/10.1177/1420326X16638912, https://doi.org/10.1177/1420326X16638912.

Magalhaes, C., Mendes, J., & Vardasca, R. (2021). Meta-analysis and systematic review of the application of machine learning classifiers in biomedical applications of infrared thermography. *Applied Sciences, 11*(2), 2076−3417. Available from https://doi.org/10.3390/app11020842.

Marani, R., Palumbo, D., Galietti, U., Stella, E., D'Orazio, T. (2016). Automatic detection of subsurface defects in composite materials using thermography and unsupervised machine learning. *IEEE 8th International Conference on Intelligent Systems (IS)*, pp. 516−521, doi: 10.1109/IS.2016.7737471.

Marinetti, Sergio, & Cesaratto, Pier Giorgio (2012). Emissivity estimation for accurate quantitative thermography. *NDT & E International, 51*, 127−134. Available from https://doi.org/10.1016/j.ndteint.2012.060.001, https://www.sciencedirect.com/science/article/pii/S0963869512000813.

Mashekova, A., Zhao, Y., Ng, E. Y. K., Zarikas, V., Fok, S. C., & Mukhmetov, O. (2022). Early detection of the breast cancer using infrared technology − A comprehensive review. *Thermal Science and Engineering Progress, 27*101142. Available from https://doi.org/10.1016/j.tsep.2021.101142.

Matrone, F., & Martini, M. (2021). Transfer learning and performance enhancement techniques for deep semantic segmentation of built heritage point clouds. *Virtual Archaeology Review, 12*(25), 73−84. Available from https://doi.org/10.4995/var.2021.15318.

Ma, L., Liu, Y., Zhang, X., Ye, Y., Yin, G., & Johnson, B. A. (2019). Deep learning in remote sensing applications: A meta-analysis and review. *ISPRS Journal of Photogrammetry and Remote Sensing, 152*, 166−177. Available from https://doi.org/10.1016/j.isprsjprs.2019.040.015, https://www.sciencedirect.com/science/article/pii/S0924271619301108.

Mačianskytė, D., & Adaškevičius, R. (2022). Automatic detection of human maxillofacial tumors by using thermal imaging: A preliminary study. *Sensors, 22*(5), 1985. Available from https://doi.org/10.3390/s22051985.

Monna, F., Rolland, T., Denaire, A., Navarro, N., Granjon, L., Barbé, R., & Chateau-Smith, C. (2021). Deep learning to detect built cultural heritage from satellite imagery. - Spatial distribution and size of vernacular houses in Sumba, Indonesia. *Journal of Cultural Heritage, 52*,

171−183. Available from https://doi.org/10.1016/j.culher.2021.100.004, https://www.sciencedirect.com/science/article/pii/S1296207421001606.

Moradi, M., Ghorbani, R., Sfarra, S., Tax, D. M. J., & Zarouchas, D. (2022). A spatiotemporal deep neural network useful for defect identification and reconstruction of artworks using infrared thermography. *Sensors*, *22*(23). Available from https://doi.org/10.3390/s22239361.

Orengo, H. A., Conesa, F. C., Garcia-Molsosa, A., Lobo, A., Green, A. S., Madella, M., & Petrie, C. A. (2020). Automated detection of archaeological mounds using machine-learning classification of multisensor and multitemporal satellite data. *Proceedings of the National Academy of Sciences*, *117*(31), 18240−18250. Available from https://doi.org/10.1073/pnas.2005583117, https://doi.org/10.1073/pnas.2005583117.

Ornek, A. H., & Ceylan, M. (2022). Medical thermograms' classification using deep transfer learning models and methods. *Multimedia Tools and Applications*, *81*(7), 9367−9384. Available from https://doi.org/10.1007/s11042-021-11852-6.

Panella, F. W., Pirinu, A., & Dattoma, V. (2021). A brief review and advances of thermographic image - processing methods for IRT inspection: A case of study on GFRP plate. *Experimental Techniques*, *45*(4), 429−443. Available from https://doi.org/10.1007/s40799-020-00414-4, https://doi.org/10.1007/s40799-020-00414-4.

Peng, X., Zhong, X., Chen, A., Zhao, C., Liu, C., & Chen, Y. F. (2021). Debonding defect quantification method of building decoration layers via UAV-thermography and deep learning. *Smart Structures and Systems*, *28*(1), 55−67. Available from https://doi.org/10.12989/sss.2021.28.10.055.

Pierdicca, R., Paolanti, M., Matrone, F., Martini, M., Morbidoni, C., Malinverni, E. S., Frontoni, E., & Lingua, A. M. (2020). Point cloud semantic segmentation using a deep learning framework for cultural heritage. *Remote Sensing*, *12*. Available from https://doi.org/10.3390/rs12061005.

Qiang Fang., Ba Diep., Nguyen Clemente., Ibarra Castanedo., Yuxia Duan., & Xavier Maldague, I. I. (2020). Automatic defect detection in infrared thermography by deep learning algorithm. *Proc.SPIE*, *11409*. Available from https://doi.org/10.1117/12.2555553.

Rasulo, A., Garrido, I., Solla, M., Lagüela, S., & Rasol, M. (2022). Review of infrared thermography and ground-penetrating radar applications for building assessment. *Advances in Civil Engineering*, *2022*5229911. Available from https://doi.org/10.1155/2022/5229911, https://doi.org/10.1155/2022/5229911.

Rodríguez-Martín, M., Fueyo, J. G., Gonzalez-Aguilera, D., Madruga, F. J., García-Martín, R., Muñóz, Á. L., & Pisonero, J. (2020). Predictive models for the characterization of internal defects in additive materials from active thermography sequences supported by machine learning methods. *Sensors*, *20*(14). Available from https://doi.org/10.3390/s20143982.

Rodríguez-Martín, M., Fueyo, J. G., Pisonero, J., López-Rebollo, J., Gonzalez-Aguilera, D., García-Martín, R., & Madruga, F. (2022). Step heating thermography supported by machine learning and simulation for internal defect size measurement in additive manufacturing. *Measurement*, *205*112140. Available from https://doi.org/10.1016/j.measurement.2022.112140, https://www.sciencedirect.com/science/article/pii/S0263224122013367.

Saeed, N., Abdulrahman, Y., Amer, S., & Omar, M. A. (2019). Experimentally validated defect depth estimation using artificial neural network in pulsed thermography. *Infrared Physics and Technology*, *98*, 192−200. Available from https://doi.org/10.1016/j.infrared.2019.030.014, https://ui.adsabs.harvard.edu/abs/2019InPhT.98.192S.

Saeed, N., King, N., Said, Z., & Omar, M. A. (2019). Automatic defects detection in CFRP thermograms, using convolutional neural networks and transfer learning. *Infrared Physics &*

Technology, *102*103048. Available from https://doi.org/10.1016/j.infrared.2019.103048, https://www.sciencedirect.com/science/article/pii/S1350449519303135.

Schalekamp, S., Klein, W. M., & van Leeuwen, K. G. (2022). Current and emerging artificial intelligence applications in chest imaging: a pediatric perspective. *Pediatric Radiology*, *52*(11), 2120−2130. Available from https://doi.org/10.1007/s00247-021-05146-0, https://doi.org/10.1007/s00247-021-05146-0.

Solla, M., Lagüela, S., Fernández, N., & Garrido, I. (2019). Assessing rebar corrosion through the combination of nondestructive GPR and IRT methodologies. *Remote Sensing*, *11*(14), 1705. Available from https://doi.org/10.3390/rs11141705.

Solla, M., Lagüela, S., Riveiro, B., & Lorenzo, H. (2013). Nondestructive testing for the analysis of moisture in the masonry arch bridge of Lubians (Spain). *Structural Control Health Monitoring*, *20*(11), 1366. Available from https://doi.org/10.1002/stc.1545.

Sperlí, G. (2021). A cultural heritage framework using a deep learning based chatbot for supporting tourist journey. *Expert Systems with Applications*, *183*115277. Available from https://doi.org/10.1016/j.eswa.2021.115277, https://www.sciencedirect.com/science/article/pii/S0957417421007089.

Susana, L., Cereijo, J., Martínez-Sánchez, J., Roca, D., & Lorenzo, H. (2014). Thermographic mobile mapping of urban environment for lighting and energy studies. *Journal of Daylighting*, *1*, 8−15. Available from https://doi.org/10.15627/jd.2014.2.

Svantner, M., Lang, V., Kohlschutter, T., Skala, J., Honner, M., Muzika, L., & Kosová, E. (2021). Study on human temperature measurement by infrared thermography. *Engineering Proceedings.*, *8*(1), 4. Available from https://doi.org/10.3390/engproc2021008004.

Tejedor, B., Lucchi, E., Bienvenido-Huertas, D., & Nardi, I. (2022). Nondestructive techniques (NDT) for the diagnosis of heritage buildings: Traditional procedures and futures perspectives. *Energy and Buildings*, *263*112029. Available from https://doi.org/10.1016/j. enbuild.2022.112029, https://www.sciencedirect.com/science/article/pii/S0378778822002006.

Usamentiaga, R., Ibarra-Castanedo, C., & Maldague, X. (2018). Comparison and evaluation of geometric calibration methods for infrared cameras to perform metric measurements on a plane. *Applied Optics*, *57*(18), D1−D10. Available from https://doi.org/10.1364/ AO.57.0000D1, https://opg.optica.org/ao/abstract.cfm?URI = ao-57-18-D1.

Usamentiaga, R., Venegas, P., Gurediaga, J., Vega, L., & Molleda Julio Bulnes Francisco, G. (2014). Infrared thermography for temperature measurement and nondestructive testing. *Sensors*, *14*(7), 12305−12348. Available from https://doi.org/10.3390/s140712305, 1424−8220.

Venegas, P., Ivorra, E., Ortega, M., & Sáez de Ocáriz, I. (2022). Towards the automation of infrared thermography inspections for industrial maintenance applications. *Sensors*, *22*. Available from https://doi.org/10.3390/s22020613.

Wei, Z., Ahmad, O., Bernd, V., & Xavier, M. (2023). Pulsed thermography dataset for training deep learning models. *Applied Sciences*, *13*(5). Available from https://doi.org/ 10.3390/app13052901.

Wu, Q., Feng, D., Cao, C., Zeng, X., Feng, Z., Wu, J., & Huang, Z. (2021). Improved mask R-CNN for aircraft detection in remote sensing images. *Sensors*, *21*(8). Available from https://doi.org/10.3390/s21082618.

Xunta de Galicia., & Turismo de Galicia (2023). Igrexa de Santiago de Cangas. Available from https://www.turismo.gal/buscador-global?langId = es_ES&q = santiago + de + cangas. (2023) Accessed 10.12.23.

Yang, J., Lin, G., Li, Q., Sun, Y., & Sun, Y. (2019). Infrared thermal imaging-based crack detection using deep learning. *IEEE Access*, *7*, 182060−182077. Available from https:// doi.org/10.1109/ACCESS.2019.2958264.

Yumnam, M., Gupta, H., Ghosh, D., & Jaganathan, J. (2021). Inspection of concrete structures externally reinforced with FRP composites using active infrared thermography: A review. *Construction and Building Materials*, *310*125265. Available from https://doi.org/10.1016/j.conbuildmat.2021.125265, https://www.sciencedirect.com/science/article/pii/S0950061821030063.

Yun-Kyu, A., Jang, K., Kim, B., & Cho, S. (2018). Deep learning-based concrete crack detection using hybrid images. *Proc. SPIE*. Available from https://doi.org/10.1117/12.2294959, 10.1117/12.2294959 1059812.

Zhang, Y., Chu, J., Leng, L., & Miao, J. (2020). Mask-refined R-CNN: A network for refining object details in instance segmentation. *Sensors*, *20*(4). Available from https://doi.org/10.3390/s20041010.

Zhang, Xin, & Dahu, Wang (2019). Application of artificial intelligence algorithms in image processing. *Journal of Visual Communication and Image Representation*, *61*, 42−49. Available from https://doi.org/10.1016/j.jvcir.2019.030.004, https://www.sciencedirect.com/science/article/pii/S1047320319300975.

Zhang, Z., Zhao, L., & Yang, T. (2021). Research on the application of artificial intelligence in image recognition technology. *Journal of Physics: Conference Series*, *1992*(3) 032118. Available from https://doi.org/10.1088/1742-6596/1992/3/032118, https://doi.org/10.1088/1742-6596/1992/3/032118.

Zhou, R., Wen, Z., & Su, H. (2022). Automatic recognition of earth rock embankment leakage based on UAV passive infrared thermography and deep learning. *ISPRS Journal of Photogrammetry and Remote Sensing*, *191*, 85−104. Available from https://doi.org/10.1016/j.isprsjprs.2022.070.009, https://www.sciencedirect.com/science/article/pii/S0924271622001836.

Analyzing non-destructive methods for building inspection and energy performance: A focus on photogrammetry and infrared thermography

Kalare Agrasar-Santiso[1], Jose Antonio Millan-Garcia[2],
Juan Pedro Otaduy-Zubizarreta[1], Abderrahmane Baïri[3], and
Alexander Martín-Garín[1]

[1]TICBE Research Group, Department of Architecture, Faculty of Engineering of Gipuzkoa, University of the Basque Country UPV/EHU, Donostia-San Sebastián, Spain,
[2]ENEDI Research Group, Department of Energy Engineering, Faculty of Engineering of Gipuzkoa, University of the Basque Country UPV/EHU, San Sebastián, Spain,
[3]Laboratoire Thermique Interfaces Environnement (LTIE), Université de Paris, EA 4415, 50 Rue de Sèvres, F-92410 Ville d'Avray, France

6.1 Introduction

In the present time, there are various methods of building inspection, when it comes to historic buildings, challenges arise due to their inherently fragile nature. This hinders the collection of standard samples for material characterization. Non-destructive testing (NDT) methods are employed to assess and examine structures without causing damage; hence, their relevance is lying in the context of historical heritage. These methods enable the acquisition of detailed information without compromising the integrity of the constructions and contribute to enhancing their energy efficiency.

The notion of architectural heritage is defined by the significance that a building holds for all humanity and for the urban identity of spaces. These are designated by the United Nations Educational, Scientific and Cultural Organization (UNESCO) under the convention concerning the Protection of the World Cultural and Natural Heritage, which was drafted in 1972 (Troi & Bastian, 2014).

The buildings considered heritage of humanity are not delimited to the classical hierarchy of heritage values, such as catholic buildings, castles, and bourgeois houses. Currently, the heritage registry integrates a wide variety of spaces and constructions, including concentration camps or industrial buildings. In fact, the protection of a building is not determined by its age or function. The cultural values

Diagnosis of Heritage Buildings by Non-Destructive Techniques. DOI: https://doi.org/10.1016/B978-0-443-16001-1.00006-1

attributed to each construction are defined through a negotiation process where experts and the community are involved. The discussion takes place internationally, with guidelines established by the International Council on Monuments and Sites, UNESCO, and the Council of Europe. In terms of practicality, it is not about the conservation of these designated values but rather taking care of the elements that possess those virtues. These include the formal qualities, the aesthetic aspect of the building, and even the traces of history. Its value is universally recognized and there is a commitment to preserving their integrity.

The commitment to conservation refers to the extension of the life of a culturally valuable building. Historic buildings play an active role in the community they belong to. They represent tangible and intangible heritage, serving as historical and cultural symbols of the present and future. Preserving these buildings is necessary not only for economic reasons, as they are a tourist attraction and a source of income, but also because they foster an emotional connection with the community. They maintain the history and cultural identity of spaces and the overall residential environment (Munarim & Ghisi, 2016). The heritage values of a building encompass symbolic meanings (identity, charisma, and so on), cultural values (aesthetic, historical, and so on), and utility values (functionality, economic value, and so on).

Lastly, it is important to mention the uniqueness of each building within cultural heritage. They are the result of construction processes that reflect the style and influences of the historical context. Each building has distinctive characteristics in terms of its architecture, ornamental details, materials used, and construction techniques. This means that the examination of each building cannot be standardized, and the development of strategies for historic buildings varies accordingly.

Urban sustainability to mitigate climate change is of great importance for the EU, a policy of rehabilitation of the current building stock is proposed, aligning with the Energy Performance of Buildings directives. The number of buildings belonging to the European World Heritage represents 30% of the total park, so the intervention in this type of buildings becomes an opportunity considering that they could reduce their energy demand by 50% and 80% through energy renovations (Diz-Mellado et al., 2021). Consequently, they are assigned a strategic role in energy saving due to the untapped energy potential.

Energy savings bring multiple benefits, including saving limited economic resources, preserving our fossil fuels (the primary energy source), and reducing CO_2 emissions. The key to achieving these savings lies in the fact that we do not consume energy itself, but energy services. Therefore it is possible to provide the same level of energy service using less energy.

The implementation of energy efficiency requires a comprehensive evaluation that involves the entire life cycle of the building. The longevity of historic buildings necessitates considering the identification of boundary conditions (Asdrubali & Desideri, 2018). As far as we refer to the evaluation of the energy performance of historic buildings, there is hardly any empirical substance due to the little information of previous analysis that these have and the exceptionality that each building represents. Therefore, to apply energy improvements, all aspects of each individual case study must be evaluated, along with their behavior, to identify elements with intervention potential.

Regarding conservation practices, there is a need for on-site building analysis to understand the characteristics of each building, not only to identify bad practices in terms of energy but also to determine suitable measures to implement. The ornamentation of facades should not be covered with external insulation, even sometimes due to certain design elements, internal insulation may not be feasible either. However, there are often areas with limited heritage value where renovation with different components could be studied.

Therefore, considering the significance of cultural heritage and the benefits of energy efficiency, it can be concluded that conservation practices must go hand in hand with the rehabilitation and analysis of historic buildings. The preservation of authenticity should be balanced with meeting current requirements, ensuring that the buildings are returned to society in optimal condition, as demanded by their conservation.

Consequently, this document will be guided by the following approach:

- Firstly, the identification of the topics that have been addressed so far will be carried out.
- A detailed exposition of the utilized non-destructive methods will be conducted.
- A study of the application of the explained methods in various case studies will be undertaken.
- Subsequently, the most significant non-destructive methods will be analyzed.
- Lastly, current trends will be explored to understand the direction toward which NDT methods are heading.

Before embarking on the aforementioned informative work, the necessity of conducting a thorough analysis of relevant databases is presented to obtain an updated perspective on non-destructive methods applied in the field of cultural heritage. By becoming acquainted with the current state of non-destructive methods, the intention is to provide an informed and meticulous approach that contributes to the dissemination and comprehension of cutting-edge practices in the field.

6.2 Research methodology

Conducting a study on historic buildings requires specialized knowledge. As mentioned earlier, on-site investigations are mandatory. This entails understanding the instrumentation required to analyze of the building. As far as measuring devices are concerned, sample collection could potentially damage the building. Therefore, keeping in mind the delicacy of the object of study, the main purpose will be to conduct a review of existing NDT ([AMG1] NDT) methods. These are techniques used to inspect the properties of materials or components of a space and/or building without compromising their integrity.

Scopus is used to access information on research, scientific articles, conferences, patents, and other documents related to the discipline of non-destructive methods and historic buildings. Its extensive coverage provides citation metrics, enabling the

Table 6.1 Queries used and number of publications obtained for the topic.

Query	Number of publications
TITLE-ABS-KEY "heritage" OR "NDT"	143,165
KEY: "heritage" OR "NDT"	55,578
TITLE-ABS-KEY "heritage" AND "NDT"	206
KEY: "heritage" AND "NDT"	69
TITLE-ABS-KEY: "cultural heritage" OR "NDT"	61325
KEY: "cultural heritage" OR "NDT"	28977
TITLE-ABS-KEY: "cultural heritage" AND "NDT"	127
KEY: "cultural heritage" AND "NDT"	41
TITLE-ABS-KEY: "heritage building" AND "NDT"	31
TITLE-ABS-KEY: "historic building"* AND "NDT"	40

Note: TITLE-ABS-KEY = Titles, Abstracts, Keywords KEY = Keywords * = plural and singular.
Source: From data of authors/Scopus.

evaluation of the importance and impact of research work. Therefore it is used for decision-making in research.

Table 6.1 aims to identify a complete overview of publications dedicated to the application of non-destructive methods in cultural heritage.

The titles, abstracts, and keywords of publications are taken into consideration to filter the documents for analysis, and keywords are used to narrow them down. Additionally, "AND" is used instead of "OR" to refine the publications related to both topics. The extracted information is divided into different queries: temporal evolution of publications over time, geographic distribution of studies, disciplinary fields, and indexed keywords. For further analysis, multiple applications are used to generate graphs to enhance the visual understanding of the data more effectively. To achieve this, Scopus filters are employed to group the temporal evolution, geographic distribution, and disciplinary field. Moreover, the VOSviewer software is used for the cooccurrence network of keywords.

6.2.1 Time evolution

The first document is published in the year 2000, and the timeline extends to the present. The significant increase in the number of publications can be observed in Fig. 6.1.

6.2.2 Geographical distribution of studies

Afterward, in Fig. 6.2 the global geographic distribution of studies can be observed. The majority of publications are from Europe, with Italy leading the way with 84 studies, followed by Spain with 39. Outside of the EU, Turkey, the United States, and China also contribute. This demonstrates that this topic is being discussed worldwide, with a particular focus on European countries.

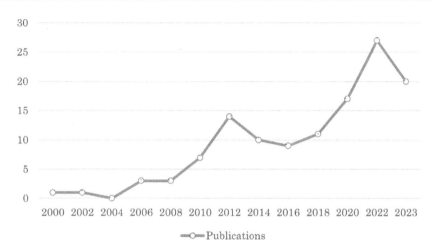

Figure 6.1 Temporal evolution of the number of publications.
Source: Own elaboration based on Scopus data.

Figure 6.2 Geographical distribution of publications from 2000 to 2023.
Source: Own elaboration based on data from Scopus and powered by Bing.

6.2.3 Disciplinary fields of study

The disciplinary field with the highest number of publications is engineering, with a total of 134 publications, representing 30% of all disciplines. These publications predominantly focus on research in machine learning, development of techniques and methodologies, and their applications. Following closely behind is materials science with 82 studies, accounting for 18% of the total. When it comes to historic buildings, many articles in this field relate to structural analysis and material properties. Additionally, a significant number of publications are associated with

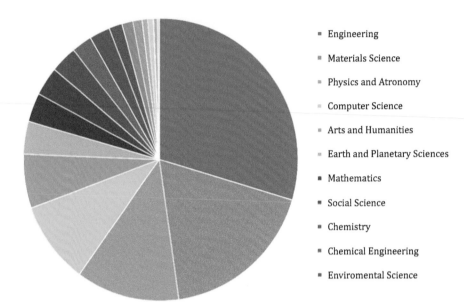

Engineering

Materials Science

Physics and Atronomy

Computer Science

Arts and Humanities

Earth and Planetary Sciences

Mathematics

Social Science

Chemistry

Chemical Engineering

Enviromental Science

Figure 6.3 Disciplinary field of publications.
Source: Own elaboration based on Scopus data.

environmental science, mathematics, physics, and other related disciplines. Fig. 6.3 compiles the main disciplinary fields of study.

6.2.4   Indexed keywords

The following graphs are based on the previously selected keywords and consider their frequency of occurrence in the database. Using the data obtained from Scopus, a cooccurrence network is generated using VOSviewer and subsequently presented in different ways. In both presented scenarios, the size of the keywords represents their level of popularity, and the proximity indicates their similarity. In one scenario, different shades of colors are used to represent thematic groups of keywords (Fig. 6.4). In the other context, the same division methodology is used, but with the purpose of distinguishing the temporal sequence of the keywords (Fig. 6.5).

In Fig. 6.4, it can be observed that the cluster with the most keywords is the existing non-destructive methods, which is represented as "Cluster 1." Therefore throughout the document, special emphasis is placed on the description of these, especially in the methods of photogrammetry and infrared thermography (IRT).

Based on the newest research findings shown in Fig. 6.5, it is concluded which are the future lines of investigation, with popularity seen in 3D, ANN, Historic Building Information Modeling (HBIM), and artificial intelligence.

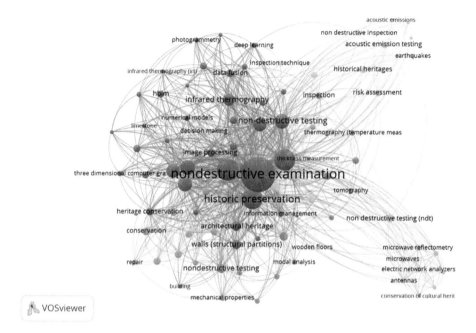

Figure 6.4 Network of cooccurrence of indexed keywords.
Source: Elaboration of the author using the VOSviewer software from the Scopus data.

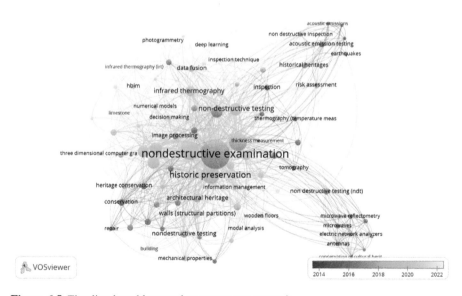

Figure 6.5 Timeline-based keyword cooccurrence network.
Source: Author's elaboration using VOSviewer software from Scopus data.

6.3   Non-destructive testing methods

There is now a wide range of methods applicable to building inspection. However, historic buildings due to their intrinsically fragile nature (traditional construction techniques, use of local materials, and so on) do not allow for the collection of samples with commonly used dimensions in new buildings for material and component characterization. Numerous aspects, such as variations in wall thickness or joints between different components require a thorough examination. It is fundamental for a correct energy analysis (Martín-Garín et al., 2020). Furthermore, energy rehabilitation must balance various requirements. In heritage buildings, the preservation of architectural character and materials is substantial. It is about preserving the invaluable cultural heritage using appropriate structural evaluation.

The use of new technologies reduces the damage to construction (Bosiljkov et al., 2010). In light of the aforementioned, a framework of different and new nondestructive techniques applicable in the development of building diagnostics should continue to be provided. These challenges must be addressed through scientific research, cutting-edge technology, and innovation in this field.

At present, various NDT methods are used, either individually or in combination with others, to examine the structure of the building and characterize the materials. The following sections provide an overview of the main NDT techniques.

6.3.1   *Digital image processing*

Digital image processing (DIP) is a comprehensive framework that involves the processing and application of information to extract pertinent data, perform in-depth analysis, determine suitable techniques for desired outcomes, monitor conservation status, generate specific maps, and facilitate remodeling endeavors, among other functions.

The utilization of DIP plays a critical role in assessing and safeguarding culturally significant heritage buildings. It offers valuable insights, intricate analysis, and visualization tools for studying, documenting, and preserving these invaluable structures. Most NDT methods mentioned subsequently rely on DIP techniques to effectively process the acquired data.

6.3.2   *Photogrammetry (P)*

Photogrammetry establishes the dimensions and location of various objects within a specific area. To achieve an accurate model, the intersection of multiple photographs is required. Additionally, results can be obtained using photographs and a digital terrain model. Photogrammetry enables the generation of outcomes without any physical contact with the structural element.

This non-destructive technique finds application in various fields such as cartography, architecture, archaeology, topography, and the production of animated media, including films and video games. Likewise, photogrammetry is used to create 3D models of objects and environments for virtual and augmented reality (Al-Ruzouq et al., 2023).

Currently, computational intelligence is not fully automated. However, photogrammetry is starting to be combined with computer vision through the incorporation of Structure from Motion (SfM) algorithms to provide 3D models in the cultural heritage sector.

There is an increasing availability of open-access technology. Software tools for reconstructing 3D models from photographs allow for a quick and realistic virtual recreation of different buildings, spaces, and artifacts. The images required for these automated applications can be obtained personally using various devices such as drones, but there is also a wealth of high-quality images available online (López et al., 2018).

6.3.3 Stereophotogrammetry (SP)

Stereophotogrammetry is a technique similar to photogrammetry, with the main difference lying in its precision. Like the previous technique, it uses two-dimensional images, but the three-dimensional information is derived from stereo images, which allows three-dimensional measurements to be obtained in more detail. The human face was the first example of a complex geometry produced in 3D using stereophotogrammetry (Krutikova et al., 2017).

This technology has been investigated using various methodologies and has been able to track characteristic points and three-dimensional maps in real time (Fu et al., 2019; Pire et al., 2017). These approaches are highly innovative in terms of time and reproduction capability. They allow the acquisition of 3D environments with great precision in just a second, demonstrating incredible potential for applications in aerial inspections that require speed. They are combined with devices such as unmanned aerial vehicles (UAVs) to obtain data from even the highest points of buildings.

6.3.4 Terrestrial aser scanning

Highly accurate technique used to capture detailed point clouds of complex geometries. It involves scanning the monument from multiple positions to conduct surveys. The process entails emitting laser beams to various points on the structure, measuring the distances and angles between these points and the laser device, and subsequently extracting 3D coordinates with respect to the scanning station. The point cloud obtained is a dense grid of data that, through data processing, can be transformed into a precise 3D geometric model.

Different sensors are categorized based on the height of the structure being examined. Phase technology scanners have a limited range but offer sharper results. On the other hand, pulse-based scanners have a greater range but provide less accurate resolutions. The final output obtained through laser scanning is comparable to that of digital photogrammetry, producing a comprehensive digital representation of the scanned object (Pallarés et al., 2021).

Terrestrial laser scanning (TLS) techniques are commonly employed to detect angular distortions in structural elements by measuring their vertical deformations.

This enables the identification of any deviations or irregularities in the geometry of the structure, providing valuable insights for analysis and assessment purposes.

6.3.5 Ambient vibration test (AVT)

Ambient vibration is the oscillation that structures usually suffer from a series of environmental actions such as wind, traffic, and human activity. Ambient vibration tests allow to determinate the stiffness of a structure by measuring the amount of movement generated by external sources (Diz-Mellado et al., 2021).

Generally, vibration tests utilize accelerometers, which are usually highly sensitive. These accelerometers can be piezoelectric, generating an electrical signal, or servo accelerometers, which rely on force balance. Depending on the accessibility of the monument, the sensor coupling to the structure is defined to achieve a greater number of results while ensuring the structural integrity is not affected (Pallarés et al., 2021).

6.3.6 Infrared thermography

Thermographic cameras are commonly used as a non-destructive method to produce images of the thermal distribution on surfaces. The data used in thermographic processes include atmospheric temperature/relative humidity, reflection temperature, and the emissivity factor of the analyzed space (Kylili et al., 2014). This occurs because the radiant heat flow is influenced by the materials and the environment surrounding the structural component.

IRT cameras, known as IRT, focus the optical radiation emitted by the object of study onto the sensor. The response of this signal is a digital image with different colors. In other words, the camera identifies the emitted electromagnetic energy and converts it into an electronic signal. Each color tone in the obtained digital image represents the temperature distribution of the object (Herraiz et al., 2020).

During the data acquisition stage, there are two possible approaches. It can be active IRT, which involves applying artificial external stimulation to the structure under study, generating internal heat flow. On the other hand, it can be passive IRT that is simply differentiated by the absence of this external source, it is based on the radiation emitted by the structure (Kylili et al., 2014).

The temperature difference of the images obtained helps to identify different thermal anomalies. These appear as a reflection of cracks, humidity, heat leaks, or any problem. In addition, the severity of the problem can be quantified with a more detailed analysis of the images, to know when to react and which changes will help to improve energy efficiency to a greater extent. The thermal imaging camera is simple and at the same time very effective (Ibarra-Castanedo et al., 2013).

6.3.7 Ground penetrating radar

The goal of this technology is to conduct in-depth investigations of structures using electromagnetic waves. It is commonly applied in cultural heritage sites to assess structural conditions. Since drilling is not an option for analyzing historical buildings, this method allows for evaluating the morphology of walls and ceilings. Additionally, it enables the identification of moisture levels, distinguishing capillary rise and dry areas.

Similarly, the ground penetrating radar (GPR) method allows for deducing the positioning and geometry of cracks in structures. The identification of these irregularities is possible due to the waveform generated by GPR technology. Depending on the intensity, the propagation time of voids can be determined through analysis of the results (Liu et al., 2023). With sufficient resolution, irregular boundaries of material heterogeneity in historical buildings can be distinguished. The transmitter and receiver are placed on the surfaces to be studied, and direct contact is not necessary, making it the most used method for delicate cases (Pallarés et al., 2021).

6.3.8   Fan pressurization method

The purpose of using the fan pressurization method is to investigate how permeable the building envelope is as far as we are concerned with unwanted airflows. Air changes between the interior and exterior spaces directly impact humidity, temperature, and consequently, energy consumption. Historical buildings generally lack sealing materials, so the presence of air leaks related to a lack of airtightness is common (Martín-Garín et al., 2020).

This test determines the amount of air leakage from a confined space under pressure differences. A fabric panel frame and a fan are placed in a specific opening, such as a door. Specialized pressure meters are used to measure the pressure differences while the fan simulates exterior wind conditions (positive pressure test) or indoor wind conditions (negative pressure test). The data obtained is analyzed to determine the air filtered by cracks, poorly sealed areas, and so on (Keefe, 2010).

6.3.9   Ultrasonic pulse velocity (UPV)

Ultrasound is used to inspect wall components. It uses short waves for structural evaluation, allowing for the generation of high-resolution images of the structures through propagating waves. Data such as duct location, thickness measurements, and fault detection can be obtained (El Masri & Rakha, 2020).

6.3.10   Heat flow meter

The measurement of heat flow allows determining the thermal transmittance of building elements, known as U-value. In modern buildings, these values are often specified. However, for historical buildings, which typically have extensive masonry and the use of traditional materials, it is important to perform in situ measurements of thermal transmittance.

The procedure for these measurements is defined by the international standard ISO 9896 (ISO, 2014). The meters have a slim design and low thermal resistance. They are equipped with temperature sensors that allow the electrical signal to be related to the heat flow through the plate. Additionally, high sensitivity is required to consider the influences of temperatures between the interior and exterior (Evangelisti et al., 2015). Temperature sensors are important for recording the ambient conditions under which data is obtained.

The size of the heat flow meter (HFM) is determined based on the structure being studied and is placed in direct contact with the surface, covering the entire sensor area. The data must be acquired at short intervals and over a period of several days to ensure reliability and to avoid dependence on specific temporary conditions (ISO, 2014).

6.4   Literature review

In the present scenario, as mentioned previously, there are several noninvasive techniques available that, when combined, facilitate the establishment of a systematic approach to assess the state of preservation and discover possible improvements that contribute to achieving greater energy efficiency in buildings. A literature review work related to NDT methods applied in heritage buildings is carried out.

In this chapter, a study of analysis of the structural behavior of the Jerónimos Monastery in Portugal is carried out (Lourenço et al., 2007). It is obtained that an iterative approach in cultural heritage buildings is necessary to obtain sufficient information and make a proper judgment. The number of measurements, which are in situ, must be sufficient to obtain adequate results, therefore it is concluded by highlighting the importance of improving NDT technologies.

In another study, GPR techniques are selected in combination with seismic techniques to know the size of the damage and columns of the Cathedral of Mallorca (Pérez-Gracia et al., 2013). The result was positive, the internal shape of several structures was inferred, in addition to the quality and damage of the stones.

The case study in another research is the Ottoman building in Urla (Kilic, 2015). Three inspection models, GPR, IRT, and visual inspection, were used to obtain information on the condition of the facilities and structure due to the absence of building documentation. The method of these inspections was valuable, especially in relation to historic buildings. The result was obtained with a combination of destructive techniques, without the need for further inspections.

In this chapter, we discuss the integrated approach in combination with visual inspection, making use of the technical tests of GPR and IRT (Kilic, 2015). It is made in situ in several historical stone masonry. The results are not reliable, only HFM measurements make it possible to use the data for standard calculations (energy labeling) as well as for accurate evaluations. The chapter is carried out in Italy but affirms scalability in European countries with similar masonry.

The following study uses IRT, electrical resistivity tomography (ERT), and GPR techniques to analyze the subsoil of the Church of San Juan Bautista, in Madrid (Martínez-Garrido et al., 2018). Wireless sensors are used for tree-type communication. The results showed that the GPR technique is the most efficient to know the profiles of underground. Automation conclusions can be obtained due to the correct evaluation of humidity.

In this study, GPR and ultrasonic measurement technique (UPV) techniques are chosen to detect the problems of instability and humidity and characterize materials in the Museum of Kariye, Chora (Yalçıner et al., 2019). It is obtained that the evaluation with GPR offers good resolution for certain criteria, but nevertheless the results cannot be associated with the properties of each material. The UPV technique offers less resolution, but the mechanical properties are identified. The combination of both can make a correct analysis.

This research analyzes the Cathedral of Seville, which has instances of damage that exist for decades (Diz-Mellado et al., 2021). Non-destructive techniques are used to know the origin of the pathologies, the data have been verified in a simulation model. It makes use of visual inspections with photographs, laser ground leveling, use of tools such as accelerometers, thermal imaging camera (IRT), and GPR. They are then analyzed with image processing and computer software. The results show that the combination of tools offers a thorough knowledge of historic buildings, because many are not documented. Therefore the collected physical, geometric, and material properties are necessary to make hypotheses.

In the case of Villa Litta Modignani (Lualdi et al., 2001), the deterioration hidden by construction material is analyzed by thermovision and georadar. Also, sonic measurements are used in combination with the georadar for morphological analysis. It was concluded that the balanced use of methods reduces costs and times and limits the need for platforms, reaffirming the cost-effectiveness of the approach.

6.5 Thermography and photogrammetry

The technological revolution achieved in recent decades has brought forth new technologies that have improved and accelerated spatial data acquisition techniques to generate accurate models of construction information. This section describes the two most influential techniques today: photogrammetry and thermography. For a precise evaluation involving the quantification of facades or the detection of thermal defects, it is necessary to have information on surface temperature and geometry.

6.5.1 Thermography

Thermography involves capturing an image using a measurement technology that makes it possible to visualize heat radiation on the surface of objects using a thermal imaging camera. Every object that has a temperature above absolute zero emits infrared radiation, which is invisible to the human eye. The thermal imaging camera measures longwave radiation in the field of view and uses the results to calculate the temperature of the average object. Thermographic measurement involves an interaction between three actors: the operator, the observed system, and the environment, where thermal exchanges occur through the different modes of heat transfer (radiation, convection, and conduction) and the generation of heat sources of various natures. To perform a quality thermographic diagnosis, it is necessary to have information about the characteristics of the camera, the composition of the observed system, and the environment, as well as basic knowledge about heat transfer.

As mentioned earlier, thermography can be active or passive. In passive thermography, the detection of abnormal temperature profiles signals potential problems, and a key element is the temperature difference relative to a reference point, commonly known as the delta value (ΔT). The value of this delta is key to identifying anomalies and detecting errors in construction. This choice reduces the influence of reflected solar radiation (CEN, 1998). It is called active thermography when the operator performs a stimulation on the observed system for the purpose of identifying system parameters through a comparative analysis of the states before and after stimulation. There are different stimulation techniques, but the most common use photons. The predominant approach to data acquisition in IRT depends on the objective that is taken on each occasion. Passive IRT is most used in the analysis of humidity, hotspots, thermal bridges, among others. On the other hand, active IRT is more common in cracking analysis, to calculate thermal diffusivity, conductivity, and so on (Garrido et al., 2020).

Thermographic inspections use thermal imaging to represent the temperature map of an object. In the field of audits and diagnostics of heritage buildings, thermography is used both qualitatively and quantitatively.

The qualitative approach (IRT) allows for the identification of the hottest and coldest points on an object compared to its surroundings, helping to detect thermal anomalies and assess their location and extent. As for quantitative infrared thermography, it is a technique that combines the use of thermal cameras and image processing to provide more precise numerical information about the thermal properties of an object, as well as spatial-temporal correlation for a deeper evaluation. Interpretations are based on realistic models (Tejedor et al., 2021; Tejedor et al., 2022).

The IRT technique originated with the invention of the first camera in the 1990s. Despite its long history, ongoing improvements are being proposed to enhance the results of this tool and expand its applications. Regarding IRT studies focused on heritage spaces, many studies analyze moisture. High moisture levels inside worship areas are often the main risk factor for degradation (Cadelano et al., 2015). The most recent publications of IRT applied in cultural heritage are studies related to crack analysis (Bisegna et al., 2014; Sfarra et al., 2019), analysis of subsurface defects (Sfarra et al., 2016), and related to the search for buried structures within heritage sites. (Bisegna et al., 2014; Carlomagno et al., 2011; Puente et al., 2018; Sfarra et al., 2019).

6.5.2 Photogrammetry

As mentioned previously, photogrammetry is a non-destructive technique that uses images of selected locations to obtain dimensions and positions of objects in space. It is performed through the intersection of photographs and the digital terrain model.

There are three recognized forms of photogrammetry. Firstly, there is analog photogrammetry, which uses mathematical models. Analytical photogrammetry aims to apply mathematical models to physical elements. Lastly, digital photogrammetry, which is the most current and commonly used method, utilizes computer software to replace analog images with digital ones. Regarding the photography process, data can be obtained either aerially (with a camera in the air) or terrestrial (with a handheld camera on a tripod) (Gómez-Zurdo et al., 2021).

Firstly, an introduction to the fundamental elements of photogrammetry is provided to understand the process. The focal length should be considered, which represents the distance between the optical center and the focal plane where the image is captured. In combination with the photographic scale, it determines the height from which the image acquisition is conducted. This is crucial to ensure accurate measurements in the digital reconstruction. Based on the Figs. 6.6 and 6.7,

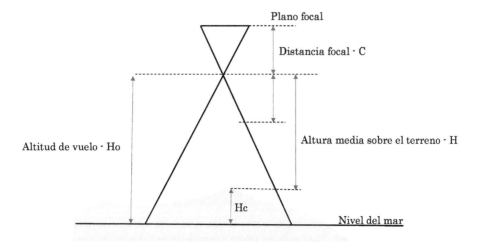

Figure 6.6 Photographic scale and camera focal length ratio.
Source: Developed by the author to visualize graphically how photogrammetry works.

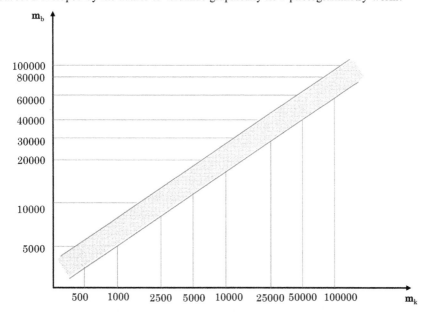

Figure 6.7 Relation between the scale of photography and that of cartography.
Source: Developed by the author to visualize graphically how photogrammetry works.

the relation between photographic scale and flight height is given with the following equation (Aber et al., 2019):

$$H = c \, m_b$$

$$H_0 = H_c + H \rightarrow M_b = \frac{1}{m_b} = \frac{c}{(H - H_c)}$$

where H is the average height above the ground and H_0 is the altimeter data, that is, the flight altitude in relation to sea level. With the altitude of the relief, the value of H_c would be known. The case study is about a vertical photograph, the projection is central. However, in most cases, the photograph is not obtained vertically, that is, the optical axis is not perpendicular to the ground. This usually happens due to lens distortions. For oblique photographs, the scale varies with magnitude and angular orientation with inclination (nadir line n) as shown in Fig. 6.8.

The relative height of each point with respect to a datum plane can be determined by the relation between parallax and height. This relationship is directly related to the elevation of a point compared to the reference level and will be higher for higher heights than for lower heights, as long as the viewing angle remains constant (Fig. 6.9) (Quirós & Rosado, 2014).

Photogrammetry relies on stereoscopy, which relies on the human ability to recognize features in relief. Two images taken from different points of view are used to achieve depth perception. On the other hand, the concept of stereoscopic parallax refers to the change of position of a point in two photographs due to the displacement of the camera during capture (Fig. 6.10).

As for the photogrammetric process, it is divided into consecutive phases that involve cartographic production. Once the photographic shot has been taken and

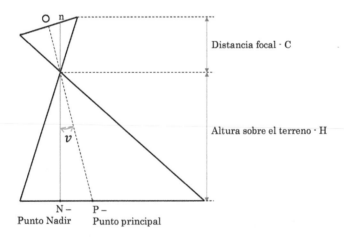

Figure 6.8 Low obliquity photography taken on completely flat terrain.
Source: Developed by the author to visualize graphically how photogrammetry works.

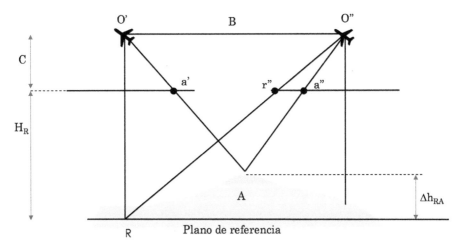

Figure 6.9 Parallax−dimension relationship.
Source: Developed by the author to visualize graphically how photogrammetry works.

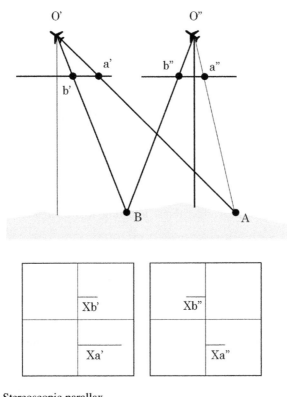

Figure 6.10 Stereoscopic parallax.
Source: Developed by the author to visualize graphically how photogrammetry works.

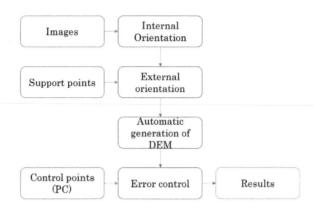

Figure 6.11 Photogrammetry process.
Source: Developed by the author to visualize graphically how photogrammetry works.

the three-dimensional coordinates of the space are entered, different products are given such as cartographic plans, elevations of digital models, and orthophotos (Fig. 6.11).

The parameters needed to define the geometry of optical paths are determined through the camera's internal and external orientations.

The internal orientation includes the focal length, lens radial distortion parameters, and the position of the principal point in the image coordinate system. The principal point is located at the center of the image and coincides with the origin of the coordinate system. This coordinate system must remain consistently stable and fixed with respect to the optical axis. In the case of digital cameras, they have a matrix of pixel cells on the image sensor that provides a Cartesian coordinate system for the images (Aber et al., 2019).

The widely used method for determining external orientation involves postsurvey reconstruction using ground control points (GCPs) with known X, Y, and Z coordinates. Theoretically, a minimum of three GCPs is required to orient a single photograph; However, in practice, multiple images are usually oriented together using least squares adjustment algorithms, allowing less than three points to be used per image.

The 3D reconstruction is performed using SfM. The SfM process consists of two stages. The structure estimation stage involves identifying 3D points in the scene and determining their spatial positions. The camera motion estimation stage calculates the position and relative orientation of the extreme cameras based on visual features and correspondences found between the images. SfM is compatible with any type of camera, but single-lens reflex cameras are most suitable.

As shown in this review, the general use of noninvasive methods has become common in research literature. The latest techniques presented for assessing cultural objects involve the combination of photogrammetry with laser scanning. Additionally, UAVs are often incorporated to reach the highest points of buildings (Gómez-Zurdo et al., 2021; Moyano et al., 2020; Tysiac et al., 2023).

6.6 Conclusions and future perspectives

The non-destructive techniques mentioned previously have an important contribution to heritage management by integrating them into tools such as Building Information Modeling (BIM) systems or artificial neural networks (ANNs) (Moropoulou et al., 2013). In addition to this, technologies are available that facilitate the acquisition of data to conduct a comprehensive analysis of the building, such as UAVs.

6.6.1 *Building Information Modeling*

BIM is a digital technology applied in construction to manage, create, and utilize information. When combined with NDT, it allows for a comprehensive and up-to-date project status presentation. The interoperability of BIM with non-destructive energy analysis methods is considered promising but may be low depending on the tool being used (Hanafi et al., 2016). It may be due to difficulties that arise in data extraction, discrepancies in files or data loss in the exchange of tools (Gao et al., 2019). Improving this combination can bring many benefits in terms of energy efficiency (Pereira et al., 2021).

The fusion of data obtained from photogrammetry and thermal cameras can be implemented with BIM thermal models to potentially increase the accuracy of building energy assessments, allowing for greater scalability and calibration. The integration of geographic information system revolutionizes energy analysis, as the detailed level of BIM combined with the properties of the built environment helps create highly precise models, as performance is influenced by environmental dynamics such as air convection and solar radiation (Murphy et al., 2009). On the other hand, with artificial intelligence through the digital twin, it is possible to develop maintenance programs with remote sensing of the study model. Currently, this information can be saved in virtual construction (BIM) models. In addition, it can be combined with new technologies such as UAVs to make a momentary review of space viable (Pereira et al., 2021).

The technique of BIM specifically applied to the study of historical buildings is known as HBIM. It emerged in 2009, attributed to Maurice Murphy (Murphy et al., 2009). In the context of historical buildings, HBIM focuses on parametric modeling based on documentation of historical materials, including historical documents, bibliographic references, photographs, and drawings. HBIM goes beyond mere graphics by semantically enriching elements, thus enhancing virtual reconstruction (Yang et al., 2013).

The first research efforts focused on building virtual models using graphic software and later enhancing them with the use of BIM for specific elements. However, current research is guided by the idea that all construction information should be centered around the digital reconstruction of existing heritage in BIM (Cursi et al., 2015).

In a study on a historical building, the possibility of integrating various research efforts to create a 3D digital archive of historical, energy, environmental, and urban planning information was established. The combination of automated SfM with photogrammetric methods was used to enhance accuracy. NDT techniques (photogrammetry, laser scanning, and IRT) were combined with the Internet of Things to achieve architectural surveys in a 3D BIM model (Delegou et al., 2019).

In the work of Santos et al. (2022), a methodology was proposed and validated to obtain a 3D HBIM model of degraded wooden structures based on data obtained through NDT. This methodology allows for an analysis of the structure's health and represents the geometry with high detail. Many studies related to HBIM aim to develop action plans. The results obtained demonstrate the potential of this methodology to automate the modeling of similar structures in the future.

On the other hand, recent studies (Udeaja et al., 2021) have used virtual reality technology in combination with the HBIM to create an interactive environment that allows the visualization and sharing of heritage-related information. The objective is to promote the preservation and sustainable management of heritage, integrating both its tangible and intangible aspects.

In addition, proposal has been made to create an integrated visualization platform to democratize access to cultural heritage by representing the HBIM model in Autodesk Revit 2020 and publishing it on a website (Rocha & Tomé, 2021).

The technique of BIM is heading toward process automation and affordability, offering key benefits such as efficiency, error reduction, cost savings, increased productivity, scalability, and improved tracking and control.

6.6.2   Artificial neural networks

ANNs, inspired by the functioning of the human brain, are computational models consisting of interconnected nodes that process and transmit information. These networks are capable of learning complex patterns and relationships from input data, making them powerful tools for data analysis and decision-making.

In the context of NDT, ANN is used to process and analyze data collected during non-destructive evaluations of materials, components, or structures. For instance, ANN can be trained to recognize anomalies in inspection data, facilitating the detection and diagnosis of potential issues. They can also automate and streamline the data analysis process, reducing time and resources required for non-destructive evaluations.

ANNs comprise processing elements known as neurons, which operate in parallel. The functions of these neurons depend on the neurons present in the "hidden layer." The network consists of input neuron layers and additional layers corresponding to each input value, where the hidden layers are located, and ultimately a neuron for each output (Fig. 6.12). The hidden layers are optimized using various techniques, such as trial and error, to determine their response to information. They typically employ activation and summation functions. A training set consists of combined input and output patterns used to train the network, enabling the automation of defect identification (Kumar et al., 2013).

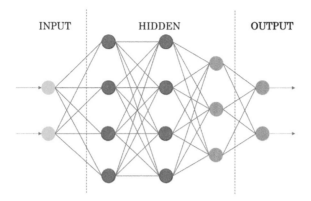

Figure 6.12 Graphically operation of neuronal layers in ANN graphical explanation illustrating the functionality of ANNs. *ANN*, Artificial neural network.

The most recent studies in deep learning (DL) have combined different NDT methods with neural networks to detect and segment defects in composite elements. However, in the field of cultural heritage, most data processing algorithms were not automated, which could lead to subjectivity on the part of the technician. To address this, it implements ANN with active thermography to automatically detect and segment defects and areas of defects in marquetry and art objects (Garrido et al., 2021). On the other hand, another study uses laser scanning and ANN to determine deformations in heritage structures with high precision.

ANNs have also been applied to the photographic microhistory (HFM) method to reduce measurement time, with promising results. However, there is a paucity of studies on the application of HFM in heritage construction, mainly due to the lack of adequate keywords in the definition of scientific works. On the other hand, it makes use of historical archives to create 3D models of lost architectural heritage, which is collected in neural networks to use as information for subsequent projects (Condorelli et al., 2020).

In summary, various DL techniques are being explored and applied in combination with thermography, HFM, photogrammetry, and other technologies for the detection, segmentation, and monitoring of defects in the context of architectural heritage.

It is concluded that, when it comes to the cultural heritage sector, relevant information regarding the construction of buildings is often overlooked. A literature review has been conducted on the tools used in the construction sector, with a focus on their applicability to cultural heritage.

Acknowledgments

This research has been carried out within the framework of the research activities of the TICBE Research Group recognized as UPV/EHU Research Group with code GIU22/001 in the call entitled *"Convocatoria de ayudas a grupos de investigación de la UPV/EHU (2022)"*.

References

Aber, J. S., Marzolff, I., Ries, J. B., & Aber, S. E. W. (2019). *Small-format aerial photography and UAS imagery. Principles, techniques, and geoscience applicationschapter 3 - principles of photogrammetry* (Second Edition). Elsevier. Available from https://doi.org/10.1016/B978-0-12-812942-5.00003-3.

Al-Ruzouq, R., Abu Dabous, S., Junaid, M. T., & Hosny, F. (2023). Non-destructive deformation measurements and crack assessment of concrete structure using close-range photogrammetry. *Results in Engineering, 18*, 101058. Available from https://doi.org/10.1016/j.rineng.2023.101058.

Asdrubali, F., & Desideri, U. (2018). *Handbook of energy efficiency in buildings* ISBN: 978-0-12-812817-6*A life cycle approach*. Imprint Butterworth-Heinemann: ELSEVIER. Available from https://doi.org/10.1016/C2016-0-02638-4.

Bisegna, F., Ambrosini, D., Paoletti, D., Sfarra, S., & Gugliermetti, F. (2014). A qualitative method for combining thermal imprints to emerging weak points of ancient wall structures by passive infrared thermography − A case study. *Journal of Cultural Heritage, 15*(2), 199−202. Available from https://doi.org/10.1016/j.culher.2013.03.006.

Bosiljkov, V., Uranjek, M., Žarnić, R., & Bokan-Bosiljkov, V. (2010). An integrated diagnostic approach for the assessment of historic masonry structures. *Journal of Cultural Heritage, 11*(3), 239−249. Available from https://doi.org/10.1016/j.culher.2009.11.007.

Cadelano, G., Bison, P., Bortolin, A., Ferrarini, G., Peron, F., Girotto, M., & Volinia, M. (2015). Monitoring of historical frescoes by timed infrared imaging analysis. *Opto-Electronics Review, 23*(1), 100−106. Available from https://doi.org/10.1515/oere-2015-0012, http://journals.pan.pl/opelre.

Carlomagno, G. M., Di Maio, R., Fedi, M., & Meola, C. (2011). Integration of infrared thermography and high-frequency electromagnetic methods in archaeological surveys. *Journal of Geophysics and Engineering, 8*(3), S93−S105. Available from https://doi.org/10.1088/1742-2132/8/3/s09.

CEN, EN 13187:1998: Thermal performance of buildings - Qualitative detection of thermal irregularities in building envelopes - Infrared method (ISO 6781:1983 modified). European Committee for Standardization (1998).

Condorelli, F., Rinaudo, F., Salvadore, F., & Tagliaventi, S. (2020). A neural networks approach to detecting lost heritage in historical video. *ISPRS International Journal of Geo-Information, 9*(5), 297. Available from https://doi.org/10.3390/ijgi9050297.

Cursi, S., Simeone, D., & Toldo, I. (2015). A semantic web approach for built heritage representation. *Communications in Computer and Information Science, 527*, 383−401. Available from https://doi.org/10.1007/978-3-662-47386-3_21.

Delegou, E. T., Mourgi, G., Tsilimantou, E., Ioannidis, C., & Moropoulou, A. (2019). A multidisciplinary approach for historic buildings diagnosis: The case study of the Kaisariani Monastery. *Heritage, 2*(2), 1211−1232. Available from https://doi.org/10.3390/heritage2020079.

Diz-Mellado, E., Mascort-Albea, E. J., Romero-Hernández, R., Galán-Marín, C., Rivera-Gómez, C., Ruiz-Jaramillo, J., & Jaramillo-Morilla, A. (2021). Non-destructive testing and finite element method integrated procedure for heritage diagnosis: The Seville Cathedral case study. *Journal of Building Engineering, 37*, 102134. Available from https://doi.org/10.1016/j.jobe.2020.102134.

El Masri, Y., & Rakha, T. (2020). A scoping review of non-destructive testing (NDT) techniques in building performance diagnostic inspections. *Construction and Building Materials, 265*, 120542. Available from https://doi.org/10.1016/j.conbuildmat.2020.120542.

Evangelisti, L., Guattari, C., Gori, P., & Vollaro, R. (2015). In situ thermal transmittance measurements for investigating differences between wall models and actual building performance. *Sustainability*, *7*(8), 10388−10398. Available from https://doi.org/ 10.3390/su70810388.

Fu, K., Xie, Y., Jing, H., & Zhu, J. (2019). Fast spatial−temporal stereo matching for 3D face reconstruction under speckle pattern projection. *Image and Vision Computing*, *85*, 36−45. Available from https://doi.org/10.1016/j.imavis.2019.02.007.

Gao, H., Koch, C., & Wu, Y. (2019). Building information modelling based building energy modelling: A review. *Applied Energy*, *238*, 320−343. Available from https://doi.org/ 10.1016/j.apenergy.2019.01.032.

Garrido, I., Erazo-Aux, J., Lagüela, S., Sfarra, S., Ibarra-Castanedo, C., Pivarčiová, E., Gargiulo, G., Maldague, X., & Arias, P. (2021). Introduction of deep learning in thermographic monitoring of cultural heritage and improvement by automatic thermogram preprocessing algorithms. *Sensors*, *21*(3), 750. Available from https://doi.org/10.3390/ s21030750.

Garrido, I., Lagüela, S., Otero, R., & Arias, P. (2020). Thermographic methodologies used in infrastructure inspection: A review—data acquisition procedures. *Infrared Physics & Technology*, *111*, 103481. Available from https://doi.org/10.1016/j.infrared.2020.103481.

Gómez-Zurdo, R. S., Martín, D. G., González-Rodrigo, B., Sacristán, M. M., & Marín, R. M. (2021). Aplicación de la fotogrametría con drones al control deformacional de estructuras y terreno. *Informes de la Construcción*, *73*(561), e379. Available from https://doi. org/10.3989/ic.77867.

Hanafi, M. H., Sing, G. G., Abdullah, S., & Ismail, R. (2016). Organisational readiness of building information modelling implementation: architectural practices. *Jurnal Teknologi*, *78*(5). Available from https://doi.org/10.11113/jt.v78.8265.

Herraiz, Á. H., Marugán, A. P., García Márquez, F. P., Papaelias, M., García Márquez, F. P., & Karyotakis, A. (2020). *Chapter 7 - A review on condition monitoring system for solar plants based on thermography. Non-destructive testing and condition monitoring techniques for renewable energy industrial assets* (pp. 103−118). Boston: Butterworth-Heinemann. Available from https://doi.org/10.1016/B978-0-08-101094-5.00007-1, https://www.sciencedirect.com/science/article/pii/B9780081010945000071.

Ibarra-Castanedo, C., Ricardo Tarpani, J., & Maldague, X. P. V. (2013). Non-destructive testing with thermography. *European Journal of Physics*, *34*(6), S91−S109. Available from https://doi.org/10.1088/0143-0807/34/6/s91.

ISO, ISO 9869-1:2014: Thermal insulation — Building elements — In-situ measurement of thermal resistance and thermal transmittance — Part 1: Heat flow meter method. International Organization for Standardization, Geneva, Switzerland. 3rd edition (2014), 36. Available from https://www.iso.org/standard/59697.html.

Keefe, D. (2010). Blower door testing. *Journal of Light Construction* (01).

Kilic, G. (2015). Using advanced NDT for historic buildings: Towards an integrated multidisciplinary health assessment strategy. *Journal of Cultural Heritage*, *16*(4), 526−535. Available from https://doi.org/10.1016/j.culher.2014.09.010, http://www.elsevier.com.

Krutikova, O., Sisojevs, A., & Kovalovs, M. (2017). Creation of a depth map from stereo images of faces for 3D model reconstruction. *Procedia Computer Science*, *104*, 452−459. Available from https://doi.org/10.1016/j.procs.2017.01.159.

Kumar, R., Aggarwal, R. K., & Sharma, J. D. (2013). Energy analysis of a building using artificial neural network: A review. *Energy and Buildings*, *65*, 352−358. Available from https://doi.org/10.1016/j.enbuild.2013.06.007.

Kylili, A., Fokaides, P. A., Christou, P., & Kalogirou, S. A. (2014). Infrared thermography (IRT) applications for building diagnostics: A review. *Applied Energy*, *134*, 531−549. Available from https://doi.org/10.1016/j.apenergy.2014.08.005, http://www.elsevier.com/inca/publications/store/4/0/5/8/9/1/index.htt.

Liu, Z., Yeoh, J. K. W., Gu, X., Dong, Q., Chen, Y., Wu, W., Wang, L., & Wang, D. (2023). Automatic pixel-level detection of vertical cracks in asphalt pavement based on GPR investigation and improved mask R-CNN. *Automation in Construction*, *146*, 104689. Available from https://doi.org/10.1016/j.autcon.2022.104689.

Lourenço, P. B., Krakowiak, K. J., Fernandes, F. M., & Ramos, L. F. (2007). Failure analysis of Monastery of Jerónimos, Lisbon: How to learn from sophisticated numerical models. *Engineering Failure Analysis*, *14*(2), 280−300. Available from https://doi.org/10.1016/j.engfailanal.2006.02.002.

Lualdi, L. B., Saisi, L., Zanzi, M., Gianinetto, & G., Roche, (2001). NDT applied to the diagnosis of historic buildings: The case of some Sicilian Churches. *III International Seminar on Structural Analysis of Historical Constructions*. 29−46.

López, F., Lerones, P., Llamas, J., Gómez-García-Bermejo, J., & Zalama, E. (2018). A review of heritage building information modeling (H-BIM). *Multimodal Technologies and Interaction*, *2*(2), 21. Available from https://doi.org/10.3390/mti2020021.

Martín-Garín, A., Millán-García, J. A., Hidalgo-Betanzos, J. M., Hernández-Minguillón, R. J., & Baïri, A. (2020). Airtightness analysis of the built heritage−field measurements of nineteenth century buildings through blower door tests. *Energies*, *13*(24), 6727. Available from https://doi.org/10.3390/en13246727.

Martínez-Garrido, M. I., Fort, R., Gómez-Heras, M., Valles-Iriso, J., & Varas-Muriel, M. J. (2018). A comprehensive study for moisture control in cultural heritage using non-destructive techniques. *Journal of Applied Geophysics*, *155*, 36−52. Available from https://doi.org/10.1016/j.jappgeo.2018.03.008, http://www.elsevier.com/inca/publications/store/5/0/3/3/3/3/.

Moropoulou, A., Labropoulos, K. C., Delegou, E. T., Karoglou, M., & Bakolas, A. (2013). Non-destructive techniques as a tool for the protection of built cultural heritage. *Construction and Building Materials*, *48*, 1222−1239. Available from https://doi.org/10.1016/j.conbuildmat.2013.03.044.

Moyano, J., Nieto-Julián, J. E., Bienvenido-Huertas, D., & Marín-García, D. (2020). Validation of close-range photogrammetry for architectural and archaeological heritage: Analysis of point density and 3D mesh geometry. *Remote Sensing*, *12*(21), 3571. Available from https://doi.org/10.3390/rs12213571.

Munarim, U., & Ghisi, E. (2016). Environmental feasibility of heritage buildings rehabilitation. *Renewable and Sustainable Energy Reviews*, *58*, 235−249. Available from https://doi.org/10.1016/j.rser.2015.12.334.

Murphy, M., McGovern, E., & Pavia, S. (2009). Historic building information modelling (HBIM). *Structural Survey*, *27*(4), 311−327. Available from https://doi.org/10.1108/02630800910985108.

Pallarés, F. J., Betti, M., Bartoli, G., & Pallarés, L. (2021). Structural health monitoring (SHM) and non-destructive testing (NDT) of slender masonry structures: A practical review. *Construction and Building Materials*, *297*, 123768. Available from https://doi.org/10.1016/j.conbuildmat.2021.123768.

Pereira, V., Santos, J., Leite, F., & Escórcio, P. (2021). Using BIM to improve building energy efficiency − A scientometric and systematic review. *Energy and Buildings*, *250*, 111292. Available from https://doi.org/10.1016/j.enbuild.2021.111292.

Pire, T., Fischer, T., Castro, G., De Cristóforis, P., Civera, J., & Berlles, J. J. (2017). S-PTAM: Stereo Parallel Tracking and Mapping. *Robotics and Autonomous Systems, 93*, 27−42. Available from https://doi.org/10.1016/j.robot.2017.03.019.

Puente, I., Solla, M., Lagüela, S., & Sanjurjo-Pinto, J. (2018). Reconstructing the Roman Site "Aquis Querquennis" (Bande, Spain) from GPR, T-LiDAR and IRT Data Fusion. *Remote Sensing, 10*(3), 379. Available from https://doi.org/10.3390/rs10030379.

Pérez-Gracia, V., Caselles, J. O., Clapés, J., Martinez, G., & Osorio, R. (2013). Nondestructive analysis in cultural heritage buildings: Evaluating the Mallorca cathedral supporting structures. *NDT and E International, 59*, 40−47. Available from https://doi.org/10.1016/j.ndteint.2013.04.014.

Quirós Rosado, E. M. (2014). *Introduction to photogrammetry and cartography applied to civil engineering.* Publications Service.

Rocha, J., & Tomé, A. (2021). Multidisciplinarity and accessibility in heritage representation in HBIM Casa de Santa Maria (Cascais) — A case study. *Digital Applications in Archaeology and Cultural Heritage, 23*, e00203. Available from https://doi.org/10.1016/j.daach.2021.e00203.

Santos, D., Cabaleiro, M., Sousa, H. S., & Branco, J. M. (2022). Apparent and resistant section parametric modelling of timber structures in HBIM. *Journal of Building Engineering, 49*, 103990. Available from https://doi.org/10.1016/j.jobe.2022.103990.

Sfarra, S., Ibarra-Castanedo, C., Tortora, M., Arrizza, L., Cerichelli, G., Nardi, I., & Maldague, X. (2016). Diagnostics of wall paintings: A smart and reliable approach. *Journal of Cultural Heritage, 18*, 229−241. Available from https://doi.org/10.1016/j.culher.2015.07.011.

Sfarra, S., Yao, Y., Zhang, H., Perilli, S., Scozzafava, M., Avdelidis, N. P., & Maldague, X. P. V. (2019). Precious walls built in indoor environments inspected numerically and experimentally within long-wave infrared (LWIR) and radio regions. *Journal of Thermal Analysis and Calorimetry, 137*(3), 1083−1111. Available from https://doi.org/10.1007/s10973-019-08005-1, http://www.springer.com/sgw/cda/frontpage/0,11855,1-40109-70-35752391-0,00.html.

Tejedor, B., Barreira, E., Almeida, R. M. S. F., & Casals, M. (2021). Automated data-processing technique: 2D Map for identifying the distribution of the U-value in building elements by quantitative internal thermography. *Automation in Construction, 122*, 103478. Available from https://doi.org/10.1016/j.autcon.2020.103478.

Tejedor, B., Lucchi, E., Bienvenido-Huertas, D., & Nardi, I. (2022). Non-destructive techniques (NDT) for the diagnosis of heritage buildings: Traditional procedures and futures perspectives. *Energy and Buildings, 263*. Available from https://doi.org/10.1016/j.enbuild.2022.112029, https://www.journals.elsevier.com/energy-and-buildings.

Troi, A., & Bastian, Z. (2014). *Energy efficiency solutions for historic buildings: A handbook.* Birkhäuser. ISBN: 9783038216469. Available from https://doi.org/10.1515/9783038216506.

Tysiac, P., Sieńska, A., Tarnowska, M., Kedziorski, P., & Jagoda, M. (2023). Combination of terrestrial laser scanning and UAV photogrammetry for 3D modelling and degradation assessment of heritage building based on a lighting analysis: case study—St. Adalbert Church in Gdansk, Poland. *Heritage Science, 11*(1). Available from https://doi.org/10.1186/s40494-023-00897-5.

Udeaja, C., Mansuri, L. E., Ncube Makore, B. C., Baffour Awuah, K. G.,Patel, D. A., Trillo, C., & Jha, K. N. (2021). Digital storytelling: the integration of intangible and tangible heritage in the city of Surat, India. *Human-Computer Interaction (HCI) International Conference.* Washington, DC, USA, PP. 12794. Available from https://doi.org/10.1007/978-3-030-77411-0_11.

Yalçıner, C. Ç., Büyüksaraç, A., & Kurban, Y. C. (2019). Non-destructive damage analysis in Kariye (Chora) Museum as a cultural heritage building. *Journal of Applied Geophysics*, *171*, 103874. Available from https://doi.org/10.1016/j.jappgeo.2019.103874.

Yang, B. Zhang, L. Zhang, W., & Ai, Y. (2013). Defects Infrared thermography non-destructive test Wind turbine blades Thermography (imaging) Non destructive testing. Non-destructive testing of wind turbine blades using an infrared thermography: A review. *ICMREE 2013 - Proceedings: 2013 International Conference on Materials for Renewable Energy and Environment*. 1. 407–410. Available from https://doi.org/10.1109/ICMREE.2013.6893694, https://www.scopus.com/inward/record.uri?eid = 2-s2.0-84910100553&doi = 10.1109% 2fICMREE.2013.6893694&partnerID = 40&md5 = d41c1834a70a0e85ce16f1ab7bb28702.

Section III

Laser Scanning and Photogrammetry

Architectural survey of built heritage using laser scanning and photogrammetry. Comparison of results in the case study of the Gothic vault of the church of San Vicente in Donostia-San Sebastián

7

José Javier Pérez-Martínez[1], Iñigo Leon[2], María Senderos Laka[2], Amaia Casado-Rezola[1], and Alexander Martín-Garín[2]
[1]TICBE Research Group, Department of Architecture, School of Architecture, University of the Basque Country UPV/EHU, Donostia-San Sebastián, Spain, [2]TICBE Research Group, Department of Architecture, Faculty of Engineering of Gipuzkoa, University of the Basque Country UPV/EHU, Donostia-San Sebastián, Spain

Abbreviations

AEC	Architecture, Engineering, and Construction
BIM	Building Information Modeling
DSLR	Digital Single-Lens Reflex
GCP	Ground Control Point
NDT	Non-destructive Testing
SfM	Structure from Motion
SLAM	Simultaneous Localization and Mapping
TLS	Terrestrial Laser Scanning
TS	Total Station
UAV	Unmanned Aerial Vehicle

7.1 Introduction

The construction industry is a behemoth, accounting for a substantial portion of the global economy. In fact, an estimated $10 trillion is spent each year on goods and services related to the sector (Barbosa et al., 2017). However, despite this significant investment, the construction industry's productivity has not kept pace with other sectors that have made notable strides. Over the past two decades, labor productivity in the construction industry has increased by a mere 1%, a figure that

Diagnosis of Heritage Buildings by Non-Destructive Techniques. DOI: https://doi.org/10.1016/B978-0-443-16001-1.00007-3

pales in comparison to the global economy's 2.8% and the manufacturing industry's 3.6% growth rates. This undeniable evidence underscores the daunting challenge that the architecture, engineering, and construction (AEC) sector faces and the pressing need to overcome it to meet society's current demands.

The AEC sector is presently undergoing a technological shift, with the emergence of various digital tools that are increasingly dominating the industry and being implemented across all three phases of the sector's value chain: design, construction, and operation and maintenance. Building information modeling (BIM) is considered to be the cornerstone of this digital transformation, as it not only replaces traditional computer-aided design as the preferred architectural design alternative but also integrates multiple functions necessary for the work processes of construction companies. This means that BIM is not limited to simple three-dimensional (3D) modeling but also enables the utilization of dimensions ranging from 4D to 7D, for applications such as planning/scheduling, cost management, sustainability analysis, and operation and maintenance (Baïri et al., 2021; Liu et al., 2015; Martín-Garín et al., 2022; Sacks et al., 2018).

In addition to the widespread implementation of BIM, other digital technologies have been acquiring an increasingly significant role in the AEC sector (Etxepare et al., 2020; Lasarte et al., 2021). For example, in recent years, there has been an enormous expansion of digital data capture technologies applied to architectural surveying that combines Terrestrial Laser Scanning (TLS), photogrammetry, both terrestrial and unmanned aerial vehicle (UAV)-assisted, and real-time kinematic satellite navigation technology (Freimuth & König, 2018; Jiang et al., 2020; Leon et al., 2020; Xiong et al., 2013). Through the application of these new technologies, 3D models can be generated in point cloud format with millimeter precision, a result that is very difficult to achieve by applying traditional architectural survey techniques. In addition, these 3D models can be easily implemented in BIM software environments, facilitating high-precision modeling that guarantees the rigor of the process.

In this sense, the built heritage sector is another area where graphic survey techniques are widely spread. The geometric complexity that usually occurs in this type of building means that traditional survey techniques do not allow achieving the necessary detail for certain tasks. The vaults of Gothic architecture are one of the outstanding examples where a series of architectonic elements of great geometric complexity coincide, thus constructively defining the generated volume. The original traces that were developed centuries ago enable us to comprehend the dimensions, shapes, and proportions that the architectural elements possess. However, documentary conservation and preservation has not always been possible and therefore makes it impossible to understand in detail the many examples of today, thus becoming a challenge for researchers in this area.

7.1.1 Church of San Vicente in Donostia-San Sebastián (Basque Country, Spain)

For a long time, traditional historiography has considered Gothic architecture after the classical period as a decadent variant, thus neglecting the vindication of the architectural values of Gothic constructions from the 14th, 15th, and 16th centuries (Goitia, 2000). An inconsideration based on two ideas: on the one hand, the perception of this architecture as a decadent and mannerist manifestation of classical Gothic, belonging to a unified panorama, which remains practically unchanged, without contributing any innovation; on the other hand, its appreciation as a past style that has been perpetuated for centuries, resisting disappearing before the innovative appearance of classical forms. After a long scientific debate, developed during the first half of the 20th century, on the structural or esthetic functionality of the rib of the Gothic vault (Martínez, 1998), at the end of the last century new research proposals emerged that recognized the value and quality of the Gothic architecture of the Modern Age throughout Europe, including the context of Castilian Gothic architecture of the 16th century. From the numerous works of José María Azcárate on late-medieval and modern Spanish Gothic architecture (de Azcárate Ristori, 1958, 1985, 1990), from the last decade of the 20th century, there has been a proliferation of authors who have delved into this area, highlighting, among others, the investigations of José Carlos Palacios, Javier Gómez Martínez, or Begoña Alonso (Martínez, 1998; Palacios Gonzalo, 2009; Ruiz, 2003). According to these authors, Hispanic Gothic architecture reached its maturity and maximum levels of creativity at the end of the 15th century, highlighting the high richness of its layouts or the sophisticated geometries of its ribs (Palacios Gonzalo, 2009).

The research actions on the late Gothic architecture in the Hispanic area that has been developed mainly during the last two decades has focused on an analysis of geometric conception, design, layout, construction, and technological development focused on direct study of the material reality of the built work. This new line of research has therefore required the characterization of the geometry of these works through precise metric and dimensional data, which has been possible to obtain by applying new digital architectural survey techniques such as the TLS and automated photogrammetry or close-range photogrammetry.

The evolutions of these new technologies have generated an important advance in the digital record of the architectural survey (Lerma García et al., 2008), including the dynamic TLS capture or the SfM (Structure from Motion) photogrammetric processing, contributing to improve the efficiency of the field work reducing the time effective data collection and allowing a greater amplitude of the working range; in short, facilitating a more efficient management of the geometric characterization within the architectural survey procedure (Grussenmeyer et al., 2010).

Terrestrial laser scanning technology allows capturing a large amount of information with millimeter precision, also being able to obtain a 360-degree panoramic image that, in combination with the point cloud, enables direct measurement of the georeferenced digital model, facilitating detailed analysis of cabinet work. However, its main disadvantage is its high economic cost, the complexity of processing the captured data and the low resolution of the images obtained.

Automated photogrammetry supposedly generates a model with a lower degree of accuracy. In addition, its application is conditioned by the homogeneity and level of lighting received by the item being captured. However, it is a cheaper application through which, in addition to the 3D point cloud model, a geometric and textured triangular mesh is obtained using high-resolution photography. This particularity allows its implementation in open and universal access software and platforms, for specific application in digital resources for the dissemination of built heritage, including augmented reality, virtual reality, and mixed reality.

The analysis at hand has focused on the church of San Vicente in the city of Donostia-San Sebastián in Gipuzkoa (Spain), specifically on one of the vaults of its main chapel Fig. 7.1. It is a characteristic work of Castilian Gothic architecture from the early 16th century. The beginning of the works dates back to 1507 (Tarifa Castilla, 2018). The work, finally executed around 1548, responds to a rectangular floor plan with three naves, an octagonal chancel, an aligned transept, and three other sections in the longitudinal direction, up to the foot of the church, with the central nave being higher than the collaterals.

As in many of the cases of similar examples, the building has a planimetry elaborated using traditional survey techniques from which it is very difficult to obtain a precision architectural survey, considering the particular complexity and irregularity of the morphology of the building elements and factories of Gothic architecture Fig. 7.2. Therefore a planimetry of these characteristics is usually insufficient for the development of research that requires a high level of geometric accuracy (Fig. 7.3).

As indicated, there are currently different technologies that allow high-precision geometric surveying of built elements, such as the TLS system. However, the use

Figure 7.1 Nadir plane of the vaults of the church of San Vicente in Donostia-San Sebastián.

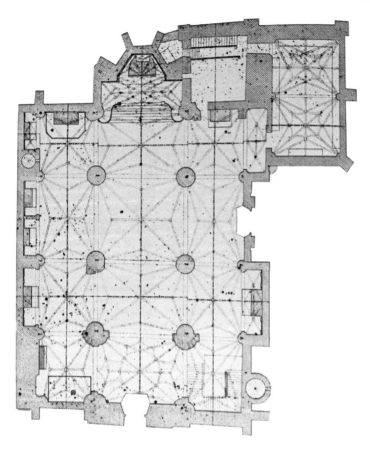

Figure 7.2 Floor plan of the church of San Vicente in Donostia-San Sebastián drawn up by Manuel Echave in 1892.
Source: Figure developed by the authors and based on Izaguirre Lacoste, M., Muñoz Baroja Peñagaricano Jesús. 1985. Monumentos Nacionales de Euskadi; Tomo II. Departamento de Cultura, Gobierno Vasco.

of this technology implies important investments that are not always possible to carry out, which supposes an important limitation for the universalization of its use. In addition, there are few studies that have made it possible to verify the quality of the results of low-cost alternative technologies such as photogrammetry. For this reason, the main objective of the research is to verify automated photogrammetry or SfM image processing as a reliable non-destructive technique (NDT) for precision and cost-effective geometric capture for the characterization of built heritage. For this purpose, it is proposed to analyze and verify the results obtained through SfM with the capture carried out both with traditional techniques based on total station (TS) and leading techniques of massive data capture such as TLS.

For this purpose, this study has focused on a section of enormous relevance in the investigation of heritage of Gothic architecture, where, for some years,

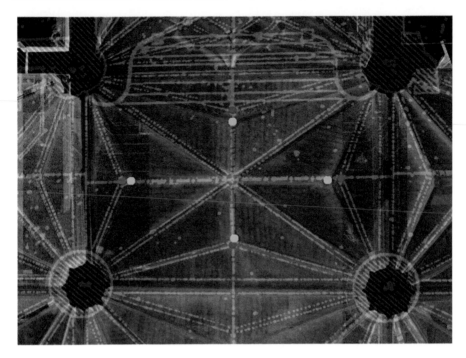

Figure 7.3 Overlay between the historical planimetry and the plan obtained from the point cloud of the laser scanning, taking the central key of the vault as the coincident point (the areas filled in black correspond to the point cloud). In red, position of the keys in the historical planimetry; in green, position of the keys in the cloud of points.

geometric and constructive analyses of ribbed vaults have been developed, which have opened a research line focused on the design and construction process of these works. Specifically, the process of geometric characterization of the Gothic ribbed vault has been taken as a case study. This is a context that presents particularly complex conditions for the application of automated photogrammetry, where the lighting of these spaces is, in most cases, a determining factor for the methodological development of this technique. The research aims to determine the level of accuracy of the results obtained, indicating the advantages and disadvantages of each of the techniques and thus allowing the choice of the most efficient workflow based on the particular conditions of the work environment.

7.2   Materials and methods

To achieve the proposed objective of the investigation, this section describes the methodology that has been carried out during the book chapter (Fig. 7.4).

It must be taken into account that the conditions of the interior spaces of the temples belonging to the analyzed typology present some drawbacks that it is

DSLR camera + SFM Photogrammetry

Total Station

Terrestrial Laser Scanning

Figure 7.4 Flowchart of the research methodology applied.

necessary to take into consideration. The problem generated by the low level of natural lighting stands out, an issue that decisively affects the application of the photogrammetric technique. This issue determines one of the theoretical foundations of the application of photogrammetric methods, such as taking photographs under a configuration of camera parameters where, for the entire series of shots, the aperture of the diaphragm and the time of exposure remain constant. Likewise, homogeneous lighting conditions are required without producing an excessive contrast between the illuminated areas and the shadow areas.

To take the photographic series based on the SfM technique, a Nikon D3400 digital single-lens reflex (DSLR) camera was used with a Nikkor 50 mm 1:1.8 lens (Fig. 7.5A), a diaphragm aperture of f/11 to limit the effects of diffraction, obtaining an exposure time of 5 seconds. Under these configuration parameters and maintaining a fixed focus, the complete series of photographic shots has been carried out.

The photographic shot combines zenithal alignment parallel to the Z axis and oblique alignment, the angle of which is determined by the possibilities offered by the physical configuration of the space around the vault section under analysis. The overhead shot has been structured taking the field of vision of each photograph as a

Figure 7.5 Instrument based surveying of the case study (A) photogrammetry, (B) total station, and (C) laser scanner.

reference, considering the parameters of average focus distance (24 m) and focal length of the lens used (50 mm). To ensure a high percentage of common points, a high level of overlap between shots has been predetermined (horizontal/vertical 90%) (Fig. 7.6). This configuration allows to establish a grid with an approximate interval of 1.5 m that determines the position of each shot. In a complementary way, oblique shots are established along the contour of the vaulted section, trying to maintain a similar degree of overlap. In this way, a total of 175 photographs have been obtained.

On the other hand, the implementation was done through the 3DF Zephyr V 7.003 software from the company 3DFLOW. The software allows the reconstruction of 3D models from digital images, standing out for its intuitive graphical interface, for following a clear workflow and for its multiple import−export options. The general workflow begins with the import of the captured images, their alignment, the subsequent generation of a sparse point cloud in a first phase, and a subsequent generation of a dense point cloud. The tool, in turn, allows obtaining the meshing and texturing of the generated element. Regarding the solutions for the quality control of the point clouds obtained, it is possible to incorporate ground control points (GCP) and reference distances that allow the correct scaling and adjustment of the model. Said GCP can be added from the images used, from the point cloud itself or obtained through topographic methods and represented through the desired reference system.

On the other hand, as previously indicated, one of the methods used to verify the results obtained through SfM has been the use of the TS.

Due to the previously mentioned lighting conditions, the correct application of those instruments that perform discrete data capture, such as the TS, can be problematic. This situation usually leads to the need to be guided by the station's laser pointer, with the consequent loss of aiming precision. Likewise, it must be taken into account that the artificial lighting conditions are available in these spaces, which are usually distinguished by two main characteristics: a level that is still

Figure 7.6 Field of view of the shot with zenithal orientation and degree of overlap.

insufficient to accurately carry out aiming through the TS viewer, of those elements located in dimly lit areas, and excessively focal overhead lighting that usually dazzles the surroundings of the luminaires, preventing the visualization of those elements located in their immediate environment. Furthermore, in those temples that preserve the wooden planks with which the burial pits distributed throughout the naves are covered, it is common for the TS, as a result of the instability of the tripod support, cannot perform any measurement due to the deviation of the compensator out of range.

Specifically, the Leica TCR407 equipment was used (Fig. 7.5B). One of the fundamental characteristics of the equipment, given the inaccessibility of the measurement points, is its ability to perform said measurement using the laser distance meter that the equipment incorporates. Regarding the accuracy of the measurements, the

Figure 7.7 Zenithal survey using the total station with GFZ3 eyepiece (A) and visualization through the eyepiece of the "tortera" or decorated part of the key (B).

equipment offers a standard deviation of 3 mm + 2 ppm for the measurement of distances without reflector according to the ISO 17123-4 standard and of 7″ (2 mgon) for the measurement of angles according to ISO 12857. On the other hand, the equipment has a 30 magnification telescope and a visual field of 1°30′. However, the need to perform readings of dome reference points made it necessary to use the GFZ3 zenithal eyepiece to allow inclined visual directions to the zenith (Fig. 7.7).

In relation to the TLS equipment, the Leica RTC360 high-speed laser scanner was used, which allows the capture of HDR spherical images. In addition, it has SLAM technology, through which information from the visual inertial system, made up of five cameras, is combined with the inertial measurements of an inertial measurement unit that allow determining the relative position and orientation between two consecutive scans. The combined use of these technologies substantially reduces field work time (Fig. 7.5C). Among the characteristics of the equipment, the scanning speed of 2×10^6 points/s, a resolution of 6 mm at 10 m, and a range of up to 130 m stand out. It has a 360-degree horizontal and 300-degree vertical field of view, requiring around 2.5 minutes to perform each high-resolution scan including the 360-degree image capture. The angular precision is 18″ and the range precision is 1.0 mm + 10 ppm. The point capture workflow is complemented by Cyclone Field software using a digital tablet or smartphone, which allows scanning data to be linked and recorded while field work is in progress. Subsequently, the registration of the point clouds during office work is carried out using the Leica Cyclone Register 360 software.

Finally, the process of lifting the vault through the previously presented instruments was complemented in situ by means of a freehand sketch (Fig. 7.8). The scheme allowed accompanying the measurements made with field annotations for the correct identification of points and thus facilitating the correct understanding later during cabinet work.

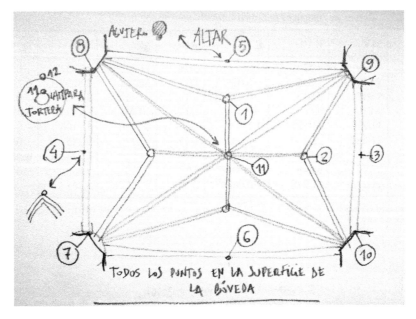

Figure 7.8 Sketch of the vault during data collection in situ.

7.3 Results and discussion

After the presentation of the research methodology, the following section presents and analyzes the results obtained through the techniques described through the case study of the church of San Vicente de Donostia-San Sebastián. The section, in turn, is subdivided into three sections with the purpose of progressively analyzing the results obtained in each of the phases.

The first one shows the results obtained through the DSLR camera and the photogrammetry carried out. The second section allows verifying the precision of the photogrammetry carried out in relation to the measurements made by TS through discrete points that will serve as GCP. As will be analyzed in the section, these points allow the reconstruction of the original model to achieve its scaling the points measured through the TS taking as reference, thus improving the precision of the model. Finally, the third section compares the results of the second scaled photogrammetric point cloud with respect to the point cloud obtained by TLS.

7.3.1 Photogrammetry

For the development of the photogrammetric model, a total of 175 photographic captures with the DSLR camera were used. The process begins with the import of the images, taking the software the camera calibration values indicated in Table 7.1. From this step, the 3D reconstruction of the images whose position and

Table 7.1 Reference distances obtained through control points.

RD	MV	BA	E	RD	MV	BA	E
D1	9.519	9.533	0.144	D10	11.473	11.437	−0.310
D2	7.870	7.866	−0.056	D11	9.818	9.829	0.114
D3	9.917	9.914	−0.033	D12	9.981	9.976	−0.051
D4	9.810	9.805	−0.044	D13	10.868	10.874	0.052
D5	11.213	11.221	0.068	D14	11.129	11.106	−0.201
D6	11.380	11.380	−0.002	D15	9.272	9.245	−0.290
D7	6.900	6.932	0.458	D16	6.004	6.024	0.323
D8	9.745	9.745	−0.003	D17	9.310	9.308	−0.021
D9	11.278	11.270	−0.068	D18	5.772	5.767	−0.078

BA, Measured value in the bundle-adjusted model (m); *E*, error (%); *MV*, measured value with total station (m); *RD*, reference distance.

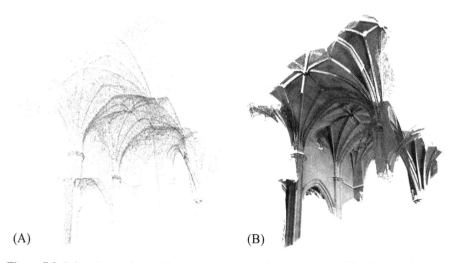

(A)   (B)

Figure 7.9 Point clouds obtained by photogrammetry from the dome of San Vicente in Donostia-San Sebastián: (A) sparse point cloud and (B) dense point cloud.

orientation are calculated based on the SfM method of the software is carried out. This process results in a first sparse point cloud together with the bounding box that allows to discard those points that are not collected by said volume. The main objective of this step is to lighten the computational load in the workflow by discarding the points that are not necessary for the dense point cloud reconstruction process.

Fig. 7.9 shows the results obtained from both point clouds (sparse and dense). To obtain these clouds, processing times of 4.32 minutes and 1 hours 11 minutes 3 seconds, respectively, are necessary. As can be seen in the figure, there is a noticeable difference in the point densities of both clouds, going from a total of 81,661 points in the sparse cloud to 26,588,237 points in the dense one. The process

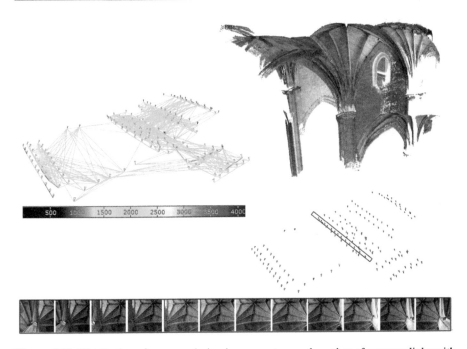

Figure 7.10 Distribution of cameras during image capture and number of cameras links with a threshold match value of 1500.

of capturing the images is reflected in Fig. 7.10 where the location and distribution of the 175 captures made from the bottom to the vault are reflected. In addition, one of the longitudinal series carried out in which the overlap between them can be seen, including an analysis between camera links. Through this study, it is possible to show the strength of the matches between the cameras during the orientation phase. A series of lines which link the cameras will be shown in the workspace according to a matching threshold defined, while their degree of intensity is indicated through a color map. On this occasion, a high threshold value of 1500 has been used to be able to distinguish the connections with the greatest weight, since a large part of them occur with lower values and if the values were reduced, the pattern of connections could not be analyzed.

7.3.2 Total station

One of the handicaps that the use of photogrammetry entails is the complexity of being able to carry out scaled models if the metadata of the photographs does not include the specific coordinates from which the images have been taken. In the case of using aerial images with UAVs, for example, the speed of the workflow increases since 3DF Zephyr automatically detects pictures taken by drones leveraging their GPS coordinates to scale and georeference the 3D model. However, for those cases where the captured images do not integrate said information, it is necessary to work with GCP to scale and georeference the model. GCPs can also be used in the model bundle adjustment to correct points and cameras by minimizing the reprojection error of points through the nonlinear minimization procedure. It is recommended to set a minimum of 3 control points as constraints for running the bundle adjustment. For improved accuracy, it is advisable to increase the number of GCPs used.

As described in Section 7.2, a Leica TCR407 TS, has been used to obtain a series of GCPs. As shown in Fig. 7.11, a total of 11 control points were taken from which 18 distances were drawn, which have been taken as reference elements. Once these points and reference distances have been taken, the bundle adjustment of the point cloud is carried out to proceed with its scaling and thus obtain a model that is more adjusted to the real dimensions of the vault under study.

As previously indicated, the bundle adjustment process is an adjustment based on a nonlinear minimization procedure and, therefore, it is prone to some parts of the model being able to adjust better than others to the real dimensions. As it can be seen that the original model of the cloud of points calculated in Section 7.3.1. faithfully reflects the geometry and proportions of the model of the real vault. However, if the dimensions of the reference distances of the model are compared

Figure 7.11 Control points acquired through the total station and implemented in the photogrammetric model using GCP and reference distances. *GCP*, Ground control point.

with the values measured through the TS, it can be observed that there is a great difference in an order of magnitude 3 times greater. As previously indicated, this result is normal since the photographs used for the photogrammetric model lack additional information related to the coordinates and therefore the software does not have any reference to the dimensions of the model. Due to this, the bundle adjustment of the model is necessary using the GCPs and the previously calculated reference distances. After the adjustment process, the software generates a new model in which the dimensions take coherent values with respect to the real values, producing only an average absolute difference of 1.143 cm ± 1.104 and an average percentage difference of 0.128% ± 0.132. Table 7.1 shows the errors obtained in each of the reference distances and as can be seen the minimum and maximum errors are between 0.02 mm and 3.56 cm, values are consistent with those obtained in other investigations (Dai et al., 2014).

7.3.3 TLS results

Through the capture of the laser scanner, 29 positionings have been carried out, both inside and outside the church, proceeding later to its linking to obtain a model of the building as a whole (Fig. 7.12). The mean overall error obtained from said processing was 0.001 m.

To make a comparison between both point cloud models (Fig. 7.13), beyond the verification of the reference distances described above, both clouds have been compared using the CloudCompare software. CloudCompare is an open-source software

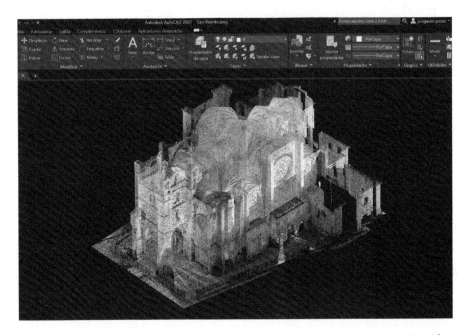

Figure 7.12 Complete point cloud of the church of San Vicente in Donostia-San Sebastián.

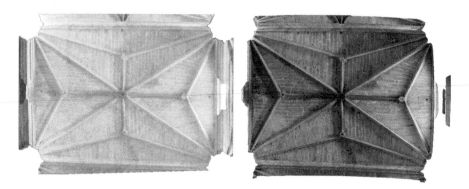

Figure 7.13 Comparison of point clouds obtained by TLS and photogrammetry from the bottom view of the vault.

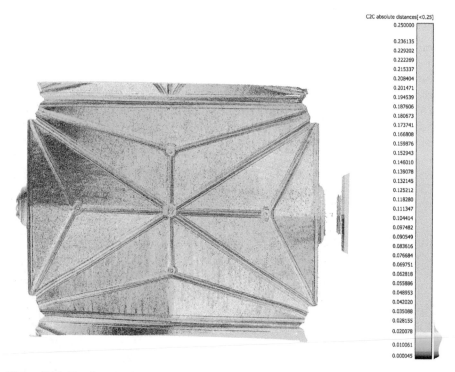

Figure 7.14 Plan image of the comparison between the distances between the photogrammetric point cloud and the reference point cloud of the laser scanning.

for 3D point cloud editing and processing. It was designed to perform a direct comparison between two dense 3D point clouds. In this way, it is intended to guarantee the adjustment of said overlap by analyzing the possible deviations between both clouds. As can be seen in the plan image of the comparison between both clouds, a high level of general overlap can be seen in Fig. 7.14.

As can be seen in the plan image of the comparison between both clouds, a high level of general overlap can be noticed. To analyze this degree of coincidence in detail, the two characteristic sections that pass through the central key are analyzed. In the case of the longitudinal section, deviations of less than 0.018 m can be seen in Fig. 7.15.

Regarding the transversal direction of the vault, the maximum deviations of the transversal section that by the central key can be quantified as less than 0.015 m (Fig. 7.16).

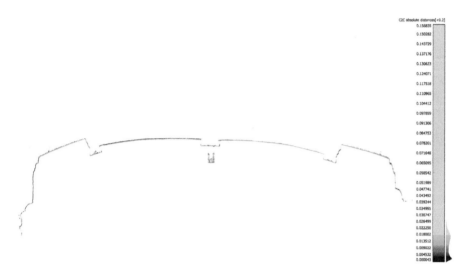

Figure 7.15 Longitudinal section of the vault through the central key. Maximum deviation of 0.018 m of the overlap between the two clouds.

Figure 7.16 Cross-section of the vault through the central key. Maximum deviation of 0.015 m of the overlap between the two clouds.

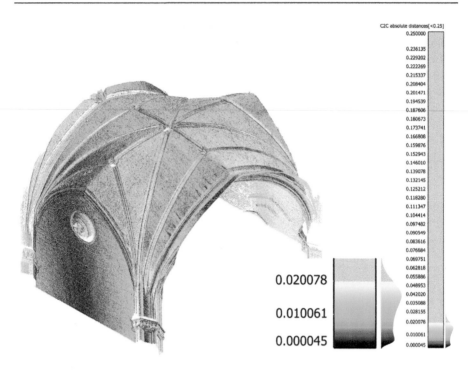

Figure 7.17 Image of the photogrammetric point cloud and representation of the distances with respect to the reference cloud of the laser scan.

After verifying the overlap between both clouds, the result offered by the comparison for the entire surface of the vaulted section allows us to determine that the distances between the points of the two clouds vary between 0.000 and 0.020 m, Fig. 7.17.

7.4 Conclusion

The objective of this study was to verify the automated photogrammetry technique or SfM image processing as a reliable non-destructive technique (NDT) for precise and cost-effective geometric capture for the characterization of built heritage. For this, the research has focused on the application of said technique in the case study of the church of San Vicente in the city of Donostia-San Sebastián, specifically in one of its Gothic vaults. The photogrammetric technique has been complemented with survey methods based on more traditional techniques such as the TS and newer techniques such as the Terrestrial Laser Scanning (TLS) that have allowed the scientific methodology of the investigation to be formed.

Through the capture of images by means of a DSLR camera and the 3DF Zephyr software, it has been possible to generate the 3D model of the case study.

The first results have shown a point cloud model that, although it approximately maintains the geometry and proportions of the original model of the vault, does not save the quantitative distance values and therefore does not allow the generation of a model that can be used in precision work. For this reason, specific points were captured through the Leica TCR407 TS to facilitate the determination of reference distances, serving as a control point system to provide the model with the necessary precision. In this way, after the implementation of these values in the photogrammetric software, it has been possible to perform the bundle adjustment of the original point cloud model, thus obtaining a higher fidelity point cloud. The results of this model have offered an average error of $0.128\% \pm 0.132$ with respect to the reference values.

On the other hand, although it is true that through the GCPs and distances, the degree of precision of the point cloud model can be analyzed, this method only allows the comparison with respect to said specific references. That is why throughout the chapter a more specific comparison of the point cloud model obtained by photogrammetry with respect to the point cloud obtained by the TLS Leica RTC360 has been developed. This process has been carried out using the open-source CloudCompare software. The two characteristic sections of the vault have been analyzed, obtaining maximum deviation values of 0.018 m for the longitudinal section and 0.015 m for the cross-section, while for the entire surface of the vault under study, the deviation is less than 0.02 m.

Therefore it could be indicated that the geometric models obtained through the SfM technique have adequate precision for use in the field of architectural survey applied to the architecture of built heritage. However, it should be noted that obtaining a model with such characteristics requires processing using GCP and comparing the values obtained with respect to said GCP or reference distances. On the other hand, future work can focus on the comparison of the results obtained through photogrammetry with photographs that incorporate GPS coordinates to facilitate the generation process of scaled point cloud models.

Acknowledgments

The authors thank 3DFLOW especially Federica Murana for the help provided through the contribution of the 3DF Zephyr Education license during the 2022/2023 academic year. This research has been carried out within the framework of the research activities of the TICBE Research Group recognized as UPV/EHU Research Group with code GIU22/001 in the call entitled "Convocatoria de ayudas a grupos de investigación de la UPV/EHU (2022)"

References

Barbosa, F., Woetzel, J., & Mischke. (2017). *Reinventing construction: A route of higher productivity*. McKinsey Global Institute.

Baïri, A., Martín-Garín, A., Alilat, N., Roseiro, L., & Millán-García, J. A. (2021). Quantification of free convection in a quarter-spherical innovative Trombe wall design. *Journal of Building Engineering*, *42*, 102443. Available from https://doi.org/10.1016/j. jobe.2021.102443.

Dai, F., Feng, Y., & Hough, R. (2014). Photogrammetric error sources and impacts on modeling and surveying in construction engineering applications. *Visualization in Engineering*, *2*(1). Available from https://doi.org/10.1186/2213-7459-2-2.

de Azcárate Ristori, J. M. (1990). *Arte gótico en España. Manuales Arte Cátedra*. El País D.L. Madrid (Spain), Madrid (Spain).

de Azcárate Ristori, J. M. (1985). *Aspectos distintivos de la arquitectura gótica española*. Cuadernos de Historia del Arte. Universidad de Extremadura, Servicio de Publicaciones, Cáceres (Spain).

de Azcárate Ristori, J. M. (1958). *La arquitectura gótica toledana del siglo XV. Artes y Artistas*. Instituto Diego Velázquez, del Consejo Superior de Investigaciones Científicas., Madrid.

Etxepare, L., Leon, I., Sagarna, M., Lizundia, I., & Uranga, E. J. (2020). Advanced intervention protocol in the energy rehabilitation of heritage buildings: A Miñones Barracks Case Study. *Sustainability*, *12*(15), 6270. Available from https://doi.org/10.3390/su12156270.

Freimuth, H., & König, M. (2018). Planning and executing construction inspections with unmanned aerial vehicles. *Automation in Construction*, *96*, 540−553. Available from https://doi.org/10.1016/j.autcon.2018.10.016.

Goitia, F. C. (2000). *Historia de la arquitectura occidental. Edad Media Cristiana En España*. CIE Inversiones Editoriales DOSSAT-2000. Madrid (Spain), Madrid (Spain).

Grussenmeyer, P., Alby, E., Landes, T., Koehl, M., Guillemin, S., Hullo, J. F., Assali, P., & Smigiel, E. (2010). Recording approach of heritage sites based on merging point clouds from high resolution photogrammetry and terrestrial laser scanning. *The International Archives of the Photogrammetry, Remote Sensing and Spatial Information Sciences*, *39*. Available from https://doi.org/10.5194/isprsarchives-XXXIX-B5-553-2012.

Jiang, W., Zhou, Y., Ding, L., Zhou, C., & Ning, X. (2020). UAV-based 3D reconstruction for hoist site mapping and layout planning in petrochemical construction. *Automation in Construction*, *113*, 103137. Available from https://doi.org/10.1016/j.autcon.2020.103137.

Lasarte, N., Elguezabal, P., Sagarna, M., Leon, I., & Otaduy, J. P. (2021). Challenges for digitalisation in building renovation to enhance the efficiency of the process: A Spanish Case Study. *Sustainability*, *13*(21), 12139. Available from https://doi.org/10.3390/su132112139.

Leon, I., Pérez, J. J., & Senderos, M. (2020). Advanced techniques for fast and accurate heritage digitisation in multiple case studies. *Sustainability*, *12*(15), 6068. Available from https://doi.org/10.3390/su12156068.

Lerma García, J. L., Van Genechten, B., Heine, E., Santana Quintero, M. (2008). *Theory and practice on terrestrial laser scanning: Training material based on practical applications*. Universidad Politecnica de Valencia Editorial. Valencia (Spain), Valencia (Spain). https://lirias.kuleuven.be/1773517?limo = 0.

Liu, H., Al-Hussein, M., & Lu, M. (2015). BIM-based integrated approach for detailed construction scheduling under resource constraints. *Automation in Construction*, *53*, 29−43. Available from https://doi.org/10.1016/j.autcon.2015.03.008, https://www.journals.elsevier.com/automation-in-construction.

Martín-Garín, A., Millán-García, J. A., Hernández-Minguillón, R. J., Prieto, M. M., Alilat, N., & Baïri, A. (2022). *Open-source framework based on LoRaWAN IoT Technology for building monitoring and its integration into BIM models. Handbook of Smart Materials, Technologies, and Devices: Applications of Industry 4.0* (Volume 1—3). Springer. Available from https://doi.org/10.1007/978-3-030-84205-5_9, https://doi.org/10.1007/978-3-030-84205-5.

Martínez, J. G. (1998). *El gótico español de la edad moderna: bóvedas de crucería Arte y Arqueología.* Secretariado de Publicaciones e Intercambio Científico, Universidad de Valladolid. Valladolid (Spain), Valladolid (Spain).

Palacios Gonzalo, J. C. (2009). *La canteria medieval: La construccion de la boveda gotica española: Ediciones Munilla-Lería.* Madrid (Spain), Madrid (Spain).

Ruiz, B. A. (2003). *Arquitectura tardogótica en Castilla: Los Rasines.* Editorial Universidad de Cantabria. Santander (Spain), Santander (Spain).

Sacks, R., Eastman, C., Lee, G., & Teicholz, P. (2018). *BIM Handbook, a Guide to Building Information Modeling for Owners, Designers, Engineers, Contractors, and Facility Managers.* John Wiley & Sons. Available from https://doi.org/10.1002/9781119287568.

Tarifa Castilla, M. J. (2018). La iglesia de San Vicente en San Sebastián: los contratos, trazas y artífices del proyecto edilicio (1507-1548). *Locus Amoenus, 16,* 71—92. Available from https://doi.org/10.5565/rev/locus.316.

Xiong, X., Adan, A., Akinci, B., & Huber, D. (2013). Automatic creation of semantically rich 3D building models from laser scanner data. *Automation in Construction, 31,* 325—337. Available from https://doi.org/10.1016/j.autcon.2012.10.006.

Intelligent recording of cultural heritage: From point clouds to semantic enriched models

Diego González-Aguilera[1], Mario Soilán[2], Alberto Morcillo[1],
Susana del Pozo[1], Lloyd A. Courtenay[3], Pablo Rodríguez-Gonzálvez[4],
and David Hernández-López[5]

[1]Department of Cartographic and Land Engineering, Universidad de Salamanca, Ávila,
Spain, [2]CINTECX, GeoTECH Research Group, Universidade de Vigo, Vigo, Spain,
[3]CNRS, PACEA UMR 5199, Université de Bordeaux, Bât B2, Allée Geoffroy Saint Hilaire,
CS50023, 7 Pessac, France, [4]Department of Mining Technology, Topography and
Structures, Universidad de León, Av. Astorga s/n, Ponferrada, Spain, [5]Institute for
Regional Development (IDR), University of Castilla La Mancha, Albacete, Spain

8.1 Overview

8.1.1 Motivation

The documentation of cultural heritage is a key aspect in the tasks associated not only with the activities of preservation and preventive conservation carried out by its stakeholders but also for its valorization, so that it can reach any citizen and thus learn about its associated history and traditions (Xiao et al., 2018). We must not forget the responsibility of owners and managers of heritage buildings and assets, so they can facilitate the accessibility to tangible and intangible information.

To advance in this cultural heritage value chain, geotechnologies play a key role in the process of recording the current state of a heritage element (Croce et al., 2019), but they must be understood as a means and not as an end in themselves. The mere application of, for example, a terrestrial laser scanner (TLS) to obtain a 3D point cloud barely generates added value in the value chain of the cultural heritage element by itself. That is why it is necessary to have a clear idea of the tasks in which the aforementioned digitization will be applied to identify the final products or tasks demanded. As an example, and without seeking to be an exhaustive list, the following points can represent real demands in the cultural heritage field:

- **Heritage Building Information Modeling (HBIM)**, which encompass 3D virtual information models of both movable and immovable architectural cultural heritage elements, allowing the performance of simulations (e.g., mechanical, structural (Alfio et al., 2022), and energetic assessment (López et al., 2018), analyses of the cultural heritage through time (Rodríguez-Gonzálvez et al., 2019), or even aspects as sustainability or maintenance (Lee et al., 2019)).

Diagnosis of Heritage Buildings by Non-Destructive Techniques. DOI: https://doi.org/10.1016/B978-0-443-16001-1.00008-5

- **Digital twins (DTs)** provide a qualitative leap with respect to HBIM by integrating together with the 3D model the information coming from the Internet of Things (IoT) devices that monitor different aspects related to the state of the cultural heritage elements (Minerva et al., 2020). Furthermore, DTs integrate some cognitive capabilities based on the codification of cultural heritage rules using artificial intelligence (AI). As a result, a virtual replica "avatar" can be achieved for working or simulating collaboratively in a remote and multidisciplinary way, allowing stakeholders to carry out predictive and prescriptive conservation tasks more efficiently and safely.
- **Augmented and mixed reality (AR/MR)**, in which the social role of cultural heritage is emphasized through which citizens can establish links with their past and traditions. In this regard, dissemination and awareness can be done through virtual museums in their different modalities (Margetis et al., 2021), highlighting AR and MR that allow hybridizing physical artifacts with digital content. While in AR there is a superimposition of digital materials on a real environment, in the MR paradigm, a new environment is generated in which physical and digital objects can interact in real time.

All the abovementioned demands have, as a common base, the availability of 3D point clouds that provide detailed, accurate, and comprehensive representations of cultural heritage elements, allowing researchers and professionals to better understand and appreciate their history and significance. However, the size and complexity of point cloud data can make it difficult to effectively analyze and interpret the underlying geometry. To extract useful information from point clouds, segmentation and classification techniques must be applied to extract semantics. These techniques, which are based on geometry and/or radiometry can help to organize, categorize, and visualize point cloud data in meaningful ways. Semantics is the conceptual model that makes it possible to display information associated with data to be able to interpret them without the need to know their true representation. The semantic dimension makes it possible to embed cultural heritage information beyond geometry and radiometry in the model. However, this operation is not trivial, since it requires the definition of an ontology, that is, a formal representation of knowledge, and the use of standardized terminology with clearly defined interrelationships (Kauppinen et al., 2008).

That is the reason why this chapter focuses on intelligent digitization, and how to provide added value to the 3D point clouds coming from the different geotechnologies currently available.

8.1.2 Geometric and radiometric approaches for cultural heritage classification: literature review

A scientometric survey carried out by Yang et al. (2022) shows a structured overview of the state of the art in the use of point clouds for cultural heritage, observing a clearly growing trend of interest in this field of research. One of the main reasons that can be related to this trend is the fact that the release of more affordable lightweight devices (e.g., handheld Light Detection and Ranging [LiDAR], smartphone LiDAR, and ultralight Unmanned Aerial Vehicle [UAV]) makes 3D data collection of cultural heritage more popular and available to more nonprofessional users. This work also studies the international collaboration on the topic, with relevant connections between large countries with strong research bodies, such as China or United States, and smaller countries with a rich cultural heritage, such as Greece, Italy, or Spain.

The differences that can be found in the treatment of three-dimensional information throughout the body of literature lie, on the one hand, in the acquisition of the data and on the other hand, in their processing and application.

While data acquisition is out of the scope of this chapter, it is of key and obvious importance for the subsequent processing of the data. 3D data can be obtained from laser scanners or photogrammetry, and sensors can be mounted on drones, ground vehicles, or static ground stations (TLS). Different sensors and acquisition equipment will produce different topologies of 3D point clouds, which will have different processing requirements. There are, however, processes that are common to all of them, such as data registration (Ghorbani et al., 2022), noise filtering (Zheng et al., 2022), or preliminary surface reconstruction (Gaiani et al., 2016).

Among the different types of processes that can be applied to a 3D point cloud to generate automated workflows, segmentation and classification play a fundamental role in the field of cultural heritage. The two terms are often confused in the literature, as they differ only in certain nuances. Point cloud segmentation is defined by Spina et al. (2011) as a way to process and organize a point cloud into meaningful subsets. Similarly, Grilli et al. (2017) highlight the importance of point cloud segmentation as a key step to identify homogeneous areas and define it as the process of grouping point clouds into multiple homogeneous regions with similar properties. In addition, they define the conceptual difference with classification, stating that classification is the step which labels those homogeneous regions grouped with a segmentation process.

Within the field of point cloud segmentation and classification, there are several popular geometry-based algorithms that are widely used among different fields, including cultural heritage (Grilli et al., 2017):

1. Edge-based segmentation: It detects the borders of different regions on a first step to group points inside their boundaries to deliver the final segments (Castillo et al., 2013).
2. Region growing (RG) segmentation: It starts from one or more points (seeds) for each segment and grows according to specific features and thresholds (Vo et al., 2015).
3. Model fitting: Primitive shapes (planes, cylinders) are fitted onto point cloud data. Points that fit to those shapes are considered as a single segment. The most popular method is Random Sample Consensus (RANSAC), although in architecture, it is usually complex to apply those methods as for such complex shapes, local descriptors may provide a better solution (Poux et al., 2016).

In fact, the calculation of geometric descriptors has shown great applicability in the field of cultural heritage. These descriptors aim to represent the features of each point with respect to its neighborhood. Their calculation allows, subsequently, to categorize sets of points based on their geometric arrangement according to the application being carried out. Capolupo (2021) investigated the accuracy of the most common geometric descriptors that can define the 3D scene structure, which are based on the eigenvalues of the 3D covariance matrix, for both aerial and terrestrial point clouds of the All Saints' Monastery of Cuti (Italy). Different mathematical relations between those eigenvalues (λ_1, λ_2, λ_3) define geometrical features such as linearity, anisotropy, sphericity, planarity, omnivariance, or curvature.

The use of geometric descriptors in cultural heritage point clouds has relevant applications, such as crack detection. Wood and Mohammadi (2021) use geometric features based on the covariance, normal, and curvatures of a voxelized point cloud to detect damaged and cracked areas in the walls of Sala degli Elementi in the Palazzo Vechio (Italy). The variation of these geometric features allows a damage evaluation process with an unsupervised learning classifier. Normal vectors, which can be obtained as well from the covariance matrix, applying principal component analysis (PCA) are a geometric feature often used for damage detection. Kim et al. (2015) computed the variation of normal vectors in respect of a reference vector to detect spalling damage of a flat surface. This strategy was enhanced with information about radiometry (color and intensity) in the work of Guldur Erkal and Hajjar (2017) to detect and quantify surface damage.

Initially, the classification of geometries was carried out based on the definition of threshold values for the different characteristics that were calculated. The use of machine learning (ML) techniques, as it is detailed in Section 8.1.3.1, improved this approach, either with supervised techniques where a classification model is trained from labeled data, or with unsupervised techniques, where the objective is to determine how the data is organized.

It is also relevant to mention that the intensity attribute of a point cloud can be relevant for different applications in cultural heritage. While the geometric properties of a point cloud are useful for shape classification or detection of cracks or deformations, the radiometric properties (i.e., color or intensity can also be included) can help to distinguish different materials, or nongeometric damage to them. It is relevant to note that the intensity attribute will only be present in point clouds acquired with LiDAR systems, as it is related to the interaction of the laser beams with the surface. In a photogrammetric point cloud, its radiometric quality will be usually given by the color, which will not be present in the case of LiDAR.

In the literature, the terms "reflectance intensity" or "pulse reflectance" are often used as a synonym for the return of one echo of a laser beam. However, value of the intensity attribute is not proportional to the surface reflectance. It is usually provided as a digital value of 8, 12, or 16 bits of resolution, and the internal algorithms for its calculation and quantization are normally not well specified by the manufacturers (Höfle & Pfeifer, 2007). That is why early research on the topic focused on intensity correction and radiometric calibration, to move from a digital, not well-specified intensity value, to one that is, at least, proportional to surface reflectance. Examples of these calibrations come from the data- and model-driven correction methods by Höfle and Pfeifer (2007) and the radiometric analysis and calibration in the work of Stilla and Jutzi (2009) and Wagner (2010). These works were developed using full-waveform airborne LiDAR, which has been used for different cultural heritage—related applications. In the study of Lasaponara and Masini (2009), the archaeological area of Monte Irsi (Italy) is surveyed using this technology, as the vegetation that covers it makes difficult its aerial analysis with other technologies. In the research work Lasaponara et al. (2011), a data processing chain is presented that is able to detect archaeological features in sparsely and densely vegetated areas.

Finally, works that combine geometric and radiometric information for the processing and classification of heritage elements can be highlighted. Alkadri et al. (2022) developed a method to detect fracture distribution and vulnerable surfaces on cultural heritage buildings' computing geometric and radiometric properties from TLS point clouds. For masonry walls, Riveiro et al. (2016) developed a point cloud segmentation algorithm based on a 2.5D approach, which created images based on the intensity attribute to isolate each block of the wall with a morphology analysis.

In summary, this section provides a state-of-the-art overview of methods for different 3D point cloud classification applications in the field of cultural heritage, with an emphasis on geometric and also radiometric techniques. The potential of new methods such as deep learning (DL), which has already taken off in other fields, is expected to be revolutionary also for cultural heritage.

8.1.3 Artificial intelligence approaches for cultural heritage classification: literature review

The turn of the century provided a significant development in point cloud classification research, especially from the perspective of supervised approaches commonly used in AI. From this perspective, the field of computer vision has been seen to have undergone a great evolution, especially since the first attempts to perform supervised point cloud classification. In the interest of clarification, the term *supervised* here refers to the process of training an algorithm to learn on data while explicitly defining the objectives of the task at hand. In the context of point clouds, this implies that algorithms are taught to assign labels to points or sets of points in a cloud, given previous experience and examples of what these elements should represent.

Adaptation of supervised approaches to point clouds, however, presents a series of challenges and issues which complicates the means in which we can train algorithms to identify features. From one perspective, most computational learning approaches require input to present a regular structure, a feature that point clouds inherently lack. For example, the position of a single point, described by its x, y, and z coordinates, does not contain information about topological structure, while remaining uninformative regarding its orientation, or its relationship with neighboring points both on a local and global scale. Similarly, the same point can be translated, thus presenting different coordinate values, yet still have the same label. Moreover, point clouds are continuously distributed in space, and thus any permutation in their ordering will provide different spatial information per point, yet still represents the same geometric entity as a whole. Similar issues can be found in the quality and completeness of the 3D model. Point clouds presenting holes, noise, and inconsistent point densities are problematic and likely to cause issues in the learning process. Finally, for algorithms to work efficiently, they must overcome complications related to invariance across classes; the identification of doors in a building, for example, should not be dependent on a specific building, yet should be generalizable across all examples of buildings.

From this perspective, most supervised approaches fall short when used to assign labels to single points when provided with their coordinate values alone. This has led to a great deal of research constructing more geometrically informative variables to describe each point, or the way coordinate values are introduced to algorithms as variables.

The present summary of research into the best means of supervising an algorithm to perform point cloud classification will be divided into a number of different, yet closely related, categories; (1) ML approaches, (2) DL approaches, and (3) graph learning (GL) approaches.

8.1.3.1   Machine Learning

In terms of ML, a number of very popular algorithms can be found applied to point cloud classification tasks. These include, but are not exclusive to, support vector machines (Golovinskiy et al., 2009; Li et al., 2016; Zhang et al., 2013), random forest (RF) (Chehata et al., 2009), k-nearest neighbor algorithms (Golovinskiy et al., 2009), as well as boosting algorithms such as Adaboost (Lodha et al., 2007), and Extreme Gradient Boosting (Carbonell-Rivera et al., 2022). In each of these cases, coordinate values are not used as input. Instead, algorithms are trained on computed geometric variables, such as; height-based, echo-based, and intensity features for LiDAR data; eigenvalues and principal components; the calculation of local planes; variables describing the density of points; color and radiometric features; as well as an array of different algorithms for the calculation of shape descriptors (Frome et al., 2004; Lamdan & Wolfson, 1988; Matei et al., 2006).

In each of these cases, authors present promising results of the training of ML algorithms on point cloud features. Chehata et al. (2009) show how RF can reach error rates as low as 6% and 7% when segmenting LiDAR point clouds of urban areas, with the highest confusion rates found when separating buildings from roads and pavements. Lodha et al. (2007), on the other hand, report error rates of under 5% and 8%, depending on the number of labels used for the classification of aerial LiDAR scans of roads, grass, buildings, and trees.

8.1.3.2   Deep learning

From the perspective of DL approaches, authors have been seen, over the years, to try and leverage the power of neural networks (NN) for the processing of point clouds for a number of different purposes (classification or otherwise). These efforts are primarily based on the success of convolutional neural networks (CNNs) when applied to computer vision tasks in image processing. CNNs, when applied to images, work by convolving a series of filters over the pixel values of the image, extracting both high- and low-level features that can be used to define the class of the object. In the case of 3D data, however, multiple challenges exist to define a generalizable version of the convolution operator, especially one that is invariant to translation and rotation, as well as the aforementioned lack of structure across the points of each cloud.

One approach to applying CNN to 3D models is the *extrinsic* approach, whereby authors attempt to analyze 3D surfaces by either projecting them as 2D images (a multiview approach, sensu (Su et al., 2015), inter alia), and then passing a typical CNN over the resultant images, or by convolving over models in the form of a voxel grid (a Volumetric approach, e.g., (Wu et al., 2015), inter alia). In the majority of these cases, point clouds are first required to be converted to meshes, prior to supervised learning and classification. One of the issues of these approaches, however, are the inability of these techniques to capture fine details, creases, and seams in the model, while the voxel-based approach is highly conditioned by the size of the voxels used. Likewise, these approaches are not invariant to shape deformations, which is a fundamental requirement in many types of 3D classification tasks (Boscaini et al., 2016; Bronstein et al., 2017).

From a different perspective, Guo et al. (2015) calculate the descriptive features of triangulated points, used as input to a CNN, which produce relatively more accurate results. Similarly, Liu et al. (2019) use a complex combination of attention-based mechanisms and autoencoder feature aggregation units to obtain high performance in point cloud classification tasks.

One of the major advances in this field comes from the use of *intrinsic* approaches that try to model 3D surfaces in terms of a non-Euclidean manifold (Bronstein et al., 2017; Masci et al., 2015). These are especially useful when the elements to classify have smoother or unclearly defined limits between the classified parts. These approaches compute highly complex functions to describe a 3D surface, before convolving over these descriptor signatures to learn the features of the model. Maron et al. (2017), for example, represent 3D surfaces by using a global seamless parametrization of a planar flat torus. Once flattened, a CNN algorithm is passed over the 2D representation of the 3D model and can thus be used to perform a number of supervised tasks. In other applications, analysts use combinations and variants of heat diffusion equations, spectral descriptors, and wave kernel signatures, to describe local and progressively global features of geometry ((Boscaini et al., 2015, 2016; Bronstein & Kokkinos, 2010; Bronstein et al., 2017; Masci et al., 2015) inter alia). The passing of these descriptors over the surface of a 3D model mimics the convolution operator as it extracts local features that can then be used as input to an NN. From this approach, this adapted version of CNN can freely move across the surface of a 3D model and provide geometric information at a finer scale than the extrinsic approaches previously defined. As opposed to the ML approaches described above, intrinsic CNN are also to capture local information, without having to consider each point indiscriminately.

This approach produces increasingly more accurate results for the classification of 3D models and, however, can often be considered computationally expensive.

8.1.3.3 Graph learning

Despite the advances made in ML and DL applications to point cloud classification, one of the most important advances in the field of supervised approaches is the integration of *graph theory* into the algorithms used for these particular tasks, fueled by the introduction of GL in 2009 (Scarselli et al., 2009). GL builds strongly from the notions proposed in the aforementioned applications on manifolds, yet it simplifies the kernel equations significantly. The notion behind GL considers points in a cloud as a function of their own position in the context of their "neighborhood." Similar to how ML applications have been applied to graphs such as in social, ecological, and citation network analyses (Scarselli et al., 2009), GL considers the entire point cloud as a graph; with points as vertices which are connected to their nearest neighbors by edges. From this perspective, the ML or DL algorithms applied can consider neighborhoods of points, as opposed to acting on each point individually.

In the context of hybrid ML and GL approaches, a number of applications can be found using notions from graph theory to describe the structural relationship of points, introduced alongside feature descriptors as input to supervised algorithms (Alkadri et al., 2022; Landrieu et al., 2017; Niemeyer et al., 2014; Vosselman et al., 2017). Nevertheless, some of the most important advances in this field are found when introducing a convolutional component to GL (known as graph convolutional networks [GCN]), which in many cases produce the best results (Bronstein et al., 2017).

A number of different methods exist to convolve an NN across a point cloud in this manner. The main idea behind this approach is to use a message passing mechanism for neighborhood aggregation (Kipf & Welling, 2017). This mechanism begins by taking a single point, before aggregating the features of neighboring points (Kipf & Welling, 2017) and moving (convolving) on to the next point in the cloud and repeating the operation. From this perspective, points are embedded as a function of themselves, and the position of their neighbors, providing a more general overview of their context in the point cloud. Each point's features can simply be defined as the coordinate values of the point, however, when combined with other hand-computed geometric features, these convolution operators can be considered a powerful insight into the local structure of the cloud.

GCNs overcome many of the issues related with point cloud classification through the presence of multiple learnable parameters included within the message passing mechanism. These parameters are fundamental in ensuring the network is invariant to the translation and rotation of a point. Moreover, the convolutional operation itself ensures that the algorithm is order invariant, without requiring data to be structured as it would be for a CNN.

Table 8.1 summarizes some of the architectures and algorithms used for point cloud classification, alongside the performance they obtained for the classification of object parts in the ShapeNetPart dataset (Yi et al., 2016).

8.1.4   Existing benchmarks and tools

To evaluate the different existing approaches in the field of semantic extractions from 3D point clouds, it is necessary to have benchmarks or datasets with a known "ground truth" to evaluate the performance of the different algorithms developed. It should not be forgotten that the variety of cultural heritage assets is an additional problem due to the complexity and diversity of each cultural element. The absence of benchmarks that account for this vast amplitude hinders the possibility of

Table 8.1 Summary of the mean intersection-over-union (mIoU) performance of different algorithms when trained for the supervised classification of object parts in the ShapeNetPart dataset.

Algorithm name	Authors (Year)	mIoU
PointNet	Qi et al. (2017a)	83.7
PCNN	Wang et al. (2018)	85.1
PointNet++	Qi et al. (2017b)	85.1
DGCNN	Wang and Lu (2019)	85.1
SpiderCNN	Xu et al. (2018)	85.3
SPLATNet	Su et al. (2018)	85.4
PointConv	Wu et al. (2020)	85.7
SGPN	Wang et al. (2019)	85.8
PointCNN	Li et al. (2018)	86.1
InterpCNN	Mao et al. (2019)	86.3
PointTransformer	Zhao et al. (2020)	86.6

Note: Mean IoU: It takes into account the IoUs of all the classes and calculates their mean. The version of the network that gets the highest validation mIoU is saved as the best version.

developing automated solutions to provide semantics to cultural heritage elements. Furthermore, the process of generating these datasets or benchmarks requires the manual classification of a considerable number of entities that must be used in the development, training, testing, and assessment of ML and DL methods and algorithms (Matrone et al., 2020).

There are numerous semantic segmentation benchmarks for 3D point clouds in the scientific literature which include a range of scene types and item classifications (Guo et al., 2021), compromising several sensor types and indoor and outdoor scenarios (De Deuge et al., 2013; Hackel et al., 2017) and also focused on specific issues such as the evaluation of 3D point cloud alignment algorithms (Korres et al., 2022), pattern detection (Lengauer et al., 2021), or the reconstruction algorithms themselves (Dulecha et al., 2020).

Currently, the number of benchmarks that addressed the semantic problem in cultural heritage is very limited, such as the one resulting from the work of Matrone et al. (2020), called ArCH (Architectural Cultural Heritage), which includes datasets and classification results for better comparisons and insights into the advantages and disadvantages of various machine and DL approaches for heritage point cloud semantic segmentation.

Also worth mentioning is the project Inception (Inclusive cultural heritage in Europe through 3D Semantic Modelling) (https://www.inception-project.eu), funded by the European Commission within the Program Horizon 2020, which addresses the data collection processes and the development of semantically enriched 3D models to achieve interoperable models able to enrich the interdisciplinary knowledge of cultural identity (Maietti et al., 2018). This approach follows the creation of a formal representation of knowledge as a hierarchy of concepts within the cultural heritage domain (namely, an ontology) and the application of the Semantic Web technology and Linked Open Data principles.

On the other hand, the tools and algorithms available to the scientific community are also another important pillar, since there are a lot of end-users who take these semantic classifications as input data. The available software or tools can be grouped according to the taxonomy proposed by (Xie et al., 2019), as follows: (1) **multiview-based approaches, founded on the creation of a set of images from 3D point clouds, on which CNNs are applied, such as SnapNet3D** (Boulch et al., 2018), Spnet (Yavartanoo et al., 2018), or View-gcn (Wei et al., 2020); (2) **Voxel-based approaches, which perform the conversion of a 3D point cloud into an ordered grid by the application of a voxelization algorithm, having this bidimensional representation as an attribute the third dimension required to apply a CNNs, being examples of this approach Voxnet** (Maturana & Scherer, 2015), Pointrcnn (Shi et al., 2019) or VoxSegNet (Wang & Lu, 2019); and (3) **Point-based, where the classification and semantic segmentation is performed by applying feature-based approaches, such as Deeppoint3D** (Srivastava & Lall, 2019), modified DGCNN (Pierdicca et al., 2020), or PointNet (Qi et al., 2017a).

However, all of them still require the critical element mentioned at the beginning of the section, a wide set of benchmarks adapted to the specific case of cultural heritage, to be able to fine-tune them to achieve an automatic and effective semantic segmentation.

8.1.5 Aim and contribution of this chapter

Deciding about the best semantic classification strategy in cultural heritage is challenging due to the complexity and variability of buildings and assets. After carefully analyzing the state of the art and especially the results obtained in the last papers focused on the semantic classification of cultural heritage buildings using DL strategies, we decided to propose an ML approach based on RF. The reason is twofold: on the one hand, we try to generalize RF over cultural heritage buildings with similar features, and on the other hand, the RF approach developed has been included in Cloud Compare, so it can be used within one of the most popular open-source point cloud processing tools in the scientific community.

In summary, the aims and contributions of this chapter are as follows:

- analyzing in depth those geometric and radiometric features that provide a better separability and thus a more efficient classification of cultural heritage buildings,
- filtering the point cloud using a novel adaptation of the anisotropic filtering,
- applying RF classifier over different cultural heritage buildings, and
- integrating the in-house tool developed within Cloud Compare open-source platform, so it can be offered to the scientific community.

8.2 Methodology

This section describes the different methods developed and implemented within the in-house tool which exploit geometric and radiometric properties of point clouds using AI. Last but not least, before applying these classification approaches, the point cloud filtering is a crucial step since it removes noise and outliers from the point cloud data. To this end, authors present in this chapter a particular adaptation of anisotropic filtering method developed by Perona and Malik (1990) since it has provided great results in denoising point clouds without causing loss of significant data.

8.2.1 Point cloud filtering

Starting with a raw 3D point cloud, one common technique for the first step of the process is the point cloud filtering, which involves removing noise and outliers from the point cloud to improve its quality. This can be accomplished through a variety of methods, such as spatial or temporal filtering. This section is focused on a particular adaptation of anisotropic filtering method developed by the authors since it has provided great results in denoising point clouds without causing loss of significant data.

This anisotropic denoising technique seeks to reduce the noise of 3D point clouds without losing relevant information. It is based on an iterative process of local spatial diffusion considering the point cloud density. Specifically, a directional multiscale Laplacian mask/convolution is applied with which intraregional smoothing is prioritized over interregional smoothing. Ultimately, and far from the expected result of a common blur technique, the edges of interest are sharpened without sacrificing the resolution of the final 3D model. This filter considers the divergence ($\nabla \cdot$), gradient (∇), and Laplacian (Δ) operators throughout (Eq. 8.1).

$$\frac{\partial \varphi}{\partial t} = \nabla \cdot \left[c\left(\|\nabla \varphi\| \nabla \varphi \right)\right] = \nabla \cdot \nabla \varphi + c\left(\|\nabla \varphi\|\right)\Delta \varphi \tag{8.1}$$

Since the Laplacian function manages to analyze the gradient of the density function, its effects could be explained by the isotropic heat diffusion equation if c is considered constant and the boundary limits are considered adiabatic. In this case, the function would be equivalent to applying a Gaussian blur. Under this assumption, information of interest and points on edges can be lost; however, if the diffusion coefficient is contemplated, edge detection occurs as a function of the gradient and the parameter k as shown in Eq. 8.2.

$$c\left(\|\nabla \varphi\|\right) = e^{-\left(\frac{\|\nabla \varphi\|}{k}\right)^2} \tag{8.2}$$

This procedure can be better explained within the 2D domain. In this case, the gradient could be described as the vector that indicates the color variation between pixels of an image. Thus, in border areas, where there is a greater color variation between regions than within regions, the gradient increases, so the function c approaches zero. If the second part of the equality (in Eq. 8.1) is forced to be zero, it would remain that $\partial \varphi/\partial t$ is also zero. This means that the image does not change at these areas. For this reason, the edges remain unmodified. In any other region than in border areas, the gradient between regions decreases and therefore the function c does not tend to zero. For these areas, the derivative of φ with respect to time is no longer zero, therefore the image does undergo a change. The variation is given by the second equality of the anisotropic diffusion equation, producing a blur in these pixels.

If instead of pixels, we think of a point cloud within the 2D domain (Fig. 8.1A), a grid could be created, of a specific resolution (on who's the results will depend), with which to be able to count the number of points contained by each cell of the grid (Fig. 8.1B).

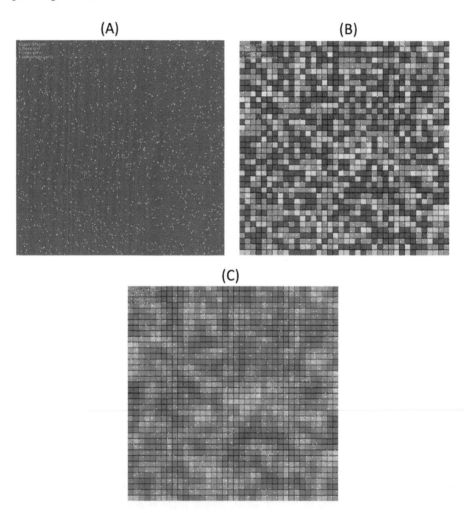

Figure 8.1 Point cloud filtering: (A) randomly 2D point cloud, (B) grid with color ramp depending on the level of intensity (density of point cloud of each cell), and (C) grid and intensity result after applying the anisotropic diffusion.

Next, if an intensity level is assigned to each 2D cell proportionally to the number of points contained in it, the case is reduced to the above, the analysis of the gradient between pixels with a certain color or intensity (Fig. 8.1B).

Thus, by applying the anisotropic diffusion equation, the smoothed result (Fig. 8.1C) would be achieved.

Since noise is in those cells with lowest point cloud density (redder colors in Fig. 8.1C) and the objective is to move them to the areas with the highest point cloud, a gradient ascent process is applied, being the gradient of each grid calculated with Eq. 8.3.

$$\nabla \varphi = \left(\frac{\partial \varphi}{\partial x}, \frac{\partial \varphi}{\partial y} \right) \tag{8.3}$$

Thus those points located in cells with greater gradient are moved toward those cells with lower gradient according to Eq. 8.4.

$$v_{n+1} = v_n + \gamma \nabla \varphi(v_n) \tag{8.4}$$

Being γ the "step size" or number of cells that each point must move in the positive gradient direction. It is about performing this process iteratively to eliminate the noise from the point cloud and that the edges of the object of interest are emphasized (Fig. 8.2).

The process described and illustrated for the 2D domain is similar for the 3D domain, where the 3D point cloud would be voxelized into cubic units (instead of grid cells) and the equations seen would be parameterized for R^3. It must be considered that the computational cost in this case is higher, and it is advised to use

Figure 8.2 2D point cloud after applying the anisotropic filter.

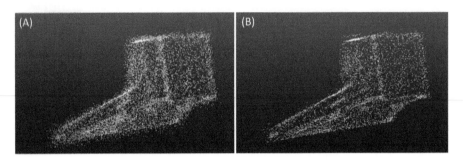

Figure 8.3 3D point cloud before (A) and after (B) applying the anisotropic filter.

several execution threads or even make use of the graphical processing unit. Fig. 8.3 shows the result of applying the anisotropic filter to a 3D point cloud.

8.2.2   Segmentation pipeline

8.2.2.1   Introduction

In this section, a point cloud segmentation pipeline based on geometric properties is presented. This pipeline is summarized in Fig. 8.4.

One important aspect of point cloud segmentation is the extraction of features from the data. This involves identifying and extracting specific local characteristics of the point cloud, for each point and its neighborhood. These features can provide valuable information about the shape and structure of the objects in the point cloud and can be used to improve the accuracy and precision of the segmentation process. Note that the concept of neighborhood plays a relevant role in this workflow, as point clouds are typically unstructured and there are not topological relationships among individual points in raw point cloud data.

A relevant algorithm for feature extraction is PCA. PCA is a dimensionality reduction method that transforms the original data into a new set of variables, known as principal components, which are uncorrelated and ordered by their importance. In the context of point cloud classification, PCA can be used to extract features from the data. By applying PCA to the point cloud, it is possible to identify the eigenvectors and corresponding eigenvalues for each point with respect to its given neighborhood. Eigenvalues are commonly used to derive point features that can be used for point cloud segmentation in combination with clustering techniques such as RG.

RG is a commonly used technique which involves grouping points in the point cloud that are spatially connected or have similar attributes. Starting from one or several seed points, the RG algorithm iteratively "grows" that seeds based on predefined attribute thresholds until no more points meet them. This can help to segment the point cloud into distinct objects or regions, allowing for more focused analysis and interpretation. For example, they can be sued to identify and separate different materials, such as stone, wood, or metal, in a cultural heritage building. They can

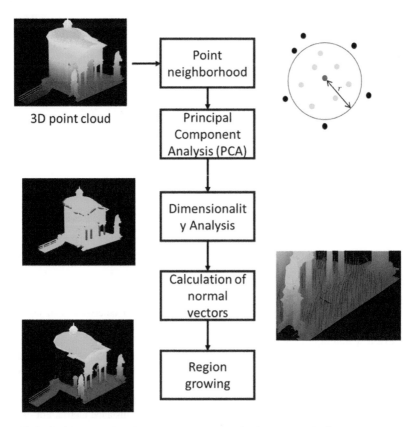

Figure 8.4 Pipeline for point cloud segmentation exploiting geometric features.

also be used to isolate cracks, holes, or other defects in the point cloud data, which can be important for conservation and restoration efforts.

These and other point cloud segmentation techniques are important tools for researchers and professionals working in cultural heritage. By applying these techniques, it is possible to gain a deeper understanding of the objects and sites being studied, and to create more detailed and informative models that can be used for a variety of purposes, including conservation, restoration, and education.

8.2.2.2 Neighborhood in a 3D point cloud

Let $\mathbf{P} = (\mathbf{p}_1, \ldots \mathbf{p}_i, \ldots \mathbf{p}_n) \mid i = 1 \ldots n$ a 3D point cloud defined as a set of 3D points, where each point can be defined as a coordinate in a 3D cartesian system $\mathbf{p}_i = (x, y, z)_i$; being $\mathbf{p}_i \in \mathbb{R}^3$.

A neighborhood of a point $\mathbf{p}_i$ in a 3D point cloud can be conceptually defined as a set of points that are *close* to $\mathbf{p}_i$. The notion *closeness*, however, is not unique and can be defined in various ways. Any formal definition will depend on the distance metric and the neighborhood search criteria.

The most common distance metrics used to determine the neighborhood of a point are Euclidean distance, Manhattan distance, and Chebyshev distance:

- The **Euclidean distance between two points** $\mathbf{p} = (p_x, p_y, p_z)$ and $\mathbf{q} = (q_x, q_y, q_z)$ in a 3D point cloud is widely known and is defined as Eq. (8.5):

$$d(\mathbf{p}, \mathbf{q}) = \sqrt{\sum_{j=1}^{n} (\mathbf{p} - \mathbf{q})(\mathbf{p} - \mathbf{q})^T} = \sqrt{(p_x - q_x)^2 + (p_y - q_y)^2 + (p_z - q_z)^2} \tag{8.5}$$

- The **Manhattan distance between two points** $\mathbf{p}$ and $\mathbf{q}$ is defined as the sum of absolute difference between the measures in all dimensions of two points (Eq. 8.6):

$$d(\mathbf{p}, \mathbf{q}) = |p_x - q_x| + |p_y - q_y| + |p_z - q_z| \tag{8.6}$$

- The **Chebyshev distance between two points** $\mathbf{p}$ and $\mathbf{q}$ is defined as the greatest of the differences along any dimension (Eq. 8.7).

$$d(\mathbf{p}, \mathbf{q}) = \max(|p_x - q_x|, |p_y - q_y|) \tag{8.7}$$

In this chapter, the concept of distance, unless specified otherwise, will correspond with the Euclidean distance.

Next, a search criteria is needed to define the neighborhood of a point. Let's assume that the distance between a point $\mathbf{p}$ and every other point in the 3D point cloud is known. Two common search criteria to define the neighborhood of $\mathbf{p}$ are k-nearest neighbor and radius search:

- The **k-nearest neighbor criterion defines the neighborhood of a point as the set of** k points that are closest to the point $\mathbf{p}$ according to the chosen distance metric. For example, if we set $k = 5$, the neighborhood of a point would be the set of 5 points in the point cloud that are closest to it.
- The **radius search criterion defines the neighborhood of a point as the set of points in the cloud that are within a specified radius of the point. For example, if we set the radius to 0.5 m, all the points that are within that distance of the point p** will be part of its neighborhood.

Summarizing, while the k-nearest neighbor criterion obtains a predefined number of neighbors, with maximum size of the neighborhood variable, the radius search criterion obtains a variable number of neighbors with a predefined size of the neighborhood.

8.2.2.3 Principal component analysis

PCA is a statistical method used to describe a group of variables of any dimensionality in terms of new, uncorrelated components ordered from highest to lowest variance with respect to the original information. Note that the variance is defined as a measure of the dispersion of a variable x_i with respect of its average $\overline{X}$ (Eq. 8.8):

$$\sigma_n^2 = \frac{1}{n} \sum_{i=1}^{n} (x_i - \overline{X}) \tag{8.8}$$

In other words, it seeks a projection of the information in a smaller or equal number of dimensions that optimizes the representation in terms of least squares.

In Fig. 8.5, a set of 2D points is represented together with its variance with respect to two different directions. The described variance corresponding to X axis (Fig. 8.5, left) is smaller than the described variance corresponding to the oblique direction (Fig. 8.5, right). PCA selects the two directions that describe the most variance, so the information can be visualized from a different reference system that presents the information on a more intuitive manner (Fig. 8.6).

It can be seen that a set of variables in N dimensions ($N = 2$ in the previous example) can be described by a number $K \leq N$ of principal components. The mathematical description of these principal components is given by two highly relevant concepts: Eigenvectors and Eigenvalues.

- An **Eigenvector defines the** direction of the principal component.
- An **Eigenvalue defines the** variance described in the direction of its corresponding eigenvector.

Each eigenvector has an associated eigenvalue, and there are as many eigenvector/eigenvalue pairs as components considered.

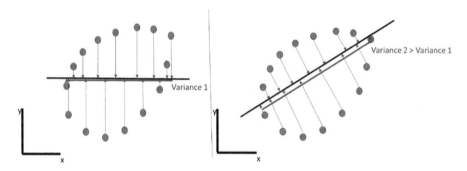

Figure 8.5 Representation of variance in a set of 2D points along different directions.

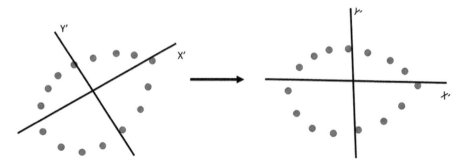

Figure 8.6 More intuitive representation of variance based on PCA. *PCA*, Principal component analysis.

The eigenvectors describe the transformation from the original coordinate system to the one defined by the principal components:

Let X be an $n \times p$ matrix with n elements of p coordinates each. In practice, PCA obtains the eigenvectors from the covariance matrix of $\overline{X} = X - \mu$, where μ is the mean of each coordinate (Eq. 8.9):

$$cov(\overline{X}) = \left(\frac{\overline{X}\,\overline{X}^T}{n-1} \right) \tag{8.9}$$

This matrix can be decomposed in p eigenvectors ($\mathbf{p}_a$) and eigenvalues (λ_a). Finally, the projections of X in the coordinate system defined by the eigenvectors $\mathbf{p}_a$ will be given by Eq. (8.10)

$$t_a = X\mathbf{p}_a \tag{8.10}$$

PCA allows us to perform two very useful analyses in point cloud processing: dimensionality analysis and calculation of normal vectors.

8.2.2.4 Dimensionality analysis

Dimensionality analysis provides a quantitative description of the shape of a point with respect to its neighborhood, or of an object/element of a point cloud, by taking into account the relative values of its eigenvalues.

Given a point $\mathbf{p}$ belonging to a point cloud, its neighborhood $p^k \backslash |p_i^k - p| < r \, \forall \, p_i^k$ as a set of points closer than a distance r to the given point, and its eigenvalues (λ_1, λ_2, λ_3) of the covariance matrix of that neighborhood corresponding to the first three eigenvectors, ordered such as $\lambda_1 > \lambda_2 > \lambda_3$, the following dimensionality features can be obtained (Eqs. 8.11−8.13):

- **Linearity** (1D):

$$a_{1D} = \frac{\sqrt{\lambda_1} - \sqrt{\lambda_2}}{\sqrt{\lambda_1}} \tag{8.11}$$

- **Planarity** (2D):

$$a_{2D} = \frac{\sqrt{\lambda_2} - \sqrt{\lambda_3}}{\sqrt{\lambda_1}} \tag{8.12}$$

- **Scatter** (3D):

$$a_{3D} = \frac{\sqrt{\lambda_3}}{\sqrt{\lambda_1}} \tag{8.13}$$

where $a_{1D} + a_{2D} + a_{3D} = 1$

8.2.2.5 Calculation of normal vectors

Normal vectors are one of the most relevant properties for many applications in point cloud processing. Mathematically, a normal vector to a point p_i is that perpendicular to the tangent plane to the surface passing through p_i.

The simplest way to define the normal vector of a point p_i belonging to a point cloud **P** is as the eigenvector associated to the third eigenvalue (λ_3) of the covariance matrix of a given neighborhood p_i.

8.2.2.6 Clustering and region growing

Clustering is a process that involves dividing data into groups (or clusters) based on their similarity. In the context of 3D point cloud processing, clustering algorithms can be used to identify distinct groups of points within the cloud that represent different objects or structures. For example, a point cloud of a scene containing multiple objects might be clustered in a way that groups together all of the points that belong to a particular object. Clustering algorithms typically work by iteratively assigning points to different clusters based on some measure of similarity, such as the distance between points.

A relevant type of clustering algorithm is RG. It is an iterative approach that involves starting with a small set of seed points and then growing the clusters by adding points from the point cloud that are similar to the points already in the cluster. This is typically done based on some measure of similarity, such as the Euclidean distance between points. RG algorithms are often used for applications such as segmentation, where the goal is to partition the point cloud into distinct regions or clusters that correspond to different objects or structures in the scene. One advantage of RG algorithms is that they can be easily customized to incorporate domain-specific knowledge or constraints, which can improve the accuracy of the resulting clusters.

Let $\mathbf{P} = (\mathbf{p}_1, \ldots \mathbf{p}_i, \ldots \mathbf{p}_n) \mid i = 1 \ldots M$ be a 3D point cloud defined as a set of M 3D points, where each point can be defined as a coordinate in a 3D cartesian system $\mathbf{p}_i = (x, y, z)_i \mid \mathbf{p}_i \in \mathbb{R}^3$. Let $\mathbf{N} = (\mathbf{n}_x, \mathbf{n}_y, \mathbf{n}_z)$ a Mx3 matrix with the normal vectors corresponding to each point $\mathbf{p}_i$.

To define an RG algorithm whose similarity is based on the normal vectors of the points, the definition of a seed point $\mathbf{p}_{\text{seed}} = (x, y, z) \mid \mathbf{p}_{\text{seed}} \in \mathbf{P}$ is needed. It can be done manually or automatically, depending on the application. Then, the growing rules should be defined. For a general case, a function $f(\mathbf{n}_1, \mathbf{n}_2)$ and a threshold n_{th} can be defined, where $\mathbf{n}_1$ is the normal vector of a point $\mathbf{p}_1$ within the grown segment (the seed, in the first iteration of the algorithm) and $\mathbf{n}_2$ is the normal vector of a neighboring point $\mathbf{p}_2$, such that:

$$\begin{cases} \text{if } f(\boldsymbol{n}_1, \boldsymbol{n}_2) < n_{th} \rightarrow \text{Add } \boldsymbol{p}_2 \text{ to the same segment than } \boldsymbol{p}_1 \\ \text{if } f(\boldsymbol{n}_1, \boldsymbol{n}_2) \geq n_{th} \rightarrow \text{Do not add } \boldsymbol{p}_2 \text{ to the same segment than } \boldsymbol{p}_1 \end{cases}$$

The algorithm can be introduced with the following pseudocode:

Inputs

- 3D point cloud $\mathbf{P}$
- normal vectors $\mathbf{N}$
- seed point $\mathbf{p}_{\text{seed}}$
- similarity function $f(\mathbf{n}_1, \mathbf{n}_2)$
- similarity threshold n_{th}
- neighborhood radius r

Process

1. The neighborhood of each point in **P is computed using a radius search (radius** r**)**.
2. The neighborhood of $\mathbf{p}_{\text{seed}}$ is selected as a set of points $\mathbf{p}_n \mid \mathbf{p}_n - \mathbf{p}_{\text{seed}} \mid < r$. The normal of $\mathbf{p}_{\text{seed}}$ is selected as $\mathbf{n}_1$.
3. The cluster of points $\mathbf{C}$ that will grow from the seed is initialized.
4. For each point in $\mathbf{p}_n$.
5. Select its normal vector $\mathbf{n}_2$
6. Compute the similarity function $f(\mathbf{n}_1, \mathbf{n}_2)$
7. Add to $\mathbf{C}$ all points that meets the similarity threshold: $f(\mathbf{n}_1, \mathbf{n}_2) < n_{\text{th}}$
8. If there are no points that meet the similarity threshold, finish the process.
9. Select the neighborhood $\mathbf{p}_n$ of new points in $\mathbf{C}$.
10. Go back to the beginning of step 4.

Output

- Cluster **C of points that meet the similarity function for the RG algorithm.**

8.2.3 Artificial intelligence classification pipeline

8.2.3.1 Random forest

RF is a nonparametric ensemble learner, meaning it helps to create more accurate results by using multiple models to come to its conclusion. As a result, RF can efficiently model nonlinear relationships, handle a large number of redundant features and prevent overfitting (Belgiu & Drăguţ, 2016). As a result, it aggregates the results of several randomized decision trees by a majority voting strategy. In particular, a bootstrap aggregating sample of the training instances is built for each tree, and each tree node is split using a user-defined number of randomly selected features. These features or descriptors constitute a step of great importance in semantic segmentation since an optimal design and combination of features along with the optimal neighborhood definition is key to success. The classification procedure thus relies on a set of feature vectors that characterize the different classes while enabling a distinct separation, that is, the classes must have a unique signature in the feature space, with sufficient differences between classes. The segments are assigned class labels based on their characteristic feature values. In particular, we analyze the following geometric and radiometric features (Table 8.2).

Table 8.2 Different features used for describing points and its neighborhood.

Features	Definitions
Sum of eigenvalues	$\Sigma\lambda_i$
Omnivariance	$(\prod\lambda_i)^{\frac{1}{3}}$
Eigentropy	$-\Sigma\lambda_i\ln(\lambda_i)$
Linearity	$\frac{\lambda_1-\lambda_2}{\lambda_1}$
Planarity	$\frac{\lambda_2-\lambda_3}{\lambda_1}$
Sphericity	$\frac{\lambda_3}{\lambda_1}$
Change of curvature	$\frac{\lambda_3}{\lambda_1+\lambda_2+\lambda_3}$
Verticality	$\left\|\frac{\pi}{2}-angle(e_i,e_Z)\right\|_{i\in(0,2)}$
Surface variation	$C_\lambda=\frac{\lambda_3}{\Sigma\lambda}$
Absolute moment	$\frac{1}{\|N\|}\left\|\Sigma(p-p_0,e_i)^k\right\|_{i\in(0,2)}$
Vertical moment	$\frac{1}{\|N\|}\Sigma(p-p_0,e_z)^k$
Number of points	$\|N\|$
Radiometric features	
Average color HSI model	$\frac{1}{\|N\|}\Sigma c$
Color variance HSI model	$\frac{1}{\|N\|}\Sigma(c-\bar{c})^2$

HSI, Hue, Saturation, Intensity.

We use the covariance-based features coming from the covariance matrix that simply derive from the eigenvalues $\lambda_1>\lambda_2>\lambda_3$ and their corresponding eigenvectors *e1*; *e2*; *e3* which are extracted from Eq. (8.14):

$$cov(N)=\frac{1}{|N|}\sum_{p\in N}(p-\bar{p})(p-\bar{p})^T \qquad (8.14)$$

where p is the centroid of the neighborhood N. From the eigenvalues, we can compute several features: *sum of eigenvalues, omnivariance, eigenentropy, linearity, planarity, sphericity, anisotropy*, and *change of curvature*, among others (Table 8.2).

Regarding radiometric features we used hue, saturation, value (HSV) color model, since it is more perceptually uniform than other color spaces. The color conversion from RGB color space to HSV color space is performed before extracting features.

Each decision tree is generated according to the following structure:

- Let N be the number of samples from the training set, of which n randomly chosen cases are considered for the construction of the tree.
- Let M be the total number of geometric and radiometric features. For each node, we randomly select m features from the total M, so that the best partition of these m variables is the one chosen to split the node. During the growth of the forest, m remains constant.
- Each tree is extended as much as possible, without pruning.

The error rate depends on the correlation between any two trees in the forest—the higher the correlation, the greater the error—and the strength of each of the individual trees: a tree with a low error rate is a strong classifier. A reduction in m reduces both correlation and strength, and an increase in m contributes to increasing them.

This error is estimated by the algorithm during processing by using "the out-of-bag," which consists of leaving out of each tree one-third of the samples chosen for the construction of the tree, so that they are used to determine the error. This also makes it possible to compute the importance of the different variables used, which is characteristic of this algorithm

Among the **advantages attributed to RF for classifying point clouds, we can point out the following:**

- It is a fast algorithm.
- It is efficient with large datasets.
- It can deal with hundreds of features (both geometric and radiometric) without excluding any.
- It provides an estimate of the most important features.
- It allows multiple decision trees to be created in parallel.
- It has an effective method for estimating missing data and maintains a good accuracy if these missing data are numerous.

The main **disadvantages are that if there is noise in the data, the algorithm overfits and overpredicts, so the resulting point cloud classifications are difficult to be interpreted. This is the main reason for which we apply this algorithm combined with an anisotropic filter previously described.**

8.2.3.2 Assessing semantic classification quality

The quality of the classification can be calculated by a number of different metrics, most of which are derived from confusion matrices. The confusion matrix takes into consideration the number of points that have been calculated as True (T) or False (F), Positives (P), or Negatives (N) (Eq. 8.15):

$$M = \begin{bmatrix} TP & FP \\ FN & TN \end{bmatrix} \tag{8.15}$$

From the confusion matrix, we can calculated the Recall (Re) and Precision (Pr) of an algorithm using Eqs. (8.16–8.17):

$$\mathrm{Re} = \frac{TP}{TP + FN} \tag{8.16}$$

$$\mathrm{Pr} = \frac{TP}{TP + FP} \tag{8.17}$$

While the *F1* score, which provides a balanced representation of the overall performance of the algorithm, can be calculated through Eq. (8.18):

$$F1 = \frac{2 \times \text{Pr} \times \text{Re}}{\text{Pr} + \text{Re}} \tag{8.18}$$

High values of *F1* are indicative of a reliable classification algorithm (*F1* = 1), while optimal performance is typically considered if *F1* > 0.8.

While other metrics exist for the evaluation of confusion matrices, *Pr*, *Re*, and *F1* are more suitable for the valuation of case studies where the number of samples per class is imbalanced (He & Ma, 2013).

Another means of evaluating the performance of a segmentation algorithm is using the Jaccard index (Jaccard, 1912), more commonly known in these contexts as the intersection-over-union (IoU) metric. This metric considers the amount of intersection divided by the size of the union of different segments of the point cloud (Eq. 8.19);

$$IoU = \frac{|x' \cap x|}{|x' \cup x|} \tag{8.19}$$

where *x'* are the predicted labels and *x* are the ground-truth labels of the different segments. IoU ranges from [0, 1], with 0 implicating poor segmentation, that is, no overlap between the predicted and the ground-truth segments, while 1 implicates reliable segmentation. The mean IoU (*mIoU*) can thus provide a balanced overview on the clarity of separation among the different segment classes.

8.3   Results and discussion

This section describes and validates the in-house tool developed for the semantic classification of point clouds in cultural heritage and integrated in Cloud Compare. In particular, we have used the ArCH dataset (http://archdataset.polito.it/), since it provides different labeled point clouds of different cultural heritage buildings acquired with different geotechnologies (e.g., LiDAR and photogrammetry). The classification toolbox is available as a GitHub repository. This can be found via the following link: https://github.com/TIDOP-USAL/RFCCToolBox. To install the library from a GitHub repository, the user must first have installed Cloud Compare (Fig. 8.7).

The tool is used by starting the Cloud Compare program and selecting the data model to work with for the point cloud classifications. In this first release, the tool offers two classification data models (Fig. 8.8): (1) the American Society of Photogrammetry and Remote Sensing model designed to work with airborne LiDAR data and (2) the ArCH data model focused on heritage buildings and

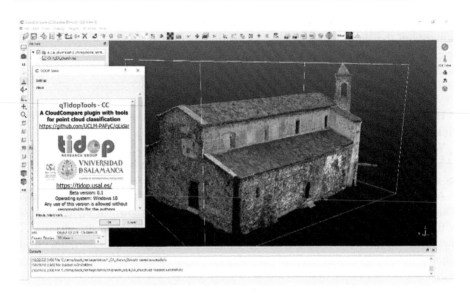

Figure 8.7 Semantic classification tool developed within Cloud Compare.

Figure 8.8 Data models included in the toolbox developed: data model for cultural heritage buildings-ArCH (left) and ASPRS data model for airborne LiDAR (right). *ASPRS*, American Society of Photogrammetry and Remote Sensing.

created by several scientific institutions (Politecnico di Torino, Università Politecnica delle Marche, FBK Trento, Italy, and INSA Strasbourg, France).

In our case, we select the ArCH data model for its validation with the semantic classification of cultural heritage buildings. It is worth noting that four special classification classes have been created, beyond the different categories created in the ArCH data model: (1) removed, (2) selected, (3) unclassified, and (4) nonclassified. The main difference between "unclassified" class and "nonclassified" class is that the former is that the user can decide to unclassify something that was already classified by the algorithm because it is failing and therefore identify false positives. The "removed" class allows to select those points that need to be removed or that we do not want to be included in the classification.

The toolbox has been designed to allow multiple classification and segmentation operations by allowing selection by different methods: (1) by circle—C, (2) by polygon—P, and (3) by rectangle—R. It also allows to change the selected part of the point cloud to any class to perform the computation of features or the training.

The RF-based classification process has been articulated in three main steps including different RF variants, so the user can compare different results (Fig. 8.9): (1) compute features, (2) training, and (3) classification using RF. Furthermore, to assess and compare the results, you can analyze the separability of the classes.

The first step is the computation of the features, which is applied to the whole point cloud and thus requires some computational effort. This step is independent of the selection of the training area. The features computed depend in the type of the RF algorithm selected (CGAL or ETHZ) (Fig. 8.10) and include geometric and radiometric features. Each feature is assigned a weight that measures its strength with respect to the other features and each feature includes the median and standard deviation.

Once the features have been computed, the second step requires the training taking a small set of points classified that perform as ground truth and the features previously computed (Fig. 8.11). Particularly, the user must provide for each label a set of known semantic points among the input data set (e.g., selecting a part of a roof, an arch, and a column). The training step computes for each feature, a range of weights, so the effect each feature has on each label is estimated. For a given

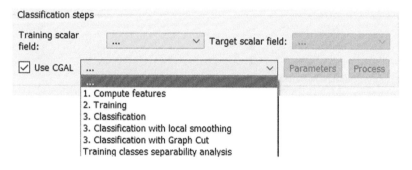

Figure 8.9 Steps required for performing the RF classification. *RF*, Random forest.

(A)

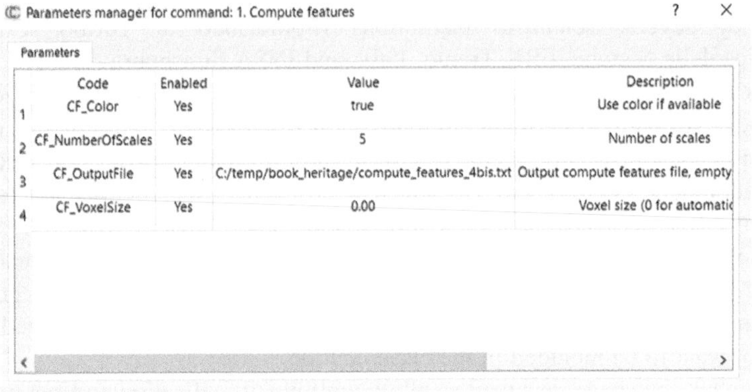

Parameters manager for command: 1. Compute features
Parameters
Code
Enabled
Value
Description
1 CF_Color Yes true Use color if available
2 CF_NumberOfScales Yes 5 Number of scales
3 CF_OutputFile Yes C:/temp/book_heritage/compute_features_4bis.txt Output compute features file, empty
4 CF_VoxelSize Yes 0.00 Voxel size (0 for automati

(B)

Parameters manager for command: 1. Compute features (no CGAL)
Parameters
Code
Enabled
Value
1 CFNCGAL_Anisotropy Yes true
2 CFNCGAL_Color Yes true
3 CFNCGAL_Density2d Yes true
4 CFNCGAL_Density3d Yes true
5 CFNCGAL_DensityKnn Yes true
6 CFNCGAL_EigenEntropy Yes true
7 CFNCGAL_EigenValue1 Yes true
8 CFNCGAL_EigenValue2 Yes true
9 CFNCGAL_EigenValue3 Yes true
10 CFNCGAL_EigenValuesSum Yes true
11 CFNCGAL_GaussianCurvature Yes true
12 CFNCGAL_Linearity Yes true
13 CFNCGAL_LocalNeighborhoodRadius Yes 0.00
14 CFNCGAL_MeanCurvature Yes true
15 CFNCGAL_MomentOrder1 Yes true
16 CFNCGAL_NormalChangeRate Yes true
17 CFNCGAL_Omnivariance Yes true
18 CFNCGAL_OutputFile Yes D:/book_heritage/qTidopTools_2023/5_SMV_chapel_1/no_cgal_color
19 CFNCGAL_PCA1 Yes true
20 CFNCGAL_PCA2 Yes true
21 CFNCGAL_Planarity Yes true
22 CFNCGAL_Roughness Yes true
23 CFNCGAL_RoughnessDirection Yes (0.0,0.0,1.0)
24 CFNCGAL_Sphericity Yes true
25 CFNCGAL_SurfaceVariation Yes true
26 CFNCGAL_Verticality Yes true

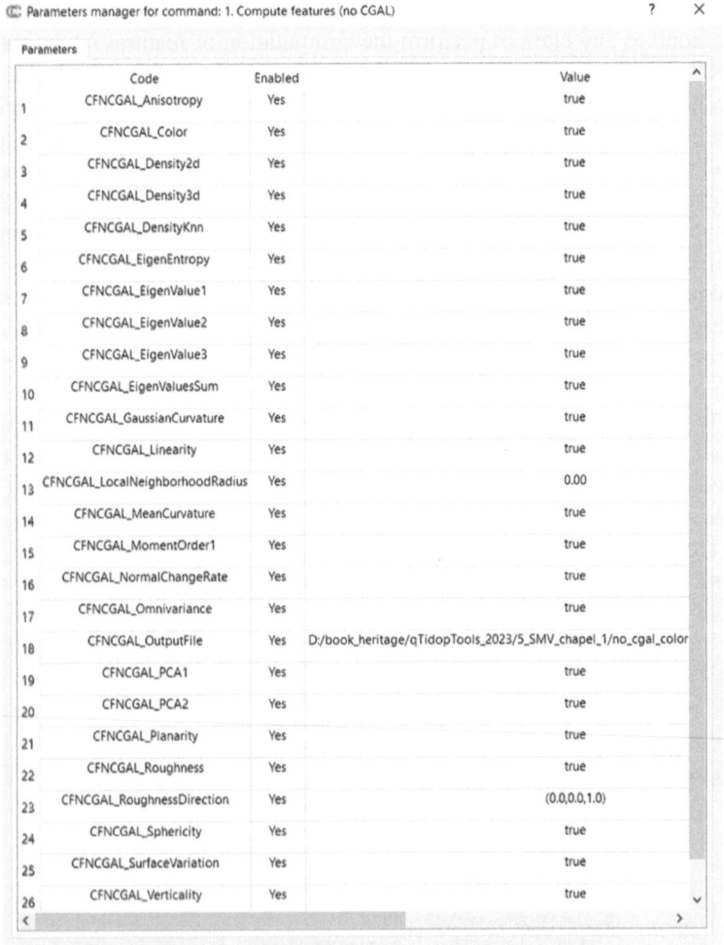

Figure 8.10 Selection of the parameters to be applied in the computation of features with CGAL (A) and with the ETHZ algorithm (B).

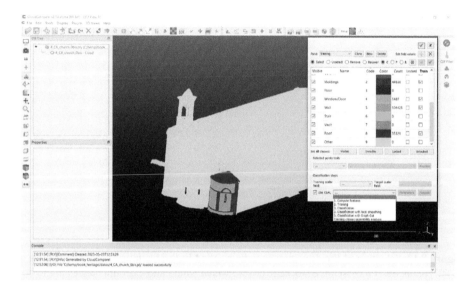

Figure 8.11 Selection of the zone to train and selection of the classes to train. Note that a new class has been created for training, called "training" and another class where the classification results are stored, called "new_classification."

weight, if a feature has the same effect on each label, it is nonrelevant for classification. The range of weights for which the feature is relevant is also estimated. For each feature, uniformly selected weight values are tested, and their effects are estimated. The *mIoU* is used to assess the quality of this set of weights and effects. The same mechanism is repeated until all features' ranges have been tested.

For performing the training, the user has to select and define the following parameters (Fig. 8.12). The training consists of calculating the random decision trees with the strategy defined by the parameters and features computed in the previous step in a way that allows it to find the optimal path to follow to correctly classify each point.

Finally, the classification based on RF is applied selecting between two different filtering strategies and their setup parameters (e.g., local smoothing, graph cut) (Fig. 8.13). The main difference between local smoothing and graph cut is the computational cost, being the former faster since it is a trade-off between quality and efficiency. For its part, graph cut provides the best quality but requires longer computation time. To speed up computations, the input domain can be subdivided into smaller subsets such that several smaller graph cuts are applied instead of a big one. The computation of these smaller graph cuts can be done in parallel. Increasing the number of subsets allows for faster computation times but can also reduce the quality of the results.

Fig. 8.14 outlines some of the results obtained with RF and the different filtering strategies.

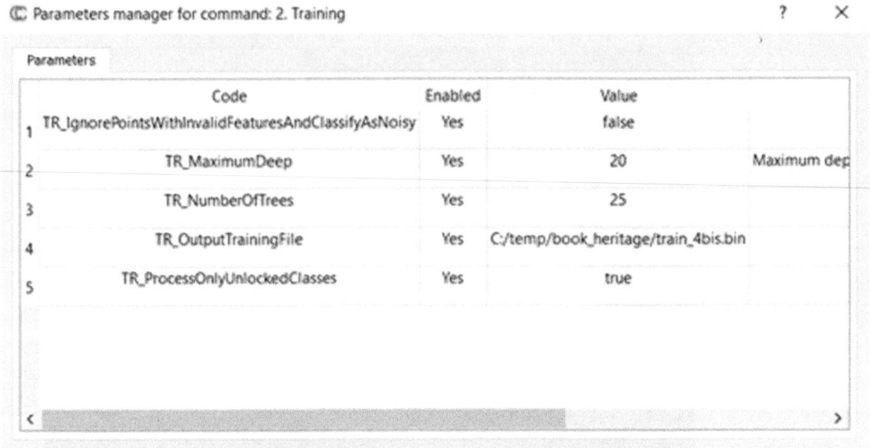

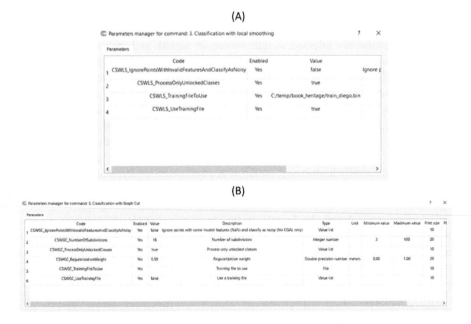

Figure 8.12 Selection of the parameters to be applied in the training. Note the presence of the parameter "ProcessOnlyUnlockedClasses" which would allow certain classes not to be classified. This is important since the majority of the classification algorithms classify the whole point cloud.

Figure 8.13 Selection of setup parameters for the RF classification using different filtering strategies: local smoothing (A) and graph cut (B). *RF*, Random forest.

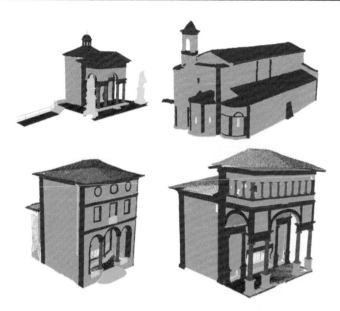

Figure 8.14 Results of the classification using RF and the toolbox developed in Cloud Compare.

8.4 Conclusion

This chapter has described in detail the importance and current state of passing from unorganized 3D point clouds to semantic enriched models. Different techniques have been tested and encapsulated within an in-house tool integrated in Cloud Compare, using different cultural heritage buildings for its validation. In particular, we developed and integrated a particular variation of the anisotropic filtering and an ML strategy based on RF which exploits the geometric and radiometric features of the point cloud.

The variation of the anisotropic filtering, due to its adaptive design, reduces the noise while preserving fine features and sharp edges of the cultural heritage buildings. This provides the best input data for performing the semantic classification based on RF. The filtering has been validated in different types of point clouds of different shapes and sizes and under various levels of noise, being applied in fully unsupervised and automatic way.

For its part, RF has demonstrated a clear potential to get a semantic point cloud classification of the different cultural heritage buildings. Furthermore, the tool offers a didactical interface where the user can select and combine the different parameters and compare and analyze the results obtained. The output provided by this toolbox can make easier the generation of BIM models, semantically separating elements in point clouds for the modeling procedure in a BIM environment. Last but not least, the RF strategy is easy to implement, and it does not require high computational cost or processing time.

References

Alfio, V. S., Costantino, D., Pepe, M., & Garofalo, A. R. (2022). A geomatics approach in scan to FEM process applied to cultural heritage structure: The case study of the "Colossus of Barletta.". *Remote Sensing, 14*(3), 664. Available from https://doi.org/10.3390/rs14030664.

Alkadri, M. F., Alam, S., Santosa, H., Yudono, A., & Beselly, S. M. (2022). Investigating surface fractures and materials behavior of cultural heritage buildings based on the attribute information of point clouds stored in the TLS dataset. *Remote Sensing, 14*(2). Available from https://doi.org/10.3390/rs14020410, https://www.mdpi.com/2072-4292/14/2/410/pdf.

Belgiu, M., & Drăguţ, L. (2016). Random forest in remote sensing: A review of applications and future directions. *ISPRS Journal of Photogrammetry and Remote Sensing, 114*, 24−31. Available from https://doi.org/10.1016/j.isprsjprs.2016.01.011.

Boscaini, D., Masci, J., Melzi, S., Bronstein, M. M., Castellani, U., & Vandergheynst, P. (2015). Learning class-specific descriptors for deformable shapes using localized spectral convolutional networks. *Computer Graphics Forum, 34*(5), 13−23, 14678659 Blackwell Publishing Ltd Switzerland. Available from https://doi.org/10.1111/cgf.12693, http://www.blackwell-synergy.com/loi/CGF.

Boscaini, D., Masci, J., Rodolà, E., & Bronstein, M. (2016). Neural information processing systems foundation Switzerland Learning shape correspondence with anisotropic convolutional neural networks. *Advances in Neural Information Processing Systems*, 3197−3205, 10495258.

Boulch, A., Guerry, J., Le Saux, B., & Audebert, N. (2018). SnapNet: 3D point cloud semantic labeling with 2D deep segmentation networks. *Computers & Graphics, 71*, 189−198. Available from https://doi.org/10.1016/j.cag.2017.11.010.

Bronstein, M. M., Bruna, J., Lecun, Y., Szlam, A., & Vandergheynst, P. (2017). Geometric deep learning: Going beyond Euclidean data. *IEEE Signal Processing Magazine, 34*(4), 18−42. Available from https://doi.org/10.1109/MSP.2017.2693418, http://ieeexplore.ieee.org/xpl/RecentIssue.jsp?punumber = 79&year = 2008.

Bronstein, M. M., & Kokkinos, I. (2010) Scale-invariant heat kernel signatures for non-rigid shape recognition. *Proceedings of the IEEE Computer Society Conference on Computer Vision and Pattern Recognition, Israel*. 1704−1711, 10636919. Available from https://doi.org/10.1109/CVPR.2010.5539838.

Capolupo, A. (2021). Accuracy assessment of cultural heritage models extracting 3D point cloud geometric features with RPAS SfM-MVS and TLS techniques. *Drones, 5*(4), 145. Available from https://doi.org/10.3390/drones5040145.

Carbonell-Rivera, J. P., Torralba, J., Estornell, J., Ruiz, L. Á., & Crespo-Peremarch, P. (2022). Classification of Mediterranean shrub species from UAV point clouds. *Remote Sensing, 14*(1). Available from https://doi.org/10.3390/rs14010199, https://www.mdpi.com/2072-4292/14/1/199/pdf.

Castillo, E., Liang, J., & Zhao, H. (2013). Point cloud segmentation and denoising via constrained nonlinear least squares normal estimates. *Mathematics and Visualization, 0*. Available from https://doi.org/10.1007/978-3-642-34141-0_13, https://www.springer.com/series/4562.

Chehata, N., Guo, L., & Mallet, C. (2009). *Laser Scanning 207−213 Airborne LiDAR feature selection for urban classification using random forests*.

Croce, V., Caroti, G., Piemonte, A., & Bevilacqua, M. G. (2019). Geomatics for cultural heritage conservation: Integrated survey and 3D modeling. *IMEKO TC4 International Conference on Metrology for Archaeology and Cultural Heritage*, MetroArchaeo. IMEKO-International Measurement Federation Secretariat Italy. 2019:271−276. 9789299008454.

De Deuge, M., Quadros, A., Hung, C., Douillard, B. (2013). Unsupervised feature learning for classification of outdoor 3D Scans. *Australasian Conference on Robotics and Automation, ACRA Australasian Robotics and Automation Association Australia*, 9780980740448. http://www.araa.asn.au/acra/.

Dulecha, T. G., Pintus, R., Gobbetti, E., & Giachetti, A. (2020). SynthPS: a Benchmark for Evaluation of Photometric Stereo Algorithms for Cultural Heritage Applications GCH. *Eurographics Workshop on Graphics and Cultural Heritage, EG GCH2020*, pp. 13−22.

Frome, A., Huber, D., Kolluri, R., Bülow, T., Malik, J. (2004). Recognizing objects in range data using regional point descriptors. *Lecture notes in computer science (including subseries lecture notes in artificial intelligence and lecture notes in bioinformatics)*. 3023, 224−237. Available from https://doi.org/10.1007/978-3-540-24672-5_18, https://www.springer.com/series/558.

Gaiani, M., Remondino, F., Apollonio, F. I., & Ballabeni, A. (2016). An advanced preprocessing pipeline to improve automated photogrammetric reconstructions of architectural scenes. *Remote Sensing*, 8(3). Available from https://doi.org/10.3390/rs8030178, http://www.mdpi.com/2072-4292/8/3/178/pdf.

Ghorbani, F., Ebadi, H., Pfeifer, N., & Sedaghat, A. (2022). Uniform and competency-based 3D keypoint detection for coarse registration of point clouds with homogeneous structure. *Remote Sensing*, 14(16), 4099. Available from https://doi.org/10.3390/rs14164099.

Golovinskiy, A., Kim, V.G., & Funkhouser, T. (2009). Shape-based recognition of 3D point clouds in urban environments. *Proceedings of the IEEE International Conference on Computer Vision*. United States. 2154−2161. Available from https://doi.org/10.1109/ICCV.2009.5459471.

Grilli, E., Menna, F., & Remondino, F. (2017). A review of point clouds segmentation and classification algorithms. *The International Archives of the Photogrammetry, Remote Sensing and Spatial Information Sciences*, XLII-2/W3(2), 339−344. Available from https://doi.org/10.5194/isprs-archives-xlii-2-w3-339-2017.

Guldur Erkal, B., & Hajjar, J. F. (2017). Laser-based surface damage detection and quantification using predicted surface properties. *Automation in Construction*, 83, 285−302. Available from https://doi.org/10.1016/j.autcon.2017.08.004.

Guo, Y., Wang, H., Hu, Q., Liu, H., Liu, L., & Bennamoun, M. (2021). Deep learning for 3D point clouds: A survey. *IEEE Transactions on Pattern Analysis and Machine Intelligence*, 43(12), 4338−4364. Available from https://doi.org/10.1109/TPAMI.2020.3005434, https://ieeexplore.ieee.org/servlet/opac?punumber = 34.

Guo, K., Zou, D., & Chen, X. (2015). 3D mesh labeling via deep convolutional neural networks. *ACM Transactions on Graphics*, 35(1), 1−12. Available from https://doi.org/10.1145/2835487.

Hackel, T., Savinov, N., Ladicky, L., Wegner, J. D., Schindler, K., & Pollefeys, M. (2017). Semantic3D. net: A new Large-scale Point Cloud Classification Benchmark. *Arxiv*. Available from https://doi.org/10.48550/arXiv.1704.03847.

He, H., & Ma, Y. (2013). *Imbalanced Learning: Foundations, Algorithms, and Applications* (pp. 1−210). United States: Wiley. Available from https://doi.org/10.1002/9781118646106.

Höfle, B., & Pfeifer, N. (2007). Correction of laser scanning intensity data: Data and model-driven approaches. *ISPRS Journal of Photogrammetry and Remote Sensing, 62*(6), 415−433. Available from https://doi.org/10.1016/j.isprsjprs.2007.05.008.

Jaccard, P. (1912). The distribution of the flora in the alpine zone. *New Phytologist, 11*(2), 37−50. Available from https://doi.org/10.1111/j.1469-8137.1912.tb05611.x.

Kauppinen, T., Väätäinen, J., & Hyvönen, E. (2008). Finland creating and using geospatial ontology time series in a semantic cultural heritage portal. *Lecture notes in computer science (including subseries lecture notes in artificial intelligence and lecture notes in bioinformatics).* 5021. 110−123, 16113349. Available from https://doi.org/10.1007/978-3-540-68234-9_11.

Kim, M. K., Sohn, H., & Chang, C. C. (2015). Localization and quantification of concrete spalling defects using terrestrial laser scanning. *Journal of Computing in Civil Engineering, 29*(6). Available from https://doi.org/10.1061/(ASCE)CP.1943-5487. 0000415, http://ascelibrary.org/cpo/resource/1/jccee5.

Kipf, T. N., & Welling, M. (2017). Semi-supervised classification with graph convolutional networks. *5th International Conference on Learning Representations, ICLR 2017 - Conference Track Proceedings International Conference on Learning Representations, ICLR* Netherlands. https://dblp.org/db/conf/iclr/iclr2017.html.

Korres, G., Eid, M., & Anagnostopoulos, C. N. (2022). Benchmark based selection of point cloud registration algorithms for cultural heritage 3D digitization. *Journal of Physics: Conference Series, 2204*(1), 012061. Available from https://doi.org/10.1088/1742-6596/2204/1/012061.

Lamdan, Y., & Wolfson, H.J. (1988) *Geometric hashing: A general and efficient model-based recognition scheme* (pp. 238−249).

Landrieu, L., Raguet, H., Vallet, B., Mallet, C., & Weinmann, M. (2017). A structured regularization framework for spatially smoothing semantic labelings of 3D point clouds. *ISPRS Journal of Photogrammetry and Remote Sensing, 132*, 102−118. Available from https://doi.org/10.1016/j.isprsjprs.2017.08.010, http://www.elsevier.com/inca/publications/store/5/0/3/3/4/0.

Lasaponara, R., Coluzzi, R., & Masini, N. (2011). Flights into the past: Full-waveform airborne laser scanning data for archaeological investigation. *Journal of Archaeological Science, 38*(9), 2061−2070. Available from https://doi.org/10.1016/j.jas.2010.10.003, http://www.elsevier.com/inca/publications/store/6/2/2/8/5/4/index.htt.

Lasaponara, R., & Masini, N. (2009). Full-waveform airborne laser scanning for the detection of medieval archaeological microtopographic relief. *Journal of Cultural Heritage, 10*(1), e78−e82. Available from https://doi.org/10.1016/j.culher.2009.10.004, http://www.elsevier.com.

Lee, J., Kim, J., Ahn, J., & Woo, W. (2019). Context-aware risk management for architectural heritage using historic building information modeling and virtual reality. *Journal of Cultural Heritage, 38*, 242−252. Available from https://doi.org/10.1016/j.culher.2018.12.010, http://www.elsevier.com.

Lengauer, S., Sipiran, I., Preiner, R., Schreck, T., & Bustos, B. (2021). A benchmark dataset for repetitive pattern recognition on textured 3D surfaces. *Computer Graphics Forum, 40*(5), 1−8. Available from https://doi.org/10.1111/cgf.14352, http://onlinelibrary.wiley.com/journal/10.1111/(ISSN)1467-8659.

Li, Y., Bu, R., Sun, M., Wu, W., Di, X., & Chen, B. (2018). PointCNN: Convolution on χ-transformed points. *Advances in Neural Information Processing Systems, 32*, 1−11. Available from https://arxiv.org/pdf/1801.07791.pdf.

Li, Z., Zhang, L., Tong, X., Du, B., Wang, Y., Zhang, L., Zhang, Z., Liu, H., Mei, J., Xing, X., & Mathiopoulos, P. T. (2016). A three-step approach for TLS point cloud classification. *IEEE Transactions on Geoscience and Remote Sensing, 54*(9), 5412−5424. Available from https://doi.org/10.1109/TGRS.2016.2564501.

Liu, X., Han, Z., Liu, Y. S., & Zwicker, M. (2019). Point2Sequence: Learning the shape representation of 3D point clouds with an attention-based sequence to sequence network. *Proceedings of the AAAI Conference on Artificial Intelligence, 33*(01), 8778−8785. Available from https://doi.org/10.1609/aaai.v33i01.33018778.

Lodha, S. K., Fitzpatrick, D. M., & Helmbold, D. P. (2007). Aerial lidar data classification using AdaBoost. 3DIM 2007 - *Proceedings 6th International Conference on 3-D Digital Imaging and Modeling*. United States. 435−442. Available from https://doi.org/10.1109/3DIM.2007.10.

López, F. J., Lerones, P. M., Llamas, J., Gómez-García-Bermejo, J., & Zalama, E. (2018). A review of heritage building information modeling (H-BIM). *Multimodal Technologies and Interaction, 2*(2). Available from https://doi.org/10.3390/mti2020021, https://www.mdpi.com/2414-4088/2/2/21/pdf.

Maietti, F., Di Giulio, R., Piaia, E., Medici, M., & Ferrari, F. (2018). Enhancing Heritage fruition through 3D semantic modelling and digital tools: the INCEPTION project. *IOP Conference Series: Materials Science and Engineering, 364*(1)012089. Available from https://doi.org/10.1088/1757-899x/364/1/012089.

Mao, J., Wang, X., & Li, H. (2019). Interpolated convolutional neural networks for 3D point cloud understanding. *International Conference on Computer Vision*, 1−10. Available from https://arxiv.org/pdf/1908.04512.pdf.

Margetis, G., Apostolakis, K. C., Ntoa, S., Papagiannakis, G., & Stephanidis, C. (2021). X-reality museums: Unifying the virtual and real world towards realistic virtual museums. *Applied Sciences, 11*(1), 338. Available from https://doi.org/10.3390/app11010338.

Maron, H., Galun, M., Aigerman, N., Trope, M., Dym, N., Yumer, E., Kim, V. G., & Lipman, Y. (2017). Convolutional neural networks on surfaces via seamless toric covers. *ACM Transactions on Graphics, 36*(4). Available from https://doi.org/10.1145/3072959.3073616, 15577368 Association for Computing Machinery Israel, http://www.acm.org/tog/.

Masci, J., Boscaini, D., Bronstein, M. M., Vandergheynst, P. (2015). *ShapeNet: convolutional neural networks on non-Euclidean manifolds.*

Matei, B. C., Tan, Y., Sawhney, H. S., Kumar, R. (2006). Rapid and scalable 3D object recognition using LIDAR data. *Proceedings of SPIE - The International Society for Optical Engineering*. United States. 6234 0277786X. Available from https://doi.org/10.1117/12.666235.

Matrone, F., Lingua, A., Pierdicca, R., Malinverni, E. S., Paolanti, M., Grilli, E., Remondino, F., Murtiyoso, A., & Landes, T. (2020). A benchmark for large-scale heritage point cloud semantic segmentation. *The International Archives of the Photogrammetry, Remote Sensing and Spatial Information Sciences, 43*(2), 1419−1426. Available from https://doi.org/10.5194/isprs-archives-XLIII-B2-2020-1419-2020, http://www.isprs.org/proceedings/XXXVIII/4-W15/.

Maturana, D., & Scherer, S. (2015). VoxNet: A 3D Convolutional Neural Network for real-time object recognition. *IEEE International Conference on Intelligent Robots and Systems*. Institute of Electrical and Electronics Engineers Inc. United States 2015:922−928, 21530866. Available from https://doi.org/10.1109/IROS.2015.7353481.

Minerva, R., Lee, G. M., & Crespi, N. (2020). Digital twin in the IoT Context: A survey on technical features, scenarios, and architectural models. *Proceedings of the IEEE,*

108(10), 1785−1824. Available from https://doi.org/10.1109/JPROC.2020.2998530, http://ieeexplore.ieee.org/xpl/RecentIssue.jsp?punumber = 5.

Niemeyer, J., Rottensteiner, F., & Soergel, U. (2014). Contextual classification of lidar data and building object detection in urban areas. *ISPRS Journal of Photogrammetry and Remote Sensing*, *87*, 152−165. Available from https://doi.org/10.1016/j.isprsjprs.2013. 11.001, http://www.elsevier.com/inca/publications/store/5/0/3/3/4/0.

Perona, P., & Malik, J. (1990). Scale-space and edge detection using anisotropic diffusion. *IEEE Transactions on Pattern Analysis and Machine Intelligence*, *12*(7), 629−639. Available from https://doi.org/10.1109/34.56205.

Pierdicca, R., Paolanti, M., Matrone, F., Martini, M., Morbidoni, C., Malinverni, E. S., Frontoni, E., & Lingua, A. M. (2020). Point cloud semantic segmentation using a deep learning framework for cultural heritage. *Remote Sensing*, *12*(6). Available from https:// doi.org/10.3390/rs12061005, https://res.mdpi.com/d_attachment/remotesensing/remote-sensing-12-01005/article_deploy/remotesensing-12-01005.pdf.

Poux, F., Hallot, P., Neuville, R., & Billen, R. (2016). Smart point cloud: Definition and remaining challenges. *ISPRS Annals of the Photogrammetry, Remote Sensing and Spatial Information Sciences*, *IV-2/W1*(2), 119−127. Available from https://doi.org/ 10.5194/isprs-annals-iv-2-w1-119-2016.

Qi, C. R., Su, H., Mo, K., Guibas, L. J. (2017). PointNet: Deep learning on point sets for 3D classification and segmentation. *Proceedings - 30th IEEE Conference on Computer Vision and Pattern Recognition, CVPR 2017*. Institute of Electrical and Electronics Engineers Inc. United States. 2017:77−85, 9781538604571. Available from https://doi. org/10.1109/CVPR.2017.16.

Qi, C. R., Su, H., Mo, K., & Guibas, L. J. (2017a). Pointnet: Deep Learning on point sets for 3D classification and segmentation. *Proceedings of the International Conference on Computer Vision and Pattern Recognition*, 1−19. Available from https://arxiv.org/pdf/ 1612.00593.pdf.

Qi, C. R., Yi, L., Su, H., & Guibas, L. J. (2017b). Pointnet++: Deep hierarchical feature learning on point sets in a metric space. *Advances in Neural Information Processing Systems*, 1−14. Available from https://arxiv.org/pdf/1706.02413.pdf.

Riveiro, B., Lourenço, P. B., Oliveira, D. V., González-Jorge, H., & Arias, P. (2016). Automatic morphologic analysis of quasi-periodic masonry walls from LiDAR. *Computer-Aided Civil and Infrastructure Engineering*, *31*(4), 305−319. Available from https://doi.org/10.1111/ mice.12145, http://www.blackwellpublishers.co.uk/journals/MICE/descript.htm.

Rodríguez-Gonzálvez, P., Campo, Á. G., Muñoz-Nieto, Á. L., Sánchez-Aparicio, L. J., & González-Aguilera, D. (2019). Diachronic reconstruction and visualization of lost cultural heritage sites. *ISPRS International Journal of Geo-Information*, *8*(2). Available from https://doi.org/10.3390/ijgi8020061, https://www.mdpi.com/2220-9964/8/2.

Scarselli, F., Gori, M., Tsoi, A. C., Hagenbuchner, M., & Monfardini, G. (2009). The graph neural network model. *IEEE Transactions on Neural Networks*, *20*(1), 61−80. Available from https://doi.org/10.1109/TNN.2008.2005605, http://ieeexplore.ieee.org/xpl/ RecentIssue.jsp?punumber = 72.

Shi, S., Wang, X., & Li, H. (2019). PointRCNN: 3D object proposal generation and detection from point cloud. *Proceedings of the IEEE Computer Society Conference on Computer Vision and Pattern Recognition*. IEEE Computer Society Hong Kong. (pp. 770−779), 9781728132938. Available from https://doi.org/10.1109/CVPR.2019.00086.

Spina, S., Debattista, K., Bugeja, K., & Chalmers, A. (2011). Point cloud segmentation for cultural heritage sites. *The Eurographics Association*. Available from https://doi.org/ 10.2312/VAST/VAST11/041-048.

Srivastava, S., & Lall, B. (2019). DeepPoint3D: Learning discriminative local descriptors using deep metric learning on 3D point clouds. *Pattern Recognition Letters*, *127*, 27−36. Available from https://doi.org/10.1016/j.patrec.2019.02.027, http://www.journals.elsevier.com/pattern-recognition-letters/.

Stilla, U., & Jutzi, B. (2009). *Waveform analysis for small-footprint pulsed laser systems* (pp. 215−234). Informa UK Limited. Available from 10.1201/9781420051438.ch7.

Su, H., Jampani, V., Sun, D., Maji, S., Kalogerakis, E., Yang, M. H., & Kautzz, J. (2018). SPLATNet: Sparse lattice network for point cloud processing. *Proceedings of the International Conference on Computer Vision and Pattern Recognition*, 1−12. Available from https://arxiv.org/pdf/1802.08275.pdf.

Su, H., Maji, S., Kalogerakis, E., Learned-Miller, E. (2015). Multi-view convolutional neural networks for 3D shape recognition. *Proceedings of the IEEE International Conference on Computer Vision*. 2015, 945−953, 9781467383912. Institute of Electrical and Electronics Engineers Inc. United States. Available from https://doi.org/10.1109/ICCV.2015.114, http://ieeexplore.ieee.org/xpl/conhome.jsp?punumber = 1000149.

Vosselman, G., Coenen, M., & Rottensteiner, F. (2017). Contextual segment-based classification of airborne laser scanner data. *ISPRS Journal of Photogrammetry and Remote Sensing*, *128*, 354−371. Available from https://doi.org/10.1016/j.isprsjprs.2017.03.010.

Vo, A. V., Truong-Hong, L., Laefer, D. F., & Bertolotto, M. (2015). Octree-based region growing for point cloud segmentation. *ISPRS Journal of Photogrammetry and Remote Sensing*, *104*, 88−100. Available from https://doi.org/10.1016/j.isprsjprs.2015.01.011, http://www.elsevier.com/inca/publications/store/5/0/3/3/4/0.

Wagner, W. (2010). Radiometric calibration of small-footprint full-waveform airborne laser scanner measurements: Basic physical concepts. *ISPRS Journal of Photogrammetry and Remote Sensing*, *65*(6), 505−513. Available from https://doi.org/10.1016/j.isprsjprs.2010.06.007.

Wang, S., Suo, S., Ma, W. C., Pokrovsky, A., & Urtasun, R. (2018). Deep parametric continuous convolutional neural networks. *Proceedings of the International Conference on Computer Vision and Pattern Recognition*, 1−18. Available from https://arxiv.org/pdf/2101.06742.pdf

Wang, Z., & Lu, F. (2019). VoxSegNet: Volumetric CNNs for semantic part segmentation of 3D shapes. *IEEE Transactions on Visualization and Computer Graphics*. Available from https://doi.org/10.1109/TVCG.2019.2896310.

Wang, W., Yu, R., Huang, Q., & Neumann, U. (2019). SGPN: Similarity group proposal network for 3D point cloud instance segmentation. *Proceedings of the International Conference on Computer Vision and Pattern Recognition*, 1−13. Available from https://arxiv.org/pdf/1711.08588.pdf.

Wei, X., Yu, R., Sun J. (2020). View-GCN: View-based graph convolutional network for 3D shape analysis. *Proceedings of the IEEE Computer Society Conference on Computer Vision and Pattern Recognition*. IEEE Computer Society China. 10636919, 1847−1856. Available from https://doi.org/10.1109/CVPR42600.2020.00192.

Wood, R. L., & Mohammadi, M. E. (2021). Feature-based point cloud-based assessment of heritage structures for non-destructive and noncontact surface damage detection. *Heritage*, *4*(2), 775−793. Available from https://doi.org/10.3390/heritage4020043, https://www.mdpi.com/2571-9408/4/2/43/pdf.

Wu, W., Qi, Z., & Fuxin, L. (2020). PointConv: Deep convolutional networks on 3D point clouds. *Proceedings of the International Conference on Computer Vision and Pattern Recognition*, 1−10. Available from https://arxiv.org/pdf/1811.07246.pdf

Wu, Z., Song, S., Khosla, A., Yu, F., Zhang, L., Tang, X., Xiao, J. (2015). 3D ShapeNets: A deep representation for volumetric shapes. *Proceedings of the IEEE Computer Society Conference on Computer Vision and Pattern Recognition*. 07−12:1912−1920, 9781467369640. Available from https://doi.org/10.1109/CVPR.2015.7298801.

Xiao, W., Mills, J., Guidi, G., Rodríguez-Gonzálvez, P., Gonizzi Barsanti, S., & González-Aguilera, D. (2018). Geoinformatics for the conservation and promotion of cultural heritage in support of the UN Sustainable Development Goals. *ISPRS Journal of Photogrammetry and Remote Sensing*, *142*, 389−406. Available from https://doi.org/10.1016/j.isprsjprs.2018.01.001, http://www.elsevier.com/inca/publications/store/5/0/3/3/4/0.

Xie, Y., Tian, J., & Zhu, X. X. (2019). *A review of point cloud semantic segmentation.* arXiv, https://arxiv.org.

Xu, Y., Fan, T., Xu, M., Zeng, L., & Qiao, Y. (2018). SpiderCNN: Deep learning on point sets with parameterized convolutional filters. *European Conference on Computer Vision*, 1−17. Available from https://arxiv.org/pdf/1803.11527.pdf.

Yang, S., Xu, S., & Huang, W. (2022). 3D point cloud for cultural heritage: A scientometric survey. *Remote Sensing*, *14*(21), 5542. Available from https://doi.org/10.3390/rs14215542.

Yavartanoo, M., Kim, E. Y., & Lee, K. M. (2018). *SPNet: Deep 3D object classification and retrieval using stereographic projection.* arXiv. https://arxiv.org.

Yi, L., Kim, V. G., Ceylan, D., Shen, I. C., Yan, M., Su, H., Lu, C., Huang, Q., Sheffer, A., & Guibas, L. (2016). A scalable active framework for region annotation in 3D shape collections. *ACM Transactions on Graphics*, *35*(6), J778. Available from https://doi.org/10.1145/2980179.2980238, http://dl.acm.org/citation.cfm?id = .

Zhang, J., Lin, X., & Ning, X. (2013). SVM-based classification of segmented airborne LiDAR point clouds in urban areas. *Remote Sensing*, *5*(8), 3749−3775. Available from https://doi.org/10.3390/rs5083749, http://www.mdpi.com/2072-4292/5/8/3749/pdf, China.

Zhao, H., Jiang, L., Jia, J., Torr, P., & Koltun, V. (2020). Point transformer. *International Conference on Computer Vision*, 16259−16268. https://arxiv.org/pdf/2012.09164.pdf.

Zheng, Z., Zha, B., Zhou, Y., Huang, J., Xuchen, Y., & Zhang, H. (2022). Single-stage adaptive multi-scale point cloud noise filtering algorithm based on feature information. *Remote Sensing*, *14*(2), 367. Available from https://doi.org/10.3390/rs14020367.

Investigating the use of 3D laser scanning to detect damaged features in heritage buildings

9

Daniel Antón[1,2], José-Lázaro Amaro-Mellado[3], and Amin Al-Habaibeh[4]
[1]Departamento de Expresión Gráfica e Ingeniería en la Edificación, Escuela Técnica Superior de Ingeniería de Edificación, Seville, Spain, [2]The Creative and Virtual Technologies Research Lab & Product Innovation Centre, School of Architecture, Design and the Built Environment, Nottingham Trent University, Nottingham, United Kingdom, [3]Departamento de Ingeniería Gráfica, Universidad de Sevilla, Seville, Spain, [4]Product Innovation Centre, School of Architecture, Design and the Built Environment, Nottingham Trent University, Nottingham, United Kingdom

9.1 Introduction

The UNESCO World Heritage Convention (United Nations Educational Scientific and Cultural Organization UNESCO, 1972) stressed the uniqueness and need to protect heritage, the valuable and irreplaceable legacy from the past of all the peoples of the world (United Nations Educational Scientific and Cultural Organization, 2019). For heritage to prevail in this rapid-developing world, continuous maintenance and conservation actions must take place. To this end, the Athens and Venice Charters (International Council on Monuments and Sites, 2004, 2011) promoted the use of modern techniques and materials in the restoration, the process of returning heritage assets to their original condition and/or spatial layout, but also consolidating them to be durable over time, thus retaining their historic integrity. To support this, digital tools have been implemented in such a way that heritage assets can be studied, understood, and virtually reconstructed from documentary sources.

9.1.1 3D scanning in cultural heritage

In recent years, as indicated by Li et al. (2023), the current practice of built heritage protection is constantly based on three-dimensional LiDAR, which accounts for light detection and ranging. 3D LiDAR, which operates in an analogous way to radar, is an active sensor that records surrounding 3D information (Huang et al., 2022). As described by Huang et al. (2022), this technology obtains the distance of target bodies by illuminating laser signals (pulses) at specific wavelengths on those targets. This is performed using the time-of-flight (ToF) principle, which modulates laser beams over time. In other words, the emitter of the device shoots a laser pulse

Diagnosis of Heritage Buildings by Non-Destructive Techniques. DOI: https://doi.org/10.1016/B978-0-443-16001-1.00009-7

at the target body, and the receiver collects the reflected laser pulse. The ToF method measures the round-trip flight time of emitted laser pulses to determine the distance between the LiDAR device and the target. Carrying out throughout the target surface, this captures 3D point cloud data (spatial data in XYZ coordinates).

The use of 3D scanning technology has been extensively investigated to detect damage in heritage buildings in high resolution. In other words, it provides detailed insight into the real state of conservation and the behavior of the assets to generate new knowledge for their life cycle. This technology produces highly accurate qualitative and quantitative information that can be used to assess complex structures under a non-destructive and efficient approach, thus supporting restoration actions for the conservation of architectural and cultural heritage. Particularly, terrestrial laser scanning (TLS) can be used to detect and monitor surface deficiencies in heritage buildings (Antón et al., 2022) using both 3D point clouds and their laser intensity data (TLS radiometric data; red, green, and blue [RGB] values depending on the reflection of the laser beam on surfaces). Combined with unsupervised classification methods for digital image processing, TLS intensity data is also useful for the detection of damage deriving from moisture content in stone materials of historic buildings (Armesto-González et al., 2010). Likewise, other researchers (Lezzerini et al., 2016) used computer-aided design (CAD) and geographic information system (GIS) software to map different stone materials of the medieval Church of St. Nicholas in Pisa, Italy, using TLS data, high-resolution images, and organoleptic data of the building surface. As a result, the facade of that historic building was characterized in terms of materials, stages, and techniques used to build it. Regarding the virtual representation of geometrical alterations for heritage buildings and archaeological sites, TLS has been used as the data source to develop semiautomatic as-built or as-is 3D modeling approaches (Antón et al., 2018, 2019) compatible with Historic Building Information Models (HBIM). In this way, 3D meshing algorithms and visual programming language implemented in CAD software packages can be used to model both structural and surface irregularities and singularities of the assets. Consequently, the recording of surfaces using this remote sensing technique can be further applied to the early detection and monitoring of the damage and changes in historic building surfaces, that is, their surface wear and deterioration and their evolution over time. This is the case of the research by Dawson et al. (2022), who monitored and detected minor changes in a historic site thanks to TLS surveys in the long term. Similarly, Lercari (2019) conducted multitemporal TLS monitoring and surface change detection at the millimetric scale from the resulting 3D point clouds of Neolithic earthen structures.

As seen earlier, the scientific community has embraced TLS because of its great capabilities. In particular, it can be used in developing countries to digitize heritage buildings and sites that are endangered (United Nations Educational Scientific and Cultural Organization UNESCO, 2023). The threats include natural disasters, climate change and agents, the expansion of urban areas, pollution, war, uncontrolled tourism (United Nations Educational Scientific and Cultural Organization UNESCO, 2023), and other human activities and behaviors such as vandalism. Besides, TLS can be combined with many other technologies to:

- Provide a more comprehensive and integrated approach to studying heritage buildings and sites for conservation: The physical and spatial characteristics of the heritage assets can be analyzed alongside their information when using 3D point cloud data on Building Information Modeling (BIM) platforms and GIS. Moyano et al. (2021) experimented with HBIM parameterization through semantic segmentation applied to a TLS point cloud dataset of one of the facades of a heritage building in Seville, Spain. Klapa and Gawronek (2022) studied the synergies between TLS and UAVs (unmanned aerial vehicles) for the creation of HBIM, in which the authors placed greater importance on the accuracy of the measurement information than on the selection of the level of detail itself. Still, HBIM open-source technology is a cutting-edge topic, and an extensive review can be found in the research paper by Diara (2022). Haznedar et al. (2023) proposed a workflow for 3D point cloud segmentation for heritage buildings using deep learning by implementing PointNet, thus improving the HBIM capabilities. Regarding GIS, Campiani et al. (2019) inserted deterioration—calculated by comparing TLS data in different periods—and environmental values of earthen walls in a Neolithic site into a GIS to relocate conservation interventions to more urgent areas. Doğan and Yakar (2018) developed a GIS to integrate and document 3D data for cultural assets in Turkey. Pepe et al. (2021) employed a scan-to-BIM method for heritage assets to create a 3D GIS model that enabled a multidisciplinary view. This was conducted by integrating TLS and close-range photogrammetry, and a BIM project was loaded into a 3D GIS, thus facilitating multiple information connections.
- Significantly improve the efficiency and accuracy of damage detection in heritage buildings: Yang et al. (2023) reviewed the use of nonartificial intelligence–based algorithms to perform 3D point cloud segmentation of cultural heritage datasets, and machine learning and deep learning methods to conduct semantic segmentation. Thanks to these approaches, damage detection can be automated. Nevertheless, those authors indicated that, among other issues, these methods are mainly limited to historic buildings, suffer from oversegmentation, lack consistency when processing data from multiple sources, and should improve for larger datasets, although progress is being made in large-scale scene segmentation (Liao et al., 2022). Dayal et al. (2019) also analyzed damage detection of heritage monuments from TLS point clouds and terrestrial optical data in India. To this end, the authors converted the 3D point cloud into a 2D dataset, leading to an "unrolled" point cloud that was employed to generate raster images. They undertook the damage detection through an approach based on geometry and a radiometric data. Alkadri et al. (2022) examined surface fractures and material behavior from TLS point cloud attribute information for a church in Java, Indonesia.
- Improve the processing and analysis of massive TLS datasets, that is, a substantial number of points in space: The research work of Liu and Boehm (2015) and Pajić et al. (2018) has addressed the management of large-scale 3D point clouds in a big data context for processing and semantic classification. Nguyen et al. (2022) researched a cost-effective and user-friendly large point cloud rendering solution based on a potential distributed computing framework for big data storing and processing (Hadoop) for distributed computing applications in civil engineering, including progress monitoring, change detection, or indoor navigation. Compared to conventional solutions, they achieved improved performance, scalability, and fault tolerance. Duchnowski and Wyszkowska (2022) modeled vertical terrain displacement from massive TLS data using M_{split} estimation, which the authors proved to be more accurate than least-square or robust M-estimation.
- Make 3D scan data of historic buildings and sites more accessible and user-friendly, especially for nonexpert users, to support the dissemination and conservation of heritage assets. This can be done by importing 3D point cloud data into immersive technologies

such as augmented reality (AR), mixed reality (MR), and virtual reality (VR). Thus Patel et al. (2021) explored the methodology of bridge inspection from TLS and AR, mainly to cover unreachable spots, from the office, not from the site. Janeras et al. (2022) employed MR to display and distribute 3D data of the Montserrat Massif (Spain) during a stability assessment, leading to better risk communication to the users. Finally, Poux et al. (2020) designed a system to implement a VR application for multidisciplinary users, which was tested in a Belgium castle from a TLS 3D point cloud.

9.1.2   Limitations of terrestrial laser scanning

In spite of the numerous benefits of TLS and future advances and its combination with other diverse technologies, this remote sensing technique is not free of limitations or challenges. These issues can be divided into two different groups: financial and operational.

9.1.2.1   Financial

- Even if the purchase of the state-of-the-art devices is not considered, 3D scanning equipment can be expensive, especially when high resolution and accuracy are needed. According to Disney et al. (2019), TLS instruments over $100,000 are beyond the reach of most researchers. In contrast, low-cost devices are in the region of $20,000, which is still a significant amount of money for many users. For those cases, the Structure-from-Motion (SfM) photogrammetric technique is the very low-cost alternative, for which the only requirement for data collection is access to a camera. Furthermore, handling and processing massive 3D point cloud data requires high-performance workstations and/or laptops.

9.1.2.2   Operational

- Time-related: Especially for large and complex heritage buildings and sites that require hundreds of stations (scan positions), TLS surveying can be a time-consuming process, although there are few other options currently available. The time that the fieldwork takes does not only depend on the (scan and image) recording period but also on the survey planning, the leveling of the tripod, and the recording of targets for subsequent registration. In addition, as seen in the work of Julin et al. (2020), TLS imaging quality also influences the total data acquisition time spent in the field since the performance of the equipment significantly differs when choosing low dynamic range or high dynamic range (HDR) imaging. Finally, the processing and analysis of 3D point cloud data may require spending considerable time depending on the computer specifications, the number of stations and points recorded, the resolution and other parameters set in the TLS survey, the intended point density, the segmentation and cleaning of the data, among many others, with a view to extracting meaningful information.
- Image quality: Research has found poor colourization of 3D point clouds depending on which TLS device is used (Julin et al., 2020), and their image quality is far inferior to what can be achieved by SfM photogrammetry. As explained by Julin et al. (2020), low image quality has an impact on the direct applicability of colored 3D point clouds to diverse cases that need to rely on visual appearance or radiometric values for object interpretation and recognition, visual analysis, or photorealism. Those authors considered that enhanced coloring of 3D point cloud data would be relevant and useful for traditional

application areas. These domains cover engineering, surveying, or cultural heritage but also emerging application fields such as virtual production in the film industry or the creation of 3D content for video games and immersive experiences.

- Accuracy: The accuracy of the 3D recording depends on numerous factors. Firstly, it depends on the TLS device used. Here, as seen in the work of Chen et al. (2018), the ability of the LiDAR sensor to measure both time and the laser beam width has an impact on the recording accuracy. Secondly, the accuracy of 3D scanning results depends on several factors such as illuminance and atmospheric conditions, surface characteristics of the building (materials, moisture content, dust, among others), number and location of stations, distance to the object, point of view, and resolution. These aspects also influence the intensity of the point cloud data, as the laser beam incidence angles and distances may differ.

With a special focus on accuracy, the scientific community has studied the effect of TLS parameters on crack width measurement in different materials and damage sizes. Nevertheless, the quantification of depth recording errors still needs to be addressed, especially in hybrid materials (Oytun & Atasoy, 2022). For their part, other researchers (Tan & Cheng, 2017) investigated the effects of specular reflection on the accuracy of the scan data. They corrected the incidence angle to eliminate the specular reflection effects for TLS intensity image interpretation and 3D point cloud representation by intensity. The authors explained that scanning at larger incidence angles avoids the influence of specular reflection effects in the intensity data of smooth surfaces. Consequently, perpendicular scanning should be avoided, as in the case of a different remote sensing technology, infrared thermography, where the camera operators themselves constitute a source of emissions that must be taken into account (Antón & Amaro-Mellado, 2021). However, as indicated by Soudarissanane et al. (2011), increasing incidence angles implies higher measurement errors, that is, lower quality in point clouds. Finally, Tan et al. (2018) studied anomalous distance measurement errors caused by target specular reflections. Their results revealed that distance measurement errors are strongly related to the original intensity values, so the correction of those errors leads to significantly improved accuracy.

In view of all the above limitations and determining factors of TLS surveys, this chapter aims to investigate the accuracy and precision of 3D laser scanning positioning to detect damage features of heritage buildings and how to enhance detection and monitoring for their preservation over time.

The rest of this chapter is organized as follows: Section 9.2 describes the case study that will be used in the investigation, details on the TLS survey, processes to carry out the geometry and accuracy analyses on 3D point clouds and meshes, and the specifications of the equipment used for data processing and analysis. Section 9.3 presents the results and their discussion, as well as the limitations of the research. The conclusion and future work are described in Section 9.4.

9.2 Methodology

This research investigates the accuracies and positioning of TLS survey for the recording of surface damages in heritage buildings. Hence, it is worth describing the selected case study for that purpose.

9.2.1   Case study: St John the Baptist Church in Nottingham, United Kingdom

This research is developed through the case study of the Church of St John the Baptist, also called Beeston Parish Church, an Anglican temple in Beeston, Borough of Broxtowe, in the city of Nottingham (Nottinghamshire County, East Midlands region in England, the United Kingdom). The site map with coordinates is shown in Fig. 9.1.

The Church of St John the Baptist was listed as a Grade II historic building by Historic England (1987) with List Entry Number: 1263823 (it was formerly listed 18.lO.49 as Parish Church of St John). This temple dates from the mid-19th century; since then, restoration interventions have taken place on its tower and bells, roofs, and interior walls (The Southwell and Nottingham Church History Project, 2021). This heritage building presents diverse surface pathologies such as patina of lichen and moss, moisture, and stone mass loss in masonry blocks. The latter is the case of a section of a wall of approximately 1.65 m (width) × 1.95 m (height) oriented to the West, with damaged blocks scattered on it, mainly in the bottom and at mid-height. The target wall is indicated in Fig. 9.2.

9.2.2   3D laser scanning

To study the TLS measurements on the aforementioned wall, a comprehensive 3D survey was first carried out throughout the exterior and interior of the building. However, the interior was not considered in this research as it was beyond its scope. The geometry of the temple was recorded using a tripod-mounted Leica Geosystems ScanStation P20 3D laser scanner (Leica Geosystems, 2012), as can be

Figure 9.1 Site map of St John the Baptist Parish Church with coordinates.
Source: Google Earth with 2D satellite imagery.

Figure 9.2 Photograph with indications St John the Baptist Parish Church: north facade and target wall.
Source: The authors, based on Google Street View.

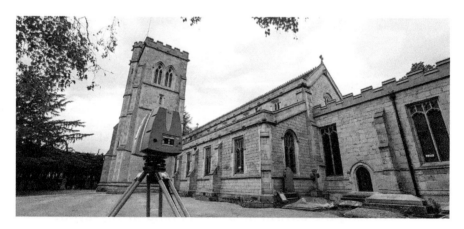

Figure 9.3 Photograph St John the Baptist Beeston Parish Church and the 3D laser scanner: wide-lens southeast perspective.

seen in Fig. 9.3. The survey parameters were as follows: the resolution was set at 6.3 mm at 10 m and 3.1 mm for closer and further stations, respectively; full scanning range and horizontal and vertical angles were established; and HDR imaging was selected.

The TLS survey was carried out according to previous planning consisting of a total of 17 stations surrounding the building. No targets were placed outside the temple given the sufficient overlap (enough points) between scans.

9.2.2.1   Registration diagnostics

Leica Cyclone 9.4 (Leica Geosystems, 2019) was used to process the scan data after collection. This allowed for creating constraints between the clouds so that the stations were registered, that is, aligned in the same coordinate system. This process was mainly automatic because of the overlap described above. However, given the numerous trees surrounding the building, the algorithm may have had difficulties registering the scans. As a result, certain scan links (constraints) needed the manual selection of three pairs of common points between the datasets for accurate alignment, which was further optimized by using the automatic tool in Cyclone. Next, the program was also used to generate the alignment report (registration diagnostics, Table 9.1), which consisted of essential data to certify and understand the accuracy of the process. In addition to specifying the stations matched, the strength of the scan links, the overlapping points, and the average errors, the report included the Root Mean Square Error (RMSE) of each constraint as per Eqs. (9.1) and (9.2) (Xu et al., 2021).

The mean absolute error of the TLS survey registration was only 0.002 m (2 mm). In relation to the RMSE, the lower it is, the better the registration result.

$$R\mu SE = \sqrt{\frac{1}{N} \cdot \sum_{i=1}^{N} \left\| P_i - q_j \right\|^2} \qquad (9.1)$$

with

$$1 \le j \le M, \qquad (9.2)$$

where p_i and q_j are, respectively, the nearest corresponding pair of points in clouds P and Q, and N and M account for the registration scales of those clouds.

9.2.3   Geometric and accuracy analysis

9.2.3.1   Assessing the terrestrial laser scanning survey layout

Once the details of the alignment have been presented, it is worth showing the selected stations to be used to assess the accuracy of the damage recording on the target wall. To do this, this research took into account the findings of Tan and Cheng (2017) to avoid specular reflection effects in 3D point cloud intensity data by scanning at greater angles and avoiding perpendicular recording, at least over short distances. At the same time, consideration was given to what Soudarissanane et al. (2011) reported on the occurrence of higher measurement errors at large angles of incidence, thus entailing lower quality in point clouds. The layout of the stations also depended on the planned positions of the scanner to avoid occlusions affecting the TLS recording of the entire exterior of the church.

Table 9.1 Registration diagnostics of the exterior TLS survey of Beeston Parish Church.

Constraint ID	Station no.	With station no.	Weight (coefficient)	Overlap (points)	Average error (m)	RMSE (m)
1	6	16	0.9218	714,800	0.003	0.023
2	7	16	0.8251	332,366	0.003	0.028
3	13	14	1.0000	319,433	0.001	0.019
4	1	2	1.0000	885,066	0.003	0.024
5	8	9	0.7213	335,033	0.001	0.026
6	3	4	1.0000	885,966	0.002	0.021
7	5	6	0.5751	487,066	0.004	0.018
8	3	15	0.7300	622,200	0.002	0.021
9	7	8	0.5048	116,033	0.002	0.029
10	15	17	0.6369	901,700	0.001	0.021
11	8	10	0.4648	214,800	0.001	0.025
12	5	7	0.4453	169,700	0.004	0.024
13	4	5	0.5345	608,433	0.001	0.018
14	6	7	0.3883	216,366	0.002	0.025
15	9	10	0.3485	463,833	0.001	0.021
16	3	5	0.4384	658,800	0.003	0.022
17	10	11	0.1964	549,266	0.004	0.022
18	1	13	1.0000	243,700	0.001	0.024
19	1	14	1.0000	319,966	0.001	0.023
20	2	3	1.0000	525,933	0.002	0.021
21	2	4	1.0000	435,766	0.001	0.021
22	2	14	1.0000	319,666	0.002	0.025
23	9	11	1.0000	513,400	0.002	0.025
24	11	12	1.0000	479,366	0.001	0.025
25	12	13	1.0000	477,533	0.001	0.025
26	12	14	1.0000	324,833	0.001	0.027

Six out of the 17 stations of the Beeston Parish Church TLS survey captured the geometry of the target wall. However, considering the abovementioned criteria, two stations were discarded for excessive incidence angle and occlusions. Therefore four stations were chosen to analyze the accuracy of the recording and damage detection on the heritage building. From the arrangement of these four remaining stations, two of them were selected as the ground truth for the analysis (stations 17 and 15). They were the closest to the target wall (A and B, respectively, in Fig. 9.4 and Tables 9.2 and 9.3), thus providing the highest resolution in the 3D point cloud and a low average error (1 mm). These stations also faced the target wall from each side.

To ease understanding, a letter was given to each station from the registration diagnostics list depending on their distance to the target wall (please see Table 9.1 and Fig. 9.4, and Table 9.2).

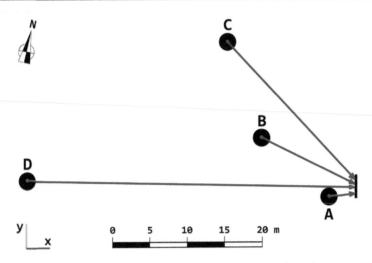

Figure 9.4 Diagram layout of selected stations for accuracy analysis, and target wall (right): top view.

Table 9.2 Reader-friendly naming of scan stations.

Station ID	17	15	3	1
Station name	A	B	C	D

Table 9.3 Distance from stations to target, approximate incidence angle, scanning resolution, and measured wall target cloud resolution.

Station name	A	B	C	D
Mean distance to target (m)	3.85	14.15	25.50	43.95
Incidence angle (degree)	70	63	42	89
Scanning resolution at 10 m (mm)	6.3	3.1	3.1	3.1
Point resolution in wall target cloud (mm)	<3	<5	<8	<14

9.2.3.2 Accuracy analysis

To analyze the recorded surface at each station, a previous segmentation process was needed to extract the target wall from the 3D point cloud. This also allowed for the removal of noise and unwanted sectors. Open-source software such as CloudCompare (Girardeau-Montaut, 2016) permits the manual creation of polygon fences to enclose the desired points and then compute the trimming.

As described earlier, the geometry of reference (ground truth) consisted of point cloud data of the target wall from stations A and B (1 mm average error in the alignment; please see Table 9.1). Therefore it is worth analyzing the accuracy of stations C and D to assess the suitability of different points of view for the detection of damage on heritage building surfaces.

Point deviation analysis

The Cloud-to-Cloud (C2C) Distance computing tool (Girardeau-Montaut, 2015) in CloudCompare software enables point deviation measurements to be conducted between two point clouds. In addition to a histogram of the distance between them, statistical data such as mean distance and standard deviation are also provided to offer insight into the scanning accuracy. This was useful to compare the clouds from stations C and D against the reference cloud to assess their accuracies, but also if combining the point clouds from those two unfavorable stations (due to excessive distance and incidence angle) significantly improves the accuracy of the scanning. The results will be shown in Section 9.3.

Accuracy of damage modeling: 3D meshing

The scientific community has demonstrated that TLS data is useful in capturing heritage building surface deficiencies. However, 3D modeling constitutes a step forward toward further analysis and simulation. 3D meshes (triangle-based 3D objects) and 3D solid models of heritage assets in CAD and BIM environments have been used for virtual reconstruction and studies to contribute to their conservation and dissemination. For these reasons, the accuracy analysis of 3D meshing should be addressed to ensure surface defects in historic buildings can be represented. In this sense, this research focuses on the following accuracy indicators:

1. Different smoothening degrees in the 3D meshing of the reference geometry (target wall cloud from stations AB, ground truth in the analysis) should be analyzed and compared with each other to find the optimal smoothening degree to model surface deficiencies in the heritage building.
2. The accuracy of clouds of the target wall from stations A and B should be verified against the reference cloud and optimal mesh (AB) to find the most suitable TLS parameters in the survey. This was carried out by running the C2C tool and the Cloud-to-Mesh Distance computing tool (Girardeau-Montaut, 2015) in CloudCompare software, which enables point deviation measurements to be conducted between a point cloud and a mesh, or those of two meshes with each other, although the vertices of one of them will be chosen instead the mesh itself.
3. Finally, it is worth analyzing the accuracy of each component of the ground truth point cloud data, A and B, against the reference 3D mesh (AB) to calculate the distortion from discretizing their geometry.

3D meshes were generated from the selected 3D point clouds using a plug-in in CloudCompare based on the Screened Poisson Surface Reconstruction algorithm developed by Kazhdan & Hoppe (2013). This requires that the point sets have their normal vectors calculated. Once this was conducted, the smoothening degree was determined by selecting the Octree depth value (level). Qualitative testing indicated that levels 9 and 10 enabled the representation of surface defects on the target wall without excessive simplification (loss of deformation details) or number of triangles (larger file size), respectively. As a consequence, both levels needed to be analyzed to determine the optimal smoothening degree for accuracy analysis through Cloud-to-Mesh and Mesh-to-Mesh distance computations.

When creating an open surface that extends beyond the point cloud edges (Neumann meshing method), it is necessary to segment the excess part so that its vertexes are not taken into account in the accuracy analysis. This was carried out by filtering mesh vertexes by the desired density values, so that outliers that were not part of the target wall geometry were removed. To do this, it was possible to set the 3D mesh density as scalar field values thanks to the aforementioned CloudCompare plug-in.

9.2.4   Equipment used

As seen in previous sections, the TLS equipment consisted of a Leica Geosystems ScanStation P20 3D laser scanner. The computer used to process and 3D mesh the 3D point cloud data and perform the analysis was a high-performance gaming laptop with an octa-core processor with hyperthreading at 2.30 GHz and a maximum turbo frequency of 4.60 GHz with 24-MB cache, 32-GB RAM DDR4 @ 3200 MHz, a 256-bit graphics card with 6144 cores @ 1245−1710 MHz and 8-GB GDDR6 dedicated memory @ 14 Gbps, and a 1TB NVMe PCIe Gen3x4 SSD (solid-state drive).

9.3   Results and discussion

This section aims to gather, interpret, and examine the outcomes of the accuracy analysis on 3D point clouds of each TLS survey station and the 3D meshes of the surface deficiencies detected on the target wall of the case study. This includes analyzing the point deviation of station clouds that derive from their incidence angle, resolution, and distance to the target against the ground truth, the reference dataset resulting from the combination of two accurate stations (A and B) for presenting more favorable surveying features.

9.3.1   Suitability of terrestrial laser scanning surveying characteristics for damage detection on heritage buildings

This subsection addresses the accuracies of both 3D point clouds and meshes by applying the approaches described in Section 9.2.

9.3.1.1   Accuracies of station clouds

Qualitatively speaking, special mention should be made of the evident difference in resolution between the 3D point cloud data from stations D and C, and the ground truth (AB) (please see Fig. 9.5). There is a greater distance between points in the former stations in comparison with the dense cloud in the more accurate cloud AB (right). Here, in the reference stations, shorter distances to the target wall and

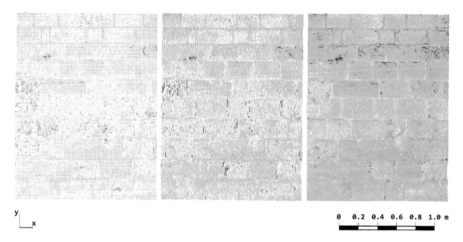

y
x

0   0.2 0.4 0.6 0.8 1.0 m

Figure 9.5 Render image target wall clouds from stations D (left), C (center), and the ground truth AB (right). Rainbow color gradient for intensity visualization: elevation view.

increased resolution set before scanning made the difference. This can also be seen in Table 9.3 with quantitative data. Considering this table, station C implied a greater incidence angle of the laser beam on the target wall, whose implications in mean distance between clouds and point accuracy (standard deviation value) will be shown below. Fig. 9.5 shows the two compared stations and the reference cloud, which is much more detailed. Because of the greater angle in station C (center), the points are not arranged in a quasi-regular grid as in station D (left). This entails lower accuracy in cloud C in comparison with a hypothetical station established at the same distance as C but with a lower angle (closer to perpendicularity) to the target wall.

The C2C Distance tool in CloudCompare yielded the following data from the point deviation analysis in clouds:

Station D against the ground truth (AB)
These are the results of the accuracy analysis of the most distant station with the most perpendicular angle and the reference cloud:
Mean distance = 0.004341 m
Standard deviation = 0.001952 m
The histogram of absolute point distances is shown in Fig. 9.6.

Station C against the ground truth (AB)
The accuracy of the far-intermediate cloud with the greatest angle (histogram of point distances in Fig. 9.7):
Mean distance = 0.002897 m
Standard deviation = 0.001245 m

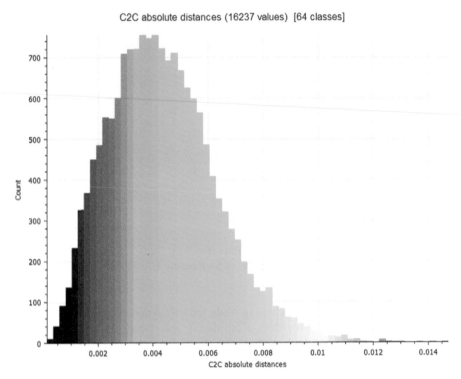

Figure 9.6 Histogram of the geometric comparison between station D and the ground truth.

Stations D and C combined against the ground truth (AB)

The accuracy of the combination between stations D and C and the histogram of point distances (Fig. 9.8) are given as follows:

    Mean distance = 0.003370 m

    Standard deviation = 0.001658 m

    Given the standard deviation values from the geometric comparison between stations D, C, and their combination (D + C) against the ground truth (AB), it can be concluded that merging the datasets of both stations improves the accuracy of the point cloud from the most unfavorable station (D, also in terms of resolution) but does not enhance their accuracy as a joint point cloud.

Stations B and A against the ground truth (AB)

Before comparing stations B and A with the reference cloud, it is worth analyzing their accuracy with each other (the histogram of their point deviation analysis is shown in Fig. 9.9).

    Mean distances = 0.001762 m

    Standard deviations = 0.000947 m

    In view of the accuracies of stations D and C against the reference dataset, stations B and A evidence a greater similarity with each other. Specifically: 59.41%,

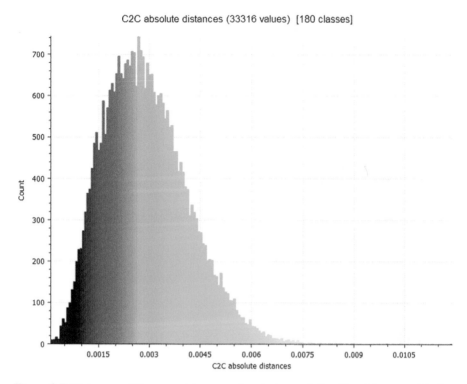

Figure 9.7 Histogram of the geometric comparison between station C and the ground truth.

39.18%, and 47.72% decrease against stations D, C, and C + D in mean distances and 51.49%, 23.94, and 42.88% decrease in standard deviation, respectively.

Besides, given the fact that stations A and B were used to build the ground truth by merging their clouds, it was expected that the comparison between them and the reference point cloud data (AB) yielded significantly low mean distances and high accuracies, as seen in the following:

Mean distances = 0.000000 m (B); 0.000001 m (A)

Standard deviations = 0.000030 m (B); 0.000042 m (A)

In these cases, given the great similarity between the clouds, the histograms are not provided. The datasets should be compared with the geometry of the reference 3D mesh (mesh AB Octree level 10, which will be validated below).

The analysis of station B against the reference mesh yielded the following data and histogram (Fig. 9.10):

Mean distance = 0.000826 m

Standard deviation = 0.000786 m

The analysis of station A against the reference mesh yielded the following data and histogram (Fig. 9.11):

Mean distance = 0.000225 m

Standard deviation = 0.000400 m

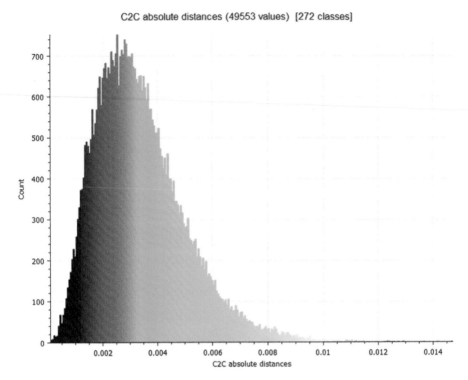

Figure 9.8 Histogram of the geometric comparison between the joint cloud from stations D and C and the ground truth.

The results of the analysis reveal that, due to the smaller distance of station A to the target wall and its slightly smaller angle of incidence, the cloud from A is twice as accurate as station B despite the fact that the scanning resolution was set at twice its value. The vertical scale in the histograms of stations B and A was constant to ease the recognition of the higher point accuracy in the latter.

9.3.1.2 Accuracies of 3D meshes

The 3D meshing algorithm was used on the reference cloud (AB) of the target wall to constitute a solid basis of geometrical data for the accuracy analysis of station clouds. Likewise, the chosen 3D meshing smoothening degree was validated for the purpose of modeling defects on heritage surfaces.

Fig. 9.12 illustrates the 3D mesh created by selecting level 10 of the Octree Depth parameter in CloudCompare both with the TLS imaging colors and showing the graded point neighboring density on its surface.

With a view to validating the meshing algorithm parameters to create the geometry in Fig. 9.12, that 3D mesh was compared against a smoother mesh that benefits from a lower polygon count, that is, a lower triangle resolution (bigger triangles), and, therefore, a lower file size.

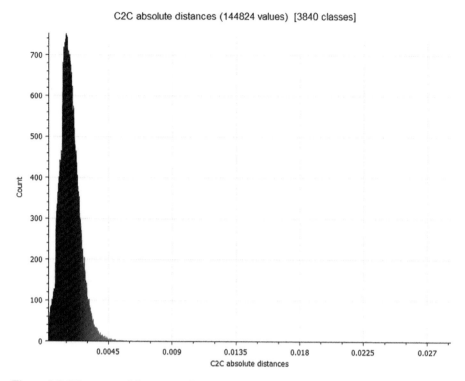

C2C absolute distances (144824 values) [3840 classes]

Figure 9.9 Histogram of the geometric comparison between stations B and A.

Smooth mesh (AB, Octree level 9) against the reference mesh (AB, Octree level 10)

In qualitative terms, the relief of the 3D meshes of the target wall is clearly lower when applying level 9 to the smoothening degree (please see the three-part Fig. 9.13), which has an impact on the representation of surface defects on heritage buildings.

Focusing on quantitative data, the accuracy of 3D mesh level 9 against level 10 (reference mesh) is given below:

Mean distance = 0.000052 m

Standard deviation = 0.000295 m

Scalar field RMS = 0.000299847 m

The histogram of their point deviation analysis is shown in Fig. 9.14.

Here, the 0.05 mm average distance between meshes and the approximately 0.3 mm errors could be accepted in favor of a lower polygon count. However, the global surface (area) and triangle surface (3D mesh resolution) of these meshes become essential aspects with regard to the reliability of the 3D representation. They can also be quantified so that the decision is based on accurate data.

Level 9:

Mesh surface = 3.18227 m^2

Average triangle surface: 7.22781 $\times$ 10^{-6} m^2

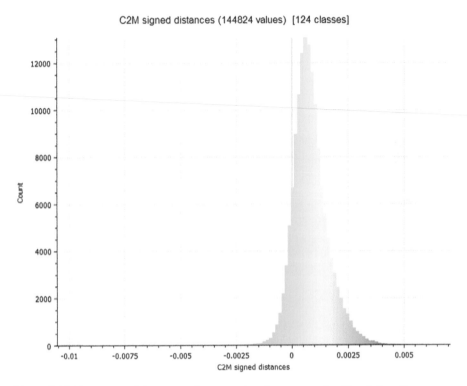

Figure 9.10 Histogram of the geometric comparison between station B and the reference mesh.

Level 10:

Mesh Surface = 3.22211 m^2

Average triangle surface: 1.70902×10^{-6} m^2

In view of these data, there is an evident loss of geometry when smoothening the meshes. In particular, the level 9 mesh loses 1.24% of geometry, and its triangle resolution significantly decreases (greater triangle size, meaning a poorer geometry) by 76.35%.

Therefore the 3D mesh generated using the Octree Depth level 10 was chosen as the reference mesh in the comparison for its ability to represent building surface deficiencies in a more accurate and representative way.

9.3.2 Limitations

In relation to the scan registration diagnostics from the TLS survey of Beeston Parish Church (Table 9.1), the reason for the alignment errors (RMSE values [from 2 to 3 mm] and average error of 4 mm in some cases) may have been the constant movement of the leaves and branches of the trees surrounding the temple.

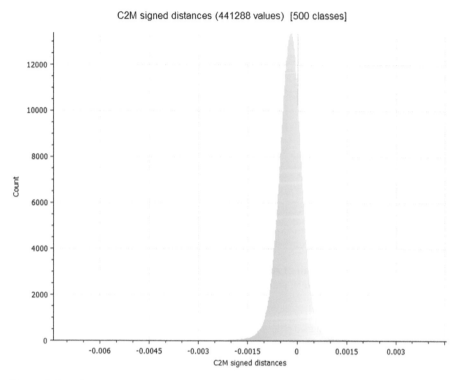

Figure 9.11 Histogram of the geometric comparison between station A and the reference mesh.

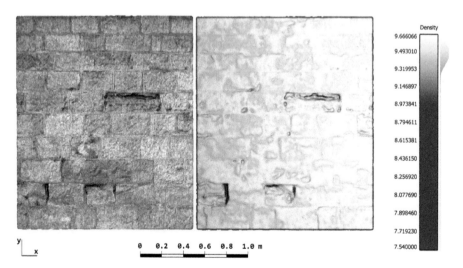

Figure 9.12 Render image Octree level 10 3D mesh of the target wall in RGB and point neighboring density visualization modes: elevation view. *RGB*, Red, green, and blue.

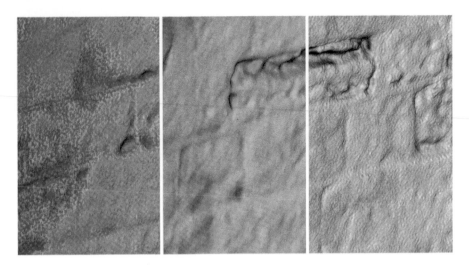

Figure 9.13 Render image detail of surface defects on 3D meshes: levels 9 and 10 overlapping (left); level 9 (center); and level 10 (right).

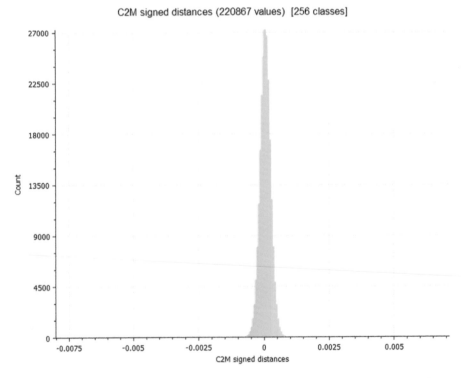

Figure 9.14 Histogram of the geometric comparison between 3D mesh level 9 and level 10 (reference mesh).

Nevertheless, as the comparison between stations A and B shows, their low alignment average error (1 mm) and great similarity (1.76 mm mean distance and 0.95 mm standard deviation) account for an accurate global registration, thus validating the accuracy analyses of point clouds and 3D meshes carried out.

The reason why there are no round distances and incidence angles between the stations and the target wall in Table 9.3 is that the experiment was not carried out under laboratory conditions but after an actual TLS survey at Beeston Parish Church for subsequent 3D modeling. Nonetheless, this does not translate into inaccurate analysis results.

9.4 Conclusion

3D laser scanning, particularly TLS, is commonly used to capture dimensions and develop 3D CAD models in buildings. In the field of cultural heritage, this technology is becoming crucial because of the need for a virtual record of buildings and to detect their surface wear and deterioration. Scientific research has demonstrated that exhaustive as-built modeling from 3D point cloud data allows for performing diverse accurate analyses of heritage buildings and sites. However, some issues in TLS surveying are the number and location of stations, the distance to the object, the point of view, and the resolution. These factors have a significant impact not only on the scanning and modeling accuracies but also on the intensity of the point cloud data for visual inspection and quantitative analyses.

This chapter investigates the accuracy and precision of TLS positioning and resolution to produce detailed models of damage features on heritage buildings. A sector of a masonry wall in an Anglican temple from the mid-19th century, St John the Baptist Beeston Parish Church in Nottingham, United Kingdom, was selected as a case study for this purpose.

This research sheds light on the accuracy that can be reached in a real TLS survey by combining different points of view and distances to target bodies. The results reveal that point cloud data of the building from a low-resolution scan station with a suitable angle to target can see its accuracy enhanced when combined with the cloud from another nearer station at a greater angle. Nonetheless, this research detected that the joint accuracy is still under that of the second station.

In addition, this chapter determined the suitable smoothening degree for a case study of such characteristics to model the geometrical alterations of the damage features on the target wall. The distortion from discretizing (triangle-based 3D meshing) the 3D point cloud geometry was analyzed and validated for the accuracy analysis of scan stations.

Finally, although it is expected that 3D scanning technologies will continue to improve, future research will delve into the analysis of additional samples of stations at different distances and incidence angles so that the scanning and modeling accuracies for geometrical alterations of surface pathologies at the millimetric scale are analyzed under laboratory conditions for the digitization and virtual reconstruction of heritage assets.

Acknowledgments

This research has been supported by funding for a postdoctoral researcher contract from the VI Plan Propio de Investigación y Transferencia of Universidad de Sevilla (reference VIPPIT-2020-II.5), Spain.

Additional support was granted by the "Live Experiential and Digital Diversification – Nottingham" (LEADD:NG) project, part-funded by the European Union European Regional Development Fund (ERDF) (reference 08R20S04177) as part of the European Structural and Investment Funds Growth Programme 2014–2020, in partnership with Midlands Engine (The Government of the United Kingdom, HM Government), University of Nottingham, and Nottingham Trent University (United Kingdom).

The authors also wish to thank the School of Architecture, Design and the Built Environment at Nottingham Trent University for access to their TLS equipment and workstations for data processing.

Special thanks to Father Wayne Plimmer and St John the Baptist Beeston Parish Church for willingly granting access to their facilities.

References

Alkadri, M. F., Alam, S., Santosa, H., Yudono, A., & Beselly, S. M. (2022). Investigating surface fractures and materials behavior of cultural heritage buildings based on the attribute information of point clouds stored in the TLS dataset. *Remote Sensing, 14*(2), 410. Available from https://doi.org/10.3390/rs14020410, https://www.mdpi.com/2072-4292/14/2/410.

Antón, D., & Amaro-Mellado, J.-L. (2021). Engineering graphics for thermal assessment: 3D thermal data visualisation based on infrared thermography, GIS and 3D point cloud processing software. *Symmetry, 13*(2), 335. Available from https://doi.org/10.3390/sym13020335, https://www.mdpi.com/2073-8994/13/2/335.

Antón, D., Carretero-Ayuso, M. J., Moyano-Campos, J., & Nieto-Julián, J. E. (2022). *Laser Scanning Intensity Fingerprint: 3D Visualisation and Analysis of Building Surface Deficiencies New Technologies in Building and Construction* (pp. 207–223). Singapore: Springer Nature Singapore. Available from https://link.springer.com/10.1007/978-981-19-1894-0_12, https://doi.org/10.1007/978-981-19-1894-0_12.

Antón, D., Medjdoub, B., Shrahily, R., & Moyano, J. (2018). Accuracy evaluation of the semi-automatic 3D modeling for historical building information models. *International Journal of Architectural Heritage, 12*(5), 790–805. Available from https://doi.org/10.1080/15583058.2017.1415391, https://www.tandfonline.com/doi/full/10.1080/15583058.2017.1415391.

Antón, D., Pineda, P., Medjdoub, B., & Iranzo, A. (2019). As-built 3D heritage city modelling to support numerical structural analysis: Application to the assessment of an archaeological remain. *remote sensing, 11*(11), 1276. Available from https://doi.org/10.3390/rs11111276, https://www.mdpi.com/2072-4292/11/11/1276.

Armesto-González, J., Riveiro-Rodríguez, B., González-Aguilera, D., & Rivas-Brea, M. T. (2010). Terrestrial laser scanning intensity data applied to damage detection for historical buildings. *Journal of Archaeological Science, 37*(12), 3037−3047. Available from https://doi.org/10.1016/j.jas.2010.060.031.

Campiani, A., Lingle, A., & Lercari, N. (2019). Spatial analysis and heritage conservation: Leveraging 3-D data and GIS for monitoring earthen architecture. *Journal of Cultural Heritage, 39,* 166−176. Available from https://doi.org/10.1016/J.CULHER.2019.020.011.

Chen, Z., Fan, R., Li, X., Dong, Z., Zhou, Z., Ye, G., & Chen, D. (2018). Accuracy improvement of imaging lidar based on time-correlated single-photon counting using three laser beams. *Optics Communications, 429,* 175−179. Available from https://doi.org/10.1016/j.optcom.2018.080.017, https://linkinghub.elsevier.com/retrieve/pii/S0030401818307004.

Dawson, P., Brink, J., Farrokhi, A., Jia, F., & Lichti, D. (2022). A method for detecting and monitoring changes to the Okotoks Erratic − "Big Rock" provincial historic site. *Journal of Cultural Heritage Management and Sustainable Development.* Available from https://doi.org/10.1108/JCHMSD-10-2021-0183/FULL/PDF, ahead-of-p (ahead-of-print).

Dayal, K. R., Tiwari, P. S., Sara, R., Pande, H., Senthil Kumar, A., Agrawal, S., & Srivastav, S. K. (2019). Diagnostic utilisation of ground based imaging and non-imaging sensors for digital documentation of heritage sites. *Digital Applications in Archaeology and Cultural Heritage, 15,* e00117. Available from https://doi.org/10.1016/j.daach.2019.e00117, https://linkinghub.elsevier.com/retrieve/pii/S2212054818300481.

Diara, F. (2022). HBIM Open Source: A review. *ISPRS International Journal of Geo-Information, 11*(9), 472. Available from https://doi.org/10.3390/ijgi11090472, https://www.mdpi.com/2220-9964/11/9/472.

Disney, M., Burt, A., Calders, K., Schaaf, C., & Stovall, A. (2019). Innovations in ground and airborne technologies as reference and for training and validation: terrestrial laser scanning (TLS). *Surveys in Geophysics, 40*(4), 937−958. Available from https://doi.org/10.1007/S10712-019-09527-X/FIGURES/9, https://link.springer.com/article/10.1007/s10712-019-09527-x.

Doğan, Y., & Yakar, M. (2018). GIS and three-dimensional modeling for cultural heritages. *International Journal of Engineering and Geosciences.* Available from https://doi.org/10.26833/ijeg.378257, https://dergipark.org.tr/en/doi/10.26833/ijeg.378257.

Duchnowski, R., & Wyszkowska, P. (2022). M_{split} estimation approach to modeling vertical terrain displacement from TLS data disturbed by outliers. *Remote Sensing, 14*(21), 5620. Available from https://doi.org/10.3390/rs14215620, https://www.mdpi.com/2072-4292/14/21/5620.

Girardeau-Montaut, D. (2015). *Cloud-to-cloud distance* https://www.cloudcompare.org/doc/wiki/index.php/Cloud-to-Cloud_Distance.

Girardeau-Montaut, D. (2016). *Open Source Project CloudCompare: 3D point cloud and mesh processing software.* http://www.danielgm.net/cc/.

Haznedar, B., Bayraktar, R., Ozturk, A. E., & Arayici, Y. (2023). Implementing PointNet for point cloud segmentation in the heritage context. *Heritage Science, 11*(1), 2. Available from https://doi.org/10.1186/s40494-022-00844-w, https://heritagesciencejournal.springeropen.com/articles/10.1186/s40494-022-00844-w.

Historic England. (1987). *Listing Church of St John the Baptist, Non Civil Parish - 1263823.* https://historicengland.org.uk/listing/the-list/list-entry/1263823.

Huang, J., Ran, S., Wei, W., & Yu, Q. (2022). Digital integration of LiDAR system implemented in a low-cost FPGA. *Symmetry, 14*(6), 1256. Available from https://doi.org/10.3390/sym14061256, https://www.mdpi.com/2073-8994/14/6/1256.

International Council on Monuments and Sites. (2004). *History of the Venice Charter.* https://www.icomos.org/venicecharter2004/history.pdf.

International Council on Monuments and Sites. (2011). *The Athens Charter for the Restoration of Historic Monuments — 1931.* https://www.icomos.org/en/167-the-athens-charter-for-the-restoration-of-historic-monuments.

Janeras, M., Roca, J., Gili, J. A., Pedraza, O., Magnusson, G., Núñez-Andrés, M. A., & Franklin, K. (2022). Using mixed reality for the visualization and dissemination of complex 3D models in geosciences—Application to the Montserrat Massif (Spain). *Geosciences, 12*(10), 370. Available from https://doi.org/10.3390/geosciences12100370, https://www.mdpi.com/2076-3263/12/10/370.

Julin, A., Kurkela, M., Rantanen, T., Virtanen, J.-P., Maksimainen, M., Kukko, A., Kaartinen, H., Vaaja, M. T., Hyyppä, J., & Hyyppä, H. (2020). Evaluating the quality of TLS point cloud colorization. *Remote Sensing, 12*(17), 2748. Available from https://doi.org/10.3390/rs12172748, https://www.mdpi.com/2072-4292/12/17/2748.

Kazhdan, M., & Hoppe, H. (2013). Screened Poisson surface reconstruction. *ACM Transactions on Graphics, 32*(3), 1–13. Available from https://doi.org/10.1145/2487228.2487237, https://dl.acm.org/doi/10.1145/2487228.2487237.

Klapa, P., & Gawronek, P. (2022). Synergy of geospatial data from TLS and UAV for Heritage Building Information Modeling (HBIM). *Remote Sensing, 15*(1), 128. Available from https://doi.org/10.3390/rs15010128, https://www.mdpi.com/2072-4292/15/1/128.

Leica Geosystems. (2012). Scanners Leica ScanStation P20 — Industry's best performing ultra-high speed scanner. http://w3.leica-geosystems.com/downloads123/hds/hds/ScanStation_P20/brochures-datasheet/Leica_ScanStation_P20_DAT_us.pdf.

Leica Geosystems. (2019) Leica Cyclone — 3D Point Cloud Processing Software HEXAGON. https://leica-geosystems.com/en-gb/products/laser-scanners/software/leica-cyclone.

Lercari, N. (2019). Monitoring earthen archaeological heritage using multi-temporal terrestrial laser scanning and surface change detection. *Journal of Cultural Heritage, 39,* 152–165. Available from https://doi.org/10.1016/j.culher.2019.040.005, https://linkinghub.elsevier.com/retrieve/pii/S1296207418307799.

Lezzerini, M., Antonelli, F., Columbu, S., Gadducci, R., Marradi, A., Miriello, D., Parodi, L., Secchiari, L., & Lazzeri, A. (2016). Cultural heritage documentation and conservation: Three-dimensional (3D) laser scanning and geographical information system (GIS) techniques for thematic mapping of facade stonework of St. Nicholas Church (Pisa, Italy). *International Journal of Architectural Heritage, 10*(1), 9–19. Available from https://doi.org/10.1080/15583058.2014.924605, http://www.tandfonline.com/doi/full/10.1080/15583058.2014.924605.

Liao, K.-Y., Lu, M.-H., Fan, Y.-C. (2022). 3D Point-cloud segmentation system based on AI model. *IEEE 4th Global Conference on Life Sciences and Technologies (LifeTech)* pp. 433–434. Available from https://doi.org/10.1109/LifeTech53646.2022.9754862 https://ieeexplore.ieee.org/document/9754862/, 978-1-6654-1904-8.

Liu, K., & Boehm, J. (2015). Classification of big point cloud data using cloud computing. *The International Archives of the Photogrammetry, Remote Sensing and Spatial Information Sciences, XL-3/W3,* 553–557. Available from https://doi.org/10.5194/isprs-archives-XL-3-W3-553-2015, https://www.int-arch-photogramm-remote-sens-spatial-inf-sci.net/XL-3/W3/553/2015/.

Li, Y., Zhao, L., Chen, Y., Zhang, N., Fan, H., & Zhang, Z. (2023). 3D LiDAR and multi-technology collaboration for preservation of built heritage in China: A review.

International Journal of Applied Earth Observation and Geoinformation, *116*, 103156. Available from https://doi.org/10.1016/j.jag.2022.103156, https://linkinghub.elsevier.com/retrieve/pii/S1569843222003442.

Moyano, J., León, J., Nieto-Julián, J. E., & Bruno, S. (2021). Semantic interpretation of architectural and archaeological geometries: Point cloud segmentation for HBIM parameterisation. *Automation in Construction*, *130*, 103856. Available from https://doi.org/10.1016/j.autcon.2021.103856, https://linkinghub.elsevier.com/retrieve/pii/S0926580521003071.

Nguyen, M. H., Yoon, S., Ju, S., Park, S., & Heo, J. (2022). B-EagleV: Visualization of big point cloud datasets in civil engineering using a distributed computing solution. *Journal of Computing in Civil Engineering*, *36*(3), 04022005. Available from https://doi.org/10.1061/(ASCE)CP.1943-5487.0001021. https://ascelibrary.org/doi/10.1061/%28ASCE%29CP.1943-5487.0001021.

Oytun, M., & Atasoy, G. (2022). Effect of terrestrial laser scanning (TLS) parameters on the accuracy of crack measurement in building materials. *Automation in Construction*, *144*, 104590. Available from https://doi.org/10.1016/j.autcon.2022.104590, https://linkinghub.elsevier.com/retrieve/pii/S0926580522004605.

Pajić, V., Govedarica, M., & Amović, M. (2018). Model of point cloud data management system in big data paradigm. *ISPRS International Journal of Geo-Information*, *7*(7), 265. Available from https://doi.org/10.3390/ijgi7070265, http://www.mdpi.com/2220-9964/7/7/265.

Patel, N., Parikh, K., & Patel, B. (2021). Bridge information modeling and AR using terrestrial laser scanner. *Reliability: Theory and Applications*, *16*(60), 17−23. Available from https://doi.org/10.24412/1932-2321-2021-160-17-23.

Pepe, M., Costantino, D., Alfio, V. S., Restuccia, A. G., & Papalino, N. M. (2021). Scan to BIM for the digital management and representation in 3D GIS environment of cultural heritage site. *Journal of Cultural Heritage*, *50*, 115−125. Available from https://doi.org/10.1016/j.culher.2021.050.006, https://linkinghub.elsevier.com/retrieve/pii/S1296207421000881.

Poux, F., Valembois, Q., Mattes, C., Kobbelt, L., & Billen, R. (2020). Initial user-centered design of a virtual reality heritage system: Applications for digital tourism. *Remote Sensing*, *12*(16), 2583. Available from https://doi.org/10.3390/rs12162583, https://www.mdpi.com/2072-4292/12/16/2583.

Soudarissanane, S., Lindenbergh, R., Menenti, M., & Teunissen, P. (2011). Scanning geometry: Influencing factor on the quality of terrestrial laser scanning points. *ISPRS Journal of Photogrammetry and Remote Sensing*, *66*(4), 389−399. Available from https://doi.org/10.1016/J.ISPRSJPRS.2011.010.005.

Tan, K., & Cheng, X. (2017). Specular reflection effects elimination in terrestrial laser scanning intensity data using phong model. *Remote Sensing*, *9*(8), 853. Available from https://doi.org/10.3390/rs9080853, http://www.mdpi.com/2072-4292/9/8/853.

Tan, K., Zhang, W., Shen, F., & Cheng, X. (2018). Investigation of TLS intensity data and distance measurement errors from target specular reflections. *Remote Sensing*, *10*(7), 1077. Available from https://doi.org/10.3390/rs10071077, http://www.mdpi.com/2072-4292/10/7/1077.

The Southwell and Nottingham Church History Project. (2021). *The Southwell & Nottingham Church History Project Beeston St John the Baptist.* https://southwellchurches.nottingham.ac.uk/beeston/hintro.php.

United Nations Educational Scientific and Cultural Organization (UNESCO). (1972). *17th General Conference of the United Nations Educational, Scientific and Cultural Organization The World Heritage Convention.* https://whc.unesco.org/en/conventiontext/.

United Nations Educational Scientific and Cultural Organization (UNESCO) (2023). *World Heritage Centre The 55 properties which the World Heritage Committee has decided to include on the List of World Heritage in danger in accordance with Article 11 (4) of the Convention. List of World Heritage in Danger.* https://whc.unesco.org/en/danger/.

United Nations Educational Scientific and Cultural Organization. (2019). *World Heritage Centre World Heritage.* https://whc.unesco.org/en/about/.

Xu, G., Pang, Y., Bai, Z., Wang, Y., & Lu, Z. (2021). A fast point clouds registration algorithm for laser scanners. *Applied Sciences, 11*(8), 3426. Available from https://doi.org/10.3390/app11083426, https://www.mdpi.com/2076-3417/11/8/3426.

Yang, S., Hou, M., & Li, S. (2023). Three-dimensional point cloud semantic segmentation for cultural heritage: A comprehensive review. *Remote Sensing, 15*(3), 548. Available from https://doi.org/10.3390/rs15030548, https://www.mdpi.com/2072-4292/15/3/548.

Comparison of results obtained by photogrammetry tools versus LED handheld scanning technique in architectural heritage. Application to plasterwork located in a world heritage site

10

Joaquín Aguilar-Camacho[1], Elena Cabrera-Revuelta[2], and Marta Torres Gonzalez[3]

[1]Department of Graphic Engineering, University of Seville, Seville, Spain, [2]Department of Mechanical Engineering and Industrial Design, University of Cadiz, Cadiz, Spain, [3]Department of Architectural Constructions II, University of Seville, Seville, Spain

10.1 Introduction

10.1.1 Background

Adequate preventive conservation measures require both physical and digital data about the current state of conservation and measures to take into consideration to ensure a correct maintenance. Thus the documentation, recording, and conservation of heritage have gained importance in recent decades. Natural and man-made disasters make three dimensional (3D) scanning projects essential for the reconstruction of buildings and cultural heritage objects damaged or lost due to earthquakes, fires, floods, or simply natural degradation over time. Furthermore, the use of non-invasive techniques (commonly known as 'NDT') in these projects makes a relevant contribution to diagnosis, maintenance, and conservation tasks and facilitates the correct and sustainable management of cultural heritage.

Traditionally, the geometric recording was complex, requiring a great deal of field and laboratory work, resulting in drawings of great value (Santofimia et al., 2022). In recent years, the geometric recording of pieces of great cultural value by means of reverse engineering techniques has been consolidated (Ham et al., 2020). The creation of 3D models allows them to be stored in databases, making them accessible even virtually (Wilson, 2023). Furthermore, these digital models enable simulations and non-destructive testing (Morena et al., 2019), including the production of replicas through 3D printing (Henriques et al., 2020), among any other possibilities. The complexity of scanning a 3D object, whether it is a building or a small object, depends

Diagnosis of Heritage Buildings by Non-Destructive Techniques. DOI: https://doi.org/10.1016/B978-0-443-16001-1.00010-3

on many factors, such as the equipment used, the location of the object or monument, the lighting conditions, and the texture of the material. Consequently, a 3D scanning project becomes more challenging due to all of the mentioned factors.

The incorporation of multi-image photogrammetry based on photographic sequences (Structure from Motion, SfM) (Fig. 10.1) and the new 3D scanners, which in a short period of time have been a true revolution in the graphic survey and digital recording of heritage, makes it possible to obtain hyperrealistic digital models. Both techniques, in addition to be noninvasive, offer highly accurate digital models, with relative simplicity, economy, and visual appeal of their results. In this way, the integrity and safeguarding of the recorded element is guaranteed. However, there are significant differences in the workflows involved in data acquisition and postprocessing, as well as in the obtained results.

The aim of this work is to compare both workflows considering not only aspects such as the methodology, the hardware and software resources, the processing time required, but also the results (i.e., density of points, graphic quality of the textures, accuracy, and level of detail of the 3D models) achieved by applying both techniques to specific architectural heritage objects.

The case study selected was a plasterwork panel that decorates the ceilings, walls, and arches of the Mudejar Palace of Pedro I in the Royal Alcazar of Seville (RAS). The RAS is traditionally considered the best example of a Mudejar construction palace preserved in the Iberian Peninsula (Torres-González, Alejandre, et al., 2021). In fact, this building is considered a World Heritage Site by the UNESCO since 1987 and these plasterworks with great historical and cultural value are under a high degree of protection (Torres-González, Alejandre, et al., 2021). Its plasterworks are characterized by their rich composition, made up of complex 3D geometric patterns (Fig. 10.2). Their geometry is mainly based on the elements of Latin and Kufic epigraphy, heraldic emblems, muqarnas, and vegetal or geometric elements. All of them are characterized by very irregular and lattice shapes with

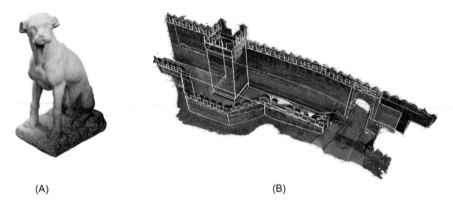

(A)   (B)

Figure 10.1 (A) Example of medium-size object (0.4x0.4x1.0 m) registered by photogrammetry; (B) example of a large-structure (20x5x5 m) registered by photogrammetry. Two different objects registered with the same technique.

Figure 10.2 Royal Alcazar of Seville. Maidens' Courtyard in the RAS. *RAS*, Real Alcázar of Seville.

many small deep holes that allow these two techniques to be subjected to an authentic intensive test, demonstrating their strengths and limitations. Therefore digitizing the plasterworks is a challenge in the preservation of the monument and suppose an indispensable resource for restorers and conservators of the building.

In recent years, the research project 'Development and assessment of models for the durability and preventive conservation of historic plasterwork from the decorative elements of the Royal Alcázar of Sevilla,' funded by the Spanish Ministry of Science, Innovation, and Universities, has made it possible to realize numerous studies on these unique elements (Aguilar-Camacho et al., 2023; Torres-González et al., 2021a,b, 2022a,b, 2023; García et al., 2021). Among the work carried out, the geometric recording of Mudejar plasterwork panels using photogrammetry (Torres-González et al., 2023) and handheld 3D scanners have been executed. Therefore this work will analyze the suitability of these two techniques for the recording of these singular pieces.

10.1.2 Literature review

In 2010, the Italian Professor Gabriele Guidi systematized and determined the state of reverse modeling (Guidi et al., 2010) constituting what could be the first step in current digital surveying. The conservation and enhancement of cultural heritage has undergone a significant improvement in recent years, as a result of the advance of digital technologies developed in the field of industrial design (Sequenzia et al., 2021). Although these techniques and technologies are now widely used, there are few highly qualified groups and SMEs (small and medium enterprises) that are researching and correctly applying digitization processes in the field of 3D scanning (Rodríguez-Navarro et al., 2022). The European Union has recently expressed its interest in this research area by funding specific projects under the Shaping Europe's Digital Future program, such as the survey on the quality of digitization of tangible cultural heritage (Loannides, 2023). This project assesses the parameters that determine the quality of a 3D cultural heritage digitization project by creating an inventory of existing formats, standards, guidelines, and methodologies used by the industry.

One of the most used technique to digitize heritage is photogrammetry. Photogrammetry is an image-based modeling technique based on projective geometry that converts 2D images into 3D information (Khries, 2021). The quality of the data obtained is highly dependent on the quality of the photographs taken, emphasizing the importance of natural or artificial lighting, exposure time, camera aperture, and the appropriate distance from which the photographs were taken (Kim et al., 2022). Also, a high percentage of overlap between the images is required.

This technique dates back to the beginning of photography and has been widely used in its aerial variant for the creation of cartography (Đorđević et al., 2022). However, in recent years, and thanks to the development of computational vision algorithms such as SfM, it has undergone a great growth (Iglhaut et al., 2019). Its application is being used in fields as diverse as engineering (Puerta et al., 2020), architecture (Barba et al., 2021), art (Cabezos Bernal et al., 2021), medicine (Talevi et al., 2023), archaeology (Howland et al., 2014), or the video game industry (Statham, 2020), among others.

Within photogrammetry, there are different levels, depending on the distance between the camera and the object. Therefore a distinction can be made between terrestrial photogrammetry, or close-range photogrammetry, which also includes photogrammetry carried out using drones (Nex & Remondino, 2019), underwater photogrammetry (Doneus et al., 2020), or aerial photogrammetry, carried out from aircrafts which has been used for several decades for the creation of accurate cartography and analysis of the territory (Monmonier, 1985).

This technique is based on human stereoscopic vision, in which the brain is able to reconstruct 3D geometric information by perceiving a scene from two slightly different angles. In this sense, the photogrammetric software use collinearity equations to create a 3D point cloud from the points detected in at least two different images.

In terms of 3D scanner technology, a distinction can be made between terrestrial laser scanners (TLS) and handheld scanners (HS). The first ones, TLS, are used for the registration of large structures, such as buildings, bridges, or ships among others. This type of laser scanners can work based on two different technologies: time of flight or phase-shift. In both cases, the instrument remains in a particular position to register geometric data around it. The work distance ranges of these devices/goes from 2 to 100 m (Dong et al., 2020; Moyano et al., 2020).

On the other hand, HS can be considered as a relatively new technology, which is particularly useful for the registration of close and small objects with complex geometry (Santofimia et al., 2022; Sevara et al., 2018). They are based on the capture of geometric information by projecting structured light onto the object. For that reason these devices are known as structured-light scanners. They project a pattern of light, whose deformation when it is projected over a surface gives geometrical information back. The pattern is projected by a stable light source, while a camera calculates the distance of every point in the field of view based on the shape of the projected pattern (Wersényi et al., 2022). This technique is widely used not only for heritage registration, but also in areas of medicine, such as plastic surgery and dentistry, among others (Thongma-Eng et al., 2022). Ambient light is a factor that significantly influences the results obtained with this type of instrument.

10.1.3 Problem statement

Nowadays, the graphic survey of heritage elements of great value involves obtaining a 3D digital model of high precision and graphic quality. The question of which technique should be chosen for this task is often raised. In many cases, this decision is linked to the available resources of the institution or enterprise in charge of carrying out this work. In this sense, the maturity reached by digital photogrammetry— thanks to the recovery of quality lenses for cameras and the development of algorithms for obtaining high-quality 3D models from photographic sequences—is considered an advanced technique for low-cost geometric registration, due to the fact that the equipment necessary to carry it out is accessible: mainly a photographic camera and specialized software. On the other hand, registration using 3D scanners incorporates increasingly higher quality in their internal systems, incrementing their speed and accuracy of capture, and reducing their size and weight, although they continue to require a significant initial investment (Eker, 2023).

Leaving aside the economic issue, another aspect to consider when deciding the most suitable technique is the time invested on data capture and processing workflows. Normally, the time needed for photogrammetric data acquisition is higher than the time needed for scanner-based registration. In addition, the photogrammetric technique requires a longer time for postprocessing.

One aspect that makes 3D scanners more user-friendly is the possibility of visualizing the results obtained in situ. This fact means that the professional in charge of carrying out the survey can be sure of having completed the data capture at the time of the field work. However, this circumstance cannot be known in photogrammetric works until the data is postprocessed. Taking into account the quality of the data obtained, both techniques offer results at a submillimeter precision. However, it is considered that the photorealistic quality of the textures is higher for photogrammetric models, due to the fact that it is based precisely on the quality of the used images (Zhao et al., 2023).

All these pros and cons make the researcher wonder which non-invasive graphic registration technique is more suitable for each occasion, and which one is more efficient in providing the best ratio between the quality of the results and the time required. Starting from the fact that both results are highly accurate, in this work, it is analyzed the performance of both techniques on a specific and complex case study.

10.2   Materials and methods

To carry out a complete comparison of photogrammetry and structured-light hand-held scan, the next workflow was implemented. First, a visual inspection was undertaken to select a panel that remains in good state of conservation, accessible from the ground floor, and located in a place where visitors to the RAS are not likely to pass through. Thus a plasterwork frieze located specifically in the entrance hall of the Palace of Pedro I, was selected (Fig. 10.3). This frieze is composed by 1.15×0.56 m panel located at a height of approximately 1.60 m and repeated systematically around the entire perimeter of the room.

As it can be observed, in this panel, there are three decorative motifs: the Kufic script is repeated along both the upper and lower parts of the panel, the sequence of different poly-lobed medallions of squared geometry can be found, and the rectangular spaces between the scripts and the medallions are composed of various decorative elements of laceries and vegetation known as 'atauriques'.

This case study is particular due to the high degree of details contained on the panel, which makes it a complex element to record. To obtain the complete geometry of the element, it was necessary to cover a large number of angles due to

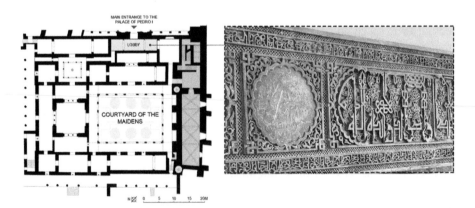

Figure 10.3 Floor plan of the Palace of Pedro I. Plasterwork panel in the entrance hall of the Palace of Pedro I.

occlusions and shadows generated by the own relief. Furthermore, the object under study is anchored to the wall so it cannot be moved; hence, the data capture was on-site. This situation complicated the data acquisition, due to factors such as the space limitations, the schedule of the RAS, or the natural or artificial light. Additionally, it must be considered that every instrument (e.g., lights, computers, wires, electrical power, auxiliary table, and stairs) had to be brought to the workplace.

10.2.1   Data acquisition

Once the plasterwork was selected, the working site was prepared for the digital data collection, and the subsequent data acquisition with both the camera and the HS. Times invested in preparing the worksite and in carrying out the data acquisition for both techniques was measured. In this section, the two main techniques used to carry out the geometric data acquisition are described, as well as their main features and the equipment required for each one.

10.2.1.1   Photogrammetric technique

During the digitization process, it was necessary to employ many different distances and points of view, paying attention to the angle between the camera and the panel, to capture precise information about the relief, specifically the depth. In addition, seven coded markers were placed on the scene, to ease the alignment of the images, as well as to facilitate the scaling of the photogrammetric model[1].

The data collection was done with a Panasonic DMC-GX80 camera equipped with a Digital Live MOS sensor whose size is 17.3×13 mm and its weighs is 426 g. This device is equipped with Wi-Fi and allows interchangeable lenses. Taking into account preliminary studies (Cabrera-Revuelta et al., 2022), a fixed lens LEICA DG NOCTICRON 42.5 mm F/1.2 ASPH Panasonic (Fig. 10.4) was selected for this work. This lens offers a high brightness, and includes an inner focusing system that enables excellent resolution and contrast. Thus photographs were recorded in maximum quality (4592×3448 pixels), obtaining a pixel size of 3.77 μm.

The camera was set to manual mode, keeping the diaphragm aperture between f/8 and f/16, ISO 100, and time of exposure 1/100 s. After taking images (Fig. 10.5), measures between coded markers were noted down. The data acquisition was organized by horizontal sweeps, considering an overlap of at least 80% between images. As it can be seen in Fig. 10.6, three correlative photos are very similar between them.

The distance between the camera and the panel has been kept lower than 1.20 m, which provides a ground sample distance (GSD) of 0.11 mm. The GSD is a fundamental concept in photogrammetry. It measures the distance between the center of two consecutive pixels, and it is very important to define the degree of detail guaranteed. This distance depends on the pixel size, the focal length, and the

---

[1] The coded markers were placed on the wall to preserve the plasterwork panel.

Figure 10.4 LEICA DG NOCTICRON 42.5 mm F/1.2 ASPH Panasonic.

Figure 10.5 Captured photographs of the plasterwork panel.

distance between the object and the camera. For the selected camera and lense, the GSD for a distance of 1 m is 0.09 mm. It means that, if photographs are taken from a distance of 1 m, the level of detail obtained is lower than 0.1 mm.

To carry out a successful data acquisition through photogrammetry, it was necessary to use some extra auxiliary instruments. First, for the photogrammetric

1st photograph   2nd photograph   3rd photograph

Figure 10.6 Three sequential photographies, offering and overlap of an 80%.

data acquisition, it was necessary to use light emitting diode (LED) lighting. Furthermore, to scale the model, a metric tape was employed and different coded markers were printed and placed where the preservation of the panel was guaranteed. To conclude, it was necessary to use instruments such as a tripod and a remote control, to avoid noisy and blurry pictures.

10.2.1.2 Handheld LED scanner technique

One of the objectives of this research is to test the versatility and performance of a high-precision HS, with LED light source, in the digitization of medium-sized heritage elements with complex geometry, as in the case of the plasterwork panels analyzed.

To obtain a high-accurated 3D model, it was used the multifunctional HS EinScan Pro 2X Plus from Shining 3D company designed for industrial use in reverse engineering and manufacturing, with a reduced weight (1.13 kg) and acceptable price−performance ratio that offers high quality and accurate results in the scanning of small and mediumsized objects.

To gain success in the data acquisition, it was necessary to use some extra auxiliary instruments such as a table and a chair, a laptop connected to the scan, and a stool, to provide high points of view. An advantage offered by this technique is the possibility to check results in real time during the data acquisition.

The scanning process takes place by emitting a LED light pattern that varies according to the scanning mode used. The optimum scanning distance is set by the manufacturer at 400 mm with a depth field of $\pm$ 100 mm and a scanning area ranging from 220×160 to 260×190 mm^2.

Scanner calibration is carried out in five different steps or positions of the calibration panel (black area with white circles). In the first step, the calibration board is placed flat. In the other steps, the calibration board is placed at an angle on a support, its position in relation to the camera being rotated 90 degrees in each step. When all five steps are completed, the software automatically calibrates the camera. Verification of the calibration process and the instructions for each step can be followed via the software interface (Fig. 10.7).

This instrument allows working in four different scanning modes: (mode 1a) handheld rapid scan mode; (mode 1b) handheld rapid scan mode with texture capture; (mode 2) handheld HD scan mode, and (mode 3) fixed scan with turntable scan mode.

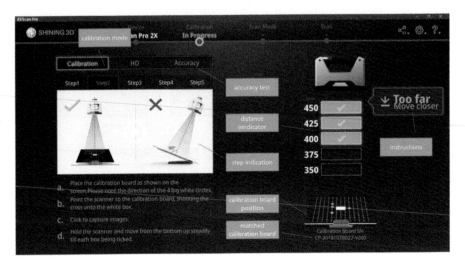

Figure 10.7 Calibration process and instructions for each step via the software interface.

Handheld rapid scan mode (1a) is the genuine scanning mode and allows fast scans (30 fps and 1,500,000 points/s) with an acceptable level of resolution (points distance 0.25—3.00 mm) and a scan accuracy of up to 0.1 mm under optimal capture conditions. This scanning mode (mode 1a) is primarily intended for objects of a certain size, or objects that cannot be moved. In addition, this mode allows the optional incorporation (mode 1b) of a digital camera (Color Pack complement) for the capture of color textures with very basic results which are not exactly the strong point of this equipment, as the obtaining of textures in the industrial field is relegated to a secondary level.

The handheld HD scan mode (2), available as an option (HD Pack), significantly improves the density of captured points (points distance 0.20—3.00 mm) and the scanning accuracy, which can reach 50 μm under optimal capture conditions, at the cost of a significant loss of scanning speed (20 fps and 1,100,000 points/s) together with the impossibility of capturing textures, being not compatible with the Color Pack.

The fixed scan with turntable scan mode (3) is designed to perform scans in optimal capture conditions (fixed mode on tripod) of small objects, guaranteeing high precision (40 μm) and resolution (points distance 0.24 mm). The turntable with coded markers, included in the Industrial Pack, are available as an option, and it can be used to facilitate the process of capturing and aligning points. This mode is not indicated for the case of study due to the dimensions and location of the plasterwork panels.

During this work (Fig. 10.8), several tests were made on the plasterwork panel to analyze the mean differences of the offered scan modes (modes: 1a, 1b, and 2) and complements available for the EinScan Pro 2X Plus scanner. Fig. 10.9 shows the differences in resolution between the analyzed scanning modes: (1a) detail of areas A and B scanned in rapid scan mode with a maximum resolution of 0.25 mm

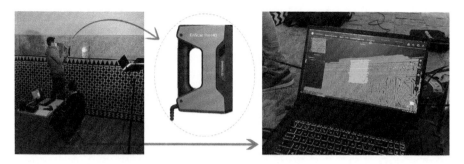

Figure 10.8 Data acquisition using handheld scan. Projection of structured-light over panel and visualization in real time of the registration by handheld scan.

and alignment by geometry; (1b) detail of areas A and B scanned in rapid scan mode with a maximum resolution of 0.25 mm, texture capture with the optional Color Pack module, and alignment by markers; and (2) detail of areas A and B scanned in HD scan mode with the optional HD Pack, maximum resolution of 0.20 mm, and alignment by markers.

The rapid scan mode (1a), aligning by geometries of the point cloud, works perfectly on objects with plenty well-defined geometry, as is the case with the plasterwork panels studied. The same cannot be said of alignment by textures (mode 1b), which is highly penalized in the scanned plasterwork panels due to its excessive homogeneity. In these cases, alignment using markers guarantees the highest levels of accuracy, but the continuity of the capture process requires a high density of markers which, with the help of the scanner control software, can be removed from the point cloud with relative ease.

On the other hand, working in the HD scan mode (2) provides an average distance between points very close to the lower end of the range specified by the manufacturer for the handheld rapid scan mode (0.20−3.00 mm). Furthermore, this mode significantly reduces its ability to align by geometry, being not possible to apply this mode to the digitization of the plasterwork panels, despite their geometric richness. Instead, it was necessary to use the marker alignment mode, which requires the use of a high density of markers (with at least four markers visible simultaneously).

Finally, for this study, the digitization of the entire plasterwork was carried out in the rapid scan mode (1a) with alignment by geometry and a resolution of 0.25 mm, as the HD scan mode had alignment problems without the use of markers and it was impossible to distribute the markers homogeneously over the entire panel due to the irregular geometry of the plasterwork. For the same reason, only the texture of a part of the plasterwork panel was digitized.

10.2.2 Data postprocessing

The following step consisted of postprocessing by using the indicated software for each technique, Agisoft Metashape Pro or EXScan Pro. A computer equipped with

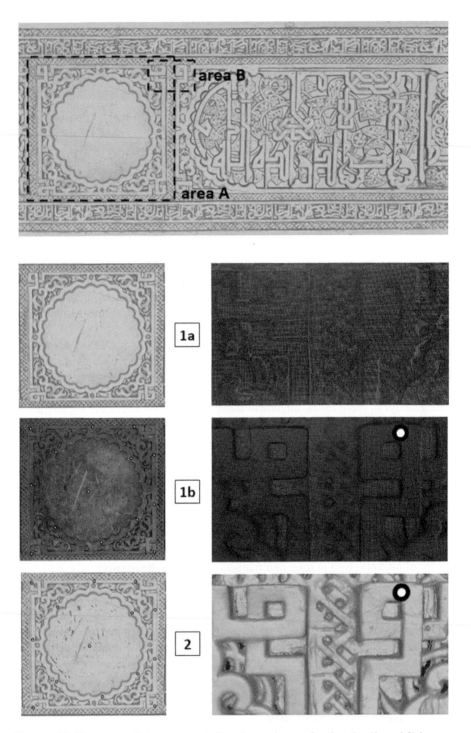

Figure 10.9 Comparison between three different scanning modes (i.e., 1a, 1b, and 2) in area A (general view of a medallion) and area B (detailed view of one latice) of the model.

an NVIDIA GeForce RTX graphic card, an Intel Core i7 processor, and a 32-GB RAM was used for postprocessing the data to ensure accuracy in graphical results. Time required for processing in both software packages was measured, and also the weights of the exported data.

10.2.2.1 Photogrammetric technique

Once data collection was completed, the set of photographs was processed using Agisoft Metashape Pro software. First of all, the images were filtered to discard those whose quality was lower than 0.5, being this value recommended by the software according with the sharpness level of the most focused part of the picture. Next, the coded markers were automatically detected, which helps the process of alignment. Different measurements between the coded markers were inserted as scale bars. After completing this process, the first step of Align Photos was carried out, selecting the Higher Accuracy. After a successful orientation of the photos, a first sparse cloud of points was generated.

Subsequently, the construction of the dense cloud. This task is really time-consuming, as this process involves generating multiple pairwise depth maps for each camera and merging them into a combined depth map (Iglhaut et al., 2019). The result is a very detailed colored cloud of points, with realistic appearance.

Next step in the workflow is the generation of the mesh. In this point, two options are available: the first one is to use the dense point cloud as source data, while the second option is to use the depth maps, already generated. This second option enables more efficient utilization of all the information generated, and it demands less resources compared to dense cloud-based reconstruction. According to the software manual, it is highly recommended for reconstructing arbitrary surface types. To conclude, the mesh was completed after building the texture (Fig. 10.10).

Once the workflow was completed, the software allows the generation of different graphic documentation, as models, textures, orthoimages, or digital elevation models.

10.2.2.2 Handheld scanner technique

The software EXScan Pro allows to configure the optimization of the contrast level of the light pattern, the scanning resolution, and the alignment mode of the point clouds and perform device calibration. In addition, elementary editing and processing tools of the point cloud are available, including cleaning, editing, meshing, closing holes in the mesh, and exporting in several formats.

Once the data capture is complete, the software generates a point cloud adapted to the previously configured resolution pattern and discards the rest of the captured points that are not necessary. Before meshing the generated point cloud, this can be edited to remove noise or unwanted elements. The edited point cloud can be saved in American Standard Code (ASC) format. Previous to the meshing phase, the software can merge point clouds from different acquisition processes in the same project.

Figure 10.10 Agisoft Metashape Pro screenshot, showing the photogrammetric process: the sparse cloud, dense cloud, mesh and textured model.

EXScan Pro_v3.7.0.3 can create watertight (closed 3D model for 3D printing) and unwatertight (nonclosed 3D model) meshes. The mesh can be created without any processing or be optimized by applying different levels of philters: simplified mesh, smoothed mesh, maximum number of triangles, filling of small holes, and suppression of small floating parts or spikes. The available meshing options also include the automatic closing of the holes created in the mesh of the model by the markers used during the acquisition process. Once the initial mesh is generated, the unclosed holes can be filled manually or automatically. The final mesh can be saved in four different standardized formats: STL, OBJ, PLY, and 3MF.

Postprocessing of the 3D model of the plasterwork panel began with cleaning up the point cloud, removing all unwanted points. An unwatertight mesh of the model was then created, without any optimization and manually closing of the mesh holes (Fig. 10.11). Finally, the model was exported in OBJ format.

10.2.3 Analysis procedure

A visual inspection of the results obtained with both techniques was carried out to check if all the geometry was properly covered. More specifically, the density of points obtained in the different clouds of points was quantified, followed by the density of the generated meshes. Finally, the qualities of the textures obtained were also visually examined.

Therefore the efficiency of both methods was compared, taking into account the following factors: data acquisition time, processing time, size of the generated files, density of the clouds of points obtained, density of the generated models, and quality of textures.

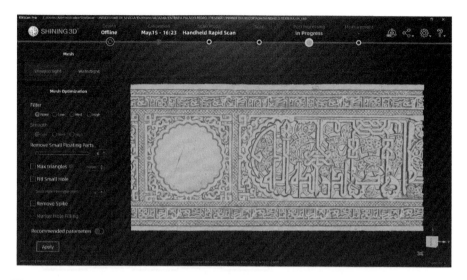

Figure 10.11 Editing and processing tools for meshing in EXScan Pro_v3.7.0.3

Additionally, a comparison of the geometrical quality of some specific parts was made. Some significant details of the plasterworks were selected to execute a more exhaustive comparison over them thanks to CloudCompare software. This means that clouds of points of specific elements of the virtual model were superimposed to analyze the degree of coincidence of the geometry using the CloudCompare software revealing the accuracy of the model. The meshes of the models obtained from both techniques were also superimposed to verify the geometric similarities.

10.3 Results and discussion

Different aspects from both workflows and models were analyzed in this section, such as processing time, weight, resolution, amount of data, and texture quality, and finally pros and cons were established to indicate when each technique is recommended.

10.3.1 3D model from digital photogrammetry

To cover the entire panel, 247 photographs were collected. The images were automatically filtered by Metashape software to discard those whose quality was lower than 0.5 units. This means that 239 photographs were finally used. After a successful orientation of the photos, a first sparse cloud of points composed of 154,494 points was generated and also the subsequent dense cloud composed of 28,572,281 points. The 3D model contained more than 14 million faces. The result of the ortho-image of the generated model is shown in Fig. 10.12.

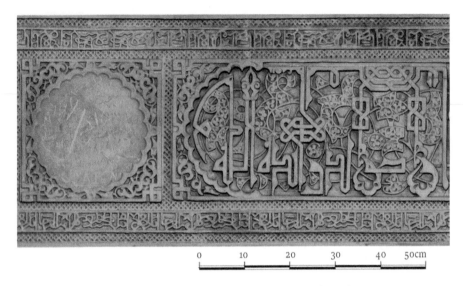

0   10   20   30   40   50cm

Figure 10.12 Orthophoto of the plasterwork panel: metric information.

Table 10.1 Processing times in Agisoft Metashape Pro and size of the generated files.

Workflow step	Quality	Time	Size
Align Photos	Medium	00 h 03 m 12 s	24.23 MB
Depth maps generation	Ultra High	18 h 15 m 00 s	5.86 GB
Dense cloud generation	Ultra High	01 h 24 m 00 s	831.87 MB
Model generation	Ultra High	03 h 04 m 00 s	10.67 GB

The set of the worksite took 15 minutes, while the data acquisition took 35 minutes; and the processing time to obtain both the dense cloud of points and the textured model was completed in 23 hours and 11 minutes. Taking these data into account, the photogrammetric process was completed after 24 hours. The point cloud generated, after cleaning the model those unwanted parts, was made up of 28,572,281 points. After completing the process, the cloud of points and the textured mesh were exported. The generated model was made up of just over 12 million faces and the result obtained is shown in Fig. 10.12. The spent time to complete the entire workflow and the sizes of the exported files are shown in Table 10.1.

According to the generated report provided by Agisoft Metashape Pro, the mean distance of the capture data was 1.02 m, and the GSD obtained is 0.0934 mm/pix. The density of points reached, computed with CloudCompare software, was 33.581 ± 10.091 points/mm^2.

10.3.2 3D model from handheld scan

The calibration of the scanner took only 6 minutes. The full scan of the plasterwork panel with a resolution of 0.25 mm and alignment by geometries (mode 1a) required the acquisition of 2,259 frames at a scanning speed of 30 fps. Five full sweeps scan of the plasterwork were taken: one orthogonal to the panel and four oblique scans from different angles, trying to capture the entire visible surface of the plasterwork. The scanning time for these five scans was 7 minutes. Additional local scans were then taken in the areas that had not been correctly captured to fill in the hard-to-capture holes and smaller gaps, which added an additional 4 minutes. The total scan time was approximately 11 minutes. For all sweeps performed, the scan distance was in a range between 350 and 500 mm. This parameter was controlled at all times by the scanner software, which warned the operator when the range was exceeded.

The point cloud generated, after cleaning those unwanted parts, was composed by 1,774,261 points. The cleaning and generation of the final point cloud took 3 minutes. One more minute was needed to save the cloud in ASC format in a 100.54-MB file. The 3D model generated from the point cloud took 4 minutes, including the time necessary to edit the mesh to manually close the gaps that were not automatically closed by the software. The generated model was made up of just over 2.8 million faces and the result obtained is shown in Fig. 10.13. The spent time to complete the entire workflow and the sizes of the exported files are shown in Table 10.2.

The density of the definitive 3D cloud of points, computed with Cloud Compare software, was 1.961 ± 0.303 points/mm^2. Furthermore, as a result of the different tests carried out with the scanner, two 3D models of a part of the plasterwork

Figure 10.13 3D Model of the plasterwork panel: 3D model of the scanned plasterwork panel (without texture).

Table 10.2 Processing times in EXScan Pro_v3.7.0.3 software.

Workflow step	Scan Resolution	Time	Size
Cleaning unwanted points	0.25 mm	00 h 03 m 09 s	166.03 MB
Definitive point cloud creation	0.25 mm	00 h 00 m 33 s	138.99 MB
Creation 3D model mesh	0.25 mm	00 h 00 m 29 s	148.41 MB
Editing 3D model mesh	0.25 mm	00 h 03 m 26 s	150.28 MB

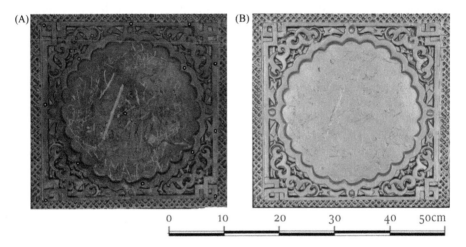

Figure 10.14 3D models of a part of the plasterwork: (A) 3D model from mode scan with texture (mode 1b); (B) 3D model from mode scan with high resolution (mode 2).

(zone A identified in Fig. 10.9) were generated: the first model was made from a scan with the optional mode texture Color Pack (mode 1b); the second model (Fig. 10.14) was made from a scan in the optional HD Prime high-resolution scan (mode 2). In both modes, the use of markers was essential to achieve the correct alignment of the cloud of points. In the processing of the model obtained with texture (mode 1b), the main difference was in the need to remap the texture of the model after closing the holes generated in the mesh. This additional workflow consumed an extra time (00 hour 01 minute 25 seconds) of postprocessing in the zone A of the plasterwork. Furthermore, in these last two models, textured and with high resolution, it was necessary to close the gaps created by the markers in the respective meshes.

10.3.3 Comparison of workflows and results between 3D models

In terms of the time invested in this work, the data capture process was similar in the photogrammetry and the 3D scanner procedure, 35 and 11 minutes, respectively, ignoring the previous times needed for site conditioning and the camera or scanner calibration. However, the postprocessing time marks a clear difference between both technologies. Meanwhile, the generation and editing of the 3D model from the point cloud obtained with the scanner took only 8 minutes, it was necessary almost an entire day to generate the 3D model from the point cloud obtained by photogrammetry. In addition, digitization by scanner provides a fundamental advantage over photogrammetry because it allows real-time preview of the data capture and redigitize on-site those incomplete areas. On the other hand, photogrammetry allows data capture to work at a greater distance from the object, compared to the limited range that the scanner allows (set by the manufacturer at 400 mm with a depth field of $\pm$ 100 mm). This question may be relevant in certain circumstances, in which the accessibility of the object may be conditioned.

After having completed the registration by the two different workflows, both models were imported into Blender software, to limit their dimensions and delete those areas that are not relevant for this work. This fact reduces the number of faces to 12,172,321 for the photogrammetric model, and to 2,835,613 for the scanned model. It may be considered that this work aims to obtain the highest quality models as possible.

The comparison of the amount of data produced by the two techniques is analyzed in Table 10.3. It can be observed that, while the photogrammetric model is more detailed, it must be considered the actual necessity of such a high quantity of data. In this sense, a scanned model is more efficient since the geometry is completely registered with a lower number of faces, and thus, with files easier to treat.

As it is shown in Fig. 10.15, both models register in a high level of detail the complexity of the geometry. To obtain a more accurate analysis about the geometrical similarity between both models, they were imported into the software CloudCompare. This software allows to align them with the tool 'Fine registration with Iterative Closest Point' (Fig. 10.16). In this process, the model obtained from the handheld scan was selected as the reference model, and so the photogrammetric

Table 10.3 Comparison between 3D model results obtained by photogrammetry and handheld scanner.

		Photogrammetry	Handheld scan
Points Cloud	Points	28,572,281	1,774,261
	Point density $\pm$ SD (points/mm^2)	33.581 $\pm$ 10.091	1.961 $\pm$ 0.303
	Size of cloud points (.ASC)	1.76 GB	0.14 GB
Mesh	Faces	12,172,321	2,835,613
	Average triangle surface (mm^2)	0.071	0.320
	Size of mesh (.OBJ)	1.28 GB	0.15 GB
	Size of self-software file	5.75 GB	13,72 GB
	Size of texture file (.JPG)	3345 KB	2910 KB

Figure 10.15 Model obtained by photogrammetry (A) and by handheld scan (B). It can be observed a denser wireframe model and a very detailed surface (A) in contrast with the smooth surface of the model (B).

Figure 10.16 Fine registration of both models Photogrammetric model is shown in red color, while the yellow is assigned to the scanned model.

model was the aligned model. This decision is due to the accuracy of metric information that scanners provide, while the photogrammetric model is scaled by introducing some scale bars by hand. Nevertheless, the scaling of the photogrammetric model was very satisfactory, since it has only differed by 0.014%, and the root mean square obtained was of 0.00068 m^2. When inspection is made in this software, it is possible to appreciate certain deviation between both models. As it can be seen, the panel obtained with the scanner is flatter, while the other model is slightly convex. The deviation is minimal, but it can be explained by the predominance of the horizontal dimension. If markers were placed on the panel, this convexity could have been absorbed. Anyway, this difference is not significant.

After this adjustment, the comparison between meshes was executed, using the tool 'compute cloud/mesh distance' from the software CloudCompare. This tool is able to compare a mesh with a point cloud or two meshes. When two meshes are compared, it is necessary to define the roles of each one: one mesh will work as the reference mesh, while the other one will work as the compared mesh. Distances from each point of a mesh to the closest point of the other mesh were calculated and showed in a scalar field. As it can be seen in Fig. 10.17, there is a high level of coincidence between both meshes, being the mean of the distances of 0.2 mm and a standard deviation of 0.8 mm.

To perform a more exhaustive comparison of the geometric similarity of elements, some specific parts are extracted from both meshes. Specifically, four different details uniformly distributed are selected, as is shown in Fig. 10.18.

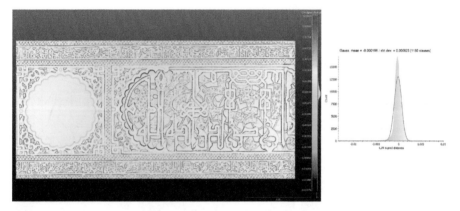

Figure 10.17 Cloud-to-mesh distances, mean and standard deviation. The mean of the distances is 0.2 mm with a standard deviation of 0.8 mm.

Figure 10.18 Four different details to be compared between models. Once the plasterwork was digitized by both techniques, four specific details are compared.

Detail A refers to a part of the medallion; Detail B is a lotus flower—shaped; Detail C is a ribbon; and Detail D is a portion of the Kufic script.

Those details were finely registered using CloudCompare, to adjust their position as much as possible. Applying the tool 'Compute cloud/mesh distance' the scalar fields of distances from meshes that are shown in are obtained. It can be observed the mean distances of those details, as well as the standard deviations (Fig. 10.19). In every case, it can be observed that the mean distance is under 0.1 mm.

The areas where the distances are greater coincide with the small recesses as well as on the side faces of the carved elements (Fig. 10.20). As it could be seen in Table 10.4, the mean distance between models is less than 0.08 mm in all cases.

To conclude with this comparison, it was observed the wireframe meshes of Detail C to appreciate the difference in the distribution of vertices in both cases.

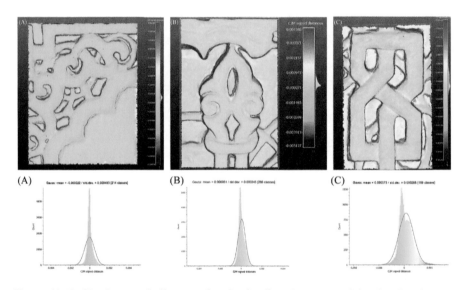

Figure 10.19 Cloud-to-mesh distances for the details. Histograms of the cloud/mesh distance applied to details A, B, and C.

Figure 10.20 Cloud-to-mesh distances for detail D. Histograms of the cloud/mesh distance applied to detail D.

Table 10.4 Mean distance between models and standard deviation of compared areas in detail.

Area in detail	Mean distance (mm)	Standard deviation (mm)
A	0.022	0.483
B	0.061	0.346
C	0.073	0.288
D	0.012	0.352

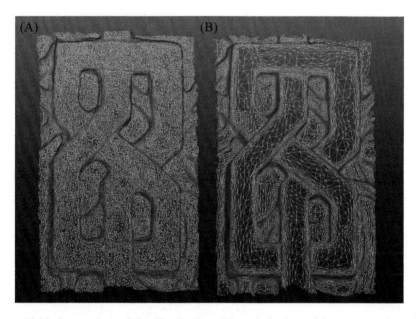

Figure 10.21 Comparison of the distribution of faces in both models: photogrammetric model (A) and scanned model (B).

As it can be seen the scanned mesh (B) is smoother when a planar surface is detected, while the amount of vertices increases on borders. In contrast, for the photogrammetric mesh (A), the density of faces is constant over the entire surface. This fact makes the scanned model more optimal than the photogrammetric one (Fig. 10.21).

Finally, it would be interesting to verify the quality of the textured models. As it can be seen in Fig. 10.22, the photogrammetric model is suitable for an appropriate dissemination of the heritage due to the higher photorealistic quality compared to the results obtained by using a HS. This is because the creation of photogrammetric models is specifically based on taking high-quality photographs, which serve not only to detect the geometry, but also to adequately capture the color of the feature (Fig. 10.23). Poor lighting conditions have been decisive in the texture quality

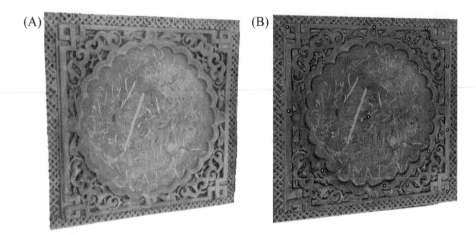

Figure 10.22 Textured model obtained from the photogrammetry (A) and from the 3D scanner technique (B). Two different methods and textures are shown.

Figure 10.23 Textured model obtained from photogrammetric technique: complete textured model of the panel.

captured by the scanner camera, with less sensitivity for capturing light than the optics of the photographic camera used. However, the HS can digitize a 3D model without texture under poor lighting conditions.

10.4   Conclusion

After a detailed comparison of the results achieved, it can first be confirmed that SfM photogrammetry and handheld scanning are perfectly suitable for the graphic digitization of elements of the cultural heritage of this type. Specifically, the performance of both techniques was compared using the graphic digitization of a Mudejar plasterwork panel with rich and complex geometric compositions in relief, and a very homogeneous texture for the case study. The 3D models, created using HSs and close-range photogrammetry, achieve a high level of accuracy and detail, providing an unprecedented graphic record and technical documentation of the plasterwork panels that adorn the walls of the RAS, helping to optimize the tasks necessary for their diagnosis, maintenance, and conservation. The data obtained thus have countless purposes, especially in the context of preventive conservation of these historic elements.

It can be seen that although the photogrammetric model is more detailed, the true need for such a high amount of data must be considered. As is well known, photogrammetry is a relatively low-cost technique that requires a high investment of time, expensive computer equipment, specialized software, and a high level of training and experience in its use to obtain good results during the postprocessing phase. However, it can be seen that photogrammetry is capable of obtaining an extra high level of detail, with very high point resolutions generating models that can be inefficient due to the difficulty in moving, editing, and working with them, as well as very high times of processed.

On the contrary, the models obtained with 3D HSs are more workable and efficient, using a lower resolution of points, particularly in those areas or flat surfaces of the model that do not require a large number of points for their correct definition. The use of this technology is less demanding in terms of on-site working conditions and the need for user training in the postprocessing phase to achieve the generation of 3D models with a high level of accuracy, while the initial investment is higher than that required for photogrammetry. In this sense, the scanned model is more efficient since the geometry is completely registered with a smaller number of faces, and therefore, generating files that are easier to treat.

Therefore, in the study carried out, the compared 3D models mainly differ, fundamentally, in the definition of the flat faces of the object; although to a lesser extent, in the small holes in the object that could not be captured and have not been defined in the cloud of points. Regarding the flat faces, we have verified that the model generated by the scanner creates a significantly smoother geometry in those areas, using fewer points for the definition of the mesh in comparison to the model generated by photogrammetry. This is not the case in the curved transition surfaces of the digitized object, in which there are hardly any differences between one model and another. Consequently, the 3D model generated from the scanner is better optimized than the one obtained by photogrammetry, as it adapts well to areas with conflicting geometric definition while saving information in areas with no geometry changes where the model photogrammetric does not optimize the information.

Regarding the small holes in the object that have not been correctly captured and must be closed later, the remeshing algorithms used by the Agisoft Metashape Pro and the EXScan Pro software worked correctly and did not show significant differences in 3D results.

Another essential advantage of digitization by scanner is the possibility of aligning the different point clouds from the geometry of the object, without the use of markers (mode 1a: classic or lighting mode in the EinScan Pro 2X Plus). The installation of these markers on patrimonial elements is not always possible and, sometimes, it is not allowed. Furthermore, the use of markers can negatively influence the final accuracy of the generated 3D model. In the case of photogrammetry, the location of these markers can condition the final accuracy of the model in those areas whose depth differs from that of the plane defined by the markers used. Nor should we forget that the use of photogrammetry entails the need to scale the 3D model, a task that can also condition its final accuracy.

Finally, it is worth noting the high quality of the textures obtained using photogrammetric techniques that, for the moment, cannot be matched with 3D scanners. This is due to the quality of the optics used by the cameras that incorporate the scanners, still very far from what the digital cameras that are commonly used in photography can achieve. In addition, the distance from which the images are captured, which is highly conditioned in HSs, is another factor to take into account in relation to the quality of the texture.

In summary, it can be concluded that SfM photogrammetry is the most appropriate technique when obtaining textured 3D digital models at low cost is a priority. On the other hand, if it is necessary to work with highly accurate models, optimizing their size to improve the workability of subsequent editions, the minimally invasive technology provided by the HS is the best choice.

References

Aguilar-Camacho, J., Granado-Castro, G., & Barrera-Vera, J. A. (2023). Optimising workflows and proposal of a standard in surveying of deco- rative heritage elements (The Royal Alcazar of Seville, Spain). *Disegnarecon, 16*(30), 1.1–1.16. Available from https://doi.org/10.20365/disegnarecon.30.2023.1, https://disegnarecon.univaq.it/ojs/index.php/disegnarecon/article/view/1130/572.

Barba, S., Ferreyra, C., Cotella, V.A., di Filippo, A., & Amalfitano, S. (2021). A SLAM integrated approach for digital heritage documentation. Culture and Computing. *Interactive Cultural Heritage and Arts: 9th International Conference, C&C 2021*, held as part of the 23rd HCI International Conference, HCII 2021, Virtual Event, July 24–29, 2021, Proceedings, Part I. Springer. 27–39.

Cabezos Bernal, P.M., Navarro, P.R., & Piqueras, T.G. (2021). Documenting paintings using gigapixel SfM photogrammetry. International Archives of the Photogrammetry, Remote Sensing and Spatial Information Sciences, (pp. 93–100).

Cabrera-Revuelta, E., Aguilar-Camacho, J., & Granado-Castro, G. (2022). *Influence of the camera lens in the photogrammetric survey of historic plasterworks. Comparative study in the Royal Alcazar of Seville. Architectural Graphics: Volume 1-Graphics for Analysis* (pp. 178−187). Springer.

Doneus, N., Miholjek, I., Džin, K., Doneus, M., Dugonjić, P., & Schiel, H. (2020). Archaeological prospection of coastal and submerged settlement sites. Re-evaluation of the Roman site complex of Vižula, Croatia. *Archaeologia Austriaca, 104*, 253−281.

Dong, Z., Liang, F., Yang, B., Xu, Y., Zang, Y., Li, J., Wang, Y., Dai, W., Fan, H., & Hyyppä, J. (2020). Registration of large-scale terrestrial laser scanner point clouds: A review and benchmark. *ISPRS Journal of Photogrammetry and Remote Sensing, 163*, 327−342.

Đorđević, D. R., Đurić, U., Bakrač, S. T., Drobnjak, S. M., & Radojčić, S. (2022). Using historical aerial photography in landslide monitoring: Umka Case Study, Serbia. *Land, 11* (12), 2073. Available from https://doi.org/10.3390/land11122282, -445X.

Eker, R. (2023). Comparative use of PPK-integrated close-range terrestrial photogrammetry and a handheld mobile laser scanner in the measurement of forest road surface deformation. *Measurement, 206*112322. Available from https://doi.org/10.1016/j.measurement.2022.112322, https://www.sciencedirect.com/science/article/pii/S0263224122015184.

García, E.V., Calero Castillo, A.I., & Bueno, A.G. (2021). *Estudio de métodos de limpieza sobre yeserías medievales.*

Guidi, G., Russo, M., & Beraldin, J. A. (2010). *Acquisizione 3D e modellazione poligonale.* Milano: Mc-Graw-Hill, ed.

Ham, N., Bae, B. I., & Yuh, O. K. (2020). Phased reverse engineering framework for sustainable cultural heritage archives using laser scanning and BIM: The case of the Hwanggungwoo (Seoul, Korea). *Sustainability., 12*(19), 8108.

Henriques, F., Bailão, A., Rocha, J., & Costa, J. (2020). Restoration of an 18th century frame: 3D modelling, printing and matching color of decorative flowers elements. *Geconservacion,* 313−322.

Howland, M. D., Kuester, F., & Levy, T. E. (2014). Photogrammetry in the field: Documenting, recording, and presenting archaeology. *Mediterranean Archaeology and Archaeometry, 14*(4), 101−108.

Iglhaut, J., Cabo, C., Puliti, S., Piermattei, L., O'Connor, J., & Rosette, J. (2019). Structure from Motion Photogrammetry in forestry: A review. *Current Forestry Reports, 5*(3), 155−168. Available from https://doi.org/10.1007/s40725-019-00094-3, https://doi.org/10.1007/s40725-019-00094-3.

Khries, H. M. (2021). Photogrammetry versus 3D scanner: Producing 3D models of museums' artifacts. *Collection and Curation.*

Kim, J., Kim, J., Jeon, K., Lee, J., & Lee, J. (2022). Development and validation of unmanned aerial vehicle photogrammetry simulator for shaded area detection, shaded area simulator validation UAV 3D modeling UAV photogrammetry UAV simulator. *Applied Sciences, 12*(9), 2076−3417. Available from https://doi.org/10.3390/app12094454.

Loannides, M. (2023). *Study on quality in 3D digitisation of tangible cultural heritage: mapping parameters, formats, standards, benchmarks, methodologies, and guidelines.* European Commission. https://digital-strategy.ec.europa.eu/en/library/study-quality-3d-digitisation-tangible-cultural-heritage. https://doi.org/10.2759/471776.

Monmonier, M. S. (1985). *Technological transition in cartography. Technological transition in cartography.* University of Wisconsin Press.

Morena, S., Barba, S., & Álvaro-Tordesillas, A. (2019). Shining 3D EinScan-pro, application and validation in the field of cultural heritage, from the Chillida-Leku museum to the archaeological museum of sarno. *International Archives of the Photogrammetry, Remote Sensing & Spatial Information Sciences.*

Moyano, J., Odriozola, C. P., Nieto-Julian, J. E., Vargas, J. M., Barrera, J. A., & León, J. (2020). Bringing BIM to archaeological heritage: Interdisciplinary method/strategy and accuracy applied to a megalithic monument of the Copper Age. *Journal of Cultural Heritage*, *45*, 303−314.

Nex, F., & Remondino, F. (2019). Preface: Latest developments, methodologies, and applications based on UAV platforms MDPI. *Drones*, *26*(3), 2504, -446X.

Puerta, A. P. V., Jimenez-Rodriguez, R. A., Fernandez-Vidal, S., & Fernandez-Vidal, S. R. (2020). Photogrammetry as an engineering design tool. In C. Alexandru, C. Jaliu, & M. Comşit (Eds.), *Product design* (pp. 41−64). TechOpen.

Rodríguez-Navarro, P., Gil-Piqueras, T., Sender-Contell, M., Giménez-Ribera, M., & De las líneas, M. (2022). *La gráfica y sus usosEstablecimiento de estándares para el levantamiento gráfico del patrimonio*. Estado del arte. In: Pedro Miguel. Jiménez Vicario, María. Mestre Martí, David. Navarro Moreno, IX Congreso Internacional de Expresión Gráfica Arquitectónica: 2−4 de junio de. Ediciones Universidad Politécnica de Cartagena, Cartagena (Spain) https://doi.org/10.31428/10317/11324.

Santofimia, E., González, F. J., Rincón-Tomás, B., López-Pamo, E., Marino, E., Reyes, J., & Bellido, E. (2022). The mobility of thorium, uranium and rare earth elements from Mid Ordovician black shales to acid waters and its removal by goethite and schwertmannite. *Chemosphere*, *307*135907.

Sequenzia, G., Fatuzzo, G., & Oliveri, S. M. (2021). A computer-based method to reproduce and analyse ancient series-produced moulded artefacts. *Digital Applications in Archaeology and Cultural Heritage*, *20*e00174.

Sevara, C., Verhoeven, G., Doneus, M., & Draganits, E. (2018). Surfaces from the visual past: Recovering high-resolution terrain data from historic aerial imagery for multitemporal landscape analysis. *Journal of archaeological method and theory*, *25*, 611−642.

Statham, N. (2020). Use of photogrammetry in video games: a historical overview. *Games and Culture*, *15*(3), 289−307.

Talevi, G., Pannone, L., Monaco, C., Bori, E., Cappello, I. A., Candelari, M., Wyns, M., Ramak, R., La Meir, M., & Gharaviri, A. (2023). Evaluation of photogrammetry for medical application in cardiology. *Frontiers in Bioengineering and Biotechnology*, *11*.

Thongma-Eng, P., Amornvit, P., Silthampitag, P., Rokaya, D., & Pisitanusorn, A. (2022). Effect of ambient lights on the accuracy of a 3-dimensional optical scanner for face scans: An in vitro study. *Journal of Healthcare Engineering*, *2022*.

Torres-González, M., Alejandre, F. J., Alducin-Ochoa, J. M., Calero-Castillo, A. I., Blasco-López, F. J., Carrasco-Huertas, A., & Flores-Alés, V. (2022). Methodology to evaluate the state of conservation of historical plasterwork and its polychrome to promote its conservation. *Applied Sciences*, *12*(10), 4814.

Torres-González, M., Alejandre, F. J., Flores-Alés, V., Calero-Castillo, A. I., & Blasco-López, F. J. (2021). Analysis of the state of conservation of historical plasterwork through visual inspection and non-destructive tests. The case of the upper frieze of the Toledanos Room (The Royal Alcázar of Seville, Spain). *Journal of Building Engineering*, *40*102314. Available from https://doi.org/10.1016/j.jobe.2021.102314, https://www.sciencedirect.com/science/article/pii/S2352710221001704.

Torres-González, M., Prieto, A. J., Alejandre, F. J., & Blasco-López, F. J. (2021). Digital management focused on the preventive maintenance of World Heritage Sites. *Automation in Construction*, *129*103813.

Torres-González, M., Revuelta, E. C., & Calero-Castillo, A. I. (2023). Photogrammetric state of degradation assessment of decorative claddings: The plasterwork of the Maidens' Courtyard (The Royal Alcazar of Seville). *Virtual Archaeology Review*, *14*(28), 110−123.

Torres-González, M., Rubio-Bellido, C., Bienvenido-Huertas, D., Alducin-Ochoa, J. M., & Flores-Alés, V. (2022). Long-term environmental monitoring for preventive conservation of external historical plasterworks. *Journal of Building Engineering, 47*103896.

Wersényi, G., Wittenberg, T., & Sudár, A. (2022) Handheld 3D scanning and image processing for printing body parts - A workflow concept and current results. *IEEE 1st International Conference on Internet of Digital Reality (IoD).* 000061−000068, https://doi.org/10.1109/IoD55468.2022.9987113.

Wilson, O. J. (2023). The 3D Pollen Project: An open repository of three-dimensional data for outreach, education and research. *Review of Palaeobotany and Palynology, 104860.*

Zhao, L., Zhang, H., & Mbachu, J. (2023). Multi-sensor data fusion for 3D reconstruction of complex structures: A case study on a real high formwork project. *Remote Sensing, 15* (5), 1264.

Exploring the accessibility of deformed digital heritage models

11

Daniel Antón[1,2], José-Lázaro Amaro-Mellado[3], Fernando Rico-Delgado[1], and Pablo Díaz-Cañete[1]

[1]Departamento de Expresión Gráfica e Ingeniería en la Edificación, Escuela Técnica Superior de Ingeniería de Edificación, Seville, Spain, [2]The Creative and Virtual Technologies Research Lab & Product Innovation Centre, School of Architecture, Design and the Built Environment, Nottingham Trent University, Nottingham, United Kingdom, [3]Departamento de Ingeniería Gráfica, Universidad de Sevilla, Seville, Spain

11.1 Introduction

Considering the enormous evolution and availability of advanced digitization and modeling technologies in recent decades, their application to the field of cultural heritage—the use of accurate reports and data was already established in the Athens Charter (International Council on Monuments & Sites, 2011)—opens up multiple, diverse possibilities such as the in-depth study of these assets, the modeling of their actual geometry for the sake of knowledge generation and preventive conservation, as well as their cataloging and dissemination through different channels.

As seen in the work by Taher Tolou Del et al. (2020) and Taher Tolou Del and Tabrizi (2020), heritage conservation involves physical aspects and semantic aspects. The former are related to the tangible condition of the assets, where advanced technologies may be applied. In contrast, the latter aspects address the assets' heritage values, meanings, messages, and concepts latent in the heritage spaces (Australia, 1999). These semantic aspects fall beyond the scope of this chapter but can be addressed with digital technologies to include narratives, social history, and anthropological mapping (Selim et al., 2022). On the other hand, in addition to experts and institutions, architectural (and archaeological) heritage conservation should involve individuals, communities, in cooperation with those specialists and governments. The physical and semantic aspects of conservation allow for identifying the heritage values and what needs to be conserved, which is mainly to be done by the cited experts. Other important requirements should be raising public awareness of the positive impact of heritage conservation on people's lives, the existence of clear conservation standards from governments, and their cooperation with private companies and other stakeholders (Taher Tolou Del et al., 2020).

Finally, in addition to heritage management for conservation, the accessibility of heritage models becomes crucial in today's increasingly inclusive society, not only

Diagnosis of Heritage Buildings by Non-Destructive Techniques. DOI: https://doi.org/10.1016/B978-0-443-16001-1.00011-5

for individuals with certain difficulties but also for the general public to experience, explore, and learn from this legacy.

11.1.1   Accessibility

The word accessibility has different acceptations, that is to say, dictionary entries (Cambridge University Press, 2023). They are mainly related to the quality of something of being able to be reached, entered, or obtained with no difficulty and to the ability of a person to reach, approach, enter, or use it. On that basis, the scientific community has addressed accessibility in diverse fields. A simple search for this term on two of the primary databases alone, Scopus (Elsevier, 2023) and Web of Science (Clarivate, 2023), yielded 236,200 (only considering the document title, abstract, and keywords) and 132,909 (all search fields) documents, respectively, a number in continuous evolution. The publications found were in the form of journal articles, review articles, books and book chapters, and conference papers, among others. They were more or less focused on the implications of accessibility, dealing with this term as a research topic or keyword. The chronological distribution and growth of indexed documents can be seen in Fig. 11.1.

Statistical data from Fig. 11.1 also reveals a substantial increase in research publications on accessibility in recent decades, especially since 2018. Considering periods every five years until 2022 (the latest full year), Table 11.1 shows the breakdown of the research outputs dealing with the topic.

This active topic has been addressed in different fields such as education (Lambert, 2020), health care (Matin et al., 2021), social inclusion (Sen et al., 2022; Woodgate et al., 2020), disabilities (Kubenz & Kiwan, 2022), transport and micro-mobility (Milakis et al., 2020), tourism (Gillovic & McIntosh, 2020), among many

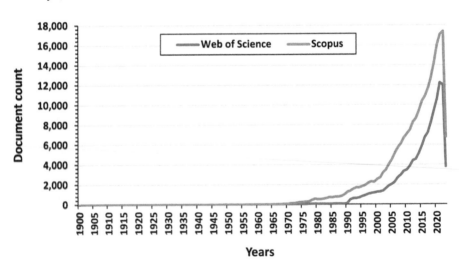

Figure 11.1 Graph. Scopus and Web of Science-indexed documents dealing with accessibility.

Table 11.1 Quantitative data on the chronological distribution and growth of indexed documents in Scopus and Web of Science from 2003 to 2022.

Database	Statistical descriptor	20 years (2003–2022)	15 years (2008–2022)	10 years (2013–2022)	5 years (2018–2022)
Scopus	Mean	9360.10	10,948.87	12,852.90	15,498.20
	Standard deviation	4334.00	3801.48	3177.80	1959.72
	Coefficient of variation	0.46	0.35	0.25	0.13
Web of Science	Mean	5523.35	6668.07	8202.30	10,495.00
	Standard deviation	3405.14	3173.56	2771.744	1711.71
	Coefficient of variation	0.62	0.48	0.34	0.16

others. Still, there are authors who provided additional meanings of the term accessibility, as in the case of the urban fields of transport and land use. Here, particularly focusing on land development from the economic and zoning points of view, Hansen (1959) identified accessibility as the potential of opportunities for interaction against the usual definition, which he explained was the intensity of the possibility of interaction. Besides, one of the acceptations described at the beginning of this section includes the word disability as an obstacle in those activities. According to the literature review conducted by Mack et al. (2021), the leading community of focus in research on accessible computing are blind or low-vision persons—it is worth noting the disproportionate frequency of this group in research (43.5%). Motor or physical disabilities and hearing impairment appear in significantly fewer publications (14.2% and 11.3%, respectively). Below 10% of the total of papers, the rest of the communities are cognitive impairments (9.1%), older adults (8.9%), autism (6.1%), other general disabilities or accessibilities (4.10%), and intellectual or developmental disability (2.8%). As a result, accessibility can be studied from different perspectives, depending on the object and the person or group.

11.1.2 Toward accessible cultural heritage in Europe

In the field of cultural heritage, by means of the Digital Agenda for Europe, the European Commission advocates the need for large-scale digitization and online accessibility of heritage assets, for which European Union (EU) Member States are invited to increase their digitization efforts (European Commission, 2011; Santos et al., 2017). At this point, it is worth synthesizing the Commission Recommendation of 27 October 2011 on the digitization and online accessibility of cultural material and digital preservation (2011/711/EU), currently in force.

Europe's Digital Agenda focuses on harnessing the benefits of information technology to boost economic growth, the creation of jobs, and quality of life for European citizens. One of the main areas covered is the digitization and preservation of European cultural memory, including prints, photographs, museum artifacts, archival documents, audio and audiovisual materials, monuments, and archaeological sites. In this way, the EU has developed a digitization and preservation strategy for cultural material and heritage assets, building on previous efforts such as the development of Europeana, the European Digital Library Archive and Museum. Digitization is critical to enabling online access to cultural materials, facilitating their use, and ensuring their preservation for future generations. In turn, this offers vast economic opportunities and is crucial for the development of Europe's cultural and creative capacities and industrial presence in this field. Nonetheless, collaborative action by Member States is required to avoid duplication of digitization and create a safer environment for companies investing in digitization technologies. Private funding and partnerships with the public sector can help cover the high costs of digitization. Likewise, digitization activities can be cofinanced by EU Structural Funds as part of projects that impact the local economy.

Materials in libraries, archives, and museums are generally protected by intellectual property rights that inspire creativity. However, the digitization and preservation of cultural materials must be done with due respect for copyright and related rights. Digitization of noncommercial works may require legal support for stakeholder-developed licensing solutions. Obtrusive watermarks and other visual protections should be avoided to allow broad access and use of public domain content. Europeana currently provides direct access to over 19 million digitized objects, with a set target of 30 million objects by 2015. Together with scientific publications achieving complex digital heritage models and data, and those achieving the availability of these assets, this is the accessibility this chapter addresses.

To ensure the survival of digitized materials, it is necessary to develop effective means of digital preservation. In this line, legal submission policies and practices should also be encouraged. Efficient cooperation among EU Member States is necessary to avoid major differences in the policies for the deposit of digital materials.

Overall, this Recommendation aims to promote the digitization and online accessibility of cultural materials, strengthen long-term preservation strategies for digital materials, and promote cooperation among countries to achieve these goals.

11.1.3 Digitization and modeling technologies for complex heritage shapes

Manual measurement instruments and processes have been left far behind with the advent of massive geometric data capture tools (Balsells et al., 2021). Without delving into the principle of operation of these remote sensing technologies here, they allow specialists to obtain hundreds of thousands and millions of data on the shape of objects within the Cartesian coordinate system (XYZ coordinates in space) (Bakker et al., 2009). In this way, complex heritage shapes can be accurately represented in a digital environment, which in turn facilitates the generation of three-dimensional (3D) models. Among others, those technologies include the LiDAR (Light Detection and Ranging)-based terrestrial laser scanning (TLS) and the often handheld structured-light scanning (SLS) for smaller objects, as well as the Structure-from-Motion (SfM) photogrammetric technique. All these techniques are capable of capturing colors to be later mapped onto the resulting 3D point clouds, which are the sets of points defining the objects' volume. In the case of SfM, this also allows for generating textured 3D meshes of the assets' geometry. Point clouds require processing to extract meaningful information from the data; this may include filtering, segmentation, registration, and feature recognition.

Once digitization and data processing are completed, the 3D modeling stage takes place. Different modeling approaches can be adopted depending on the intended accuracy, level of detail (usually referred to as LoD), or level of development (usually LOD, considering geometry and information), from an abstracted, theoretical models to geometrically complex shapes (as-built or as-is models). Thus computer-aided design (CAD) tools or Building Information Modeling (BIM) technology can be used to produce 3D models of heritage assets. However, to avoid manual 3D modeling and those approaches that do not fully represent the real condition of the assets, the virtual reconstitution should be based on 3D point cloud data (Logothetis et al., 2015). Regardless of the technology used to capture the geometry, there are algorithms capable of discretizing it and calculating 3D meshes that fit the shape of the object, with lower or greater degrees of simplification and smoothening (Antón et al., 2019). One of the most widely used algorithms is Kazhdan and Hoppe's Screened Poisson Surface Reconstruction (Kazhdan & Hoppe, 2013), usually integrated into open-source software as a plug-in. The Historic Building Information Modeling (HBIM) methodology, the application of BIM to heritage or historic buildings (H), can be enriched when importing as-built or as-is geometries (based on TLS or SfM data) to define the complex shapes of assets (Liu et al., 2023).

In view of the above, the European Commission's Expert Group on Digital Cultural Heritage and Europeana (DCHE Expert Group) developed a set of guidelines for 3D digitization of cultural heritage assets, called the basic principles and tips for 3D digitization of cultural heritage (European Commission's Expert Group on Digital Cultural Heritage & Europeana DCHE Expert Group, 2020). This document gathers a series of benefits of 3D digitization for heritage, such as aiding conservation, reproduction, research, education, and exploration, creative, or tourism-related purposes, safeguarding of at-risk tangible cultural heritage, or avoiding direct handling of assets. However, in words by the Expert Group, it should be noted that, by itself, this process does not prevent risks to cultural heritage, and it is by no means a replacement of physical preservation. Besides, 3D digitization by itself does not imply digital preservation in the long term. Therefore 3D digitization must be accompanied by good practices and policies for prevention, management, and restoration. These guidelines also highlight the possibility of virtual access to nonreachable, inaccessible, cultural heritage, especially for visually impaired individuals by enabling tactile experiences.

11.2   Research aim

This chapter is intended to explore the current and recent research on the accessibility of digital heritage models that represent the real condition of the assets. In other words, the focus is on those publications that have achieved deformed models, which include geometrical alterations following as-is or as-built modeling approaches, thus avoiding ideal geometries.

11.3   Methodology

This exploratory research aims to investigate relevant scientific literature on the accessibility of deformed digital heritage models in the last 5 years (from 2018 to present). Given their extensive use and reputation in providing access to a diverse range of scholarly literature across various disciplines, the Scopus and Web of Science databases were the primary sources for identifying those publications. The methodology adopted was as follows:

1. Keyword selection: Appropriate keywords, or different combinations of them, were used to search a wide range of documents closely aligned with the research topic. Those keywords did not have to be exclusively the keywords used for indexing the publications in the databases, but also terms included in their abstracts: as-is, as-built, accessible, accessibility, repository, TLS, scanning, photogrammetry, point cloud, 3D model, modeling, complex shapes, geometry, heritage, digital, virtual, virtual reality (VR), disabilities, or tour.
2. Database search: The selected keywords were used to search in the Scopus and Web of Science databases, as mentioned before. Additional search using the artificial intelligence (AI)-based Consensus App (Olson et al., 2022) obtained further publications—some of them are already indexed in Scopus and/or Web of Science. To do this, the prompt "accessibility of digital heritage models" was used to search for research on the topic, and the processes mentioned next were also conducted for this secondary search.
3. Publication filtering: Once a number of documents were found, a manual filtering process was conducted to assess the relevance and fit of each publication to the research topic. The filtering process involved examining the titles, abstracts, and keywords of the identified publications.
4. Inclusion and exclusion criteria: Specific criteria were applied during the filtering process. Publications that were deemed unrelated or did not sufficiently address the research topic were excluded, while those directly related to the topic or providing valuable insights were included.
5. Data synthesis: Finally, once documents were filtered and selected, data synthesis was carried out to show the current and recent perspectives on the accessibility of heritage models considering this modeling approach, and to what extent published digital models represent the real condition of the assets.

11.4   Accessibility of deformed digital heritage models

As seen earlier, this research refers to "deformed" models as those that take into consideration the surface geometrical alterations of heritage assets.

Prior to 3D reconstruction, Grilli and Remondino (2019) explained that the heterogeneous remote sensing 3D point clouds need to be segmented and classified to better characterize, describe, and interpret the study object, and that those processes can be automated and accelerated with the aid of machine learning. At this stage, the different specialists involved in the conservation process can automatically annotate 2D textures of heritage objects and see them mapped onto 3D geometries for a better understanding.

Having processed the data, 3D reconstruction methods enable mesh-based geometries consisting of triangulated surfaces fitting the point clouds (Antón et al., 2018, 2019; Bassier et al., 2016; Sani et al., 2022). 3D entities can also be produced using NURBS (nonuniform rational basis-splines) with a lower or greater degree of simplification or LoD for their integration into HBIM projects (Barazzetti et al., 2015; Murphy et al., 2013; Rodríguez-Martín et al., 2022). In this context, the scan-to-BIM workflow is still semiautomatic, despite the great efforts by the scientific community. Recent documents aim to improve the automation of HBIM from point clouds using AI, which was recently reviewed by Cotella (2023). Those 3D reconstruction approaches based on TLS, SLS, or SfM data that produce 3D meshes (Fig. 11.2, center) significantly differ from ideal or simplified 3D modeling in terms of accuracy (Fig. 11.2, left and right, respectively), which, in fact, has a severe impact on structural behavior simulations.

In view of both the research context and the given definitions of the term accessibility in Section 11.1.1, the authors of the present study found the need to explore how accessible deformed 3D models of heritage sites, buildings, and assets are.

3D digitization and restitution, augmented reality (AR), mixed reality (MR), and VR enable scientific knowledge, digital storytelling, and access (in a more or less immersive experience) to both natural and cultural heritage through a 3D database with relevant information of the asset for the users to understand their history and specifications (Ioannides et al., 2017; Maietti et al., 2017). In this way, Kidd and McAvoy (2019) discussed the increasing interest on immersive experiences for heritage organizations and their funders, driven by challenges and emerging technologies. These authors emphasized the need for better evidence on the role of immersive experiences in achieving goals such as visibility, audience appeal, participation, engagement, and revenue, which benefit not only the heritage assets in question but also the institutions and public involved. Immersive experiences in heritage models and assets have implications for accessibility, requiring consideration of inclusive design and features to accommodate diverse abilities and preferences. Here, the possibilities for physical interaction (one of the definitions of accessibility reviewed in this chapter) decrease when using the generally exclusive VR. Not devoid of challenges such as the cost of the equipment and the specialization needed to develop the immersive content, this technology often requires a separate space to be experienced by the users and is intended for the use of a single person per headset, an individual with the capacity of bearing the unpleasant physical symptoms involved. To a certain extent, this diminishes how accessible these immersive technologies are. However, they offer users the opportunity to actively engage with and become part of the heritage site, establish personal connections to

Figure 11.2 Modeling approaches. Digital reconstitution of a column in the Archaeological Ensemble of Baelo Claudia, in Tarifa, Cádiz (Spain): (left) ideal model; (center) as-is model; and (right) simplified model.

Source: From Antón, D., Pineda, P., Medjdoub, B., & Iranzo, A. (2019). As-built 3D heritage city modelling to support numerical structural analysis: Application to the assessment of an archaeological remain. *Remote Sensing*. 11 (11) (2019), 1276–1276, https://www.mdpi.com/2072-4292/11/11/1276, https://doi.org/10.3390/rs11111276.

it, and acquire knowledge (learn) through the immersive experience (Bekele & Champion, 2019).

Putting aside the immersive technologies, the accessibility of heritage should be manyfold. Not being preliminary related to digital matters, the ease of access (the lack of barriers) starts from the actual physical access to historic buildings. Kose (2022) analyzed the accessibility challenges in a set of Japanese historic buildings, mainly posed by the strict Japanese regulations on the protection of cultural properties, which hinder the changes needed to comply with accessibility standards in new built buildings. Here, the use of digital technologies could support reform proposals to persuade the governments involved. Similarly, from a university teaching perspective, the contribution of Vardia et al. (2018) consisted in developing design proposals to raise awareness of the importance of accessibility of heritage sites. Using the heritage site of Jantar Mantar in India as a case study, these authors defended the universal design as a show of sensitivity and compassion for diversity and respect for vulnerable populations, but also as a great opportunity to leverage an economic return from an increase in visitors to a previously inaccessible site.

In that line, Jiménez Martín et al. (2022) posed an interesting question that led to their research on the accessibility of historic city centers: What is the point of heritage if it cannot be known and visited? Considering accessibility and universal design principles, the authors analyzed the urban fabric and pedestrian itineraries as crucial spaces in the search for good accessibility practices and challenges. Being applicable to other historic centers, their research yielded essential aspects to be considered when undertaking actions on pedestrian routes in those urban heritage environments, encompassing mobility, location, orientation, understanding, and more. This research could be supported, again, by as-is digital models as a valuable basis for the study of heritage toward its accessibility.

Having introduced the implications of digital technologies for the accurate representation of heritage buildings and sites and the importance of ensuring access to both the physical and digital assets, the following classification gathers relevant contributions by the scientific community to the topic on the accessibility of deformed digital heritage models.

11.4.1 Disabilities

This chapter found a review article on the weight of disabilities on research, at least on accessible computing. Although majorly focusing on visual impairments, other physical limitations and conditions are present in research. Here, the heritage field has also addressed the accessibility of digital heritage models from the perspective of disabilities.

Pérez et al. (2020) addressed the barriers faced by people with disabilities, particularly in relation to archaeological heritage buildings and environments. They developed a VR experience specifically designed for individuals in wheelchairs to evoke realistic sensations on them while virtually touring archaeological sites. Focusing on the monumental building of Cancho Roano, a significant site in Spanish Protohistory, the VR application aimed to promote social integration by granting access to inaccessible cultural heritage for people with disabilities. The authors adopted two strategies to enhance the experience: TLS-based models to visualize the site accurately and a novel VR system that integrated motion capture, visualization, a motion platform with haptic rollers, and a workstation. The application was intended to undergo testing with wheelchair users and general users to evaluate its realism and the perceived value of accessing otherwise inaccessible areas of the site. By seeking to raise awareness about the challenges faced by wheelchair users in such environments, the research endeavored to provide inclusive and immersive experiences, thus enabling individuals with disabilities to engage with and appreciate cultural heritage.

Marín-Nicolás and Sáez-Pérez (2022) created and deployed a physical accessibility assessment tool tailored to the specific needs of heritage buildings by following BIM methodology. They focused on heritage buildings in collaboration with the Regional Government of Murcia, Spain. Different parameters were considered: types of buildings (religious, military, and civil buildings); area of analysis (access, door, moving walkway, and furniture); and the type of disability (list of barriers and levels of accessibility). The tool not only detected barriers but also analyzed their impact on user activities in different spaces and elements of the buildings. It assessed accessibility levels based on the main qualities of the building and identified areas to be improved. The tool provided accurate information for specific groups of people with disabilities, allowing for prioritized actions and their planning. It is also worth highlighting the benefits of BIM in analyzing and managing barrier removal. The research offered new possibilities for UD while promoting inclusive design in heritage buildings by integrating accessibility considerations into BIM tools.

Regarding visually impaired individuals, Scianna and Filippo (2019) designed a rapid prototyping approach to improve cultural heritage accessibility. The authors compared the effectiveness of a set of 3D models for guiding blind and visually impaired people in the tactile use of models of monuments. Thus a sample of disabled people helped define some 3D printing parameters to obtain a more fruitful product.

Using a different technology, Lo Valvo et al. (2020) explored the potential impact of convolutional networks trained to recognize monuments using smartphones as mediation instruments. This enabled users to access associated content and offered new forms of engagement for visually impaired individuals. Additionally, computer vision can support autonomous mobility by identifying predefined paths in heritage sites, bridging the gap between the digital and real world. The paper presented two key components: improving accessibility for visually impaired and mobility-restricted users and facilitating access to digital content associated with real monuments through smartphone-based virtual navigation. The authors envisioned personalized service instances that catered to specific user needs, such as voice descriptions, simplified content for children, and enlarged targets for the elderly, thereby enhancing the overall tourist experience.

In an effort to enhance accessibility to cultural heritage objects for the visually impaired, Montusiewicz et al. (2022) used 3D scanning, modeling, and 3D printing techniques. This addressed the limitations faced by contemporary museums in accommodating this group. The digitization of cultural artifacts enabled the creation of virtual museums and touch-accessible 3D replicas. As an illustration, the authors presented a case where prototype copies of museum objects from the Silk Road were developed, catering specifically to tactile recognition by blind individuals. These replicas, characterized by their lightweight nature, allowed for the identification of intricate details and the reading of Braille texts. By integrating Braille descriptions onto the surfaces of the objects, a multifunctional approach was established, thus enabling tactile exploration of the objects' shapes and access to accompanying textual content. The effectiveness of the chosen methodology, technology selection, and 3D printing with fused filament fabrication was demonstrated in the production of manipulable and lightweight 3D models. These versatile replicas can be used in various contexts, including museum exhibitions, home printing, conservation training, and archival purposes. This allowed for expanding accessibility and facilitating the sharing of cultural heritage objects.

Zada (2021) contributed to the development of innovative cartographic solutions for blind and partially impaired students in Kurdistan, addressing the challenges they face in utilizing maps. The study implemented a geographic information system (GIS) to create digital interactive tactile maps, also with the possibility of being updated, allowing users to customize data representation through touch displays. The effectiveness of those maps in improving text and space memorization among visually impaired Kurdish students was examined. The research aimed to fill the gap in providing effective multimodal communication maps for various demographic groups in Kurdistan. It explored the application of interactive tactile maps not only for visually impaired individuals but also for other populations. The authors believed their results were beneficial for blind and low-sighted users in easily accessing geographical information through assistive technologies, and it enhanced the learning experience and exposure to the latest information and technological advancements for students.

Closing the visual impairment disability, Rossetti et al. (2018) deployed low-cost interactive 3D models to enhance the understanding of architectural details in cultural sites. The focus was on designing digital models, printing, and assembling tangible models, in addition to creating an interactive and self-explanatory system. Eight participants with visual disabilities, including four blind and four severely visually impaired individuals, were involved in evaluating the impact of the proposed method. The results revealed positive feelings and perceptions from participants.

Finally, regarding the enhancement of disability rights, particularly for individuals with motor disabilities, Kocaman and Ozdemir (2020) proposed a comprehensive framework by using advancements in geographic information science (GIScience). This framework integrated GIS, volunteered geographic information, and citizen science to address location-based challenges faced by individuals with motor disabilities. They also explored the relationship between law and geography, emphasizing the structural injustices caused by geographies and the efforts by the United Nations to promote the cited disability rights. The authors noted that establishing a disability-oriented Spatial Data Infrastructure is crucial to ensure compatibility at both national and international levels. By implementing this framework, factors enabling or disabling motor disabilities in specific geographic contexts were identified, which led to innovative solutions and evidence-based policy making.

11.4.2   Multidisciplinary collaboration (collaborative platforms, geographic information system, and HBIM)

Pinto Puerto (2018) considered digital resources as rule breakers in relation to how data are understood, accessed, and used to build knowledge and support management and dissemination processes. In this way, the researcher and his team focused their efforts on enabling an interoperable database infrastructure that would allow various disciplines to operationalize the data and visualize phenomena from a spatio-temporal point of view. Their proposal involved using BIM and GIS, digital systems that have evolved the way architecture and land data are read, organized, selected, and analyzed.

López et al. (2018) reviewed the implementation of HBIM in the cultural heritage sector for modeling and managing architectural elements. The work highlighted the importance of digital 3D models for remote planning and conservation projects. Combining BIM and GIS tools, along with auxiliary software, enables the semiautomatic modeling of graphical and semantic data. The authors also described that HBIM libraries facilitate interdisciplinary collaboration and knowledge exchange, although challenges such as the lack of international libraries and shape recognition algorithms remain. By the time the research was conducted, there was a need to advance BIM platforms and create a universal HBIM library accessible to various experts in the field. The latter has been partly achieved as seen in the multidisciplinary collaborations in HBIM projects.

Regarding digital preservation, Champion and Rahaman (2019) highlighted the limited visibility and sustainability of 3D digital heritage models despite their potential to convey the importance of preserving cultural artifacts and intangible heritage. Through an examination of digital heritage papers, they revealed a lack of active promotion and preservation of these models as scholarly resources, hindering the evolution of the discipline and impeding public engagement in heritage preservation. The problem stems from inadequate infrastructure and the absence of integrated links to 3D assets in academic publication and dissemination systems. To address these issues, the authors proposed recognizing 3D models as scholarly resources and establishing guidelines for creating and maintaining a robust infrastructure. They emphasized the need to incorporate the dynamic and environmental aspects of built heritage into digital models to effectively communicate the principles of research and sustainability. By adopting an evolving scholarly digital ecosystem approach, the authors believed that both the scholarly publication system and digital heritage projects can better fulfill their goals of knowledge dissemination, preservation, and sustainability.

Nishanbaev (2020) drew attention to the progress made in 3D surveying and web technologies for the digital preservation and sharing of cultural heritage. The author introduced a methodology and web repository that integrated maps, 3D models, and geospatial data for long-term archiving and visualization on the web. According to the researcher, the use of free and open-source content management systems enhanced flexibility and reusability. Remaining challenges were also addressed, including metadata and semantic interoperability, as well as interlinking 3D models with relevant knowledge bases. Overall, this research contributed to the development of web repositories for 3D cultural heritage models.

Croce et al. (2020) referred to the growing trend in cultural heritage research toward digital information systems that combine geometric representations of artifacts with meaningful tags. This process, known as semantic annotation, enhances digital architectural heritage models by linking geometric representations to knowledge-related information. The study compared traditional 2D mapping methods with more recent approaches, for example, HBIM techniques and collaborative reality-based platforms. The advantages of semantic annotations over 2D mapping were explored, and ongoing research focused on constructing a formalized knowledge base, transferring annotations between representation systems, and automating the annotation process. These advancements aimed to reduce human involvement, improve the detection and labeling of architectural elements in 3D models, streamline annotation procedures, and facilitate the exchange and dissemination of research results in the field of heritage architectural assets.

Messaoudi et al. (2018) focused on addressing the challenge of integrating diverse data and creating a unified information model for the conservation and restoration of historical monuments. They developed a correlation pipeline that combines semantic, spatial, and morphological dimensions of built heritage using an ontological model. By spatially representing and semantically classifying information about material, alteration phenomena, and conservation state, they provided conservation experts with a practical means to explore and record scientific observations. The pipeline was successfully tested on a church in France, demonstrating its ability to correlate and retrieve various types of information. The proposed domain ontology model enabled the annotation and monitoring of stone degradation phenomena, facilitating decision-making for heritage experts. Further advancements include in situ data acquisition and correlation using mobile devices to provide real-time results. The integration of components and the concept of an "informative continuum" contribute to the interconnected representation of spatialized and semantically enriched photographs, enriching the knowledge base on heritage building degradation.

Pepe et al. (2021) implemented a novel Scan-to-BIM method in the cultural heritage field, specifically focusing on a rock church in Grottaglie, Italy. Through an integrated survey using TLS and SfM, they constructed a 3D-GIS model of the structure, enabling architectural and historical analysis. They developed a procedure in Rhinoceros software to accurately model the objects from the point cloud, resulting in a high-quality BIM project. Importing this BIM into a 3D-GIS environment allowed for the connection of multiinformation to each identified element.

This method overcome limitations in BIM software and facilitated the parameterization and management of objects, even those with complex and irregular surfaces. The use of unique codes for each layer within the structure enhanced the relational database and enabled the association of multiple information to each element.

In the field of archaeology, Pietroni et al. (2023) participated in the e-Archeo project, which aimed to enhance knowledge of Italian archaeological sites by using digital technologies (GIS, Digital Elevation Models, and 3D surveys) to restore their three-dimensionality, highlight cultural convergence, and promote accessibility. It fostered collaboration between public entities, research bodies, and private industries, resulting in the development of multimedia solutions and applications. The project had a positive impact, leading to increased cultural offerings, educational programs, and the integration of project narratives into civic education initiatives and VR experiences. Its goal was to raise awareness and foster a greater appreciation for Italy's cultural heritage.

Malinverni et al. (2019) conducted a multidisciplinary study at the Umm ar-Rasas archaeological site in Jordan, focusing on enhancing knowledge of the polychrome mosaic floor in the Church of Saint Stephen. They collected a vast amount of data, including archaeological investigations and geomatic surveys, and organized it in a geo-database for information exchange. Through the use of GIS and multimedia applications, they created virtual experiences and disseminated information to experts and visitors, including those with disabilities. This innovative approach improved accessibility, provided immersive cultural experiences, and facilitated the sharing of knowledge across different platforms.

Finally, the scientific community has also used BIM, a well-known methodology of the Architecture, Engineering, and Construction sector, for documentation and data sharing purposes in archaeology. Among other achievements, Cortés-Sánchez et al. (2018) documented and characterized cave surfaces using TLS and SLS techniques, and stone engravings by means of SfM and SLS at a millimetric scale. Data accessibility allowed the different specialists of that multidisciplinary team to consult and manage the as-is models of the cave sectors in BIM for diverse studies in the research. Similarly, Moyano et al. (2020) worked on the A-BIM (Archaeological Building Information Model) of La Pastora tholos in Valencina de la Concepción, Seville (Spain), a funerary megalithic monument of the Copper Age. The A-BIM of the tholos granted access to the data by the different specialists, including the archaeologist and the BIM operator. On the other hand, by adopting the Scan-to-BIM approach, these authors carried out the parametric modeling of archaeological elements using TLS data and Kazhdan and Hoppe's 3D reconstruction algorithm. This produced 3D meshes defining the original shapes and were later converted into BIM morph elements so that Boolean operations trimmed the thickness excess in the created BIM wall. However, the model failed to achieve the full as-is condition, since the wall geometry in both the corridor and the chamber was clearly affected by the modeling method followed. The upward extrusion Boolean operation, with the morph mesh as the splitting object, resulted in a loss of relief on the wall. For their part, Marini et al. (2022) 3D surveyed and reconstructed a complex-shaped archaeological site, the San Giorgio Cave in Usini (Italy), for documentation purposes. However, the authors did not explicitly describe how they achieved the desired "public archaeology," as no link to the 3D model was provided to facilitate accessibility. This suggests that there is a potential benefit to accessibility, as it should be inherent in the production of digital models of heritage assets. Furthermore, the authors recommended that accessibility of the data to archaeologists should occur through the point cloud rather than the 3D model. This is nothing but missing the opportunity to generate knowledge from actual geometry.

11.4.3   Open access, repositories, and virtual museums

According to Benchekroun and Ullah (2021), sharing access to 3D heritage scanning data is an important obstacle to the creation of a complete pipeline for this sort of material. In their work, these authors developed a flexible and accessible approach for the creation, processing, and web hosting of digital heritage materials, which was exemplified through a case study conducted in partnership with the La Mesa Historical Society (LMHS). By showcasing different 3D scanning pipelines, they demonstrated that even inexperienced practitioners can successfully produce high-quality 3D data using their methods. The authors also created an interactive online Web3D platform to share the captured data, allowing individuals who cannot physically visit the site to experience it virtually. Furthermore, they aimed to empower local stakeholders, including the LMHS, by enabling them to conduct their own scans using the accessible 3D digital recording pipelines. In their research, the authors compared the output of their pipelines with professional quality 3D scanning tools, aiming to bridge the accessibility gap in equipment. The authors believed that their methodologies would not only benefit fields such as digital archaeology and heritage conservation in practical terms but would also broaden access, control, and action over the creation and dissemination of digital heritage assets. Moreover, the authors predicted increased public engagement in the documentation and visualization of cultural heritage, ultimately generating more interest and care for humanity's valuable heritage.

While most of her ideas are still valid today, Keene (2008) highlighted the evolving expectations for museums, which were urged to become more responsive to public wishes and make better use of their collections. She revealed that expert reports advocate for a shift toward excellence, inclusivity, and approaches that focus on the users. It is also emphasized that museums have the opportunity to be innovative and creative in reinventing their services, particularly by engaging people directly with collections. Overcoming barriers to innovation, such as economic factors and cultural conservatism, is crucial. The sector should prioritize community intelligence, build alliances, and tailor services to local needs. In addition, the fact that only a small number of artifacts in the collecting institutions are accessible to the public indicates the need to improve the accessibility to information about cultural heritage in line with modern requirements. By embracing these changes, museums can shape a future that offers meaningful experiences and challenges the perception of collections as elitist, ultimately fostering greater public engagement and interest.

In that sense, García-Bustos et al. (2022) developed a virtual museum to make Paleolithic art more accessible to public, since this sort of heritage is usually prioritized for conservation and research rather than for touristic purposes. To do this, the authors digitized sectors in caves and open-air stations with valuable features using SfM to obtain textured 3D meshes. Following a simplification process to ease data handling, they designed the digital exhibition environment to store the digitized content and made it accessible online in the Sketchfab platform to (1) democratize this (in the words by the authors) important cultural manifestation for understanding human history and (2) inspire the creation of more digital resources related to this art form.

Foo et al. (2023) explored the use of digital documentation methods, including 3D scanning and photogrammetry, to create 3D models of a heritage shophouse. They evaluated each method for its strengths and weaknesses in producing accurate models and photorealistic images. The study emphasized the importance of comprehensive documentation for conservation and highlighted the potential of 3D models for initial assessment and data sharing with stakeholders. The authors suggested that a combination of methods may be necessary to capture intricate architectural features in an effective manner. The implementation of 3D digital documentation offers advantages, but challenges remain, including the need for expertise in both digital documentation and heritage conservation. The authors proposed a multidisciplinary approach and the creation of a searchable central repository for historical architectural elements to enhance conservation efforts and information sharing. They claimed that more research is needed in this field to improve and expand the use of 3D digital documentation for cultural heritage preservation.

Albertini et al. (2022) developed PROTEUS, a VR tool integrated with an online repository for 3D models (both 3D reconstructed and handcrafted), metadata, and chemical analyses in digital humanities. According to the authors of this research, PROTEUS enables seamless transition between macroscopic and microscopic worlds, offering user-friendly and immersive experiences for researchers to manipulate objects, test hypotheses, and visualize metadata interactively. It fosters collaboration, knowledge sharing, and multidisciplinarity, bridging chemistry and cultural heritage. They believed that software's flexibility, compatibility, and future expansion plans, such as temporal analysis and multiuser sessions, should enhance its utility for scholars. Thus PROTEUS would represent a significant step toward democratizing science through VR.

Delving into 3D repositories, Gothandaraman and Muthuswamy (2021) developed a symmetry detection and analysis methodology for a more effective search of similar 3D complex objects of cultural heritage.

Halabi et al. (2022) coordinated archaeological exercises, data management, digital object representation, and spatial analysis. They proposed a framework that allowed intuitive exploration of 3D reconstructed artifacts and provided detailed information on specific points of interest within a 3D-GIS repository. This pioneering work in Qatar focused on archiving and recording all archaeological data from the Murwab site, shedding light on the social and monetary relations within the ancient community. The implementation of HTML5, Cesium, and WebGL allowed large-scale web-based 3D geospatial visualization without requiring browser plug-ins or additional applications. The potential applications of this concept ranged from attracting tourists to museums through 3D archaeological data access to 3D printing of artifacts for a tactile experience and creating a cloud-based map with a simulation of the surrounding environment.

Antlej (2022) proposed the creation of a CAD repository for modeled elements and 3D scanned features of the K67 Kiosk, a sample of modernist modular architecture by the Slovenian architect and designer Saša J. Mächtig. This research enabled systematic documentation of the heritage asset and audience engagement through VR, MR and AR, 3D printing, and serious games. Nishanbaev (2020) developed a methodology and a web repository to integrate maps, 3D models, and geospatial data, enabling long-term web archiving and visualization of geo-located 3D digital cultural heritage models, thus fostering interest and tourism. The author addressed challenges in data integration, interoperability, and visualization, while acknowledging the expansion potential of this methodology to other domains. In addition, the author discussed two major technical challenges, focusing on metadata and semantic interoperability and the interconnection of 3D model parts with relevant information. Despite remaining challenges, the author's work contributed to enhancing 3D digital cultural heritage preservation and engagement on the web.

Following the 3D digitization and reconstruction of the Inquisition Prisons and Archaeological Excavation Area of the Palazzo Chiaramonte-Steri monumental complex, Scianna et al. (2023) developed a web app to create a single digital platform for the enhancement, use, and management of that case study.

Finally, Zheliazkova et al. (2015) focused their research on nonspecialist users who do not necessarily have access to expensive equipment. The aim was to bring as-built BIM for heritage (HBIM) to a wider audience. The authors presented a parametric-assisted method for reconstructing and creating BIM projects of built heritage using low-cost technologies and open-source software. They focused on the 3D reconstruction of the ceiling of the Albergo Diurno "Venezia" in Milan (Italy), an architecturally significant heritage site. The research highlighted the development of custom algorithms to automatically rebuild the complex geometry from 3D meshes into an NURBS-based 3D model. The proposed methodology streamlined data elaboration, improved accuracy depending on the quality of the input dataset, and allowed for different BIM levels of detail.

11.4.4 Immersive experiences

The scientific community has explored visually appealing and immersive settings for different purposes. Examples of those are the work by Kang et al. (2022), who presented a physics-based simulation of the deformation of soft-bodies using a low-cost RGB-D (Red, Green, Blue-Depth) sensor. With the potential to be applied to different interactive AR and VR environments, the authors' realistic outputs were useful for topological deformations. Halder & Afsari, (2022) designed an inspector assistant robot to conduct real-time construction verifications in an immersive environment.

In a way to communicate built cultural heritage, HBIM and other 3D reconstruction methods allow for producing immersive experiences to ensure accessibility to users, experts, and institutions. An example of this is the work by Argiolas et al. (2022), who emphasized the widespread use of HBIM methodology in managing architectural heritage, particularly the Scan-to-BIM process using laser scanning and photogrammetry. They also highlighted the integration of HBIM models into game engines for immersive virtual tours and interactive experiences. The paper focused on developing a virtual tour of the Jesuitical Complex of Santa Croce in Cagliari, adopting HBIM methodology and game engines to enhance understanding and communication of historic architecture. The advantages of informative models' directly transposable to development software and the ability to expand user interactions are evident.

Kowalski et al. (2023) developed methodological protocols for enhancing and preserving built heritage through digital reconstruction, particularly using VR-based models. They focused on a 19th-century case study, the coastal battery in the Gdańsk port area. Using TLS, SfM, and GIS, together with historical research, the process involved 3D digitization, comparison of historical records and maps to the present state, and 3D reconstruction of the lost heritage. This enabled an immersive and accurate experience of historical changes through VR. The contribution of this methodology is twofold: It aids in preserving incomplete or altered heritage sites and opens possibilities for interactive heritage education and engagement through modern technologies, for example, VR, thus offering fresh insights into the past.

Giovannini and Bono (2023) developed a methodological workflow to create virtual environments of replicated physical spaces. These immersive and virtual experiences enabled social interaction and augmented content exploration. The authors focused on the temporary "Phygital Exhibition" at the Sordevolo Passion Museum in the Church of Santa Marta, Italy. The digitization was performed via SfM using an unmanned aerial vehicle, and the resulting 3D model was stored in SketchFab to ensure user accessibility. The study showcased that methodology involving the creation of a digital twin to highlight the transformation of scenography over time. The authors demonstrated how virtual scenes can be generated, personalized, and shared on the web. This approach should enhance visitor experience, provide content customisation, encourage collaboration, and offer cost-effective alternatives to traditional exhibitions, ultimately broadening audience reach and offering diverse content experiences.

Jouan et al. (2022) proposed a virtual environment prototype to create a built heritage immersive experience of the Collegiate Church of Saint-Jean in Belgium. The technologies they used included 3D laser scanning, photogrammetry, and 3D models and employed the 3D Unity game engine. The result facilitated not only the tasks of the experts dealing with the difference stages of the conservation project but also the views of the nonexpert stakeholders in the maintenance issues and previous states of existence.

Benardou and Droumpouki (2022) brought together multiple works on immersive experiences related to built heritage. Topics addressed included digital immersion to analyze built heritage as an enabler of equity in urban sites, experiences in a concentration camp, or unearthing the Rosewood massacre.

Spallone et al. (2023) pursued accessibility to built heritage via 3D reconstruction and AR−VR interactive visualization. The case study was Piffetti's Library in Villa Della Regina Museum in Turin, Italy. The interdisciplinary team comprised experts from art history, digital surveying, 3D modeling, and digital cultural heritage solutions, who combined 3D laser scanning and photogrammetry to recreate the library's original structure on an accurate basis. The AR application allowed visitors to experience the library's original location, while a detailed VR model provided a comprehensive online experience. This project aimed to engage users, making history accessible through immersive technologies. Being ongoing, the project strives to offer interactive elements and connections between Villa della Regina and Quirinale, enhancing the understanding of historical artifacts and their significance. The project's approach integrates historical, architectural, and technological aspects to bridge the gap between cultural heritage and modern accessibility.

Pavelka jr. et al. (2023) achieved a comprehensive SfM 3D reconstruction of the church of St. Panteleimon, near Skopje (North Macedonia). However, in relation to the development of the VR experience from the case study, the authors missed the opportunity to reach as-is geometries as they focused on the simplified (ideal) HBIM of the temple.

Additionally, it is worth including freely accessible examples of immersive content based on as-is or as-built 3D reconstructions from remote sensing data. Although they are not indexed in scientific databases, the following contributions were made within the framework of nationally and internationally funded research and development (R&D) projects [United Kingdom, and the EU], other international collaborations, or contracts with private institutions. Given that they are available online on YouTube, the accessibility to the general public, experts, institutions, and businesses is ensured.

Another interesting contribution is that by Odyssey Visual Media (2019) video on YouTube, an impressive and realistic panoramic virtual tour of the as-is model of some Pyramids of Egypt.

Funded by the National Lottery Heritage Fund, a TLS survey of St Barnabas Catholic Cathedral in Nottingham, United Kingdom, enabled a simplified structural as-is model to support future analysis and restoration of the building (Antón, 2020). In this line, the European Regional Development Fund's part-funded project entitled "Live Experiential and Digital Diversification − Nottingham" (LEADD:NG) made it possible to survey and create simplified as-is models of important landmarks in that city. First, the Anglican St John the Baptist Beeston Parish Church was selected as a case study of the application of remote sensing and modeling digital technologies to deformed heritage buildings, thus enabling fully rendered 2D (Antón, 2022b) and 360-degree (spherical, panoramic) (Antón, 2022a) virtual tours to explore and disseminate the temple. Second, Ye Olde Trip to Jerusalem, claimed to be the oldest pub in England (CE 1189), shows numerous deformations due to the course of time. A game engine−based immersive experience was created, for which a virtual tour showing its exploration was made available to the public (Antón, 2022c). More recently, a team of researchers from Universidad de Sevilla (Spain) applied non-destructive techniques from the architectural field to a different heritage typology to study it and make it accessible to the public and the institution involved. Thus they carried out the SfM digitization and virtual exploration of the Virgin Nuestra Señora del Socorro's procession cape, of the Hermandad del Amor in Seville (Antón, 2023). Here, the as-is 3D reconstruction enabled the analysis, digital preservation, and dissemination of this embroidery textile asset, a religious heritage piece of a great value.

11.5   Conclusion

Cultural heritage is a source of knowledge and enjoyment for all people, including those with visual disabilities. However, these sites are usually not accessible or do not offer adapted resources for exploration. One way to improve the inclusion and participation of blind or visually impaired individuals in the cultural field is through the use of 3D models that reproduce architectural or artistic elements of interest. These models, based on precise geometries and representative of the current conservation status of heritage assets, allow for detailed study, restoration actions, preventive conservation, and exploration for dissemination purposes. Thus it is essential to make this legacy accessible so that all people, regardless of their physical or mental condition, can enjoy it.

This research has identified a number of relevant studies that achieve as-built geometries and make those models accessible. However, other publications do not demonstrate achieving that level of accuracy completely. Nevertheless, there is a great effort by the scientific literature to create and/or apply collaborative platforms in which stakeholders can share information.

The main limitation of this research is the overview (exploration) of the existing literature, which, while ensuring reasonable coverage of relevant publications, does not constitute a fully comprehensive and extensive literature review as such. Therefore a more thorough search for publications in these and other databases, considering additional criteria, will be undertaken in future research.

Acknowledgments

This research has been supported by funding for a postdoctoral researcher contract from the VI Plan Propio de Investigación y Transferencia of Universidad de Sevilla (reference VIPPIT-2020-II.5), Spain.

Artificial intelligence disclosure

During the preparation of this work, the author(s) used (1) Consensus App and (2) ChatGPT to (1) locate publications that may have been overlooked, or those not indexed in Scopus or WoS, and (2) edit text to improve readability and language of the work, and not to analyze or draw insights from data or publications' content. After using this tool/service, the author(s) reviewed and edited the content as needed and take(s) full responsibility for the content of the publication.

References

Albertini, N., Baldini, J., Pino, A. D., Lazzari, F., Legnaioli, S., & Barone, V. (2022). PROTEUS: An immersive tool for exploring the world of cultural heritage across space and time scales. *Heritage Science*, *10*(1), 71. Available from https://doi.org/10.1186/s40494-022-00708-3, https://heritagesciencejournal.springeropen.com/articles/10.1186/s40494-022-00708-3.

Antlej, K. (2022). Digital heritage interpretation of modernist modular architecture: The K67 Kiosk. In C. Bartolomei, A. Ippolito, & S. H. T. Vizioli (Eds.), *Digital Modernism Heritage Lexicon*. Cham: Springer. 978-3-030-76239-1. Available from https://doi.org/10.1007/978-3-030-76239-1_23.

Antón, D. (2020). 2020/// 2020 2 26 *YouTube St Barnabas Cathedral, Nottingham, UK - 3D model virtual tour (flythrough and walkaround animation)*. Unpublished content St Barnabas Cathedral, Nottingham, UK - 3D model virtual tour (flythrough and walkaround animation) https://irep.ntu.ac.uk/id/eprint/43139.

Antón, D. (2022a) 2022/// 2022 12 9 *YouTube Ye Olde Trip to Jerusalem, England's oldest pub (fly-through of VR immersive experience)*. Unpublished content Ye Olde Trip to Jerusalem, England's oldest pub (fly-through of VR immersive experience) http://irep.ntu.ac.uk/id/eprint/47661/.

Antón, D. (2022b) 2022/// 2022 3 9 *YouTube Beeston Parish Church (Nottingham) - rendered 3D model and video tour*. Unpublished content Beeston Parish Church (Nottingham) - rendered 3D model and video tour http://irep.ntu.ac.uk/id/eprint/45979/.

Antón, D. (2022c) 2022/// 2022 3 14 *YouTube 360/VR mobile devices - Beeston Parish Church (Nottingham) - rendered 3D model and 360 video tour*. Unpublished content 360/VR mobile devices - Beeston Parish Church (Nottingham) - rendered 3D model and 360 video tour http://irep.ntu.ac.uk/id/eprint/45980/.

Antón, D. (2023) 2023/// 2023 2 10 *YouTube Manto de Salida de Nuestra Señora del Socorro - Fotogrametría y exploración virtual realista*. Unpublished content Manto de Salida de Nuestra Señora del Socorro - Fotogrametría y exploración virtual realista http://irep.ntu.ac.uk/id/eprint/48368/.

Antón, D., Medjdoub, B., Shrahily, R., & Moyano, J. (2018). Accuracy evaluation of the semi-automatic 3D modeling for historical building information models. *International Journal of Architectural Heritage*, *12*(5), 790−805. Available from https://doi.org/10.1080/15583058.2017.1415391, https://www.tandfonline.com/doi/full/10.1080/15583058.2017.1415391.

Antón, D., Pineda, P., Medjdoub, B., & Iranzo, A. (2019). As-built 3D heritage city modelling to support numerical structural analysis: Application to the assessment of an archaeological remain. *Remote Sensing*, *11*(11), 1276. Available from https://doi.org/10.3390/rs11111276, https://www.mdpi.com/2072-4292/11/11/1276.

Argiolas, R., Bagnolo, V., Cera, S., & Cuccu, S. (2022). Virtual environments to communicate built cultural heritage: A HBIM based virtual tour. *The International Archives of the Photogrammetry, Remote Sensing and Spatial Information Sciences*, *XLVI-5/W1*,

21−29. Available from https://doi.org/10.5194/isprs-archives-XLVI-5-W1-2022-21-2022, https://isprs-archives.copernicus.org/articles/XLVI-5-W1-2022/21/2022/.

Australia ICOMOS. (1999). *The Burra Charter: the Australia ICOMOS Charter for Places of Cultural Significance*. Australia ICOMOS Incorporated 2000, Burwood. Available from: https://australia.icomos.org/wp-content/uploads/BURRA_CHARTER.pdf.

Bakker, W. H., Feringa, W., Gieske, A. S. M., Gorte, B. G. H., Grabmaier, K. A., Hecker, C. A., Horn, J. A., Huurneman, G. C., Janssen, L. L. F., Kerle, N., van der Meer, F. D., Parodi, G. N., Pohl, C., Reeves, C. V., van Ruitenbeek, F. J., Schetselaar, E. M., Tempfli, K., Weir, M. J. C., Westinga, E., & Woldai, T. (2009). *Principles of Remote Sensing* (4th, p. 591). Enschede, The Netherlands: The International Institute for Geo-Information Science and Earth Observation (ITC). Available from https://webapps.itc.utwente.nl/librarywww/papers_2009/general/principlesremotesensing.pdf.

Balsells, C. M., Besora, J. M. L., Jover, A. C., & Pla, S. C. (2021). Register of dry stone domes. Simplified method for point clouds. *Nexus Network Journal*, 23(2), 493−506. Available from https://doi.org/10.1007/s00004-020-00533-w, https://link.springer.com/10.1007/s00004-020-00533-w.

Barazzetti, L., Banfi, F., Brumana, R., Gusmeroli, G., Previtali, M., & Schiantarelli, G. (2015). Cloud-to-BIM-to-FEM: Structural simulation with accurate historic BIM from laser scans. *Simulation Modelling Practice and Theory*, 57, 71−87. Available from https://doi.org/10.1016/j.simpat.2015.06.004, https://linkinghub.elsevier.com/retrieve/pii/S1569190X15000994.

Bassier, M., Hadjidemetriou, G., Vergauwen, M., & Roy, N. V. (2016). Implementation of scan-to-BIM and FEM for the documentation and analysis of heritage timber roof structures. In M. Ioannides, E. Fink, A. Moropoulou, M. Hagedorn-Saupe, A. Fresa, G. Liestøl, & P. Grussenmeyer (Eds.), *Els Verstrynge, digital heritage. progress in cultural heritage: Documentation, preservation, and protection: 6th International Conference, EuroMed*. Nicosia, Cyprus: Springer International Publishing. 978-3-319-48496-9. Available from https://doi.org/10.1007/978-3-319-48496-9_7.

Bekele, M. K., & Champion, E. (2019). A comparison of immersive realities and interaction methods: Cultural learning in virtual heritage. *Frontiers in Robotics and AI, 6*. Available from https://doi.org/10.3389/frobt.2019.00091, https://www.frontiersin.org/article/10.3389/frobt.2019.00091/full.

Benardou, A., & Droumpouki, A. M. (2022). *Difficult heritage and immersive experiences* (p. 182) London: Routledge. Available from https://www.taylorfrancis.com/books/9781003200659.

Benchekroun, S., & Ullah, I.I. T. (2021 1−9). Preserving the past for an uncertain future. *The 26th International Conference on 3D Web Technology*. ACM, New York, NY, USA, Available from: https://doi.org/10.1145/3485444.3507684, https://dl.acm.org/doi/10.1145/3485444.3507684.

Cambridge University Press. (2023) 2023/// 2023 1 7 *Cambridge Dictionary Accessibility*. Unpublished content Accessibility. https://dictionary.cambridge.org/dictionary/english/accessibility.

Champion, E., & Rahaman, H. (2019). 3D digital heritage models as sustainable scholarly resources. *Sustainability*, 11(8), 2425. Available from https://doi.org/10.3390/su11082425, https://www.mdpi.com/2071-1050/11/8/2425.

Clarivate. (2023). 2023/// 2022 6 1 *Web of Science Unpublished content*. Web of Science https://www.webofknowledge.com/.

Cortés-Sánchez, M., Riquelme-Cantal, J. A., Simón-Vallejo, M. D., Giráldez, R. P., Odriozola, C. P., Román, L. C., Carrión, J. S., Gómez, G. M., Vidal, J. R., Campos,

J. J. M., Delgado, F. R., Julián, J. E. N., García, D. A., Martínez-Aguirre, M. A., Barredo, F. J., & Cantero-Chinchilla, F. N. (2018). Pre-Solutrean rock art in southernmost Europe: Evidence from Las Ventanas Cave (Andalusia, Spain. *PLoS One, 13*(10), e0204651. Available from https://doi.org/10.1371/journal.pone.0204651, https://journals.plos.org/plosone/article/file?id = 10.1371/journal.pone.0204651&type = printable.

Cotella, V. A. (2023). From 3D point clouds to HBIM: Application of artificial intelligence in cultural heritage. *Automation in Construction, 152*, 104936. Available from https://doi.org/10.1016/j.autcon.2023.104936, https://linkinghub.elsevier.com/retrieve/pii/S0926580523001966.

Croce, V., Caroti, G., De Luca, L., Piemonte, A., & Véron, P. (2020). Semantic annotations on heritage models: 2D/3D approaches and future research challenges. *The International Archives of the Photogrammetry, Remote Sensing and Spatial Information Sciences, XLIII-B2-2*, 829−836. Available from https://doi.org/10.5194/isprs-archives-XLIII-B2-2020-829-2020, https://isprs-archives.copernicus.org/articles/XLIII-B2-2020/829/2020/.

Elsevier, B. V. (2023). 2023/// 2022 6 1 *Scopus Unpublished content*. Scopus https://www.scopus.com/.

European Commission (2011) 2011/// 2022 9 15 *Official Journal of the European Union* L283 This report is structured according to the actions to be implemented by all Member States as recommended by the Commission in its Recommendation of 24 August 2006 on digitisation and online accessibility of cultural material and digital preservation and the related Council Conclusions of 13 November 2006. The deadline for submission of Member States' reports was 29/2/2008. 39-45 Commission recommendation of 27 October 2011 on the digitisation and online accessibility of cultural material and digital preservation The European Commission Brussels Commission recommendation of 27 October 2011 on the digitisation and online accessibility of cultural material and digital preservation https://eur-lex.europa.eu/eli/reco/2011/711/oj2011/711/E.

European Commission's Expert Group on Digital Cultural Heritage and Europeana (DCHE Expert Group). (2020). 2020/// 2022 11 15 *Shaping Europe's digital future. Basic principles and tips for 3D digitisation of cultural heritage*. https://digital-strategy.ec.europa.eu/en/library/basic-principles-and-tips-3d-digitisation-cultural-heritage.

Foo, C. P., Wong, C. W., Ng, Y. H., & Jacosalem, R. (2023). Exploring digital documentation for shophouses in Singapore. *The International Archives of the Photogrammetry, Remote Sensing and Spatial Information Sciences, XLVIII-M-2*, 579−586. Available from https://doi.org/10.5194/isprs-archives-XLVIII-M-2-2023-579-2023, https://isprs-archives.copernicus.org/articles/XLVIII-M-2-2023/579/2023/.

García-Bustos, M., Rivero, O., Bustos, P. G., & Mateo-Pellitero, A. M. (2022). From the cave to the virtual museum: accessibility and democratisation of Franco-Cantabrian Palaeolithic art. *Virtual Archaeology Review, 14*(28), 54−64. Available from https://doi.org/10.4995/var.2023.17684, http://polipapers.upv.es/index.php/var/article/view/17684.

Gillovic, B., & McIntosh, A. (2020). Accessibility and inclusive tourism development: Current state and future agenda. *Sustainability, 12*(22), 9722. Available from https://doi.org/10.3390/su12229722, https://www.mdpi.com/2071-1050/12/22/9722.

Giovannini, E. C., & Bono, J. (2023). Creating virtual reality using a social virtual environment: Phygital Exhibition at the Museum Passion in Sordevolo. *The International Archives of the Photogrammetry, Remote Sensing and Spatial Information Sciences, XLVIII-M-2*, 669−676. Available from https://doi.org/10.5194/isprs-archives-XLVIII-M-2-2023-669-2023, https://isprs-archives.copernicus.org/articles/XLVIII-M-2-2023/669/2023/.

Gothandaraman, R., & Muthuswamy, S. (2021). Virtual models in 3D digital reconstruction: Detection and analysis of symmetry. *Journal of Real-Time Image Processing, 18*(6),

2301–2318. Available from https://doi.org/10.1007/s11554-021-01115-w, https://link. springer.com/10.1007/s11554-021-01115-w.

Grilli, E., & Remondino, F. (2019). Classification of 3D digital heritage. *Remote Sensing, 11* (7), 847. Available from https://doi.org/10.3390/rs11070847, https://www.mdpi.com/ 2072-4292/11/7/847.

Halabi, O., Al-Maadeed, S., Puthern, M., Balakrishnan, P., & El-Menshawy, S. (2022). 3D GIS interactive visualization of the archaeological sites in Qatar for research and learning. *International Journal of Emerging Technologies in Learning (iJET)., 17*(01), 160–178. Available from https://doi.org/10.3991/ijet.v17i01.25933, https://online-journals.org/index.php/i-jet/article/view/25933.

Hansen, W. G. (1959). How accessibility shapes land use. *Journal of the American Institute of Planners, 25*(2), 73–76. Available from https://doi.org/10.1080/01944365908978307, http://www.tandfonline.com/doi/abs/10.1080/01944365908978307.

Halder, S., & Afsari, K. (2022). Real-time Construction Inspection in an Immersive Environment with an Inspector Assistant Robot. In T. Leathem, W. Collins, & A. J. Perrenoud (Eds.), *ASC2022. 58th Annual Associated Schools of Construction International Conference* (pp. 389–379). The Associated Schools of Construction, Atlanta, Georgia. https://doi.org/10.29007/ck81.

International Council on Monuments and Sites. (2011). 2011/// 2019 6 16 *The Athens Charter for the Restoration of Historic Monuments – 1931.* Unpublished content. https://www. icomos.org/en/167-the-athens-charter-for-the-restoration-of-historic-monuments.

Ioannides, M., Magnenat-Thalmann, N., & Papagiannakis, G. (2017). *Mixed reality and gamification for cultural heritage* (1, p. 594). Cham: Springer International Publishing. Available from http://link.springer.com/10.1007/978-3-319-49607-8.

Jiménez Martín, D., Saiz, A. R., & Escudero, M. A. A. (2022). *Transforming our world through universal design for human developmenturban accessibility in world heritage cities. Accessibility Considerations in Pedestrian Routes in Historic City Centres Studies in Health Technology and Informatics.* IOS Press. Available from https://ebooks. iospress.nl/doi/10.3233/SHTI220879.

Jouan, P., Moray, L., & Hallot, P. (2022). Built heritage visualizations in immersive environments to support significance assessments by multiple stakeholders. *The International Archives of the Photogrammetry, Remote Sensing and Spatial Information Sciences, XLVI-2/W1,* 267–274. Available from https://doi.org/10.5194/isprs-archives-XLVI-2-W1-2022-267-2022, https://isprs-archives.copernicus.org/articles/XLVI-2-W1-2022/267/2022/.

Kang, D., Moon, J., Yang, S., Kwon, T., & Kim, Y. (2022). Physics-based simulation of soft-body deformation using RGB-D data. *Sensors, 22*(19), 7225. Available from https://doi. org/10.3390/s22197225, https://www.mdpi.com/1424-8220/22/19/7225.

Kazhdan, M., & Hoppe, H. (2013). Screened Poisson surface reconstruction. *ACM Transactions on Graphics, 32*(3), 1–13. Available from https://doi.org/10.1145/ 2487228.2487237, https://dl.acm.org/doi/10.1145/2487228.2487237.

Keene, S. (2008). *Collections for people. Museums' stored collections as a public resource.* London: UCL Institute of Archaeology. Available from https://discovery.ucl.ac.uk/id/ eprint/13886/1/13886.pdf.

Kidd, J., & McAvoy, E. N. (2019). Immersive experiences in museums, galleries and heritage sites: a review of research findings and issues. *Creative Industries' Policy and Evidence Centre (PEC).* Available from https://orca.cardiff.ac.uk/id/eprint/128879.

Kocaman, S., & Ozdemir, N. (2020). Improvement of disability rights via geographic information science. *Sustainability, 12*(14), 5807. Available from https://doi.org/10.3390/ su12145807, https://www.mdpi.com/2071-1050/12/14/5807.

Kose, S. (2022). Transforming our world through universal design for human development how can we ensure accessibility of cultural heritage? Toward better utilization of existing assets in japanese context. In I. Garofolo, G. Bencini, & A. Arenghi (Eds.), *Studies in Health Technology and Informatics*. IOS Press. Available from http://doi.org/10.3233/SHTI220871.

Kowalski, S., La Placa, S., & Pettineo, A. (2023). From archives sources to virtual 3D reconstruction of military heritage – The case study of Port Battery, Gdańsk. *The International Archives of the Photogrammetry, Remote Sensing and Spatial Information Sciences, XLVIII-M-2*, 885–893. Available from https://doi.org/10.5194/isprs-archives-XLVIII-M-2-2023-885-2023, https://isprs-archives.copernicus.org/articles/XLVIII-M-2-2023/885/2023/.

Kubenz, V., & Kiwan, D. (2022). "Vulnerable" or systematically excluded? The Impact of Covid-19 on disabled people in low- and middle-income countries. *Social Inclusion, 11* (1). Available from https://doi.org/10.17645/si.v11i1.5671, https://www.cogitatiopress.com/socialinclusion/article/view/5671.

Lambert, S. R. (2020). Do MOOCs contribute to student equity and social inclusion? A systematic review 2014–18. *Computers & Education, 145*, 103693. Available from https://doi.org/10.1016/J.COMPEDU.2019.103693, 103693.

Liu, J., Azhar, S., Willkens, D., & Li, B. (2023). Static terrestrial laser scanning (TLS) for Heritage Building Information Modeling (HBIM): A systematic review. *Virtual Worlds, 2*(2), 90–114. Available from https://doi.org/10.3390/virtualworlds2020006, https://www.mdpi.com/2813-2084/2/2/6.

Lo Valvo, A., Garlisi, D., Giarré, L., Croce, D., Giuliano, F., & Tinnirello, I. (2020). A cultural heritage experience for visually impaired people. *IOP Conference Series: Materials Science and Engineering, 949*(1), 012034. Available from https://doi.org/10.1088/1757-899X/949/1/012034, https://iopscience.iop.org/article/10.1088/1757-899X/949/1/012034.

Logothetis, S., Delinasiou, A., & Stylianidis, E. (2015). Building information modelling for cultural heritage: A review. *ISPRS Annals of Photogrammetry, Remote Sensing and Spatial Information Sciences, II-5/W3*, 177–183. Available from https://doi.org/10.5194/isprsannals-II-5-W3-177-2015, http://www.isprs-ann-photogramm-remote-sens-spatial-inf-sci.net/II-5-W3/177/2015/.

López, F., Lerones, P., Llamas, J., Gómez-García-Bermejo, J., & Zalama, E. (2018). A review of heritage Building Information Modeling (H-BIM. *Multimodal Technologies and Interaction, 2*(2), 21. Available from https://doi.org/10.3390/mti2020021, http://www.mdpi.com/2414-4088/2/2/21.

Mack, K., McDonnell, E., Jain, D., Wang, L.L., Froehlich, J.E., Findlater, L. (2021 1–18). What do we mean by "accessibility research"? *Proceedings of the 2021 CHI Conference on Human Factors in Computing Systems*. ACM, New York, NY, USA, Available from: https://doi.org/10.1145/3411764.3445412, https://dl.acm.org/doi/10.1145/3411764.3445412.

Maietti, F., Giulio, R. D., Balzani, M., Piaia, E., Medici, M., & Ferrari, F. (2017). *Digital memory and integrated data capturing: Innovations for an inclusive cultural heritage in Europe through 3D semantic modelling. Mixed reality and gamification for cultural heritage* (pp. 225–244). Cham: Springer International Publishing. Available from http://link.springer.com/10.1007/978-3-319-49607-8_8.

Malinverni, E. S., Pierdicca, R., Stefano, F. D., Gabrielli, R., & Albiero, A. (2019). Virtual museum enriched by GIS data to share science and culture. Church of Saint Stephen in Umm Ar-Rasas (Jordan). *Virtual Archaeology Review, 10*(21), 31. Available from https://doi.org/10.4995/var.2019.11919, https://polipapers.upv.es/index.php/var/article/view/11919.

Marini, I., Caradonna, C., Melis, M. G., & Nardinocchi, C. (2022). Terrestrial laser scanning for 3D archaeological documentation. The prehistoric Cave of Sa Miniera de Santu Josi (Sardinia, Italy). *Journal of Physics: Conference Series*, *2204*(1), 012030. Available from https://doi.org/10.1088/1742-6596/2204/1/012030, https://iopscience.iop.org/article/10.1088/1742-6596/2204/1/012030.

Marín-Nicolás, J., & Sáez-Pérez, M. P. (2022). An evaluation tool for physical accessibility of cultural heritage buildings. *Sustainability*, *14*(22), 15251. Available from https://doi.org/10.3390/su142215251, https://www.mdpi.com/2071-1050/14/22/15251.

Matin, B. K., Williamson, H. J., Karyani, A. K., Rezaei, S., Soofi, M., & Soltani, S. (2021). Barriers in access to healthcare for women with disabilities: A systematic review in qualitative studies. *BMC Women's Health*, *21*(1), 1−23. Available from https://doi.org/10.1186/S12905-021-01189-5/TABLES/3, https://link.springer.com/article/10.1186/s12905-021-01189-5.

Media, O.V. (2019) 2019/// 2023 1 21 *YouTube pyramids of Egypt virtual tour* | VR 360° Travel Experience Unpublished content. https://www.youtube.com/watch?v = mOuvAJRknXk.

Messaoudi, T., Véron, P., Halin, G., & Luca, L. D. (2018). An ontological model for the reality-based 3D annotation of heritage building conservation state. *Journal of Cultural Heritage*, *29*, 100−112. Available from https://doi.org/10.1016/j.culher.2017.05.017, https://linkinghub.elsevier.com/retrieve/pii/S1296207417304508.

Milakis, D., Gebhardt, L., Ehebrecht, D., & Lenz, B. (2020). *Is micro-mobility sustainable? An overview of implications for accessibility, air pollution, safety, physical activity and subjective wellbeing*. Handbook of Sustainable Transport (pp. 180−189). Edward Elgar Publishing. Available from https://www.elgaronline.com/view/edcoll/9781789900460/9781789900460.00030.xml.

Montusiewicz, J., Barszcz, M., & Korga, S. (2022). Preparation of 3D models of cultural heritage objects to be recognised by touch by the blind—Case studies. *Applied Sciences*, *12*(23), 11910. Available from https://doi.org/10.3390/app122311910, https://www.mdpi.com/2076-3417/12/23/11910.

Moyano, J., Odriozola, C. P., Nieto-Julián, J. E., Vargas, J. M., Barrera, J. A., & León, J. (2020). Bringing BIM to archaeological heritage: Interdisciplinary method/strategy and accuracy applied to a megalithic monument of the Copper Age. *Journal of Cultural Heritage*, *45*, 303−314. Available from https://doi.org/10.1016/j.culher.2020.03.010, https://linkinghub.elsevier.com/retrieve/pii/S1296207420300467.

Murphy, M., McGovern, E., & Pavia, S. (2013). Historic Building Information Modelling − Adding intelligence to laser and image based surveys of European classical architecture. *ISPRS Journal of Photogrammetry and Remote Sensing*, *76*, 89−102. Available from https://doi.org/10.1016/j.isprsjprs.2012.11.006, https://linkinghub.elsevier.com/retrieve/pii/S0924271612002079.

Nishanbaev, I. (2020). A web repository for geo-located 3D digital cultural heritage models. *Digital Applications in Archaeology and Cultural Heritage*, *16*, e00139. Available from https://doi.org/10.1016/j.daach.2020.e00139, https://linkinghub.elsevier.com/retrieve/pii/S221205481930058X.

Olson, E., Salem, C., Ali, F., Van Welie, M., Brett, N. (2022). 2022/// 2023 1 6 *Consensus*. Unpublished content. https://consensus.app/#.

Pavelka jr., K., Kuzmanov, P., Pavelka, K., & Rapuca, A. (2023). Different data joining as a basic model for HBIM − A case project St. Pataleimon in Skopje. *The International Archives of the Photogrammetry, Remote Sensing and Spatial Information Sciences*, *XLVIII-5/W*, 85−91. Available from https://doi.org/10.5194/isprs-archives-XLVIII-5-W2-2023-85-2023, https://isprs-archives.copernicus.org/articles/XLVIII-5-W2-2023/85/2023/.

Pepe, M., Costantino, Da, Alfio, V. S., Restuccia, A. G., & Papalino, N. M. (2021). Scan to BIM for the digital management and representation in 3D GIS environment of cultural heritage site. *Journal of Cultural Heritage, 50,* 115−125. Available from https://doi.org/10.1016/j.culher.2021.05.006, https://linkinghub.elsevier.com/retrieve/pii/S1296207421000881.

Pérez, E., Merchán, P., Merchán, M. J., & Salamanca, S. (2020). Virtual reality to foster social integration by allowing wheelchair users to tour complex archaeological sites realistically. *Remote Sensing, 12*(3), 419. Available from https://doi.org/10.3390/rs12030419, https://www.mdpi.com/2072-4292/12/3/419.

Pietroni, E., Menconero, S., Botti, C., & Ghedini, F. (2023). e-Archeo: A pilot national project to valorize italian archaeological parks through digital and virtual reality technologies. *Applied System Innovation, 6*(2), 38. Available from https://doi.org/10.3390/asi6020038, https://www.mdpi.com/2571-5577/6/2/38.

Pinto Puerto, F. S. (2018). La tutela sostenible del patrimonio cultural a través de modelos digitales BIM y SIG como contribución al conocimiento e innovación social. *revista PH, 93,* 27−29. Available from https://doi.org/10.33349/2018.0.4125, http://www.iaph.es/revistaph/index.php/revistaph/article/view/4125.

Rodríguez-Martín, M., Sánchez-Aparicio, L. J., Maté-González, M. Á., Muñoz-Nieto, Á. L., & Gonzalez-Aguilera, D. (2022). Comprehensive generation of historical construction CAD models from data provided by a wearable mobile mapping system: A case study of the Church of Adanero (Ávila, Spain). *Sensors, 22*(8), 2922. Available from https://doi.org/10.3390/s22082922, https://www.mdpi.com/1424-8220/22/8/2922.

Rossetti, V., Furfari, F., Leporini, B., Pelagatti, S., & Quarta, A. (2018). Enabling access to cultural heritage for the visually impaired: An interactive 3D model of a cultural site. *Procedia Computer Science, 130,* 383−391. Available from https://doi.org/10.1016/j.procs.2018.04.057, https://linkinghub.elsevier.com/retrieve/pii/S1877050918304101.

Sani, N. H., Tahar, K. N., Maharjan, G. R., Matos, J. C., & Muhammad, M. (2022). 3D reconstruction of building model using UAV point clouds. *The International Archives of the Photogrammetry, Remote Sensing and Spatial Information Sciences, XLIII-B2-2,* 455−460. Available from https://doi.org/10.5194/isprs-archives-XLIII-B2-2022-455-2022, https://isprs-archives.copernicus.org/articles/XLIII-B2-2022/455/2022/.

Santos, P., Ritz, M., Fuhrmann, C., Monroy, R., Schmedt, H., Tausch, R., Domajnko, M., Knuth, M., & Fellner, D. (2017). Mixed reality and gamification for cultural heritage acceleration of 3D mass digitization processes: Recent advances and challenges. In M. I. N. Magnenat-Thalmann, & G. Papagiannakis (Eds.), *Mixed reality and gamification for cultural heritage.* Cham: Springer International Publishing. Available from http://doi.org/10.1007/978-3-319-49607-8_4.

Scianna, A., & Filippo, G. D. (2019). Rapid prototyping for the extension of the accessibility to cultural heritage for blind people. *The International Archives of the Photogrammetry, Remote Sensing and Spatial Information Sciences., XLII-2/W15,* 1077−1082. Available from https://doi.org/10.5194/isprs-archives-XLII-2-W15-1077-2019, https://isprs-archives.copernicus.org/articles/XLII-2-W15/1077/2019/.

Scianna, A., Gaglio, G. F., & Guardia, M. L. (2023). Augmented virtual accessibility of CH: The web navigation model of inquisition prisons. *The International Archives of the Photogrammetry, Remote Sensing and Spatial Information Sciences, XLVIII-M-2,* 1443−1447. Available from https://doi.org/10.5194/isprs-archives-XLVIII-M-2-2023-1443-2023, https://isprs-archives.copernicus.org/articles/XLVIII-M-2-2023/1443/2023/.

Selim, G., Jamhawi, M., Abdelmonem, M. G., Ma'bdeh, S., & Holland, A. (2022). The Virtual Living Museum: Integrating the multi-layered histories and cultural practices of Gadara's archaeology in Umm Qais, Jordan. *Sustainability, 14*(11), 6721. Available from https://doi.org/10.3390/su14116721, https://www.mdpi.com/2071-1050/14/11/6721.

Sen, K., Prybutok, G., & Prybutok, V. (2022). The use of digital technology for social well-being reduces social isolation in older adults: A systematic review. *SSM - Population Health*, *17*, 101020. Available from https://doi.org/10.1016/j.ssmph.2021.101020, https://linkinghub.elsevier.com/retrieve/pii/S2352827321002950.

Spallone, R., Russo, M., Teolato, C., Vitali, M., Palma, V., & Pupi, E. (2023). Reconstructive 3D modelling and interactive visualization for accessibility of Piffetti's Library in The Villa Della Regina Museum (Turin). *The International Archives of the Photogrammetry, Remote Sensing and Spatial Information Sciences, XLVIII-M-2*, 1485−1492. Available from https://doi.org/10.5194/isprs-archives-XLVIII-M-2-2023-1485-2023, https://isprs-archives.copernicus.org/articles/XLVIII-M-2-2023/1485/2023/.

Taher Tolou Del, M. S., Sedghpour, B. S., & Tabrizi, S. K. (2020). The semantic conservation of architectural heritage: the missing values. *Heritage Science*, 8(1), 70. Available from https://doi.org/10.1186/s40494-020-00416-w, https://heritagesciencejournal.springeropen.com/articles/10.1186/s40494-020-00416-w.

Taher Tolou Del, M. S., & Tabrizi, S. K. (2020). A methodological assessment of the importance of physical values in architectural conservation using Shannon entropy method. *Journal of Cultural Heritage*, *44*, 135−151. Available from https://doi.org/10.1016/j.culher.2019.12.012, https://linkinghub.elsevier.com/retrieve/pii/S1296207419305497.

Vardia, S., Khare, A., & Khare, R. (2018). Studies in health technology and informatics universal access in heritage site: A case study on Jantar Mantar, Jaipur, India. *Studies in Health Technology and Informatics*, *256*. Available from https://doi.org/10.3233/978-1-61499-923-2-67, https://ebooks.iospress.nl/publication/50548.

Woodgate, R. L., Gonzalez, M., Demczuk, L., Snow, W. M., Barriage, S., & Kirk, S. (2020). How do peers promote social inclusion of children with disabilities? A mixed-methods systematic review. *Disability and Rehabilitation*, *42*(18), 2553−2579. Available from https://doi.org/10.1080/09638288.2018.1561955, https://www.tandfonline.com/doi/full/10.1080/09638288.2018.1561955.

Zada, A. A. K. (2021). Novel cartographical designs for blind and partially impaired students in Kurdistan. *Proceedings of the ICA*, *4*, 1−3. Available from https://doi.org/10.5194/ica-proc-4-56-2021, https://ica-proc.copernicus.org/articles/4/56/2021/.

Zheliazkova, M., Naboni, R., & Paoletti, I. (2015). WIT transactions on the built environment. In C. Brebbia, & S. Hernández (Eds.), *Structural Studies, Repairs and Maintenance of Heritage Architecture XIVA parametric-assisted method for 3D generation of as-built BIM models for the built heritage* (vol. 153). Southampton: WIT Press, WIT Transactions on The Built Environment, vol. 153: Structural Studies, Repairs and Maintenance of Heritage Architecture XIV.

Sensor integration for built heritage diagnostics: From aerial and terrestrial photogrammetry to simultaneous localization and mapping technologies

12

Ilaria Trizio[1], Giovanni Fabbrocino[1,2], Adriana Marra[1],
Marco Giallonardo[1], Alessio Cordisco[1], and Francesca Savini[1]
[1]Institute for Construction Technologies - Italian National Research Council, L'Aquila, Italy, [2]Department of Biosciences and Territory, StreGa Lab, University of Molise, Campobasso, Italy

12.1 Introduction

The chapter deals with the role of integrated, multiscale, survey in the knowledge, diagnostics, and documentation of the built heritage. It offers an overview of the design and execution of survey campaigns performed by using tools deployed by the most advances of digital technologies.

Indeed, the analyses performed on the built heritage with the aim of knowing in detail the geometric-constructive characteristics that determine its statics and the state of conservation of its surfaces and structures, as well as its more formal features, have experienced an increase in the application of digital techniques in recent years. This technological impetus has favored the use of non-destructive tools for the knowledge of the current and conservation state of the built heritage, which is closely linked to technological development and provides the market with tools that are not only increasingly high-performing but also more accessible in economic terms.

In addition to modern non-destructive investigation techniques with the use of special instruments such as thermography or tomography, there are those for the digitization of historical built heritage and cultural heritage in the broadest sense. These techniques favor analytical and cognitive processes that are fundamental and preparatory to implementing multiple actions, ranging from protection and conservation to fruition and valorization. The focus here is directed primarily at identifying operational ways of exploiting these advances in digital technologies to address the complex issue of historic built environment conservation and preservation. This resulted in questions on the methods and procedures to be implemented for the

Diagnosis of Heritage Buildings by Non-Destructive Techniques. DOI: https://doi.org/10.1016/B978-0-443-16001-1.00012-7

accurate digital reproduction of architectural artifacts, addressing an extremely current issue both in the field of research and in the operational field.

The potential of digitization of the artifacts' current state is most significant when working in emergency conditions, in fragile contexts, or on damaged artifacts that are subject to particular morphological and typological conditions. Digital documentation assumes a central role in these cases since it provides a digital replica of the real artifact to be obtained, which becomes an integral part of the analysis process during diagnostic investigations. In particular, it is essential in emergency contexts, such as those resulting from traumatic events such as wars and earthquakes that compromise the built heritage by leaving it in a critical condition, it becomes essential to identify procedures capable of facilitating the processes of analysis and design of the interventions to be implemented to conserve and reconstruct artifacts. Photorealistic digital models thus become powerful tools that can facilitate safe analysis and become containers of useful information for data sharing between the several professionals involved in the historic building conservation process.

In these cases, even field data acquisition operations, the first step in the digitization of the built environment, are characterized by an underlying complexity and must respond to requirements strictly related to the safety and security of the operators, so it is essential to consider this aspect when designing the appropriate operational procedures.

The protocols defined within highly specialized disciplines and the recourse to the main geomatic and topographical techniques, together with the presence on the market of increasingly high-performance expeditious surveying tools, facilitate this step. The use of these instruments for the acquisition of point clouds is supported by surveying methodologies that provide digital products to be obtained in a short time, which can be used for on-site inspections, study, and accurate assessment of the state of conservation.

The possibility of obtaining in a short time point clouds with high levels of precision, namely, with a dispersion of points in the order of a centimeter, has favored the massive use of 3D surveying in multihazard contexts. In this case, expeditious instrumental surveying represents a concrete solution since it can combine the needs of documenting the heritage in all its complexity with those of safeguarding and securing the operators. Many studies demonstrate the effectiveness of data acquired through expeditious systems in the surveying of emergency contexts for the material and structural knowledge of damaged cultural heritage to plan the interventions to be carried out (Chiabrando et al., 2017; Di Stefano et al., 2020).

Among the main expeditious systems for acquisition, there are sensors that use light radiation and differ according to the light used. These are distinguished into active sensors when the light is directly involved in the acquisition process, as in the case of laser systems, and passive sensors when the light is natural, as in the case of photogrammetric techniques (Russo et al., 2011).

The application of laser scanning procedures—terrestrial and mobile—and photogrammetric ones—terrestrial, aerial, or with a 360-degree camera—of archaeological sites and complex urban and architectural areas, namely, historical buildings, parks, gardens, and ornamental elements (Barba et al., 2020; Chiabrando et al., 2017; Rabbia et al., 2020; Teppati Losè et al., 2020), has demonstrated the effectiveness of these techniques and procedures in the field of built heritage digitization. Indeed, the different acquisition techniques provide point clouds that can be used to produce two- and three-dimensional data, such as semantic models that facilitate the exchange of information between several professionals. At the same time, thanks also to the photorealistic textures obtained with photogrammetry, these can be the starting point for qualitative and quantitative assessments of the current condition of artifacts.

Recognizing the role and potential of the two expeditious surveying technologies, briefly described and analyzed in detail later, the group's research is aimed at identifying an optimal procedure for their integration to obtain high-precision point clouds that, as raw data, can be used for the creation of semantic models and high-resolution orthomosaics capable of supporting the knowledge and diagnosis process at the architectural scale. The application of multisensor, image and range-based techniques by researchers from different disciplinary fields has demonstrated the reliability of indoor and outdoor integrated survey procedures (Keitaanniemi et al., 2021; Malinverni et al., 2018; Sammartano & Spanò, 2018; Spanò, 2019) and the accuracy of multiscalar data acquired in the context of different types of cultural heritage (Bronzino et al., 2019; Patrucco et al., 2019; Pulcrano et al., 2019; Rinaudo & Scolamiero, 2021). The use of integration of data derived from these different sensors is useful not only in emergency conditions, but also in more common environmental and typological contexts. Indeed, particular conditions given by the characteristics of the built heritage, such as the complexity of multilayered buildings or the difficulty of reaching some artifacts, not only those located on high ground and neglected, but also historical buildings located in narrow alleyways, cause a lack of data when the acquisition procedures are considered singularly. At the same time, the processed data are inhomogeneous, and they can be used if properly integrated.

Starting from this premise, the chapter discusses the results obtained by validating the clouds derived from the surveys with simultaneous localization and mapping (SLAM) and photogrammetry, both terrestrial and aerial. Following this validation process, the importance of integrated surveying is argued by discussing the procedure proposed for its achievement and analyzing a collection of explanatory cases addressed by the authors within national and international research programs.

Technological development has made the digitization of heritage possible and has contributed to profoundly modify the methods of acquisition, processing, and sharing of heterogeneous data underlying the processes of knowledge and analysis of the built heritage. Several studies have shown how it is possible to acquire important information regarding the state of conservation of surfaces and materials and the state of health of structures with the aid of techniques based on active and passive sensors that allow for the recording of both geometric-formal information as well as appearance and color data. Among the active sensor survey techniques, the one based on laser scanning technology represents a versatile solution for acquiring large amounts of data with acceptable survey times.

A laser scanner is a technological device consisting of a laser distance meter combined with a set of high-precision mechanical equipment, capable of detecting the position of a large number of points with high speed and accuracy. The result of the acquisition is a set of points scattered in space in a more or less regular manner, which is commonly called "point cloud." Each of these points is characterized by coordinates (x, y, z) referring to a local reference system and other information, such as the reflectance, color, and normal direction of each individual point (Bornaz, 2006).

The terrestrial laser scanner allows to be recorded reality with very high precision thanks to the ability to detect the geometry of the surface with millimeter accuracy (Bartolucci, 2009). One of the strengths of this technology is the ability to carry out surveys of large surfaces with accuracy and speed, thus allowing to identify and detect particular types and forms of degradation (not always easily distinguishable with the naked eye). In fact, although always guided by the operator, the laser scanner (LS) survey is more automated than traditional visual and manual methods, as it reduces errors due to human nature and certainly increases the degree of objectivity (Bartolucci, 2009; Teza et al., 2009). At the same time, being a contactless technique, it guarantees adequate levels of operator safety (Tang et al., 2007). Technological progress has also favored the development of instruments that make use of sensors and rapid systems that favor expeditious surveying. In particular, among the most widely used expeditious systems are mobile laser scanners that, thanks to SLAM technology, allow the rapid acquisition of large amounts of data and the generation of point clouds from which geometric and spatial information can be obtained with accuracies between 1 and 5 cm (Chudá et al., 2020; Hou & Chio, 2021). These characteristics make this technology highly productive and, therefore, particularly suitable in the field of surveying for diagnostics.

Clouds obtained by LS technology, both terrestrial and mobile, are functional for the material and structural knowledge of historical buildings and cultural heritage (Charron et al., 2019; Chiabrando et al., 2017; Di Stefano et al., 2020). Furthermore, these can be applied in different areas of built analysis ranging from visual inspections to geometric evaluation and surface conformity verification. Very significant is the use of this kind of technology in the field of structural engineering applied to existing bridges, not only historical ones, which are often difficult to survey because they are related to road systems in use and often not easily accessible. Many research groups have employed point clouds to analyze the geometric characteristics and identify any unevenness on the surfaces (Bo et al., 2018; Tang et al., 2007), but also for deformation analysis, as in the striking case of the Pathein suspension bridge in Myanmar (Kitratporn et al., 2018), in which the inclination of the tower, the inclination of the hook, and the deflection of the bridge truss were analyzed. The laser survey technique is optimal for the integrated evaluation of data such as changes in geometry caused by underground mining (Skoczylas et al., 2016). What has been said so far shows the potentiality of the tool in the field of diagnostic analysis of existing infrastructures, but the literature offers many insights into the application of the procedures to historic buildings. In fact, laser scanning is also used for the analysis of the deterioration of structures through the identification and survey of deterioration and for the evaluation of geometric variations of structures, covering surfaces and decorations over time. The constant recording by laser technology of the artifacts and the comparison between them makes it possible to assess the changes that have occurred to the geometry of the structure over the period between different sets of measurements. On the other hand, digital photogrammetry techniques, both terrestrial and aerial, are characterized by the ability to acquire large amounts of data, guaranteeing accurate measurements with significantly reduced processing times compared to previous years, thanks to the development of computer vision.

This technique is based on image-based acquisition systems that make it possible to detect an object from digital image sets of the same and to return a high-density point cloud with high accuracies of the order of centimeter (Gomarasca, 2004). This cloud, through the semiautomated procedure at the base of the photogrammetric process, makes it possible to generate a very high-resolution 3D photogrammetric model that preserves the RGB data of the digital images and whose applications lend themselves to the survey of artifacts on an architectural scale or archaeological remains, as well as many emergencies that characterize the broader cultural heritage. Furthermore, thanks to the development and cost reduction of unmanned aerial vehicle (UAV) technologies, this technique becomes even more valuable for the survey of large objects, such as the survey of suburban areas or cities, facilitating the acquisition of geometric-formal data of architectural complexes even in emergency situations. The potentiality of using UAV photogrammetry is also particularly suitable for the analysis of infrastructures and structures that are difficult to access (Jordan et al., 2018; Teppati Losè et al., 2020; Ulvi, 2021). This is the case with the application of digital photogrammetry to perform surface degradation analyses of bridges and viaducts (Mistretta et al., 2019; Schweizer et al., 2018; Teodoro & Araujo, 2016; Valença et al., 2012), but also for the morpho-typological mapping of surface degradation forms of the historical built heritage (De Fino et al., 2022; Torres-González et al., 2023).

Technological progress is directing research toward the automated analysis of degradation patterns on orthomosaics resulting from this process using the object-based image analysis technique with which degraded areas of surfaces can be identified and classified (Zollini et al., 2020). This procedure also makes it possible to quantify georeferenced metric information, such as the width and length of cracks and the extent of degraded areas. This allows quickly periodic inspections to assess the evolution of degradation and plan the urgency of conservation or maintenance work.

These characteristics make digital photogrammetry one of the most widely used indirect surveying techniques in several fields, including heritage diagnostics as it is an excellent solution for visual inspections to accurately assess the state of conservation of artifacts. In conclusion, the results of the photogrammetric process make it possible to obtain low-cost three-dimensional models with a high degree of detail in both mesh and color data and allow the creation of photorealistic orthomosaics with a high pixel resolution, which are extremely functional for the knowledge path. What has been said suggests the potentiality of integrating the two techniques, which becomes the way forward to have the right compromise between the degree of detail, acquisition and processing times, and manageability of data for diagnostic analysis. The products resulting from the integration of the acquisitions made through laser and photogrammetric surveying techniques are increasingly used in the generation of 3D information digital models that become functional for diagnostic analysis and data management for knowledge and conservation of the assets (Banfi et al., 2020; Conti et al., 2022; Luleci et al., 2021; Savini et al., 2022; Tucci et al., 2022). With respect to the extensive literature available on these topics, the research group has oriented its work toward the integration of products derived from SLAM acquisition with those derived from photogrammetric surveying, in line with the current work of other research groups (Abate et al., 2023; Elamin et al., 2022; Kamnik et al., 2020; Kartini & Saputri, 2022; Rachmawati & Kim, 2022; Shang & Shen, 2018). The strength of the work is the validation of the individual clouds obtained with the two techniques and then the integration of these. This process is facilitated by the registration of the clouds using a satellite positioning system based on GNSS (global navigation satellite system) technology. In this regard, the scientific literature is full of examples where GNSS surveying is often used for the accurate identification of the position of ground control points (GCPs) used for photogrammetric surveying operations (Ellum & El-Sheimy, 2006; Kajzar et al., 2011; Lo et al., 2015; Valença et al., 2012) or for position control of UAVs for aerial photogrammetry or Lidar surveys (Chiabrando et al., 2017; Forlani et al., 2019; Freimuth & König, 2015; Mader et al., 2016; Shetty & Xingxin Gao, 2019; Suzuki et al., 2016).

12.2 Materials and methods

The expeditious survey of the historic built heritage carried out with the aid of SLAM and photogrammetric techniques permits to obtain in a short time a large quantity of geometric-spatial data that form the basis for the generation of complex products such as digital models or high-resolution orthomosaics, which are useful for diagnostic analyses. As demonstrated by the scientific literature, the integration of technologies facilitates the optimization of information according to time and need. In this regard, a procedure aimed at integrating clouds from multisensor surveys through the use of the GNSS satellite system is discussed. Specifically, the point of clouds obtained with mobile laser based on SLAM technology and those derived from the photogrammetric procedure were processed by integrating them with coordinates, geographical or cartographic, acquired with a Zenith 16 receiver by Geomax, used in real-time kinematic (RTK) mode by receiving from GPS, GLONASS, BEIDOU, and GALILEO satellites. This step is possible through correct planning of the acquisition campaigns, aimed at facilitating integration, and that requires for each survey phase the positioning in the field of topographic control points materialized through the use of targets on which mobile markers have been placed. These must be clearly visible during the acquisition phases with the multisensors; in particular, the targets should be present in the photos captured with the drone and acquired during the SLAM survey through the positioning of the instrument on the point. This step favors the georeferencing and alignment of the data acquired through the two different survey methodologies using these points as GCPs and check points, thanks to the associated coordinates.

The laser surveys were carried out with the ZEB Horizon mobile scanner by GeoSLAM (GeoSLAM, 2022), which permits the acquisition of good-quality point clouds, thanks to the inertial measurement unit inside it and the SLAM technology (Barba et al., 2019; Barrera-Vera & Benavides-López, 2018; Rinaudo & Scolamiero, 2021; Sammartano & Spanò, 2018). The laser has a range of 100 m and is capable of returning scans of indoor and outdoor environments in a short time, with an accuracy ranging from 1 to 3 cm.

For the optimal acquisition of scans with the ZEB Horizon, closed loops with an average duration of 10 minutes are preferred. This acquisition procedure facilitates cloud alignment by minimizing errors along the trajectory and increasing the rigidity of the system. The data acquired in the field are then processed through the GeoSLAM Hub cloud computing service.

For the photogrammetric survey procedure, the UAV remote-controlled acquisition techniques, through the use of a Dji Mini 2 drone with a 2/3″ CMOS sensor, were applied. In addition, the terrestrial proximity photogrammetry was implemented using a Nikon D610 and Canon EOS 4000D. These techniques were complemented by the acquisition of photographic data using a Ricoh Theta Z1 360° cam. The proprietary software Agisoft Metashape was used to process the data.

The clouds obtained with the two laser and photogrammetric technologies, integrated thanks to the GNSS system, were compared separately before proceeding to the processing phases aimed at merging the two clouds to create a three-dimensional model and the related two-dimensional data. This phase permits an assessment of the integrability degree of the survey techniques adopted and thus confirms the validity of the survey. In particular, comparisons between the different clouds were made using the Cloud Compare software through the cloud-to-cloud (C2C) tool (Fig. 12.1).

This uses the "Nearest Neighbor" algorithm which is based on a mathematical model of the "Hausdroff" distance type (Batur et al., 2020; Rockafellar & Wets, 2009) to calculate the Euclidean distance between each point of the cloud used as a reference and the nearest point of the other (Bronzino et al., 2019; Cloud Compare, 2022). This distance value, called "C2C absolute distance," is calculated for all points in the dataset to evaluate the mean value (avg. distance) and the root mean square error (RMSE) in a Gaussian distribution. The C2C tool was used by entering the parameters for the reference cloud among those produced with the mobile laser and those for the cloud to be compared among those derived from the photogrammetric process. In addition, the parameter for the maximum distance was set and 0.5 m was set as the maximum value for all comparisons. This "max distance" parameter indicates the maximum distance value of the points not to be compared (Fig. 12.2).

The choice of the maximum value at 50 cm was determined by the properties of the techniques adopted, which return clouds with precision values in the order of

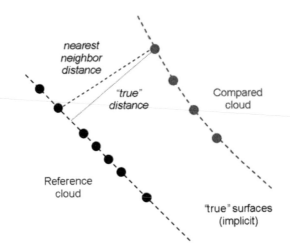

Figure 12.1 Sample distance calculation diagram.
Source: From https://www.cloudcompare.org/doc/wiki/index.php/Distances_Computation.

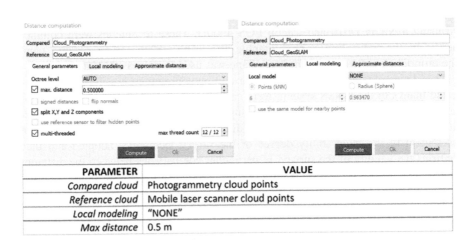

PARAMETER	VALUE
Compared cloud	Photogrammetry cloud points
Reference cloud	Mobile laser scanner cloud points
Local modeling	"NONE"
Max distance	0.5 m

Figure 12.2 Main setting parameters of the C2C tool in the Cloud Compare software and set values. *C2C*, Cloud-to-cloud.

centimeters. Therefore the maximum distance value set in this way makes it possible to exclude the points of the compared cloud having anomalous distances from the reference cloud.

The results obtained from the comparison procedure confirm the potential of cloud integration, highlighting the effectiveness of SLAM acquisition in the context of the expeditious survey of urban contexts and artifacts at an architectural scale. The point clouds resulting from the procedure described earlier provide the possibility of obtaining mesh models that can be used for creating digital replicas of the investigated artifacts that are useful for sharing information, carrying out analyses, and planning the actions to be undertaken. This confirms the benefits of implementing informative three-dimensional models, such as parametric models, in the context of data management and collaboration between different specialists in the field of in-depth knowledge of heritage for its conservation.

12.3 Application to case studies

Several experiments were carried out on selected case studies to test the effectiveness of the procedure aimed at obtaining an integrated survey, with the aim of acquiring the information necessary to understand the qualities, geometric and formal, of the urban and building fabric and artifacts on an architectural scale, and then to generate documents useful for diagnostic analyses of the built heritage. The case studies are characterized by different typologies and have been permitted to investigate different aspects and issues from time to time. The experiments involved the survey of extensive areas, although impervious, where the union between nature and buildings is very clear, as in the case of the "Pagliare di Fontecchio," up to complex artifacts that require a careful digitization phase of spaces and environments preparatory to the production of two-dimensional drawings aimed at mapping thematic analyses, as in the case of Palazzo San Giorgio in Campobasso. But also complex cases characterized by multihazards for which expeditious surveys permit safe operation in emergency conditions both on an urban scale, as in the historical center of Civitacampomarano, and on the architectural scale of a single architectural complex, such as the Church of Santa Maria e Pietro in Castello, in the territory of L'Aquila.

12.3.1 The "Pagliare di Fontecchio" AQ

In the collection of application cases on which the procedure discussed before was tested, moving from the scale of the territory to that of the individual architectural artifact, a special case is the "Pagliare di Fontecchio" (Fig. 12.3).

The "Pagliare" represents an example of a built landscape specific to the L'Aquila area, characterized by the presence of the mountainous massifs of the inner Apennines. It is in fact a highland settlement located on the Sirente Velino mountain massif, characterized by a vernacular architecture mostly of spontaneous origin, closely connected with the surrounding natural environment. This temporary village, built with local limestone ashlars by local shepherds and farmers, was used until the last century by the inhabitants of Fontecchio during the summer months for the cultivation of the land at high altitudes and grazing and sheep breeding. In this example, the particular nature of the site, located in an inaccessible mountainous area, especially during the winter period, conditioned the design of the survey. In fact, it was decided to acquire as much data as possible in a single survey

Figure 12.3 Panoramic view of the "Pagliare di Fontecchio."

Figure 12.4 Overview of the "Pagliare di Forntecchio" SLAM point cloud. *SLAM,*
Simultaneous localization and mapping.

campaign, set in early spring, before the thick vegetation of the summer months made survey operations even more problematic. Despite the problems caused by the particular characteristics and extension of the site, the results of the integrated survey applied to a territorial scale, and briefly described later, proved satisfactory.

The laser scanning acquisition was carried out with 4 closed loops and the individual clouds were aligned in the proprietary software and then merged into a single cloud of approximately 147 million points (Fig. 12.4).

The photogrammetric survey resulted in a dense cloud of approximately 200 million points that was processed to obtain a solid model consisting of approximately

13 million faces, which was subsequently textured (Fig. 12.5). From the model, orthomosaics were generated with a ground sample distance (GSD) of 1.83 cm/pix.

The clouds were aligned with each other using the GNSS receiver, which acquired the points with an average satellite coverage of about 14 satellites, achieving an average accuracy of Hrms = 0.016 m and Vrms = 0.024 m. A comparison of the two clouds in the present case study reveals a different situation from the other cases. In particular, when observing the mean value and the RMSE relative to the absolute distance between the two clouds, it can be seen that the values are of the order of 10 cm compared to the mean values found in the other cases, of the order of 4 cm (Fig. 12.6).

Figure 12.5 Textured photogrammetric model in a Metashape software view.

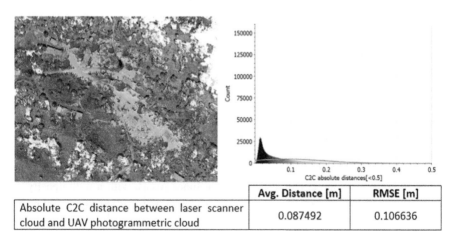

	Avg. Distance [m]	RMSE [m]
Absolute C2C distance between laser scanner cloud and UAV photogrammetric cloud	0.087492	0.106636

Figure 12.6 Pagliare settlement: point value trend of C2C distance, histogram with Gaussian curve and related statistical values. *C2C*, Cloud-to-cloud.

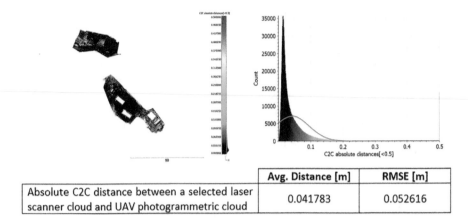

	Avg. Distance [m]	RMSE [m]
Absolute C2C distance between a selected laser scanner cloud and UAV photogrammetric cloud	0.041783	0.052616

Figure 12.7 Portion of the Pagliare settlement: point value trend of C2C distance, histogram with Gaussian curve and related statistical values. *C2C*, Cloud-to-cloud.

This increase can be attributed to various factors, such as, for example, the presence of vegetation, identifiable in red in the scalar fields, or the lower resolution of the large-scale photogrammetric survey carried out at a high distance from the ground (approx. 60 m). These effects are considerably reduced if vegetation areas are excluded from the comparison, an operation that yields an average C2C distance value and mean square deviation in the order of 4–5 cm, thus aligning the error of the two techniques. Observing the scaled fields for the portions with little vegetation, one can see a homogeneous trend in the absolute distance values C2C (Fig. 12.7).

This remark, together with the statistical values within the range of the accuracies of the two techniques, confirms the comparability of the surveys and thus their integration, especially for the elements of architecture, although they were also surveyed using a spatial approach.

12.3.2 The town of Civitacampomarano (CB)

The test of previously developed integrated survey procedure was subsequently tested at the urban scale, using as a test case a small urban center characterized by a territory that must be constantly monitored as it is subject to periodic natural hazards. The application concerns the medieval village of Civitacampomarano, a small municipality located in the province of Campobasso in the inland area of the Molise Apennines (Fig. 12.8).

The town reflects the characteristics of the Inner Areas, with a high density of cultural and historical assets in a complex geological and geomorphological environment subject to natural hazards as well as constant depopulation. Among the risks that most afflict the village, a relevant role is played by the hydrogeological one associated with the complex geomorphological and geotechnical texture of the areas located in the Apennine chain. The town stands on an arenaceous wall

Figure 12.8 Panoramic view of the town of Civitacampomarano in Molise region.

characterized by the presence of active and quiescent landslides that have damaged historic buildings on the ridge. The landslide and flood hazard have made it
necessary to ban an entire area of the historic center from public access. The need
to acquire as much data as possible on the consistency of the assets to be preserved within the red zone in a short period, without endangering the operators,
also conditioned the choices during the survey design phase. These conditions oriented in fact the survey operations toward the integration of the two expeditious
techniques described earlier, namely, SLAM and UAV photogrammetry. In some
portions, the built-up area was again the subject of an expeditious survey on an
architectural scale aimed at supporting the knowledge phases preparatory to the
drafting of specific conservation and protection proposals. The SLAM survey was
carried out through 7 closed loops aligned in the proprietary software that
returned a complete cloud of a portion of the village of approximately 36 million
points (Fig. 12.9).

In parallel, photogrammetric campaigns were conducted with UAVs for the digitization and acquisition of the colorimetric data of the village (Fig. 12.10). The photogrammetric process resulted in a dense cloud of approximately 60 million points
and a mesh model of approximately 4 million faces.

For some buildings particularly damaged by the last landslide phenomenon in
2016, and for some particular buildings of value, the aerial campaign was supplemented with a photographic survey from the ground that allowed us to obtain high-
resolution orthomosaics with a GSD of 2.27 mm/pix, on which investigations were
carried out to map the disruption and damage (Fig. 12.11).

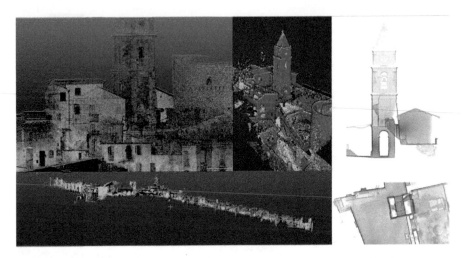

Figure 12.9 SLAM point cloud and views in proprietary software. *SLAM*, Simultaneous localization and mapping.

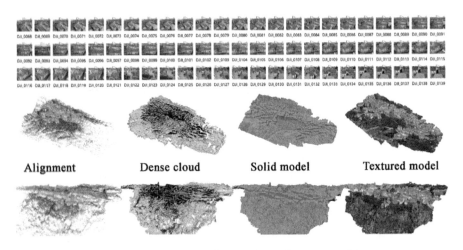

Figure 12.10 Views of the photogrammetric model at various stages of the process.

A comparison of the available clouds was carried out between the mobile laser scanner and photogrammetric clouds relating to the portion of the bell tower (Fig. 12.12).

Observing the statistical values for absolute distance, one can see that they are within the centimeter range, in line with the precision of the techniques adopted for the survey. Furthermore, a critical reading of the maps describing the point distribution of the absolute distance C2C shows a regular distance trend, with the exception of portions characterized by external objects, as in the case of the car obliterating a section of the tower.

Figure 12.11 Photogrammetric model of the Casa Pepe building and detail of the orthomosaics with the highlighted lesions.

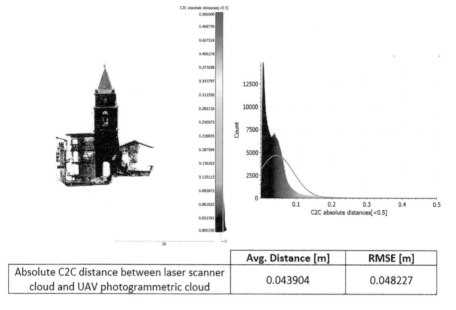

	Avg. Distance [m]	RMSE [m]
Absolute C2C distance between laser scanner cloud and UAV photogrammetric cloud	0.043904	0.048227

Figure 12.12 Bell tower of Civatacampomarano: point value trend of C2C distance, histogram with Gaussian curve and related statistical values. *C2C*, Cloud-to-cloud.

The outputs resulting from the integrated survey constitute the starting point from which the geological and geotechnical aspects of the area and the results of the damage analysis of the buildings observed on a predictive basis can be conveyed. The digital model obtained from the integrated survey and enriched with data from multidisciplinary investigations could be a crucial building block to improve the management process for the protection and conservation of the historic center and also as a starting point for the design choices of the restoration process of some valuable artifacts in the village. One of the subjects of the expeditious survey was in fact the bell tower of S. Maria Maggiore, the original gateway to the feudal village, and the house of Gabriele Pepe, an Italian politician and literary, a small palace built at the end of the 18th century.

12.3.3   The San Giorgio Palace

Continuing to test the integrated survey procedure and moving from the urban to the architectural scale, a significant example is that of San Giorgio Palace, the current seat of the town hall of Campobasso, the administrative center of the Molise region, on the edge of the historic center, in the area that had already undergone a reorganization of its spaces since the end of the 18th century, due to the role that the town was beginning to play at that time (Fig. 12.13).

The palace, 19th century in its features, has a floor plan that occupies an entire block and is characterized by an elegant three orders facade, with a large portico in the first order and a portal framed by ionic columns supporting a balcony in line with the clock.

Behind the building, the original grounds of the building became a public green area. An in-depth analysis of the artifact was performed in parallel with the historical research, through the recognition of published sources and the study of unpublished ones kept in the municipal administration archives. Specifically, a photographic and thermographic survey was carried out using rapid visual screening techniques and the collection of data through the compilation of first-level inspection forms for the material-structural understanding and the definition of the current state of affairs with the analysis of damage and surface degradation. The starting point of the cognitive phase was the digital survey integrating SLAM and terrestrial photogrammetry for the geometric restitution of the building, which also made it possible to validate the existing graphic drawings. The location of the building, which stands in the city center, in an urban block within the road network constantly traversed by vehicular traffic, and the particular function of the building, the

Figure 12.13 Panoramic photo of San Giorgio Palace in Campobasso.

seat of the municipality of Campobasso, conditioned the choices of the survey project also in this case.

In particular, it was considered useful to carry out the laser scanner survey of the interior and exterior spaces using SLAM technology to acquire in a short time the volumetric-spatial complexity of the building's spaces characterized by an extremely articulated shape.

With this tool, the geometric survey of the building's interior spaces and exterior vertical surfaces was carried out by processing the acquired data using proprietary software that allowed the single acquisitions to be merged into a single cloud. The result of the processing is a 3D point cloud of the 17 loops carried out for the survey, which digitally reproduces the spatial geometry of the building with centimeter precision (Fig. 12.14).

From the 3D cloud, through a sequence of appropriate slices, it was possible to extract sections and plans, which were subsequently digitized in the computer-aided design (CAD) environment (Fig. 12.15).

In the case of the photogrammetric survey, a digital terrestrial survey was carried out from images acquired in the field with a Canon EOS 4000D digital camera. The products resulting from the photogrammetric process made it possible to detect the external geometry of the building and to return it in the form of a digital 3D model, of approximately 60 million faces, with centimeter precision, containing RGB color information (Fig. 12.16).

The texturing phase of the model and the subsequent export of two-dimensional photo plans obtained with a GSD of 1.82 mm/pix, which were later digitized in a

Figure 12.14 San Giorgio Palace (CB); 3D point cloud surveyed with Geoslam Zeb Horizon.

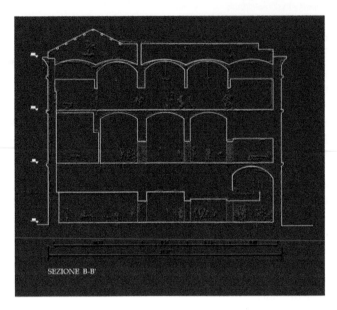

Figure 12.15 Section digitized in CAD environment from the point cloud recorded by the Geoslam.

Figure 12.16 Photogrammetric model of San Giorgio Palace: dense cloud, solid model, and textured model.

Figure 12.17 Orthomosaic of main elevation and digitization in CAD environment.

CAD environment, made it possible to understand the construction details of the surveyed object (Fig. 12.17).

To assess the two clouds, a comparison was made between the one obtained from the mobile laser scanner and the one resulting from the digital terrestrial photogrammetry process considering the portions of the building relating to the external envelope. The comparison generates a mapping of scalar fields associated with the point values of C2C distance, characterized by warm colors for higher values and cold colors for lower ones. A regular distribution over the application case is evident when looking at this mapping. Furthermore, the statistical parameters for the absolute distance show values in the order of magnitude of approximately 5 cm, in line with the instrumental precision of the mobile laser (Fig. 12.18). Also in this example, the postprocessing of the highly reliable data, acquired with the expeditious survey, made it possible to obtain a series of elaborates useful for thematic mapping aimed at conservation and maintenance projects of the building by the competent administration.

12.3.4   The Church of Santa Maria and Pietro in Castello, Fagnano Alto (AQ)

Finally, continuing to test the integrated survey procedure at the architectural scale, a significant example of a different type of building is the Church of Santa Maria e Pietro in Castello, a hamlet in the municipality of Fagnano Alto in the province of L'Aquila (Fig. 12.19). The building is located outside the fortified walls of the village, the ancient castle of medieval origin. The complex is characterized by a series of buildings leaning against each other, which over time have altered the appearance of the 13th-century church. The religious building has a rectangular tripartite

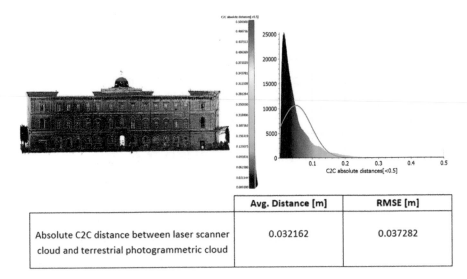

	Avg. Distance [m]	RMSE [m]
Absolute C2C distance between laser scanner cloud and terrestrial photogrammetric cloud	0.032162	0.037282

Figure 12.18 San Giorgio Palace: point value trend of C2C distance, histogram with Gaussian curve and related statistical values. *C2C*, Cloud-to-cloud.

Figure 12.19 Panoramic photo of Church of Santa Maria e Pietro in Castello, Fagnano Alto (AQ).

plan and shows clear signs of remodeling over the centuries, with Romanesque walls supplemented by others dating from the 18th-century reconstruction. Significant intervention following the 1703 earthquake helped give the church its current Baroque appearance, including the interior decorations. The building was severely damaged by the 2009 earthquake and is currently still uninhabitable, so here too the survey project was geared toward favoring expeditious survey campaigns, mainly aimed at preserving the safety of the operators. In fact, the building is heavily damaged and has an important cracking framework and large portions of the building have been secured using wooden support works, which led to favoring expeditious survey campaigns that saw the integration of SLAM and photogrammetric technologies.

The SLAM acquisition campaign was carried out with 8 closed loops, paying more attention to the rooms adjoining the church that are difficult to access. The integration of the various clouds returned a point survey representative of the entire architectural complex of 266 million points (Fig. 12.20).

The photogrammetric survey was conducted externally using a drone and integrating the dataset with ground acquisitions. The procedure resulted in a point cloud of approximately 159 million points from which a mesh model with 10 million faces was generated and subsequently textured (Fig. 12.21).

The two clouds were compared with the C2C algorithm, which returned mean distance values and the standard deviation on the Gaussian distribution of the order of centimeters.

From the mapping on the cloud of the scalar fields (Fig. 12.22), ranging from blue for lower values to red for higher ones, a considerable distance emerges at the points in the photogrammetric cloud corresponding to areas absent in the laser cloud (red areas).

Aware of the inaccuracy of the data in these portions of the building caused by the lack of data acquisition in the upper part of the building, a new comparison was carried out by excluding these areas from the analysis. With this newly performed analysis, it can be seen that the values for the absolute distance C2C drop

Figure 12.20 Geoslam cloud point of the Church.

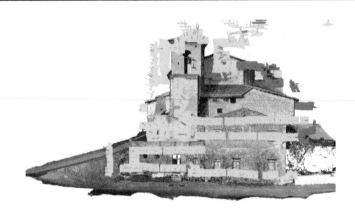

Figure 12.21 View of the photogrammetric model with reference photos.

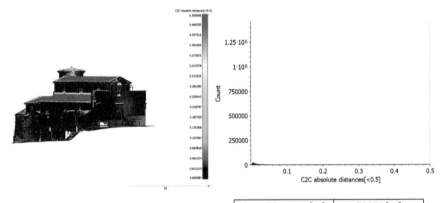

	Avg. Distance [m]	RMSE [m]
Absolute C2C distance between laser scanner cloud and integrated photogrammetric cloud	0.192525	0.215958

Figure 12.22 Church of Santa Maria e Pietro: point value trend of C2C distance, histogram with Gaussian curve and related statistical values. *C2C*, Cloud-to-cloud.

considerably and are around the accuracy values of the two techniques. In addition, the distribution maps of the scalar fields no longer show areas where discordant trends between the two clouds are evident (Fig. 12.23).

A three-dimensional model of the artifact was generated starting from the integrated survey, which is the starting point for diagnostic analyses. These included a stratigraphic reading of the masonry parameters conducted from high-resolution orthomosaics created with a GSD of 2.23 mm/pix. Also in this case, the general reliability of the survey and the postprocessing of good-quality data made it possible to obtain photo plans to be used as a basis for thematic readings. In fact, the work is mainly aimed at identifying noncoeval masonry at identifying noncoeval

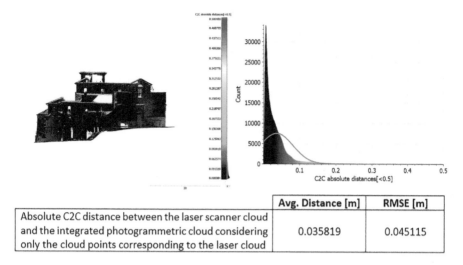

	Avg. Distance [m]	RMSE [m]
Absolute C2C distance between the laser scanner cloud and the integrated photogrammetric cloud considering only the cloud points corresponding to the laser cloud	0.035819	0.045115

Figure 12.23 Church of Santa Maria e Pietro: improved comparison with the point assessment carried out only on the portions present on both clouds.

masonry septa and the quality of the masonry that may affect the structural preservation of the building in different ways and therefore orientate the choices of those who will have to carry out the restoration project of the building.

12.4 Discussion

The application of the procedure for integrating the clouds obtained from different sensors to the case studies described earlier has produced significant data that allows us to validate the effectiveness of the expeditious survey carried out with the two techniques. It proves functional in the different contexts analyzed, ranging from territorial to architectural scales, with particular attention to multihazard situations. In general, the point clouds from GeoSLAM are more deficient in the summit part and information-rich in the ground connections of buildings and facades. On the other hand, those acquired with the aereophotogrammetric survey are very detailed in the summit part and noisier and more deficient in the lower parts further away from the sensor. Moreover, both techniques have different performances according to the contexts to be surveyed, being very functional for the territorial scale in terms of costs, acquired data, and precision. Although this confirms what is already present in the state of art, the comparison procedure between the SLAM and photogrammetric clouds highlights the effectiveness of integration through the use of the GNNS positioning system. It is possible to support this assertion after a trial conducted on the case study of the Church of Santa Maria e Pietro in Castello (AQ), which included a manual alignment phase carried out by an operator within the Cloud Compare software and an automated alignment phase using points

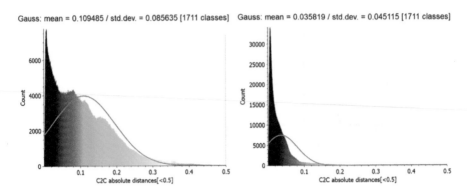

Figure 12.24 Histogram showing the results of manual alignment (left) and automated alignment using GNSS points (right). *GNSS*, Global navigation satellite system.

acquired with GNSS. The comparison between the two alignments showed a gap between the two operating modes, with a greater distance between the points of the hand-aligned clouds (Fig. 12.24).

The manual process was carried out with an average performing Workstation (Intel(R) Xeon(R) CPU E5−1650 v3 @ 3.50 GHz 3.50 GHz), with operations completed in about 10−15 minutes. Alignment is significantly affected by the operator's manual dexterity, and the C2C distance may vary depending on personal skill and working time. In this regard, a further test was conducted in which the operations were perfected with an overall total time of about 25−30 minutes, resulting in an alignment with shorter distances and an error similar to that obtained with the automated process on GNSS points in about 1 hour. The results obtained therefore allow us to confirm the effectiveness of the individual techniques, which returned accuracies in line with instrumental performance, particularly when combined with appropriate planning of the acquisition campaign using GNSS instrumentation, which significantly reduces work time.

The optimal results of the comparison between the GNSS-aligned clouds showed a consistent trend in the scalar fields associated with the C2C distance and a minimum deviation between them, calculated in terms of mean distance and RMSE, with centimeter accuracies for both techniques. These values pointed toward the cloud integration phase after a preliminary editing phase necessary to eliminate redundant and superfluous data.

The integration procedure confirms the application potential of the multisensor survey and its effectiveness in a multiscalar context. In fact, the cases discussed represent the settlement complexity due to the presence of historical buildings in close relationship with the surrounding landscape and artifacts; this is the result of the culture of the peoples who have inhabited the territory over the centuries. The historic built environment is the material result of human actions that have used construction techniques and technologies, considering the material resources of the territories. The combination between materiality and immateriality, landscape and building, empirical knowledge and architecture is present in many national and international areas, such as many parts of Europe with inland mountainous and with a rich intensity of built heritage assets.

The expeditious survey with different instruments, which are now more accessible in terms of cost and easy to use thanks to semiautomatic processes, and the procedure used for aligning and merging the clouds, carried out after careful comparative analysis, return point clouds that are functional for multiple knowledge analyses applied to diverse contexts, as demonstrated in the international scientific literature (Crescenzi & Llopis, 2023; Karaki et al., 2022; Parenti et al., 2022; Stellacci & Condorelli, 2022).

The integrated clouds are, therefore, the starting point for the generation of three-dimensional models and two-dimensional drawings such as sections, plans, and photo plans generated in an automated manner. The outputs of the digitization process of the built heritage can be used for cognitive analyses such as the correct interpretation of architectural forms and the peculiarities of the building fabric, architectural archaeology investigations aimed at assessing the consistency of the masonry and the identification of construction phases, and diagnostic investigations aimed at understanding the state of conservation of the surfaces and masonry itself. Furthermore, these models, in addition to becoming the ideal starting point for the detailed thematic analyses, can be used for the creation of three-dimensional models through the consolidated Scan to BIM procedures (Barba et al., 2021; Costantino et al., 2021; Godinho et al., 2020) or become an integral part of complex information systems capable of linking data and facilitating intervention operations for the conservation of historic buildings.

The results presented here, aimed at discussing the comparison of clouds and their reliability enhanced by the GNSS system, enrich the scientific panorama that includes multiple studies involving laser surveying procedures integrated with photogrammetric ones.

The proposed procedure constitutes the starting point, and it would be desirable to test its effectiveness by expanding the cases of investigation to other subject areas or for other purposes, such as the natural phenomena observation in geology and monitoring fields.

12.5 Conclusion

The research presented in this chapter is intended to increase the panorama already present in the scientific literature to standardize operative procedures for the correct digitization of the built heritage that respects the optimal levels of precision for representation on an architectural scale. The application of multisensor systems in the survey of morphologically and typologically complex areas and buildings represents the most effective and suitable solution for the speedy acquisition of multiple pieces of information for documentation. These become functional for linking the collected data with multidisciplinary approaches to defining conservation and safety strategies for the built heritage. On the other hand, to obtain surveys with a larger scale of detail, such as those aimed at digitizing more or less complex architectural elements, decorations, or ornamental apparatuses, it is possible to integrate the recorded data with other types of sensors, such as precision laser scanners. In this way, it is possible to obtain digital models that can be used in multilevel modality, facilitating the simultaneous consultation of several data with different accuracies.

The models, suitably edited with the results of the research, become rich in meaning and become the medium of information, favoring the sharing of the acquired data not only with the technical and scientific community but also with a broader public, that is, the authorities involved in the management and conservation processes. In this context, the use of new digital technologies, with the development of information systems based on virtual or augmented reality or digital replicas of artifacts, has brought great benefits and advantages in the field of knowledge for the identification of strategies for the maintenance, conservation, and valorization of historic buildings. The development of semantic models is also intended to provide tools for the preservation and fruition of heritage and territory from the perspective of e-government and e-conservation.

References

Abate, N., Ronchi, D., Vitale, V., Masini, N., Angelini, A., Giuri, F., Minervino Amodio, A., Gennaro, A. M., & Ferdani, D. (2023). Integrated close range remote sensing techniques for detecting, documenting, and interpreting lost Medieval settlements under canopy: The case of Altanum (RC, Italy). *Land*, *12*(2), 310. Available from https://doi.org/10.3390/land12020310.

Banfi, F., Previtali, M., & Brumana, R. (2020). Towards the development of a cloud-based BIM platform and VR apps for complex heritage sites subject to the risk of flood and water level changes. *IOP Conference Series: Materials Science and Engineering. 949* (1), 012105, https://doi.org/10.1088/1757-899x/949/1/012105.

Barba, S., di Filippo, A., Cotella, V. A., & Ferreyra, C. (2021). BIM reverse modelling process for the documentation of Villa Rufolo in Ravello. *DISEGNARECON*, *14*(26). Available from https://doi.org/10.20365/disegnarecon.26.2021.1, http://disegnarecon.univaq.it/ojs/index.php/disegnarecon/article/view/808/500.

Barba, S., di Filippo, A., Limongiello, M., & Messina, B. (2019). Integration of active sensors for geometric analysis of the Chapel of the Holy Shroud. *The International Archives of the Photogrammetry, Remote Sensing and Spatial Information Sciences*, *XLII-2/W15*(2), 149−156. Available from https://doi.org/10.5194/isprs-archives-xlii-2-w15-149-2019.

Barba, S., Parrinello, S., Limongiello, M., & Amico. (2020). *D-SITE, Drones - Systems of Information on culTural hEritage. For a spatial and social investigation*. Pavia University Press.

Barrera-Vera, J. A., & Benavides-López, J. A. (2018). Handheld mobile mapping applied to historical urban areas. *DISEGNARECON*, *11*(21). Available from http://disegnarecon.univaq.it/ojs/index.php/disegnarecon/article/view/489/335.

Bartolucci, D. (2009). Principi di laser scanning 3D. Dario Flaccovio Editore.

Batur, M., Yilmaz, O., & Ozener, H. (2020). A case study of deformation measurements of Istanbul land walls via terrestrial laser scanning. *IEEE Journal of Selected Topics in Applied Earth*

Observations and Remote Sensing, 13, 6362−6371. Available from https://doi.org/10.1109/JSTARS.2020.3031675, http://ieeexplore.ieee.org/xpl/RecentIssue.jsp?punumber = 4609443.

Bo, W., Zaw, K. M., & Kyaw, K. (2018). Investigation on the existing geometry of Pathein Suspension Bridge. *International Journal of Trend in Scientific Research and Development, Volume-2*(Issue-6), 118−124. Available from https://doi.org/10.31142/ijtsrd18436.

Bornaz, L. (2006). *Principi di funzionamento e tecniche di acquisizione. Atti del corso La tecnica del laser scanning terrestre.* CISM.

Bronzino, G. P. C., Grasso, N., Matrone, F., Osello, A., & Piras, M. (2019). Laser-visual-inertial odometry based solution for 3D heritage modeling: The Sanctuary of the Blessed Virgin of Trompone. *The International Archives of the Photogrammetry, Remote Sensing and Spatial Information Sciences, XLII-2/W15*(2), 215−222. Available from https://doi.org/10.5194/isprs-archives-xlii-2-w15-215-2019.

Charron, N., McLaughlin, E., Phillips, S., Goorts, K., Narasimhan, S., & Waslander, S. L. (2019). Automated bridge inspection using mobile ground robotics. *Journal of Structural Engineering, 145*(11), 04019137. Available from https://doi.org/10.1061/(asce)st.1943-541x.0002404.

Chiabrando, F., Spanò, A., Sammartano, G., & Teppati Losè, L. (2017). UAV oblique photogrammetry and lidar data acquisition for 3D documentation of the Hercules Fountain. *Virtual Archaeology Review, 8*(16), 83. Available from https://doi.org/10.4995/var.2017.5961.

Chudá, J., Hunčaga, M., Tuček, J., & Mokroš, M. (2020). The handheld mobile laser scanners as a tool for accurate positioning under forest canopy. *The International Archives of the Photogrammetry, Remote Sensing and Spatial Information Sciences, XLIII-B1−2020*(1), 211−218. Available from https://doi.org/10.5194/isprs-archives-xliii-b1-2020-211-2020.

Cloud Compare. (2022). *User manual.* Version 2.6.1. http://www.cloudcompare.org/doc/qCC/CloudCompare%20v2.6.1%20-%20User%20manual.pdf.

Conti, A., Fiorini, L., Massaro, R., Santoni, C., & Tucci, G. (2022). HBIM for the preservation of a historic infrastructure: The Carlo III bridge of the Carolino Aqueduct. *Applied Geomatics, 14*, 41−51. Available from https://doi.org/10.1007/s12518-020-00335-2, http://www.springerlink.com/content/1866-9298/.

Costantino, D., Pepe, M., & Restuccia, A. G. (2021). Scan-to-HBIM for conservation and preservation of Cultural Heritage building: the case study of San Nicola in Montedoro church (Italy). *Applied Geomatics.* Available from https://doi.org/10.1007/s12518-021-00359-2, http://www.springerlink.com/content/1866-9298/.

Crescenzi, C., & Llopis, J. (2023). The "Cuevas de Los Moros" in Bocairente (ES) on the integrated expeditious survey. *The International Archives of the Photogrammetry, Remote Sensing and Spatial Information Sciences, XLVIII-M-2−2023*, 447−452. Available from https://doi.org/10.5194/isprs-archives-xlviii-m-2-2023-447-2023.

De Fino, M., Porcari, V. D., Scioti, A., Guida, A., & Fatiguso, F. (2022). Polychrome majolica of Apulian domes: History, technique, pathology and conservation. *Vitruvio, 7*(2), 34−45. Available from https://doi.org/10.4995/vitruvioijats.2022.18684, https://polipapers.upv.es/index.php/vitruvio/article/download/18684/15589.

Di Stefano, F., Chiappini, S., Piccinini, F., & Pierdicca, R. (2020). *Integration and assessment between 3D data from different geomatics techniques. Case Study: The ancient city walls of San Ginesio (Italy). Communications in computer and information science* (1246, pp. 186−197). Italy: Springer Science and Business Media Deutschland GmbH. Available from http://www.springer.com/series/7899, https://doi.org/10.1007/978-3-030-62800-0_15, 18650937.

Elamin, A., Abdelaziz, N., & El-Rabbany, A. (2022). A GNSS/INS/LiDAR integration scheme for UAV-based navigation in GNSS-challenging environments. *Sensors, 22* (24). Available from https://doi.org/10.3390/s22249908, http://www.mdpi.com/journal/sensors.

Ellum, C. M., & El-Sheimy, N. (2006). New strategies for integrating photogrammetric and GNSS data. *International Archives of the Photogrammetry, Remote Sensing and Spatial Information Sciences - ISPRS Archives* 16821750 36 103−108 International Society for Photogrammetry and Remote Sensing Canada. http://www.isprs.org/proceedings/XXXVIII/4-W15/.

Forlani, G., Diotri, F., Cella, U. M. d., & Roncella, R. (2019). Indirect UAV strip georeferencing by on-board GNSS data under poor satellite coverage. *Remote Sensing, 11*(15), 1765. Available from https://doi.org/10.3390/rs11151765.

Freimuth, H., & König, M. (2015). Generation of waypoints for UAV-assisted progress monitoring and acceptance of construction work. *Conference: Conference on Construction Applications of Virtual Reality (CONVR).* 10.

GeoSLAM (2022). *ZEB Horizon.* Available from: https://geoslam.com/solutions/zeb-horizon/

Godinho, M., Machete, R., Ponte, M., Falcão, A. P., Gonçalves, A. B., & Bento, R. (2020). BIM as a resource in heritage management: An application for the National Palace of Sintra, Portugal. *Journal of Cultural Heritage, 43*, 153−162. Available from https://doi.org/10.1016/j.culher.2019.11.010, http://www.elsevier.com.

Gomarasca, M. A. (2004). *Elementi di geomatica.* ASITA.

Hou, K., & Chio, S. (2021). Plane-based range calibration method for Geoslam Zeb-Horizon handheld lidar instrument. *Proceedings of the International Symposium on Remote Sensing (ISRS), Virtual Conference* 26−28.

Jordan, S., Moore, J., Hovet, S., Box, J., Perry, J., Kirsche, K., Lewis, D., & Tse, Z. T. H. (2018). State-of-the-art technologies for UAV inspections. *IET Radar, Sonar and Navigation, 12*(2), 151−164. Available from https://doi.org/10.1049/iet-rsn.2017.0251, http://www.ietdl.org/IET-RSN.

Kajzar, V., Doležalová, H., Souček, K., & Staš, L. (2011). Aerial photogrammetry observation of the subsidence depression near Karviná. *Acta Geodynamica et Geomaterialia, 8* (3), 309−317. Available from http://www.irsm.cas.cz/abstracts/AGG/03_11/12_%20Kajzar.pdf, Czech Republic.

Kamnik, R., Nekrep Perc, M., & Topolšek, D. (2020). Using the scanners and drone for comparison of point cloud accuracy at traffic accident analysis. *Accident Analysis and Prevention, 135.* Available from https://doi.org/10.1016/j.aap.2019.105391, http://www.sciencedirect.com/science/journal/00014575.

Karaki, A. A., Bibuli, M., Caccia, M., Ferrando, I., Gagliolo, S., Odetti, A., & Sguerso, D. (2022). Multi-platforms and multi-sensors integrated survey for the submerged and emerged areas. *Journal of Marine Science and Engineering, 10*(6). Available from https://doi.org/10.3390/jmse10060753, https://www.mdpi.com/2077-1312/10/6/753/pdf?version = 1653902700.

Kartini, G. A. J., & Saputri, N. D. (2022). 3D modelling of Boscha Observatory with TLS and UAV integration data. *Geoplanning, 9*(1), 37−46. Available from https://doi.org/10.14710/geoplanning.9.1.37-46, https://ejournal.undip.ac.id/index.php/geoplanning/article/view/41615.

Keitaanniemi, A., Virtanen, J. P., Rönnholm, P., Kukko, A., Rantanen, T., & Vaaja, M. T. (2021). The combined use of slam laser scanning and TLS for the 3D indoor mapping. *Buildings, 11*(9). Available from https://doi.org/10.3390/buildings11090386, https://www.mdpi.com/2075-5309/11/9/386/pdf.

Kitratporn, N., Takeuchi, W., Matsumoto, K., & Nagai, K. (2018). Structure deformation measurement with terrestrial laser scanner at Pathein Bridge in Myanmar. *Journal of Disaster Research*, *13*(1), 40−49. Available from https://doi.org/10.20965/jdr.2018. p0040, https://www.fujipress.jp/main/wp-content/themes/Fujipress/pdf_subscribed.php.

Lo, C. F., Tsai, M. L., Chiang, K. W., Chu, C. H., Tsai, G. J., Cheng, C. K., El-Sheimy, N., & Ayman, H. (2015). The direct georeferencing application and performance analysis of UAV helicopter in GCP-free area. *The International Archives of the Photogrammetry, Remote Sensing and Spatial Information Sciences*, *XL-1/W4*(1), 151−157. Available from https://doi.org/10.5194/isprsarchives-xl-1-w4-151-2015.

Luleci, F., Li, L., Chi, J., Reiners, D., Cruz-Neira, C., & Catbas, F. N. (2021). Structural health monitoring of a foot bridge in virtual reality environment. *Procedia Structural Integrity*, *37*(C), 65−72. Available from https://doi.org/10.1016/j.prostr.2022.01.060, http://www.journals.elsevier.com/procedia-structural-integrity, 24523216 Elsevier B.V. United States.

Mader, D., Blaskow, R., Westfeld, P., & Weller, C. (2016). Potential of UAV-based laser scanner and multispectral camera data in building inspection. *The International Archives of the Photogrammetry, Remote Sensing and Spatial Information Sciences*, *XLI-B1*, 1135−1142. Available from https://doi.org/10.5194/isprs-archives-xli-b1-1135-2016.

Malinverni, E. S., Pierdicca, R., Bozzi, C. A., & Bartolucci, D. (2018). Evaluating a slam-based mobile mapping system: A methodological comparison for 3D heritage scene real-time reconstruction. *2018 IEEE International Conference on Metrology for Archaeology and Cultural Heritage, MetroArchaeo 2018 − Proceedings*. Institute of Electrical and Electronics Engineers Inc. Italy. 9781538652763 265−270. http://ieeexplore.ieee.org/xpl/mostRecentIssue. jsp?punumber = 9081813. https://doi.org/10.1109/MetroArchaeo43810.2018.13684.

Mistretta, F., Sanna, G., Stochino, F., & Vacca, G. (2019). Structure from motion point clouds for structural monitoring. *Remote Sensing*, *11*(16), 1940. Available from https://doi.org/10.3390/rs11161940.

Parenti, C., Rossi, P., Soldati, M., Grassi, F., & Mancini, F. (2022). Integrated geomatics surveying and data management in the investigation of slope and fluvial dynamics. *Geosciences*, *12*(8), 293. Available from https://doi.org/10.3390/geosciences12080293.

Patrucco, G., Rinaudo, F., & Spreafico, A. (2019). Multi-source approaches for complex architecture documentation: The "Palazzo Ducale" in Gubbio (Perugia, Italy). *The International Archives of the Photogrammetry, Remote Sensing and Spatial Information Sciences*, *XLII-2/W11*(2), 953−960. Available from https://doi.org/ 10.5194/isprs-archives-xlii-2-w11-953-2019.

Pulcrano, M., Scandurra, S., Minin, G., & di Luggo, A. (2019). 3D cameras acquisitions for the documentation of cultural heritage. *The International Archives of the Photogrammetry, Remote Sensing and Spatial Information Sciences*, *XLII-2/W9*(2), 639−646. Available from https://doi.org/10.5194/isprs-archives-xlii-2-w9-639-2019.

Rabbia, A., Sammartano, G., & Spanò, A. (2020). Fostering Etruscan heritage with effective integration of UAV, TLS and SLAM-based methods. *2020 IMEKO TC-4 International Conference on Metrology for Archaeology and Cultural Heritage*. International Measurement Confederation (IMEKO), Italy, pp. 322−327.

Rachmawati, T. S. N., & Kim, S. (2022). Unmanned aerial vehicles (UAV) integration with digital technologies toward Construction 4.0: A systematic literature review. *Sustainability (Switzerland)*, *14*(9). Available from https://doi.org/10.3390/su14095708, https://www.mdpi.com/2071-1050/14/9/5708/pdf?version = 1652089446.

Rinaudo, F., & Scolamiero, V. (2021). Comparison of multi-source data, integrated survey for complex architecture documentation. *The International Archives of the Photogrammetry,*

Remote Sensing and Spatial Information Sciences, XLVI-M-1-2021(M-1-2021), 625–631. Available from https://doi.org/10.5194/isprs-archives-xlvi-m-1-2021-625-2021.

Rockafellar, R. T., & Wets, R. J. B. (2009). *Variational analysis. 317.*

Russo, M., Remondino, F., & Guidi, G. (2011). Principali tecniche e strumenti per il rilievo tridimensionale in ambito archeologico. *Archeologia e Calcolatori, 22.*

Sammartano, G., & Spanò, A. (2018). Point clouds by SLAM-based mobile mapping systems: accuracy and geometric content validation in multisensor survey and stand-alone acquisition. *Applied Geomatics, 10*(4), 317–339. Available from https://doi.org/10.1007/s12518-018-0221-7, http://www.springerlink.com/content/1866-9298/.

Savini, F., Marra, A., Cordisco, A., Giallonardo, M., Fabbrocino, G., & Trizio, I. (2022). A complex virtual reality system for the management and visualization of bridge data. *SCIRES-IT, 12*(1), 49–66. Available from https://doi.org/10.2423/i22394303v12n1p49, http://www.sciresit.it/.

Schweizer, E. A., Stow, D. A., & Coulter, L. L. (2018). Automating near real-time, post-hazard detection of crack damage to critical infrastructure. *Photogrammetric Engineering and Remote Sensing, 84*(2), 75–86. Available from https://doi.org/10.14358/PERS.84.2.75, http://docserver.ingentaconnect.com/deliver/connect/asprs/00991112/v84n2/s14.pdf?expires = 1518097275&id = 0000&titleid = 72010567& checksum = 764E47502B1A0E5CA10298089 2497EE1.

Shang, Z., & Shen, Z. (2018). *Real-time 3D reconstruction on construction site using visual SLAM and UAV. Construction Research Congress 2018: Construction Information Technology - Selected Papers from the Construction Research Congress 2018* (2018-, pp. 305–315). United States: American Society of Civil Engineers (ASCE). Available from https://doi.org/10.1061/9780784481264.030, 9780784481264.

Shetty, A., & Xingxin Gao, G. (2019). Adaptive covariance estimation of LiDAR-based positioning errors for UAVs. *NAVIGATION, 66*(2), 463–476. Available from https://doi.org/10.1002/navi.307.

Skoczylas, A., Kamoda, J., & Zaczek-Peplinska, J. (2016). Geodetic monitoring (TLS) of a steel transport trestle bridge located in an active mining exploitation site. *Annals of Warsaw University of Life Sciences − SGGW. Land Reclamation, 48*(3), 255–266. Available from https://doi.org/10.1515/sggw-2016-0020.

Spanò, A. (2019). Rapid Mapping methods for archaeological sites. *2019 IMEKO TC4 International Conference on Metrology for Archaeology and Cultural Heritage,* MetroArchaeo 2019. IMEKO-International Measurement Federation Secretariat Italy. 9789299008454 25–30.

Stellacci, S., & Condorelli, F. (2022). Remote survey of traditional dwellings using advanced photogrammetry integrated with archival data: The case of LISBON. *Int. Arch. Photogramm. Remote Sens. Spatial Inf. Sci., XLIII,* 893–899. Available from https://doi.org/10.5194/isprs-archives-XLIII-B2-2022-893-2022.

Suzuki, T., Takahashi, Y., & Amano, Y. (2016). Precise UAV position and attitude estimation by multiple GNSS receivers for 3D mapping. *29th International Technical Meeting of the Satellite Division of the Institute of Navigation,* ION GNSS 2016 2 1455–1464 https://doi.org/10.33012/2016.14621 9781510834101. Institute of Navigation Japan.

Tang, P., Akinci, B., & Garrett, J. H. (2007). Laser scanning for bridge inspection and management. *IABSE Symposium Report. 93* (18), 17–24.

Teodoro, A. C., & Araujo, R. (2016). Comparison of performance of object-based image analysis techniques available in open source software (Spring and Orfeo Toolbox/ Monteverdi) considering very high spatial resolution data. *Journal of Applied Remote Sensing, 10*(1), 016011. Available from https://doi.org/10.1117/1.jrs.10.016011.

Teppati Losè, L., Sammartano, G., Chiabrando, F., & Spanò, A. (2020). Challenging multi-sensor data models and use of 360 images. The Twelve Months Fountain of Valentino park in Turin. IOP Conference Series: Materials Science and Engineering. *949* (1), 012060. Available from https://doi.org/10.1088/1757-899x/949/1/012060.

Teza, G., Galgaro, A., & Moro, F. (2009). Contactless recognition of concrete surface damage from laser scanning and curvature computation. *NDT & E International*, *42*(4), 240−249. Available from https://doi.org/10.1016/j.ndteint.2008.10.009.

Torres-González, M., Revuelta, E. C., & Calero-Castillo, A. I. (2023). Photogrammetric state of degradation assessment of decorative claddings: The plasterwork of the Maidens' Courtyard (The Royal Alcazar of Seville). *Virtual Archaeology Review*, *14*(28), 110−123. Available from https://doi.org/10.4995/var.2023.18647, http://polipapers.upv.es/index.php/var/index.

Tucci, G., Conti, A., & Fiorini, L. (2022). *Rilievi e modelli 3D come strumenti diagnostici per lo studio e l'interpretazione di superfici complesse ai fini dell'intervento di restauro.* San Carlo de'Barnabiti: Il Consolidamento del Sistema Voltato: dalla Documentazione al Cantiere di Restauro. 8.

Ulvi, A. (2021). Documentation, three-dimensional (3D) modelling and visualization of cultural heritage by using unmanned aerial vehicle (UAV) photogrammetry and terrestrial laser scanners. *International Journal of Remote Sensing*, *42*.

Valença, J., Júlio, E. N. B. S., & Araújo, H. J. (2012). Applications of photogrammetry to structural assessment. *Experimental Techniques*, *36*(5), 71−81. Available from https://doi.org/10.1111/j.1747-1567.2011.00731.x.

Zollini, S., Alicandro, M., Dominici, D., Quaresima, R., & Giallonardo, M. (2020). UAV photogrammetry for concrete bridge inspection using object-based image analysis (OBIA). *Remote Sensing*, *12*(19), 3180. Available from https://doi.org/10.3390/rs12193180.

Section IV

Heritage Building Information Modeling (BIM) and Digital Twins (DTs)

From a multidisciplinary analysis to HBIM: Tools for the digital documentation of historical buildings

13

Francesca Savini[1], Adriana Marra[1], Giovanni Fabbrocino[1,2], and Ilaria Trizio[1]
[1]Institute for Construction Technologies - Italian National Research Council, L'Aquila, Italy, [2]Department of Biosciences and Territory, StreGa Lab, University of Molise, Campobasso, Italy

13.1 Introduction

This chapter deals with the role of digital technologies applied to the knowledge, documentation, and multidisciplinary analysis of the built environment with specific reference to those constructions characterized by historical and/or architectural value. This is a relevant topic involving a wide area of knowledge ranging between human sciences (i.e., history, history of architecture, and archaology) and hard sciences (i.e., physics and engineering) and that finds a balance in the area of the architecture.

Historical constructions, and more in general all the built environment, reflect the expertise of the builders and of the architects at the time of the construction and are generally the manifestations of empirical knowledge and the result of a trial and error process performed directly in the field.

It is indeed well known that the developments of physical mathematics and structural mechanics are achievements of the last 15 decades. The modern comparison between the demand—the internal actions due to the loads—and the capacity—associated with the shape and the dimensions of the members as well as with the strength of materials—indeed dates back to the last century and marks the difference between the engineered structures and those built according to empiric rules.

This classification, however, is not relevant whenever the preservation and the structural maintenance of historical structures are concerned, particularly due to the increased consciousness of the risk associated with natural hazards and the demand of safety at different scales (i.e., regional and urban). This is the case of those areas classified as "peripheral" of "inner" in European countries, like the ones recently classified in Italy.

They are regions characterized by a network of minor centers having a high historical and environmental value, but also exposed to high natural hazards. It is well

Diagnosis of Heritage Buildings by Non-Destructive Techniques. DOI: https://doi.org/10.1016/B978-0-443-16001-1.00013-9

known that minor centers located in the Italian Apennines are caskets of archaeological, historical, and architectural treasures, whose preservation and refurbishment are key issues in the socioeconomical balance of the nation, but also challenging technical tasks to be accomplished in a fully multidisciplinary framework.

Herein, attention is primarily paid to the knowledge and to a historical and technical characterization of the constructions and to the operational procedures for the exploitation of the advances in the field of the digital technologies. In particular, a collection of explanatory cases analyzed by the authors in the framework of national and international research programs are discussed in a way that the features of the most relevant phases, that is, data processing and management in a digital environment, implementation of semantic and critical models of artifacts, are analyzed. In such a context, the Building Information Modeling (BIM) environment offers a really integrated and powerful tool for the maintenance, diagnosis, and management of the historic and architectural heritage making possible the transition to the so-called Historical Building Information Modeling (HBIM).

The challenges faced in documenting the built heritage are closely related to its complexity due to the intrinsic peculiarities and heterogeneity of the geometric-formal and typological characteristics of the artifacts. It is also important to emphasize that each artifact is the result of the evolution of human ability and the relationship between the works of man and nature. Indeed, in the creation of an artifact and its preservation over time, one can recognize environmental, economic, and social factors such as those that have contributed to defining its formal or decorative and structural aspects.

These have also influenced and oriented conservation strategies, in some cases reiterating maintenance interventions over the centuries, which have ensured survival to the present day, in other causing the abandonment of a building or whole village. On the other hand, the interaction between man and nature has also affected the choice of places for the establishment of settlements, which were often built in relation to specific natural resources, such as the presence of fertile land, communication routes, or waterways. In addition, each artifact also includes aspects related to the sphere of the immaterial, which fulfills the needs of an ideological-cultural nature and which are the result of choices conditioned by the will and empirical knowledge of our ancestors, which led them to design and construct buildings and infrastructures according to the rules of art and therefore with different criteria and needs from current ones.

All these factors contribute to assigning a value to the artifacts that are an integral part of our man-made landscape (churches, fountains, towers, stately homes, and so on) and, for this reason, they are recognized as a significant part of the cultural heritage that, as such, must be studied, protected, and transmitted to all citizens and future generations. These aspects underline the complexity of the built heritage, whether it belongs to the sphere of architecture in use or that of archaeological remains, and, as a consequence, the complexity in the analysis and research phases. Therefore it becomes crucial to implement a detailed cognitive analysis that is not only preparatory to all conservation activities, but also to those aimed at enhancing and communicating the cultural heritage. Before planning any conservative intervention, even if it is just a consolidation for safety, it is essential to analyze the heritage and collect data that can provide a full understanding of the construction and evolutionary history of architectural complexes by identifying the construction stages, life phases, and postdeposition transformations. From this perspective, the integration of the methodologies provided by different disciplines to ensure a correct knowledge of the artifacts appears to be advantageous. Moreover, these single artifacts, characterized by a constructive complexity, often interconnected with the evolutionary history of sites (such as scattered settlements, historic centers, or large cities) but also environments and landscapes in which they are embedded, become the main, namely, direct, source of knowledge since they are capable of recording, as true archives, all traces of the past.

These issues become more complex when the heritage is located in multihazard contexts, such as the territories of the inner areas investigated by the research group. In these cases, the analysis of the heritage and its documentation is aimed at providing a rational overview that can support the decision-making process by assessing the hazard of natural events and their potential consequences on the objects to be protected. In emergency contexts, with valuable buildings undamaged by natural disasters, a well-designed planning process capable of supporting the recovery phases is required. The need for surveys and analyses aimed at acquiring data for in-depth knowledge in these contexts was demonstrated by the experience of the postearthquake reconstruction of the areas neighboring the city of L'Aquila.

Indeed, the built heritage of the city of L'Aquila and 56 smaller municipalities in its territory were heavily damaged by the earthquake that hit the Abruzzo region in 2009. To repair the damage, while preserving the area's building peculiarities and environmental and identity values, specific reconstruction plans have been identified and adopted for each municipality to plan the most suitable interventions to be implemented to repair the damage and recover the identity features of the historic built heritage (Fico et al., 2019; Marra et al., 2019). The level of detail achieved by the plan regulations in the different experiences was not uniform although the results reached by the process still in progress have highlighted the effectiveness and quality of these tools. However, the information gathered during the cognitive phase of the historic centers led to the identification of the building techniques and materials typical of the area in consideration, the characteristics of the urban-building structure, and the typologies and peculiarities of the built environment. This process was not supported by a precise digitalization of the data associated with a digital replica of the artifacts. It was only with the acceptance of the reconstruction procedures that the Special Offices were able to record specific information in databases designed to collect information on the buildings and to display, both analytical and graphic, information on the reconstruction progress in a GIS platform.

The procedure here described, subsequently validated on several artifacts located in the inner areas of Abruzzo and Molise characterized by different risks, such as the seismic and hydrogeological ones, becomes extremely functional to operate in these contexts since it permits the documentation of the specific characteristics of the single objects in relation to those of the surrounding territory, increasing knowledge with integrated investigations between hard sciences and humanities. Indeed, it allows to record the vulnerable elements of the area and their relationship with the built heritage in a single digital environment.

This premise highlights the amount of information that results from such an analytical approach and the importance of systematizing, organizing, and accessing the data collected with non-destructive multidisciplinary analyses performed on heritage. This also implies issues related to data integration and the need to define common lexicons and operational protocols for data management and use. In this regard, research, both at the nationally and internationally level, is directed toward the extensive use of digital and three-dimensional visualization of artifacts as privileged documentation tools as they are capable of visually and topologically correlating the multiple data acquired. These approaches, unquestionably useful for heritage conservation activities, range from mesh models obtained from the digital survey to NURBS (nonuniform rational B-spline) models, from models accessible in GIS3D and in Game Engine for developing Virtual Reality systems. The validity of each approach has been tested over the years by the L'Aquila research team, which identifies the parametric process of Heritage Building Information Modeling as the connecting tool between different disciplines and different languages used by them.

Within the framework of the issues involved in the three-dimensional representation of the historic built environment, the subject of HBIM has found more and more space in the last 10 years, highlighting the advantages of applying the BIM paradigm to the historic built environment. The great potential of HBIM as a nodal tool between the various thematic readings, such as archaeology, state of preservation, and structural monitoring, initially recognized within a few European countries, is now spreading widely internationally (Al-Bayari & Shatnawi, 2022; Khan et al., 2022; Santini et al., 2022). Of great importance has been, and continues to be, within the process the correct design of the data architecture behind which there is often a refined ontological decomposition work (Acierno et al., 2017; Gigliarelli et al., 2022) that is transversal to the various applications that have followed over time. In fact, parametric modeling of historic buildings was first recognized as a convenient way to support decision-making processes through restoration project management (Continenza et al., 2016; Fiorani, 2017; Oreni et al., 2017), representing, in fact, an effective tool for managing reconstruction sites in the areas of the 2009 earthquake disaster crater (Brumana et al., 2021; Brumana, Della Torre, Previtali, et al., 2018; Trizio et al., 2019), despite the presence of certain limitations that have often led to a critical review of the tool (Della Torre, 2017). The methodology was subsequently also tested in the field of historical heritage diagnosis moving toward Diagnosis-Aided Historic Building Information Modeling and Management (Bruno, & De Fino, & Fatiguso, 2018) and subsequently toward monitoring, leading to digital twins (Marra, Trizio, Fabbrocino, & Savini, 2021; Massafra et al., 2022; Tan et al., 2022), models to be developed to deal with automated collection and allow for the constant monitoring of potential risks by means of automated procedures. HBIM has been used effectively not only for the documentation and conservation of historical buildings but also in relation to archaeological heritage, in which the potential of HBIM procedures has also been tested before in other fields, starting with free and open source software solutions (Diara & Rinaudo, 2021; Diara, 2022), which were then further explored in different applications (Oostwegel et al., 2022). The next step saw the further use of parametric models generated using HBIM technology for the structural assessment of existing artifacts (Abbate et al., 2020; Croce et al., 2022; Moyano et al., 2022; Pepe et al., 2020; Zouaoui et al., 2022), and for the recovery of industrial archaeology artifacts (Currà et al., 2022) or experiments on the comfort of cultural heritage (Meoni et al., 2022), or on specific artifacts such as historical infrastructures (Marra, Trizio, Fabbrocino, & Savini, 2021). The experiments have further advanced to the point of extending the potential of parametric modeling to entire historic centers through the use of City Information Modeling for the definition of a workflow that quickly enables the acquisition, modeling, and structural analysis of finite elements of urban aggregates (La Russa et al., 2021; Saccucci & Pelliccio, 2018). Finally, among the most recent developments and experiments are those related to the use and valorization of artifacts through mixed reality tools (Banfi, Brumana, et al., 2022; Banfi, Roascio, et al., 2022) and model generation through Visual Programming Language (VPL) procedures (Liberotti & Gusella, 2023).

The abovementioned summary emphasizes the potential of the procedure and its versatility not only according to the different types of heritage but also in relation to the field of applications such as archaeological data management, application to restoration and structural monitoring, favoring actions for the physical preservation of artifacts through the definition of intervention priority classes, and the creation of predictive models. And, as demonstrated in the literature, the tool is equally effective and performant in the valorization phase of the built heritage, as it is possible, by using the implemented solutions, to reconstruct the evolutionary history not only to protect the artifact but also to communicate it to the people through digital and BIM integration with Virtual Reality techniques.

13.2   Materials and methods

The main topics of the chapter deal with the historic built heritage analysis and interpretation through the integration of multidisciplinary data with the digital representation of artifacts with the aim of defining operational procedures that favor both material and memory conservation. In this context, the experiments performed over the years have made it possible to develop, test, and validate an operational procedure for multidisciplinary analysis, the integration of data in a digital environment, and their three-dimensional representation with the creation of HBIM models suitable for the management of the historic built heritage (Fig. 13.1). This moves from the integration of the results of multidisciplinary research by bringing together the data produced by different professionals from the sphere of the hard sciences, such as the architect, the restorer, and the engineer, with those collected through the application of methods and tools from the humanities ones. This holistic approach

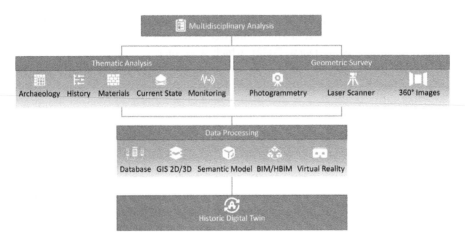

Figure 13.1 Workflow of the process.

ensures a comprehensive understanding of the traces found on buildings, such as stratigraphic sequences resulting from evolutionary phases, stylistic-functional features, or technical-structural aspects that determine their stability. In addition, the integration of different know-how and the use of diagnostic and archaeometric techniques to analyze the state of preservation, characterize materials, and identify construction techniques and technologies guarantees an in-depth knowledge of the historic built heritage.

Among the humanistic disciplines that deal with the analysis and interpretation of the built heritage, a central role is entrusted to archaeology as it is capable of implementing different procedures by applying the stratigraphic method to masonry. Starting from this process, codified in the complex discipline that has taken the name of "Archaeology of buildings," "*Archeologia dell'architettura,*" and "*Arqueología de la arquitectura*" (Azkarate Garai-Olaun et al., 2002; Azkarate Garai-Olaun, 2010; Brogiolo, 1997; Mannoni & Boato, 2002), it is possible to identify the several portions that constitute masonry, called stratigraphic masonry units (SUM), on the basis of their characteristics and discontinuities. These correspond to anthropic actions of construction or destruction, or natural collapses, that have affected the artifact over time. The correlation of the SUM with the phases (construction, intervention, and maintenance) permits the definition of a relative chronology, namely, relations of anteriority, posteriority, and contemporaneity between the SUM for reconstructing, therefore, the historical sequence and constructive evolution. Integration with historical and typological data then indicates an absolute chronology, which means associating the action that led to the construction or destruction of a wall with a specific historical period. With this in mind, it is possible to analyze the different construction periods: the building of the artifact and the choices made for its interaction with the environment and the life phase of the artifact with its collapse and additions or, in some cases, its abandonment.

In this perspective, archaeological interpretation becomes a prerequisite for any intervention in the built heritage, both ordinary and extraordinary maintenance (Brogiolo & Cagnana, 2012; Brogiolo, 2007; Doglioni, 2002; Francovich, 1988), and, like a common thread, connects the various methods adopted for heritage knowledge.

It is obvious that the archaeological data cannot be considered as an isolated entity, decontextualizing it from the building and its structural and conservation characteristics. Therefore the thematic analyses, collected by carrying out scientific investigations and applying different tools, aim to characterize the degradation of the surfaces, and the structural damage, also by identifying historical cracks and the relative measures adopted to mitigate them, the masonry instability correlated to physical stresses produced by the surrounding terrain or the erosive action of wind or water on the materials and binders with the consequent mechanical weakening of the structures.

Reliable procedures, such as periodic visual inspections, diagnostic analyses and monitoring, and efficient analysis tools capable of supporting decision-making processes and implementing predictive conservation and safety interventions, should be defined to observe changes and prevent risks to which historical artifacts are exposed. These procedures and related tools must be able to record, connect, and disseminate the contributions provided by the different professionals involved in the processes of safety and safeguarding of the historic built heritage, in line with the guidelines set out in internationally and nationally shared recommendations and regulations (ICOMOS, 2003; ISO 13822,822, 2010; Moro & Neri, 2010; NTC, 2018). The tools and methods developed in recent years in the field of diagnostics and conservation(Bonora et al., 2020; Bruno & Fatiguso, 2018; Karoglou et al., 2019; Moropoulou et al., 2018) underline the importance of detecting and relating the parameters, both qualitative and quantitative, that describe the historical building heritage. On the other hand, structural monitoring has undergone a technological advancement in recent years that has led to the definition of systems capable of monitoring the environmental and health condition of artifacts in real time. In particular, thanks to structural monitoring systems (SHMs), it is possible to measure and process a large amount of data related to different physical parameters, referring both to the environmental conditions of buildings and global and local parameters (deformations, rotations, displacements, and natural and modal frequencies) of structural and nonstructural components of the built heritage, and that can be returned as dynamic data, which are updated in real time, or static, and therefore referred to a specific observation range of the structure and the environment (De Stefano et al., 2016; Mesquita et al., 2017; Rainieri et al., 2019).

By underlining the importance of preferring integrated analytical approaches on a multidisciplinary basis, the need to manage the output of this holistic process, namely, the multiple data that are heterogeneous for the information organization, such as geometric-formal data, historical, and typological, those relating to the state of conservation and the characterization of the materials and the subsequent identification of the construction phases, is recognized. But the heterogeneity does not end in the information content but also in the formal aspect of the collected data, which ranges from alphanumeric information recorded in catalogs, files, and tables, to photographic and graphical two-dimensional and three-dimensional ones. These issues have led toward the identification of procedures for the systematization of data aimed at facilitating the digital documentation of the historic built environment through the recording, the three-dimensional reproduction of artifacts, the definition of data management systems, and the topological linking of these to digital replicas. Therefore, through the integration of relational databases with numerical and parametric modeling processes, the importance of an information structure that allows for the management and sharing of data and that can facilitate the planning of any interventions and the subsequent monitoring and maintenance phases is highlighted. The use of ICT applied to cultural heritage has been favored by the rapid development of technology and its increasing accessibility. In this context, the increasing use of ICTs in the field of documentation, analysis, and management of cultural heritage has resulted in the development of new methods of acquiring, storing, relating, and interpreting data, as well as fruition and valorization through the creation of digital replicas of real objects, easily accessible through different software and applications. Digitization characterizes the proposed procedure, which aims to document and manage data in a digital environment to favor the processes of conservation, communication, and valorization of the historical and archaeological built heritage. The main challenge that the workflow faces is the standardization of the different operational steps to guarantee the interoperability of the outputs, starting from the identification of common lexicons for the recording of alphanumeric and three-dimensional data that guarantee the correct exchange modalities among the several professionals involved in analyzing the historical built heritage. Concerning standardization in the creation phase of digital replicas, representation sciences have played a key role by defining methodological and operational processes for the artifacts' digitization and their reconstruction in a digital environment.

Digital reconstructions of the built heritage are nowadays realized using HBIM processes. The parametric models resulting from these processes can be realized using different procedures: the first exploits the interoperability of parametric software with 3D modeling software, such as freeform surfaces, and the second is that of Scan-to-BIM. Specifically, the first modeling procedure offers the possibility of importing geometric entities into the parametric environment through the open file format IFC—Industry Foundation Classes (BuildingSMART, 2022; Gerbino et al., 2021), which allows the exchange of an information model, without loss or distortion of data and information. An alternative is to obtain a parametric digital model from polysurfaces obtained through software for 3D modeling of sculptured surfaces and NURBS elements. Interoperability between the two systems is ensured by the ACI SAT file format that allows the import of single elements into the parametric environment (Marra, Trizio, Fabbrocino, 2021, 2021; Trizio et al., 2021). However, these elements are lacking in information, so in the import phase, it is necessary to associate the entities with the specific families provided by the parametric modeling software so that the geometric content can be enriched with specific information, also with the added help of shared parameters that increase the information level of the whole project. The Scan-to-BIM procedure has already been widely used and tested in the parametric modeling of existing buildings and represents a process of reverse engineering that uses data acquired from laser scans or photogrammetric techniques to reconstruct, in a digital environment, the elements that characterized historical artifacts (Allegra et al., 2020; Rocha et al., 2020). This process was more efficient in the operations of element decomposition and information management and returns an as-built model that is capable of collecting and integrating, thanks to predefined or ad hoc designed parameters, heterogeneous information, including that acquired from monitoring systems. To improve data management within the modeling environment and to facilitate data management and correlation with other information systems, a semantic deconstruction of artifact elements on the basis of construction, structural, geometric, and typological characteristics was carried out. The purpose of this decomposition is to create libraries of parametric elements that are easily adaptable to the different existing structures, on the one hand, and aligned to the semantic deconstruction of the historical building for the creation of a specific ontology, on the other.

The synthesized workflow facilitates the creation of HBIM digital models that represent the methodological connection useful to achieve shared knowledge (Fig. 13.2). This procedure, certainly articulated and transdisciplinary, can become a best practice for recovering damaged buildings and the preservation of historical

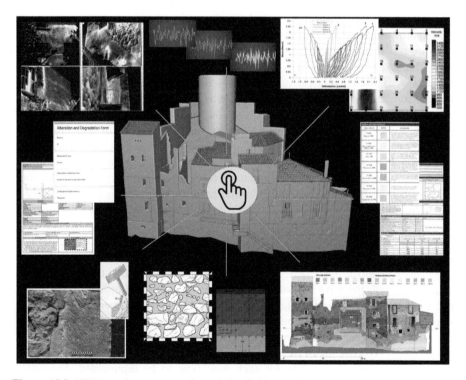

Figure 13.2 HBIM environment as a knowledge linking tool.

constructions for future generations. At the same time, it also becomes functional for the transmission of cultural heritage in a broad sense, making it accessible to a large public and easily intelligible for the community.

13.3 The HBIM of built heritage: Potential and application

The results of thematic analyses collected with the methods described in the previous paragraph facilitate the knowledge of the heritage for conservation but also become functional to represent its state in a digital environment. This process favors the definition of complex BIM models that reach high levels of definition and that can facilitate practitioners in the identification of decision-making solutions involved in built heritage conservation processes.

13.3.1 Modeling of stratigraphic masonry units and management in HBIM

The data collected from archaeological analyses applied to historical masonry allows for the collection of a great deal of information that needs to be systemized, and experiments in the management of this type of data in a parametric environment are the focus of the research of a number of groups (Banfi, Roascio, et al., 2022; Diara, 2022; Trizio & Savini, 2020). This analysis provides evidence of the uniqueness of each historical artifact and the importance of semantic modeling starting from an ontological separation, not only into the structural and formal parts that make up a building, but also into the nonstructural parts that are the result of voluntary interventions made on the masonry over time. Starting from the results of the stratigraphic research, after a process of critical survey and archaeological interpretation that allows the SUMs to be classified into periods of activity, the three-dimensional modeling phase of the building can be structured in relation to the construction and maintenance phases. From the point of view of the geometric representation of the masonry entity, the digitization procedures envisage considering the real characteristics of each unit, renouncing the approximation of the data downstream of the critical reading of the form. In this way, geometries are defined with a high Level of Geometry (LOG) development consistent with those defined at a regulatory level for new constructions. This is achieved through the identification of margins, with the auxiliary support of the point cloud and orthomosaics edited with the critical survey, which are highly reflective of reality to ensure the correct adherence of neighboring masonry in the virtual model. This modeling approach was adopted to generate entities directly in the parametric environment from a generic "wall" type by modifying its structure in identity, materials, and appearance by entering specific parameters for each wall type to be represented. Individual units were then created using the "wall" family—chosen according to the type defined by the analyses—which was modified according to the profiles and limits of the different SUMs identified (Fig. 13.3).

This modeling procedure can also be carried out using nonparametric software and importing the georeferenced entities with respect to a local system in relation to the building and respecting proximity relationships between the SUMs. An example is the import of solids modeled with Sketchup software, validated in metric form and without problems of constraints between digital entities (Fig. 13.4).

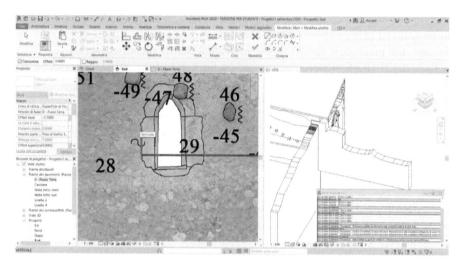

Figure 13.3 Modeling procedure in Revit software.

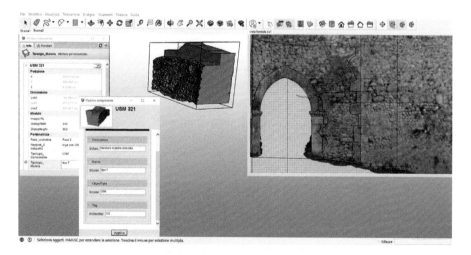

Figure 13.4 Modeling procedure in Sketchup software.

The complete model divided into its SUMs was imported as a "mass" into the parametric environment, placing it exactly in the virtual space of the building and preserving its geometric features. Each imported "mass" was subsequently transformed into a parametric element, changing the element to "wall" with the "wall from surface" command to enrich the entity not only with quantitative but also qualitative information. Again, as in modeling with the native tools, entities were chosen on the basis of characteristics, with the previously modified walls being selected for each unit.

Similarly, a parametric model representative of the geometric-formal and constructive peculiarities of the artifact from the results of the archaeological analysis can be realized using software for 3D modeling of sculpted surfaces (freeform). In this case, the parametric model of the individual SUMs identified is realized starting from the NURBS model implemented on the basis of the critical analysis of the profiles returned by the point clouds acquired with the 3D survey and extrapolated through different software such as Cloud Compare and subsequently processed in a digital environment on the basis of the stratigraphic analysis of the masonry. This type of modeling returns individual closed polysurfaces that are imported, thanks to the interoperability of parametric software with 3D freeform surface modeling software, into the BIM environment via the ACIS SAT format (Fig. 13.5).

In the element import phase, specific families are also defined, consistent with those within the parametric modeling software (Revit), to which the entities belong. Another fundamental aspect is aimed at defining not only the geometric parameters for all features of the architectural element, but also the descriptive ones, through the topological connection to the three-dimensional graphic representation of the individual SUMs of the data. To this end, thematic abacuses, and apposite schedules, had to be defined and implemented and related to each other with the

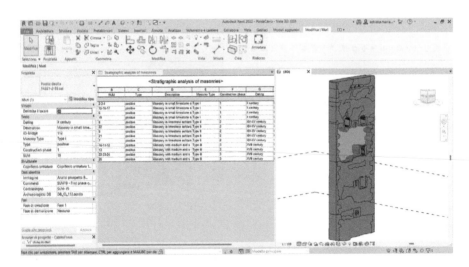

Figure 13.5 Importing the NURBS model into the parametric software. *NURBS*, Nonuniform rational B-spline.

addition of data from the archaeological analysis. The digitization of geometric and formal information was therefore augmented with the creation of specific parameters that made it possible to achieve an excellent Level of Information (LoI) for each particular element and the complex as a whole.

Thus the separate elements of the parametric model are enriched with the information collected during the analysis phase; this is possible through the use of "type parameters" that allow both to associate the elements with relational databases and to enrich the categories with the archaeological and architectural information that is collected. Specifically, three main parameters have been defined that are connected to each other, the one identifying the SUM determined through the stratigraphic analysis of the masonry, the one relating to the type of US (type), which depends on whether the elements belong to a single construction phase or show evident signs of voluntary anthropic transformations, and finally, the parameter that describes in detail the materials and construction techniques (description). From these, three further parameters were designed to be updated as new information becomes available. These parameters make it possible to associate general and detailed data with the separate entities and to represent the elements according to the type of masonry (masonry type), the construction sequences of the individual SUMs (construction phase), and the chronological arc to which the SUMs belong (dating).

13.3.2 Parametric modeling of decay forms potential and application cases

Similarly, the results of the noninvasive diagnostics were represented within the HBIM model by recurring to the creation of appropriate parameters related to the families representing the forms of degradation and damage. We propose the modeling of respective defects that may afflict the built heritage, standardized by type (vegetation, crust, erosion, and so on), through the use of "adaptive families." In this way, each pathology recorded with the diagnostic analysis was modeled by following its contours and identifying 27 adaptive points that approximate its surface (Fig. 13.6).

Finally, for each defect, shared parameters were defined, both identifying and describing pathologies, and hyperlinks were created that relate the element to a database on the cloud, thus allowing consultation of the deterioration and alteration analytical sheet directly in the parametric environment (Fig. 13.7). The model's interoperability with databases and other existing systems is ensured by the creation of specific parameters structured according to the rules of unique relationships using primary and secondary keys. In this way, it is possible to collect and manage multidata acquired over time and update them with new instances.

A different modality was put in place for the representation of problems related to certain types of artifacts, such as debris carried by the current that clogs the arches of bridges, leaning against the piers and compromising their preservation and safety. For this purpose, the solid model obtained from a specially

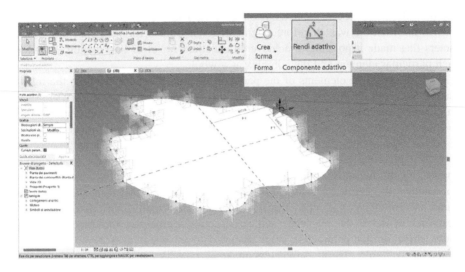

Figure 13.6 Parametric modeling of decay shapes.

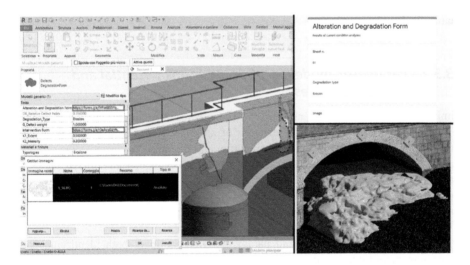

Figure 13.7 Conservation condition analysis and data management in BIM. *BIM*, Building Information Modeling.

postprocessed photogrammetric survey was imported into the parametric environment as a local entity. Again, it was possible to link the information in the form of a link to the data sheet. For this particular type of historic building, descriptive instance parameters were defined for the pathologies of degradation (defect), the area, and the coefficients of its extension and intensity (K1-Extent and K2-Intensity) in accordance with the Italian guidelines for the safety of existing bridges (MIT, 2020). The latter parameters are linked within the project to the defect weight parameter (G-Defect weight), which allows the Relative and Absolute Defect Index (DR—Relative Defect Index and DA—Absolute Defect Index) of the aforementioned standard to be evaluated. In addition, to fully exploit the potential of the system and facilitate operational procedures, hyperlinks have been set up with the interventions to be implemented. These are accessed through specific labels, modeled in BIM, in which information identifying the type of intervention to be carried out and those relating to the actions to be implemented have been parameterized.

Some practical applications to artifacts located in the L'Aquila area have made it possible to validate the procedure for the representation in a parametric environment of the different information acquired in the knowledge and analysis process to support the identification of strategies to be implemented for the restoration and conservation of historical artifacts (Fig. 13.8). The abovementioned discussion illustrates the potential of detailed modeling of diagnostic results as the use of HBIM has made it possible to identify methodological approaches capable of guiding professionals in their design choices by analyzing the characteristics of the structure as a whole before intervening.

Exploiting the conceptualization capacity of BIM, which allows the georeferencing of nongraphical information (Acierno et al., 2017; Fiorani, 2017) by correlating it with the individual parts that make up the model itself, it has been possible to obtain a model with great precision in its representation from a geometric point of view, which guarantees the management of data by many actors involved in the various phases (diagnostic analysis, historical-evolutionary analysis, and intervention design). The model's potential is not only limited to the design phase but will be further appreciated in those of the actual restoration site with economic analysis and cost calculation, as demonstrated by similar experiments in the reconstruction

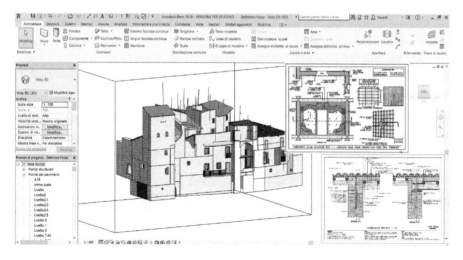

Figure 13.8 HBIM model of the Castle of Fossa for the management of the building restoration project.

of monuments in L'Aquila city (Brumana et al., 2017; Brumana, Della Torre, Oreni, et al., 2018). Similarly, the model will be functional for the management of the restored asset, monitoring the state of the structures and energy efficiency.

The undoubtedly undisputed potential of the tool in the joint operations of planning concrete actions for restoration intervention is not the only one that characterizes the tool developed from the illustrated procedure. In this regard, the use of digital facilitates predictive analysis also for the evaluation of the results of restoration work using virtual restoration techniques.

To create a functional BIM model for virtual restoration and decision-making processes, it is possible to adopt the procedures and workflow described here, taking care to define not only different project phases but also project variables that can help to identify the most suitable solution with respect to the objectives of the intervention. The design alternatives are defined starting from the parametric model of the actual state, created from the critical analysis of the acquired data, and enriched with libraries of parametric families developed to correctly represent the peculiarities of the artifacts. Within this model, the criticalities detected during the various multidisciplinary analyses must be represented, thus correctly representing the pathologies through the adaptive family already mentioned, but also the solutions to be implemented to solve the problems identified. This information can be contained in specific labels that contain the most salient information gathered from the knowledge pathway, illustrating not only the type of intervention to be implemented according to the support but also the individual steps related to the specific actions to be implemented and always guaranteeing a link to descriptive sheets of the intervention. The design options inserted in the parametric environment must be defined considering the construction features of the building and the territorial context in which it is located to identify the solution that best meets the social and cultural needs of the area. The model is then integrated with new parametric families, which remove certain forms of decay identified during the analysis phase or which constitute furnishing elements through which the forms of past eras can be reproposed, and with new elements which allow the integration of new stratigraphies of the structural elements downstream of the consolidation and restoration work. To realize a true virtual restoration and verify the different choices of materials made, it is necessary to insert high-resolution photo plans, obtained by meticulous editing of those representative of the actual state, to simulate the cleaning, restoration, and integration of the stone and plaster surfaces.

The experiments carried out in this field confirm that the HBIM model, representative of both the current conditions and those possible following restoration work on an artifact (Fig. 13.9), represents a useful tool for sharing information and design proposals with a view to the transdisciplinary nature of the processes linked to the built heritage.

13.3.3   The modeling of the monitoring system and the correlation with data

The HBIM models are effective tools within the context of built heritage management, not only because they ensure the gathering of heterogeneous information concerning the morphological and dimensional features of the artifact but also because

Figure 13.9 Restoration design hypotheses managed in a virtual environment.

they can be related to different monitoring systems that allow for intervention in the structures at the right time with proper actions that guarantee their protection and safety with the minimum impact. Therefore it is necessary to relate the digital replicas to monitoring systems that control the state of health of the artifact and the surrounding environment conditions in real time, ensuring the gathering and analysis of data collected by a sensor network within the digital environment, thus pushing toward the creation of historical digital twin (HDT) (Marra, Trizio, Fabbrocino, 2021, 2021). The HDT is conceived as a digital replica of the main qualitative, dimensional, and structural features of a physical object that can improve the management and maintenance process, as well as the ones of conservation and valorization of the built heritage, thanks to the one-to-one connection with other data acquisition and management platforms. Indeed, the relation between the digital and physical system and the sensor networks allows identifying accurately the element that could be damaged and the residual time before the failure occurs. However, it is necessary to ensure this correlation with the resorting of Internet of Things technologies and complex algorithms based on machine and deep learning. Although further experiments and efforts are required to obtain the direct link between the dynamic data and the digital replica of the artifact, as well as to create an automatic relationship between the architectural model and the ones developed for specific analysis or other monitoring systems, the procedure previously illustrated has provided for the correlation between the HBIM model and the "static" data acquired by an SHM.

In particular, a virtual sensor able to represent the accelerometers commonly used in dynamic monitoring was modeled to make information available in the parametric environment. The virtual sensor replicates in its morphology a force-balance accelerometer available in commerce; it has been characterized by a group of shared parameters (*specification* and *geometry*) that enable the sharing of technical and geometric features of the element used in the monitoring system. A further group of parameters (*sensor output data*) contains the ID of sensors, the main data recorded, and, finally, the access to a cloud that collects data acquired from single sensors, so as to be able to use information also in a parametric environment and to compare them with those resulting from the other multidisciplinary analyses performed on the artifact (Fig. 13.10).

Finally, to test and verify the effectiveness of the virtual sensor in the design of monitoring systems for buildings of public property located in areas exposed to seismic risk, a label has been added to define the directions according to the sensor location for the acquisition of natural frequencies and vibration modes of the structure. The localization of the sensors in the BIM environment facilitates the definition of the monitoring system layout and the assessment of the interferences between the existing architectural and plant components and the units of acquisition and processing.

Although the virtual sensor and the filled parameters are similar to those of a force-balance accelerometer, it should be noted that the model and the parameters themselves can easily adapt to any type of data acquisition system used in the monitoring field.

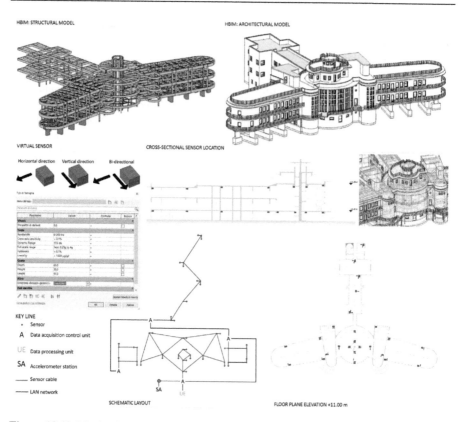

Figure 13.10 Monitoring system designed and managed in BIM environment. *BIM*, Building Information Modeling.

13.4 Discussion and final remarks

The last 30 years have been characterized by a strong increase in the use of "new technologies" in various spheres of everyday life: from primary services for citizens (identifiable with e-government actions) to those involving the sphere of leisure. This process has been fostered by the considerable development of software and hardware that has given a broad target group of users access to increasingly high-performance and user-friendly devices and programs, that is, intuitive and easy to use, but also by the growing awareness of the benefits of going digital. The built heritage has been fully affected by this process and, in fact, protection processes involving administrators (such as superintendencies, municipalities, and provinces) and professionals (such as engineers, architects, and restorers) have been extensively renewed by resorting to new technologies for in-depth knowledge of the asset in the diagnostic phase. This has oriented the work of the research group toward the development of a new paradigm, that of e-conservation, which includes the development of digital tools and best practices that can facilitate and support public administrations in the management, maintenance, and conservation phases of the existing heritage.

In association with the now-established tools for non-destructive diagnostics, predictive systems based on digital techniques have been developed in accordance with technological progress. Starting from the mapping of deterioration and the state of conservation of assets on digital models structured according to semantic bases and resorting to catalographic approaches (such as the ICCD ones and developed according to systemic architectures with an ontological basis), to arrive at complex information systems capable of correlating the information derived from structural monitoring (known by the acronym SHM) with virtual replicas for the definition of real digital twins.

This scenario opens up new issues related to the different digital literacy of a multitarget audience such as public administrations both in large cities and small towns in inland areas. Therefore research must necessarily contribute to the development of digital procedures by aligning the issue of cultural heritage conservation, understood in the broadest sense, with e-governance techniques, but also question the methods to be implemented to support the bodies in charge of the management and protection of the built heritage toward this transition.

References

Abbate, E., Invernizzi, S., & Spanò, A. (2020). HBIM parametric modelling from clouds to perform structural analyses based on finite elements: A case study on a parabolic concrete vault. *Applied Geomatics, 14*, 1–18.

Acierno, M., Cursi, S., Simeone, D., & Fiorani, D. (2017). Architectural heritage knowledge modelling: An ontology-based framework for conservation process. *Journal of Cultural Heritage, 24*, 124–133. Available from https://doi.org/10.1016/j.culher.2016.09.010, http://www.elsevier.com.

Al-Bayari, O., & Shatnawi, N. (2022). Geomatics techniques and building information model for historical buildings conservation and restoration. *Egyptian Journal of Remote Sensing and Space Science, 25*(2), 563–568. Available from https://doi.org/10.1016/j.ejrs.2022.04.002, http://www.elsevier.com/wps/find/journaldescription.cws_home/723780/description#description.

Allegra, V., Di Paola, F., Lo Brutto, M., & Vinci, C. (2020). Scan-to-BIM for the management of heritage buildings: The case study of the Castle of Maredolce (Palermo, Italy). *The International Archives of the Photogrammetry, Remote Sensing and Spatial Information Sciences, 2020*(2), 1355–1362. Available from https://doi.org/10.5194/isprs-archives-xliii-b2-2020-1355-2020, **XLIII-B2-**.

Azkarate Garai-Olaun, A., Caballero Zoreda, L., & Quirós Castillo, J. A. (2002). Arqueología de la Arquitectura: definición disciplinar y nuevas perspectivas. *Arqueología De La Arquitectura, 1*, 7–10.

Azkarate Garai-Olaun, A. (2010). Archeologia dell'Architettura in spagna. *Archeologia dell'architettura*, 17–28, **XV**.

Banfi, F., Brumana, R., Landi, A. G., Previtali, M., Roncoroni, F., & Stanga, C. (2022). Building archaeology informative modelling turned into 3D volume stratigraphy and extended reality time-lapse communication. *Virtual Archaeology Review*, *13*(26), 1−21. Available from https://doi.org/10.4995/VAR.2022.15313, http://polipapers.upv.es/index. php/var/index.

Banfi, F., Roascio, S., Paolillo, F. R., Previtali, M., Roncoroni, F., & Stanga, C. (2022). Diachronic and synchronic analysis for knowledge creation: Architectural representation geared to XR building archaeology (Claudius-Anio Novus Aqueduct in Tor Fiscale, the Appia Antica Archaeological Park). *Energies*, *15*(13). Available from https://doi.org/10.3390/en15134598, https://www.mdpi.com/1996-1073/15/13/4598/pdf?version = 1655986892.

Bonora, A., Fabbri, K., & Pretelli, M. (2020). Widespread difficulties and applications in the monitoring of historical buildings: The case of the realm of venaria reale. *Heritage*, *3* (1), 128−139. Available from https://doi.org/10.3390/heritage3010008, https://www. mdpi.com/2571-9408/3/1/8/pdf.

Brogiolo, G. P. (2007). *Dall'archeologia dell'architettura all'archeologia della complessità*. 38.

Brogiolo, G. P. (1997). *Dall'analisi stratigrafica degli elevati all'Archeologia dell'Architettura*. *Archeologia dell'architettura*. II.

Brogiolo, G. P., & Cagnana, A. (2012). Archeologia dell'Architettura. Metodi e Interpretazioni. *All'insegna del Giglio*.

Brumana, R., Della Torre, S., Oreni, D., Cantini, L., Previtali, M., Barazzetti, L., & Banfi, F. (2018). *Lecture notes in computer science (including subseries lecture notes in artificial intelligence and lecture notes in bioinformatics). SCAN to HBIM-post earthquake preservation: informative model as sentinel at the crossroads of present, past, and future*. Italy: Springer Verlag. Available from https://www.springer.com/series/558. 11196. https://doi.org/10.1007/978-3-030-01762-0_4, 16113349 39-51.

Brumana, R., Della Torre, S., Oreni, D., Previtali, M., Cantini, L., Barazzetti, L., Franchi, A., & Banfi, F. (2017). *International Archives of the Photogrammetry, Remote Sensing and Spatial Information Sciences - ISPRS Archives, . HBIM challenge among the paradigm of complexity, tools and preservation: The basilica Di collemaggio 8 years after the earthquake (L'aquila)* (42, pp. 97−104). Italy: International Society for Photogrammetry and Remote Sensing, 10.5194/isprs-archives-XLII-2-W5-97-2017 168217502. Available from http://www.isprs.org/proceedings/XXXVIII/4-W15/.

Brumana, R., Della Torre, S., Previtali, M., Barazzetti, L., Cantini, L., Oreni, D., & Banfi, F. (2018). Generative HBIM modelling to embody complexity (LOD, LOG, LOA, LOI): surveying, preservation, site intervention—the Basilica di Collemaggio (L'Aquila). *Applied Geomatics*, *10*, 545−567. Available from https://doi.org/10.1007/s12518-018-0233-3.

Brumana, R., Stanga, C., Previtali, M., Landi, A. G., & Banfi, F. (2021). *Lecture notes in computer science (including subseries lecture notes in artificial intelligence and lecture notes in bioinformatics), Wrap-up synthesis model from high-quality HBIM complex models, and specifications, to assess built cultural heritage in fragile territories (Arquata Del Tronto, Earthquake 2016, the Church of St. Francesco, IT)* (12642, pp. 100−111). Italy: Springer Science and Business Media Deutschland GmbH. Available from https://doi.org/10.1007/978-3-030-73043-7_9, https://www.springer.com/ series/558.

Bruno, S., De Fino, M., & Fatiguso, F. (2018). Historic Building Information Modelling: performance assessment for diagnosis-aided information modelling and management. *Automation in Construction*, *86*, 256−276. Available from https://doi.org/10.1016/j.autcon.2017.11.009.

Bruno, S., & Fatiguso, F. (2018). Building conditions assessment of built heritage in historic building information modeling. *International Journal of Sustainable Development and Planning*, *13*(1), 36−48. Available from https://doi.org/10.2495/SDP-V13-N1-36-48, http://www.witpress.com/journals/sdp.

BuildingSMART. (2022). *Industry Foundation Classes* (IFC). https://www.buildingsmart.org/standards/bsi-standards/industry-foundation-classes/.

Continenza, R., Trizio, I., Giannangeli, A., & Tata, A. (2016). HBIM for restoration projects: Case-study on San Cipriano Church in Castelvecchio Calvisio, Province of L'aquila, Italy. *Disegnarecon*, *9*(16), 15.1-15.9. Available from http://disegnarecon.univaq.it/ojs/index.php/disegnarecon/article/view/163/133.

Croce, P., Landi, F., Puccini, B., Martino, M., & Maneo, A. (2022). Parametric HBIM procedure for the structural evaluation of heritage masonry buildings. *Buildings*, *12*(2), 194. Available from https://doi.org/10.3390/buildings12020194.

Currà, E., D'Amico, A., & Angelosanti, M. (2022). HBIM between antiquity and industrial archaeology: Former Segrè Papermill and Sanctuary of Hercules in Tivoli. *Sustainability*, *14*(3), 1329. Available from https://doi.org/10.3390/su14031329.

Diara, F. (2022). HBIM open source: A review. *ISPRS International Journal of Geo-Information*, *11*(9), 472. Available from https://doi.org/10.3390/ijgi11090472.

Diara, F., & Rinaudo, F. (2021). ARK-BIM: Open-source cloud-based HBIM platform for archaeology. *Applied Sciences*, *11*(18), 8770. Available from https://doi.org/10.3390/app11188770.

Doglioni, F. (2002). Ruolo e salvaguardia delle evidenze stratigrafiche nel progetto e nel cantiere di restauro. *Arqueología de la Arquitectura*, *1*, 113. Available from https://doi.org/10.3989/arq.arqt.2002.10.

Fico, R., Genitti, C., Frezzini, F., Marra, A., Sabino, A., & Martino, G. (2019). PdR adottati nei Comuni del Cratere. *Modalità di approccio alla riqualificazione del costruito*, *154*.

Fiorani, D. (2017). *La modellazione della conoscenza nel restauro: uno sviluppo per il BHIMM Built Heritage Information Modelling/Management BHIMM*. Galazzano: Edizioni IMRead.

Francovich, R. (1988). Archeologia e restauro dei monumenti. *All'Insegna del Giglio*.

Gerbino, S., Cieri, L., Rainieri, C., & Fabbrocino, G. (2021). On BIM interoperability via the IFC standard: An assessment from the structural engineering and design viewpoint. *Applied Sciences*, *11*(23), 11430. Available from https://doi.org/10.3390/app112311430.

Gigliarelli, E., Cangi, G., & Cessari, L. (2022). Rilievo per la modellazione e la gestione informativa HBIM. *Approccio multicomponente per l'analisi strutturale e il restauro di edifici storici. Archeologia e Calcolatori*, *33*, 135–155.

ICOMOS. (2003). *Principles for the Analysis, conservation and structural restoration of architectural heritage*. Unpublished content. https://www.icomos.org/images/DOCUMENTS/Charters/structures_e.pdf.

ISO 13822. (2010) ISO 13822 ISO TC98/SC2 *Geneva bases for design of structures – Assessment of existing structures*.

Karoglou, M., Delegou, E. T., Labropoulos, K., & Moropoulou, A. (2019). *NDT investigation of ISIA building in Florence Springer Proceedings in Materials* (pp. 133–144). Greece: Springer Nature. Available from http://www.springer.com/series/16157, http://doi.org/10.1007/978-3-030-25763-7_10.

Khan, M. S., Khan, M., Bughio, M., Talpur, B. D., Kim, I. S., & Seo, J. (2022). An integrated HBIM framework for the management of heritage buildings. *Buildings*, *12*(7). Available from https://doi.org/10.3390/buildings12070964, http://www.mdpi.com/journal/buildings.

Liberotti, R., & Gusella, V. (2023). Parametric modeling and heritage: A design process sustainable for restoration. *Sustainability*, *15*(2), 1371. Available from https://doi.org/10.3390/su15021371.

Mannoni, T., & Boato, A. (2002). Archeologia e storia del cantiere di costruzione. *Arqueología de la Arquitectura*, *1*, 39. Available from https://doi.org/10.3989/arq.arqt.2002.5.

Marra, A., Sabino, A., Fico, R., Pecci, D., Gualtieri, R., Mannella, A., & Fabbrocino, G. (2019). *Atti del XVIII Convegno ANIDIS L'Ingegneria Sismica in Italia, . Unpublished content Conservation of historical minor centers hit by L'Aquila earthquake in the context of the reconstruction process* (13, pp. 160−170). Pisa University press.

Marra, A., Trizio, I., Fabbrocino, G., & Savini, F. (2021). Digital models for e-conservation: The HBRIM of a bridge along the Aterno River. *SCIRES-IT, 11*(2), 83−96. Available from https://doi.org/10.2423/i22394303v11n2p83, http://www.sciresit.it/.

Marra, A., Trizio, I., & Fabbrocino, G. (2021). *Lecture notes in civil engineering, . Digital tools for the knowledge and safeguard of historical heritage* (156, pp. 645−662). Italy: Springer Science and Business Media Deutschland GmbH. Available from http://www.springer.com/series/15087, https://doi.org/10.1007/978-3-030-74258-4_41.

Massafra, A., Predari, G., & Gulli, R. (2022). *International Archives of the Photogrammetry, Remote Sensing and Spatial Information Sciences - ISPRS Archives, . Towards digital twin driven cultural heritage management: A HBIM-based workflow for energy improvement of modern buildings* (46, pp. 149−157). Italy: International Society for Photogrammetry and Remote Sensing. Available from http://www.isprs.org/proceedings/ XXXVIII/4-W15/, https://doi.org/10.5194/isprs-archives-XLVI-5-W1-2022-149-2022.

Meoni, A., Vittori, F., Piselli, C., D'Alessandro, A., Pisello, A. L., & Ubertini, F. (2022). Integration of structural performance and human-centric comfort monitoring in historical building information modeling. *Automation in Construction, 138,* 104220. Available from https://doi.org/10.1016/j.autcon.2022.104220.

Mesquita, E., Arêde, A., Silva, R., Rocha, P., Gomes, A., Pinto, N., Antunes, P., & Varum, H. (2017). Structural health monitoring of the retrofitting process, characterization and reliability analysis of a masonry heritage construction. *Journal of Civil Structural Health Monitoring, 7*(3), 405−428. Available from https://doi.org/10.1007/s13349-017-0232-9, http://rd.springer.com/journal/volumesAndIssues/13349.

MIT. (2020). *Unpublished content Guidelines on risk classification and man-agement, safety assessment and monitoring of exist-ing bridges. Decree of Ministry of Infrastructure. Italian Ministry of Infrastructures and Trans-portations.* High Council of Public Works. Available from https://www.mit.gov.it/comunicazione/news/mit-approvate-le-linee-guida-per-la-sicurezza-dei-ponti.

Moropoulou, A., Avdelidis, N., Karoglou, M., Delegou, E., Alexakis, E., & Keramidas, V. (2018). Multispectral applications of infrared thermography in the diagnosis and protection of built cultural heritage. *Applied Sciences, 8*(2), 284. Available from https://doi.org/10.3390/app8020284.

Moro, L., & Neri, A. (2010). *Linee guida per la valutazione e riduzione del rischio sismico del patrimonio culturale: allineamento alle nuove Norme tecniche per le costruzioni: circolare* 2 dicembre 2010 *Ministero per i beni e le attività culturali* (26). Gangemi, Roma: Segretariato generale.

Moyano, J., Carreño, E., Nieto-Julián, J. E., Gil-Arizón, I., & Bruno, S. (2022). Systematic approach to generate Historical Building Information Modelling (HBIM) in architectural restoration project. *Automation in Construction, 143,* 104551. Available from https://doi.org/10.1016/j.autcon.2022.104551.

NTC. (2018). *Updating of technical standards for construction.* G.U. 42. Ministero per i Beni e le Attività Culturali. Segretariato Generale, Roma

Oostwegel, L. J. N., Jaud, Š., Muhič, S., & Malovrh Rebec, K. (2022). Digitalization of culturally significant buildings: ensuring high-quality data exchanges in the heritage domain using OpenBIM. *Heritage Science, 10*(1). Available from https://doi.org/10.1186/s40494-021-00640-y, http://www.springer.com/materials/journal/40494.

Oreni, D., Brumana, R., Della Torre, S., & Banfi, F. (2017). Survey, HBIM and conservation plan of a monumental building damaged by earthquake. *The International Archives of the Photogrammetry, Remote Sensing and Spatial Information Sciences* (5), 337−342. Available from https://doi.org/10.5194/isprs-archives-xlii-5-w1-337-2017, XLII-5/W1.

Pepe, M., Costantino, D., & Restuccia Garofalo, A. (2020). An efficient pipeline to obtain 3D model for HBIM and structural analysis purposes from 3D point clouds. *Applied Sciences, 10*(4), 1235. Available from https://doi.org/10.3390/app10041235.

Rainieri, C., Gargaro, D., & Fabbrocino, G. (2019). *Hardware and software solutions for seismic SHM of hospitals. Springer Tracts in Civil Engineering* (pp. 279−300). Italy: Springer. Available from http://springer.com/series/15088, http://doi.org/10.1007/978-3-030-13976-6_12.

Rocha, P., Mateus, L., Fernández, J., & Ferreira, V. (2020). A Scan-to-BIM methodology applied to heritage buildings. *Heritage, 3*(1), 47−67. Available from https://doi.org/10.3390/heritage3010004.

La Russa, F. M., Galizia, M., & Santagati, C. (2021). Remote sensing and city information modeling for revealing the complexity of historical centers. *The International Archives of the Photogrammetry, Remote Sensing and Spatial Information Sciences*, 367−374. Available from https://doi.org/10.5194/isprs-archives-xlvi-m-1-2021-367-2021, **XLVI-M-1-2021** (M-1-2021).

Saccucci, M., & Pelliccio, A. (2018). Integrated BIM-GIS system for the enhancement of urban heritage. IEEE International Conference on Metrology for Archaeology and Cultural Heritage, MetroArchaeo 2018−Proceedings, Institute of Electrical and Electronics Engineers Inc. Italy. (pp. 222−226). 9781538652763 http://ieeexplore.ieee.org/xpl/mostRecentIssue.jsp?punumber = 90818130.

Santini, S., Borghese, V., Micheli, M., & Orellana Paz, E. (2022). Sustainable recovery of architectural heritage: The experience of a worksite school in San Salvador. *Sustainability, 14*(2), 608. Available from https://doi.org/10.3390/su14020608.

De Stefano, A., Matta, E., & Clemente, P. (2016). Structural health monitoring of historical heritage in Italy: some relevant experiences. *Journal of Civil Structural Health Monitoring, 6*(1), 83−106. Available from https://doi.org/10.1007/s13349-016-0154-y, http://rd.springer.com/journal/volumesAndIssues/13349.

Tan, J., Leng, J., Zeng, X., Feng, D., & Yu, P. (2022). Digital twin for Xiegong's architectural archaeological research: A case study of Xuanluo Hall, Sichuan, China. *Buildings, 12*(7), 1053. Available from https://doi.org/10.3390/buildings12071053.

Della Torre, S. (2017). *Un bilancio del progetto BHIMM Modellazione e gestione delle informazioni per il patrimonio edilizio esistente* (pp. 10−16).

Trizio, I., Brusaporci, A., Tata, A., & Ruggieri. (2021). Advanced digital technologies for built heritage survey and historical analysis. *Tirant Humanidades: Valencia*, 621−635.

Trizio, I., & Savini, F. (2020). Archaeology of buildings and HBIM methodology: Integrated tools for documentation and knowledge management of architectural heritage. *IMEKO TC-4 International Conference on Metrology for Archaeology and Cultural Heritage 84−89. International Measurement Confederation (IMEKO), Italy.*

Trizio, I., Savini, F., Giannangeli, A., Boccabella, R., & Petrucci, G. (2019). The archaeological analysis of masonry for the restoration project in HBIM. *The International Archives of the Photogrammetry, Remote Sensing and Spatial Information Sciences* (2), 715−722. Available from https://doi.org/10.5194/isprs-archives-xlii-2-w9-715-2019, **XLII-2/W9.**

Zouaoui, M. A., Djebri, B., & Capsoni, A. (2022). From point cloud to HBIM to FEA, the case of a Vernacular architecture: Aggregate of the Kasbah of Algiers. *Journal on Computing and Cultural Heritage, 14.*

Automating built heritage inspection using unmanned aerial systems: A defect management framework based on heritage building information modeling (HBIM)

14

Botao Li[1], Tarek Rakha[1], Russell Gentry[1], Danielle S. Willkens[1], Junshan Liu[1,2], and Javier Irizarry[1]
[1]Georgia Institute of Technology, Atlanta, GA, United States, [2]Auburn University, Auburn, AL, United States

14.1 Introduction

14.1.1 Background

Within historic structures, building defects and deterioration become common and inevitable due to natural disasters, human interventions, and other environmental factors. To optimize the preservation of built heritage's architectural, historical, cultural, social, political, and material value, careful conservation requires strategies that are minimally invasive or disruptive (Zhang & Dong, 2021), as well as especially when collecting an exhaustive understanding of the building's health. The decisions related to built heritage preservation are complicated, usually because of limited antecedent documentation and the upgrading and expanding of criteria. Tejedor et al. (2022) have emphasized that all restoration-related actions for built heritage should have an in-depth knowledge of its diagnosis. Detecting building defects on heritage buildings is critical for damage diagnosis and prognosis. Without a clear picture of a building's condition, structurally and aesthetically, it is extremely difficult to apply the proper historic preservation treatment. In addition to conforming to building codes, historic buildings can also be subject to local, national, or international preservation standards. Therefore consistent and effective monitoring of building defects is essential to ensuring the building health and sustainability of historic structures. Traditional building inspection depends on human labor, which is task-intensive and time-consuming. As one of the widespread non-destructive testing (NDT) tools, unmanned aerial systems (UAS, a.k.a drones) with great mobility and capability of different sensing gears are a promising alternative

Diagnosis of Heritage Buildings by Non-Destructive Techniques. DOI: https://doi.org/10.1016/B978-0-443-16001-1.00014-0

for carrying out building inspections effectively (Masri & Rakha, 2020). Li et al. (2022) have classified four common types of NDT UAS based on their detection ability and temporal conditions matched for different building defect types. UAS path planning is a well-developed field where current applications allow users to select target areas and overlapping rates. Then, the software can automatically generate flight paths to cover targeted areas, also named drone mapping tasks. Traditionally, the mapping tasks for buildings are either from the top view planar with downward and oblique photo capture or from the elevation view's vertical planar with a camera perpendicular to the elevation. More advanced controller systems accept GPS coordinate information as input and compute UAS paths based on predefined interest spots. This technique demonstrates the vast advantage of using UAS to capture the whole building's geometric profile and the considerable potential of automating building defects' inspection.

In the past decades, Building Information Modeling (BIM) has demonstrated massive benefits in resource-saving and collaboration between stakeholders, which became a successful adoption in the Architecture, Engineering, Construction, and Operation (AECO) industries for the whole life cycle management of new buildings. With excellence at information management, BIM methodology has brought interest to existing buildings as digital as-built documentation (Volk et al., 2014). Creating the digital twin of the existing building allows easier access and management to involved stakeholders and real-time reflection without on-site requirements. Using the semantic BIM model as a long-term building defect monitoring method can support intelligent data storage, timely maintenance inspection, and further interaction and analysis among stakeholders. The process that specifies the conversion of physical historic buildings into digital 3D semantic-enriched information models is typically described as Heritage BIM or Historic BIM (HBIM). Differing from traditional BIM starting from the design phase and building components' level of detail (LoD) increasing with progress, HBIM begins with a building survey that leans on reality capture technology such as laser scanning and image photogrammetry. By collecting necessary geometric and nongeometric information with reality capture tools, the BIM modeler can construct a BIM model of the historic building. The LoD relies on the accuracy of captured data and the agreements of the project goal.

14.1.2 Problem statement

The discrepancy between BIM and HBIM and the complex nature of the historic building have raised barriers to automate the scan-to-BIM process and data repository issues. The Industry Foundation Classes (IFC) are data standards currently utilized in the open BIM environment; however, it requires a fitting extension to represent building defects and condition assessment data semantically. Diara and Rinaudo (2020) state that the classification of IFC is critical for informative systems as the current standard is specific to the modern AECO industry, and concerns should be expanded to heritage assets as well. The appropriate relationship between building elements and building defects is essential to develop data schema as an

IFC extension, which relies on an in-depth discussion of the ontology and taxonomy of building defects. In addition, the scope of building defects varies greatly, from minor crack-gap or spot spalling to wall tilting or floor deformation, which also impact the information represented in the BIM environment.

By utilizing the autonomous nature of UAS, drones can automatically execute building defect−orientated flights of collecting historic building health information and keeping tracking and updating data in the central HBIM model. To consistently monitor historic building health, a proper data repository and building defect classification method in a BIM environment is a prerequisite to ensure long-term monitoring and timely maintenance inspection. The collected information should be traceable, allowing data inquiry to track building health progress. Great at dynamic information management, the HBIM approach is the closest to fit building defects management that can associate with semantic building components and assist later intervention of decision-making. The recursive procedures of monitoring building defects by UAS and consistently updating information in the BIM model can be concluded as this research proposed framework and specified use case. Automating the whole use case requires enhancement in BIM data schema, UAS flight planning, and data processing. Those three significant improvements can guarantee information flow consistently in the BIM platform, informing all stakeholders of historic buildings' health conditions timely. Having accurate and situational awareness of historic building health is crucial for preservation. It promises further preservation steps and demonstrates robots' potential to achieve digital twin application of historic buildings. Except for building defects and historic building health concerns, the automation potential can be expanded to more features such as energy efficiency analysis and virtual reality (VR) tours.

14.2   Literature review

14.2.1   HBIM process

BIM is widespread in AECO industries with many benefits and potentials, such as expediting the design process, supporting smooth information exchange, and assisting facilities management. The BIM process allows managing building assets information throughout the whole building life cycle. Building elements in the BIM environment are semantically defined as "smart objects" which can link numerical data, text, images, and so on, with predefined entities and relationships behind them. Nowadays, BIM has spanned its interests in cultural heritage as a digital documentation and management platform. Compared to the traditional BIM concentrating on new buildings, HBIM focuses on creating 3D intelligent models and information management for documenting, preserving, and restoring historically significant structures. The BIM model allows data to be analyzed for historic preservation after creating the complete 3D model with embedded information and attributes. Although the goal is promising, the reconstruction of the virtual historic building model is still challenging because the built heritage elements are usually heterogeneous, complex, and irregular, and their ontologies are hardly represented in current BIM libraries (López et al., 2018). The generation of a digital HBIM model can be concluded as three stages: data acquisition, data processing, and BIM model making (Liu, 2022). The initial data acquisition necessitates the most performant noninvasive reality capture methods and techniques, such as laser scanning and photogrammetry, to collect highly accurate building information with as-is conditions (Rocha et al., 2020). Conventional analog techniques, such as hand measuring, which relies on manual procedure and human labor participation, have become supplemental methods to reality capture for data acquisition because they require more time and economic input and have high risks when dealing with hard-to-access areas. The current data processing workflows mainly focus on geometric information, such as generating point clouds from captured data which can assist in digital reconstruction for the BIM modeler. However, more qualitative analyses are needed to support building health diagnoses and extract nongeometric information directly.

When implementing the HBIM methodology, not only do the traditional BIM stakeholders need to acquire new expertise, but the historian and the heritage facilities manager would also require extra training and knowledge to understand the HBIM models. The fundamental obstacles stem from the distinctions in conflicting aims between BIM and HBIM. BIM systems focus on building design from scratch, usually using a solo software platform, such as Autodesk Revit, with standardized object families for new buildings (Liu & Willkens, 2021). However, the HBIM methodology begins with the reality capture of existing buildings and typically involves the use of multiple software platforms (Rolin, 2019). Limited knowledge of ancient construction techniques and the complicated nature of historic buildings usually cause the BIM environment to need to create more semantic historical building elements and define appropriate entities' relationships in the data structure. It is a vital problem of the automatic scan-to-BIM process, making current digital modeling rely primarily on human labor assistance. Furthermore, integrating massive external information with the HBIM model could be challenging for a BIM modeler who is usually working on modern buildings.

Most of the published research focuses on the geometric modeling of HBIM, such as parameterizing historical building elements (Dore & Murphy, 2017) and modeling irregular shapes from the point cloud datasets (Barazzetti et al., 2015). The current scan-to-BIM process can reach a semiautomated level where a limited number of objects can be automatically extracted and modeled from point cloud data. Banfi (2019) has demonstrated a Grades of Generation algorithm that automatically generates vaulted systems in the BIM model based on point cloud data. Although BIM methodology emphasizes semantic data and information system management, very limited research concentrates on nongeometric information of HBIM modeling, where the database design should be discussed more (Bruno & Roncella, 2019). Besides, Barontini et al. (2021) summarized three significant difficulties that must be addressed for HBIM: a lack of standards, a lack of interoperability, and the intrinsic complexity of the data. Therefore those issues are prerequisites of the HBIM platform to further develop as a holistic information management system with high interoperability.

14.2.2 HBIM data schema

The three-dimensional HBIM model aims to manage various kinds of information and analysis derived from the scan-to-BIM process, requiring high interoperability that can be handled by different software and applications (Banfi, 2020). A proper data standard is an essence of securing smooth data exchange and information interoperability among involved stakeholders. Schlenger et al. (2022) have argued that a well-designed data schema is crucial for representing complex data in a transparent manner with high reliability when the aim is achieving a digital twin application that can reflect as-built information. The current data model in an open BIM environment, IFC, is an international data standard developed by buildingSMART International for planning, design, construction, and management (Sacks et al., 2018). It is mainly created for modern AECO industries and is insufficient for historic building condition assessment and appropriately storing building defect-related information as a long-term document solution. Historic buildings emphasize detecting building anomalies and tracking their potential damage progress before taking intervention actions. Defining an appropriate classification of building defects is essential for developing data schema solutions in BIM. To extend the information capability of the IFC, solutions can be made by changing the current IFC data schema or adding a custom IFC property set for condition assessment. Modifying the IFC schema requires approval by buildingSMART, and the modification should be widely accepted internationally as common data since end users cannot change IFC classes. Customizing property set that follows IFC specification is an alternative way to store heritage building defects data by linking property sets to current IFC classes, which the end user can make. Alavi et al. (2022) have demonstrated adding condition assessment as IFC property sets for building elements and building systems. However, their research starts from a functional view, mainly including malfunction defects on certain building elements. The limited malfunctioning building defects cannot accommodate the complicated nature of heritage building defects. In addition, heritage building defects can appear in any location of building elements. The taxonomy of building defects and its relationship with building elements will be presented with the data model and LoD representations in this research, served with the proposals of modifying IFC classes or adding property sets as data solutions to support long-term monitoring framework.

Traditionally, the building elements' LoD in the BIM model are increasing with building life cycle progress. When it comes to the operation phase, the LoD grows higher, making the BIM model reflect as-built conditions. As designs are refined and construction starts, the number of building elements with qualitative information in the BIM model increase. LoD 100 represents simple bounding box geometry in the initial design process; LoD 400 describes building elements at the fabrication-ready level; LoD 500 is in as-built condition after finishing construction. Using LOD to describe the richness of information of the BIM model is the standard for modern AECO industries and ensures interoperability among stake-holders. Nevertheless, HBIM encounters different approaches and situations that directly process the after-built stage of the existing building and the as-built infor-mation, which becomes a considerable difficulty for digital documentation to achieve the jump from LoD 000 to LoD 500. Since higher LoD modeling largely relies on time-consuming labor effort for BIM modelers, the ideal solution for the digital model making of heritage buildings should compromise the economic efforts and project requirement goal. Besides, building condition assessment requires fur-ther discussion and an advanced approach for the HBIM model to support building health condition visualization. Conventional condition assessment and anomalies information are documented in an excel sheet or as images to share with involved stakeholders, which has the potential risk of loss and lacks real-time update. For example, Barontini et al. (2021) utilized a patch-type illustration method to mark building elevation drawing where different color path with unique ID links anoma-lies information as a separate form. The conventional approach needs to be revised to support dynamic change information and be embedded in the BIM model to sup-port 3D data visualization further. In addition, since building defects can be tiny scope areas with multiple tiny defects, modeling a tiny scale level of anomalies with a vast number of defects in a BIM model is laborious. The conflict arose from the difficulty of high LOD modeling, and the nature of building defects' LoD being 500 reflects as-built information. The proper representation of condition assessment in BIM needs extensive discussion, which this research will tackle. As the goal aims at long-term monitoring and data storage in BIM, the entities' relationship between building defects and building elements will also be tackled, serving as the basis of extended IFC data schema for HBIM development.

14.2.3 Unmanned aerial systems as non-destructive testing tools

Originally adopted by military use over 150 years ago, UAS are now enhanced massively and popular for civilian uses, such as traffic surveillance, monitoring the environment, built environment inspection, imaging, and mapping (Gheisari et al., 2020). Rakha and Gorodetsky (2018) state that the usage of UAS in the age of climate change has substantially advanced over traditional auditing methods and approaches. Existing software and mathematical principles aiming to automate building performance inspections by utilizing drones offer many alternatives for data processing, analysis, and visual representation with decreased manual effort. With the ability to integrate NDT tools, UAS have demonstrated a unique perspec-tive on building health inspection, comparing mobility and affordability to tradi-tional methods of human labor (Masri & Rakha, 2020). Equipped with an optical camera, UAS-based visual inspection can capture high-definition photographic information that is not obvious to human eyes from ground level (Ruiz et al., 2021). Infrared cameras, LiDAR-based, and ultrasonic-based UAS have a bright future of autonomously conducting building performance surveillance jobs.

Khodeir et al. (2016) integrate HBIM as a guiding principle and conservation approach for sustainable heritage building retrofitting, suggesting a framework into five main phases: initiation, planning, implementation, monitoring, and assessment. Initiation, as the essential part of heritage value determination and preservation problem formulation, can impact subsequent phases of orderly and preflight planning of drones. Preflight planning is a vital process that can cause deviation in the accuracy of captured data. Two limitations need further consideration and will be elaborated on next: environmental impact and mechanism effect. When capturing outdoor reality using drones, the critical environmental factors include schedule, weather, cloud condition, and flight map (Febro, 2020). When selecting a suitable type of UAS for execution, the flight purpose and the detection ability of different building defects are the considerations of mechanism effect (Li et al., 2022). Although the automation of drone path planning is relatively well advanced on mapping task that focuses on comprehensive information collection, the operation that specifies this research-proposed use case still needs to improve. Building defect—orientated path planning, differing from traditional mapping tasks, focuses on collecting local area's information that solely contains building defects, which aims at tracking the health of detected building defects in a more frequent way. When collecting information from individual local area, this path planning also includes utilizing UAS sensor position with multiple adjustments to collect sufficient information for building defect analysis. Besides, the path should make UAS arrive at the desired location efficiently and consider optimizing battery and energy usage. For example, the UAS scanning strategy for a large area of penetrating dampness should differ from a small area of spot spalling in capture scope, number of captures, and environmental features. Meanwhile, higher level autonomous drone planning could be enhanced by cooperating with weather condition thresholds to execute drone flight under the best temporal conditions.

UAS path planning necessitates first identifying the project goal and area of interest, then developing the plan, including mapping area, path pattern, and overlapping rate. To process image photogrammetry for the whole building profile, image overlapping is a vital feature to ensure final results. UAS usually execute lawnmower-pattern paths aligning with the building's footprint from the top view and elevation profile from the vertical planar views, which can take successive images with assured overlap. When collecting data for specific areas, the waypoints, which are introduced to describe virtual reference points containing GPS coordinates and altitude information, are the locations where UAS hover and usually take a series of actions (Rakha & Gorodetsky, 2018). A high-quality capture requirement for interest areas should consider camera positions and targeted geometries since inappropriate camera angles can lead to wrong object recognition and inaccurate 3D reconstruction (Ibrahim et al., 2022). This issue becomes more apparent when dealing with heritage buildings that usually contain many complex and irregular geometries. In addition, building defect information collection can be more complicated since the scope of building defects could be as tiny as a minor crack within a brick or as large as the whole masonry wall tilting. Further investigation should be conducted based on building defects' typology and monitoring goals to produce better automatic flight planning.

14.2.4 Data processing

Four common types of NDT UAS are optical cameras, infrared cameras, LiDAR scanners, and ultrasonic testing drones. Although the captured data types can be different, the majority of UAS being used are camera-based, which is relatively low affordance. Computer vision (CV) technology has become a trend as a data processing method. CV is an interdisciplinary discipline that teaches computers how to analyze and comprehend visual input; since the past few decades, the amount of photographs and videos captured in the built environment has increased exponentially (Lin & Golparvar-Fard, 2020). The two types of essential information for HBIM model making are graphic and nongraphic data. With the structure from motion technique−based photogrammetry process getting more and more mature, the software for transferring a set of 2D images to estimate the 3D structure of a scene becomes reliable for extracting geometric information. Although terrestrial laser scanning (TLS) already demonstrates the ability to deliver high-accuracy point clouds (Xu et al., 2018), the high expense and mobility limitation of TLS made the alternative way of using UAS popular. The reliability of LiDAR scanner drones in extracting geometric information still needs further validation and research.

The visual data has the advantages of transparency and communication efficiency over text description. Visual data provides a different approach for human eyes to interpret information easily. Nevertheless, examining a large number of images to identify building defects by human eyes is a repetitive task and labor-intensive work. Recent studies have shifted to CV-based machine learning (ML) and artificial intelligence (AI) applications to classify and quantify damages in built heritage (Mishra, 2021). Adopting ML approaches should be a preliminary step to classifying point cloud data as the future goal of the automated generation of the HBIM model (Croce et al., 2021). Mishra (2021) has summarized diverse ML techniques that can be applied for structural health monitoring and superficial damage detection of heritage buildings. Those ML techniques can detect anomalies timely and effectively, which ensures intervention action for the longevity of built heritage.

14.2.5 Safety considerations and standards

The excellent mobility and affordability of UAS have obtained prevalent recognition. At the year-end of 2019, the US Department of Transportation (U.S. Department of Transportation, 2019) revealed that the number of registered UAS had already reached nearly 1.5 million, with 160,000 registered remote pilots. The increasing utilization of UAS can offer multiple benefits for built heritage. However, UAS also present hazards that expose heritage property and even human lives to a range of safety issues. The inherent risks of UAS include high-speed rotating blades, overheated batteries, mechanical failures, or loss of power (Namian, Khalid, Wang, & Turkan, 2021). Besides, incorrect operation and complex environments add risks of causing damage. Some implementations are developed within UAS to mitigate inherent risks, which include equipping them with airbag systems

(Alizadeh et al., 2014), collision avoidance systems (Yasin et al., 2020), and mounting protective cages (Namian, Khalid, Wang, & Turkan, 2021).

Regulation is an alternative way of easing the potential risks of UAS. Various countries started to enact regulations for the civilian use of UAS with the drone's fast development in the past few decades. Gheisari et al. (2020) have summarized the UAS regulation of five countries or areas regarding aircraft approval and registration, remote pilot certification, autonomous operation, night operation, and so on. According to the survey conducted by Namian, Khalid, Wang, and Kermanshachi (2021), distraction is deemed one of the top safety threats in operation. On the one hand, a VR-based UAS flight training simulator is introduced to improve drone pilot performance during building inspection (Albeaino et al., 2022). On the other hand, the autopilot software and autonomous UAS flight have a promising perspective of assisting UAS operations and reducing human errors. Even though different countries and areas have diverse requirements and specifications, they have strict prerequisites for autonomous operation: autonomous flight of UAS is forbidden by the United Kingdom, Europe, and Chile, while it is allowed in the United States and Brazil only with keeping the UAS within visual line-of-sight (VLOS). Maintaining UAS in the operator's visual field can ensure real-time UAS operation monitoring of related safety concerns. Keeping UAS within VLOS can circumvent crossing lines with manned aircraft and avoid crashes with hard-to-detect obstacles by the UAS, such as natural plants and electrical wires, so the remote pilot can take control in emergencies. However, more research on balancing the decision between automatic obstacle avoidance and human-controlled condition when encountering emergency conditions still needs further validation. Meanwhile, the regulations of autonomous operation should be widely discussed and updated by considering the promising future of autonomous UAS usage. In addition, active training, education, and management for reality capture staff are essential to ensure the proper operation of UAS flights.

14.3 Materials and methods

14.3.1 Workflow process

To automate the use case—historic building inspection by UAS, this research first proposed a workflow that identified key stakeholders and essential data exchanges in this framework. Business Process Model and Notation (BPMN) is used to describe the automation workflow from beginning to end by a flowchart approach. The flowchart can clearly illustrate the entire sequence of business activities and information flows required to complete a process, which can assist people with sufficient detail to enable precise implementation. Understanding the whole process and problems in detail is critical before proposing potential automation solutions. As a means of visual communication, the use case described in a BPMN diagram can be easier to understand than traditional narrative text from a Model View Definition (Sacks et al., 2018). In addition, enhanced-entity relationship (EER) model will be utilized as a data model to describe the ontology of building defects and the interrelated building elements. The EER model can classify things of interest and specify existing relationships between entities. In software engineering, an EER model defines the data and information structure that can be implemented in a database, which serves as a basis to extend the IFC schema solution.

Figuring out the involved stakeholders and step-by-step procedures in actual use case workflow is the very first step for automating the proposal and optimizing data flows. The stakeholders in this process include the steward, the drone pilot, the BIM modeler, and the AEH (architect, engineer, and historian) group (Fig. 14.1). The steward role is usually acted by the asset owner, government, or nonprofit organization that manages, funds, and aims to preserve the built heritage site. The drone pilot executes drone flights to capture reality information, collect building defect data, and ensure the automated UAS within VLOS, which is required in current regulations. Although in the current scan-to-BIM process, the architect undertakes the BIM modeler's work since the architect is fluent in BIM modeling when designing a new building. With emphasis shifting on modeling the existing building, the BIM modeler's role is proposed to obtain extra training and knowledge to finish intricate geometric modeling and nongeometric information registration for historical buildings in the BIM environment. Separating the role of the architect and the BIM modeler can let the architect focus on later retrofit design aspects while the BIM modeler can manage built asset information within the HBIM model. The AEH group, as a collaborative team in this process, identifies the area of interest which requests more information capture and a higher LoD that can impact flight planning on data collection. Formal training and certification for stakeholders are required, especially for vulnerable historic buildings, and any poor

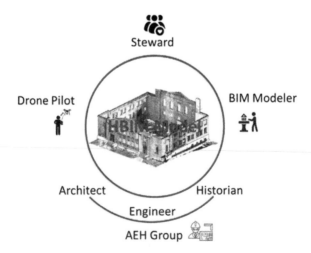

Figure 14.1 Involved stakeholders in a heritage monitoring project. The HBIM model is centralized by steward, drone pilot, BIM modeler, architect, engineer, and historian. *BIM*, Building Information Modeling.

performance might result in irreparable damage. In the United States, federally funded historic preservation projects require workers to follow the secretary of the Interior Standards and meet the qualifications to complete a project (U.S. Department of the Interior, 1983). Besides, giving the certified drone pilot the knowledge of built heritage would benefit flight planning and performance in data capture.

The BPMN diagram (Fig. 14.2) elaborates on the proposed workflow involving key stakeholders and essential data exchanges. Building typology and use cases vary, so the steward should establish the explicit goal first. Then a preliminary on-site meeting is required for most stakeholders to set up first-time drone flight planning. The on-site meeting offers the AEH group the opportunity to examine the building closely and raise the area of interest and concern, which leads to different LoD in the HBIM model and will further inform drone flight planning. Flight planning includes predefined flight paths, time schedules, cloud conditions, and environmental considerations. Flight planning is a critical document that ensures the quality and safety of first-time data capture, which is also essential for later adaptive monitoring flight use. The drone pilot will execute first-time reality capture using UAS with high-resolution cameras, the most common drone type used in historic preservation sites. Photogrammetry-based software processes the drone-captured images to generate a point cloud that assists HBIM model making and building defects information registration. Another option is to adopt LiDAR UAS to collect built heritage information and generate a point cloud directly. However, due to its high expense and new launching time, more research and studies are needed to validate LiDAR drones' utilization. The point cloud data and drone footage are evaluated by the AEH group as a first-time remote inspection so that they can place more data collection requests on a specific area and inform subsequent

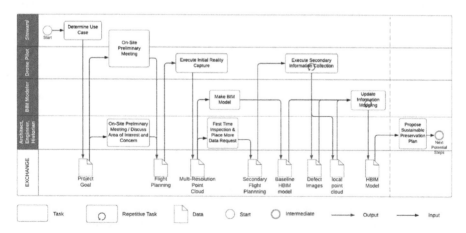

Figure 14.2 BPMN workflow of historic building defects management using UAS and BIM platform. The BPMN workflow streamlines each stakeholder's task and specifies the essential exchange data as required output or input for every task. *BIM*, Building Information Modeling; *BPMN*, Business Process Model and Notation; *UAS*, Unmanned aerial systems.

recursive flight planning. Because several latent anomalies, such as moving components and thermal leaks, require additional inspections to help verify. It may require different times of the day and even different seasons of the year to produce reliable and comprehensive information. With the secondary flight beginning as recursive tasks, the UAS with specific gears may be needed to collect detailed and specific building defect information, images, and local point clouds as the AEH group requests. The flight path will be shifted to a building defect−orientated path. The BIM modeler then registers and updates the building defect information into the comprehensive HBIM model, which can be shared with different stakeholders and prepared for the next potential steps of sustainable preservation and intervention plans.

Although current information registration in BIM is manually conducted by the BIM modeler, Dynamo inside of Revit has shown prosperous promise of automatically registering and exporting information by developing data input and output algorithms. Besides, the state-of-the-art literature has developed automatic object recognition and building defect analysis on captured images using AI and ML approaches. Once the technology becomes mature, an AI-based building defect recognition algorithm should be introduced to assist the BIM modeler in registering building anomalies information into the HBIM model.

14.3.2 Data model—entities relationship of building defect

On the one hand, more mature and adaptive AI approaches are needed to support automatic information extraction. On the other hand, the new extended data model for building defects and UAS executions is vital as a proper data container to support information interoperability in BIM environments and assist long-term historic building health monitoring goals. The EER model has been utilized to demonstrate the data relationships among building defects with other related entities in this use case (Fig. 14.3). From left to right, the whole entities are organized by four categories, building—building defect—connected features—UAS. The primary entities, building defect, UAS, and flight planning, are highlighted and centralized by other entities.

Building defects can be categorized into two subclasses: global building defect (GBD) and local building defect (LBD), which are disjoint constraint relationships and has own building defect type. GBD reflects angular, locational, and shaped faults in single or numerous building components. LBDs, on the other hand, are anomalies that are particularly attached to and within the confines of a particular building component (Li et al., 2022). Besides, building defect has attributes that subclass entities can inherit. Every building element can have a chance to contain zero to multiple building defects. Building element, which has several attributes to store information, such as unique ID, location, and materiality, is already well created and designed in the IFC data model. The materiality of the building element can affect the validity of some GBD and LDB types, so the building element and building defect have a shared attribute—materiality, which sets a restricted relationship. The impacting scope of building defects makes their graphical representation

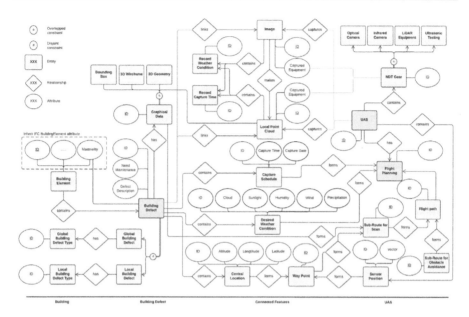

Figure 14.3 EER data model of building defect. The primary entities: building defect, UAS, and flight planning, are highlighted and centralized by other entities. *EER*, Enhanced-entity relationship; *UAS*, unmanned aerial systems.

different in the BIM environment, where GBD aligns with the whole building component. At the same time, LBD is usually a small scope area attached to a partial area of the building component or is attached to multiple building components. By considering the complexity of the building defect's modeling will take, this research proposes four subclasses of building defect's graphical data (GD): bounding box, 3D wireframe, 3D geometry, and local point cloud. Those four representation methods are overlapped constraint relationships, which will be displayed with examples and LoD in Section 14.3.4.

To execute a successful automated UAS scanning, flight planning is necessary to ensure high-quality data capture, which contains several considerations. The building defect has three primary ways to inform fight planning: first, the capture schedule acts as a time gap to set regulatory monitoring plan for BD; second, the desired weather condition plays an essential role in forming flight planning due to that weather conditions could affect the captured data quality and performance; third, the flight path is mainly dependent on the location of BD. In the flight path of tracking BD's progress, all those factors are included in the data model: the central location of BD forms the waypoint where the UAS can hover and do a series of operations; the sensor position with the vector that indicates distance and direction comprises waypoint; the flight path contains the subroute for the scan that drone scan single one BD with different perspectives and locations, which is defined BD's graphic data dimension, and the subroute for obstacle avoidance such as buildings and plants.

The building defect can link zero to multiple images or zero to multiple local point clouds, which play as visual proof of building defect. The image and local point cloud can be directly captured by UAS with different NDT gear, and they have shared attributes, and captured equipment, to document what NDT gear is used. In addition, the proper selection of NDT gear influences the validation of visual proof of building defects by considering the detection ability of the sensor. The NDT gear has four subclasses with overlapped constraint relationships since one drone can equip one to multiple sensors, allowing more comprehensive information to be captured simultaneously. The captured multiple images have the ability to be processed by photogrammetry software to generate a local point cloud.

Meanwhile, both the image and local point cloud data contain record weather conditions and record capture time, providing the retrospective ability to examine and improve the next automated UAS flight planning. Because the dissimilar natural property of building defects can influence the optimal schedule and desired weather conditions, which will further affect the captured data quality. Having recorded information can help revise and adjust the desired weather condition and capture the schedule.

14.3.3   Data container—Industry Foundation Classes data schema or extended property set

As an emerging technology, UAS provide alternative methods to capture building data which then can be registered in a BIM environment as the future digital twin use tool. The digital twin usually imply the frequent and highly-interwined data transition between physical obejct and digital model where the digital model can reflect physical object's information in real time. To automate the data consistently and constantly interwining process, the current data schema, IFC, in an open BIM environment is inadequate to support condition assessment and building defects storage. One way to achieve that is by proposing modifications to the current IFC data schema, which usually takes a long and strict process by smartBuilding International. Considering HBIM's fast development that utilizes BIM applications for built heritage, the direct modification of IFC is urgent. The other way is extending the IFC data schema by proposing property sets of building defects, which the end user can customize. In both ways, UAS-related features are partly selected for the data schema proposals because constructing a UAS data structure in IFC would be repetitive, similar to rebuilding a drone platform. By taking the current advanced UAS simulation and execution platforms into account, only essential features of UAS flight planning are needed.

Updating the current data schema can ensure seemly bridging data connections in any applicable BIM software and enrich the current data repository, rather than coming up with an ad-hoc data schema which may harm data consistency. Previous research (Li et al., 2022) proposed the IfcBuildingDefect concept, but after this chapter had discussed the comprehensive EER data model, the IfcBuildingDefect-orientated data schema modification proposal was updated (Fig. 14.4). Leaning on

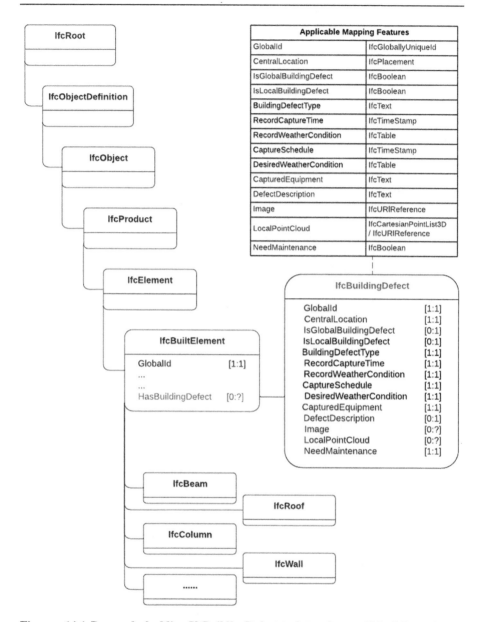

Figurew 14.4 Proposal of adding IfcBuildingDefect to data schema. IfcBuildingDefect is proposed as new IFC class with graphical and nongraphical information attached. *IFC*, Industry Foundation Classes.

the well-defined IFC data schema, IfcBuildingDefect as a new IFC class is proposed to link with IfcBuiltElement as an attribute. The GD of ifcBuildingDefect can be the bounding box, 3D wireframe, and precise 3D geometries, which will be further discussed with LoD in the next section. The IfcBuiltElement is a subclass of

IfcElement, which can be traced all the way up to IfcRoot, where the subclass inherits attributes from all upper classes. This modification guarantees that all subclasses of IfcBuiltElement, such as IfcBeam, IfcColum, IfcRoof, and IfcWall, are capable of containing building defect information because every historic building element may have zero to multiple building defects as the research discussing on EER data model. At the same time, one sole IfcBuildingDefect can be attached to zero to multiple IFC classes since, for example, one single crack can span from floor to wall and ceiling. The priority step to construct the IfcBuildingDefect data schema is to follow IFC specification, which can be achieved by mapping the proposed attributes in the EER data model with current IFC data types. This way can utilize existing well-defined IFC classes. The applicable mapping features generally show that IFC data types are eligible to assist this modification. Setting a well-defined building defect class can assist in smooth data analysis in BIM software and better inform conservation decision-making. However, adding a new IfcBuildingDefect class to the current data schema requires that data must be commonly used internationally and go through a strict revision procedure, which could be a long process.

In addition, adding a custom property set is the common and easiest way for end users to extend the IFC data schema. By doing so, users should well define their property set on the name of the property set, applicable entities, attribute's name, property type, and data type (Fig. 14.5). The PropertySet_BuildingDefect is eligible to map the current IFC property type and data type to develop. The research proposes that after defining PropertySet_BuildingDefect, the property set as a nongraphic information container should be combined with IfcShapeRepresentation, which performs as the building defect's graphic data container, to represent the building defect data. Both property set and shape representation are linked with IfcBuiltElement as an attribute, but the limitation is no strong bond between the shape representation and the building defect's property set, which has a chance to harm data consistency and damage analysis when using different BIM software. Besides, not standard as IFC class, property set can be modified by any user at any stage, thus it becomes hard to control the same data structure among various stakeholders and software.

14.3.4 Computational modeling

This section aims to provide the guideline on computational modeling of building defects in both graphical and nongraphical aspects, aided by the Level of Description (LoDes) table. Geometric modeling of historic building defects in BIM could be challenging due to its irregular geometry. Non-GD (NGD) exchange can be automated by predefined algorithms under the prerequisite of having a well-defined data structure. The complete data process and registration procedure are proposed for future reference by considering current application development (Fig. 14.6): initially, applying UAS execution to collect building health information and second, using AI and CV techniques to process UAS-captured data and extract building defects information. For example, the edge detection algorithm is widely adopted for identifying

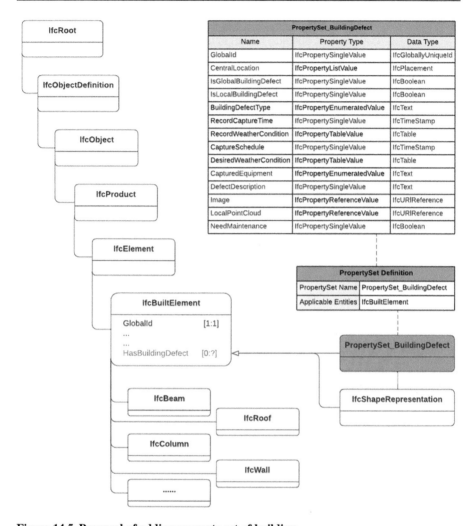

Figure 14.5 Proposal of adding property set of building defect. PropertySet_BuildingDefect needs combine IfcShapeRepresentation to represent building defect information completely.

surface cracks, which assists in quick localization and validation; third, Dynamo scripts register and update building defect information in the HBIM model. Dynamo, as a plug-in of Autodesk Revit, is a visual programming platform that enables users to construct building information workflow by graphical scripts, which can be used for nongraphic data registering and updating in the HBIM model; fourth, it provides opportunities for involved stakeholders and especially engineers to make intervention decision based on the severity of historic building health; and last, the building defect information and locations help initiate next automated UAS flight planning and execution to achieve a monitoring loop process.

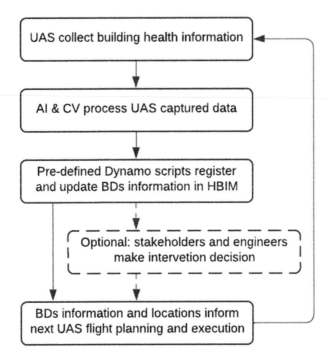

Figure 14.6 Complete data automation process. The complete data processing is a procedure loop that can keep check and update building health information on BIM model. *BIM*, Building Information Modeling.

Traditional LoD was originally created for modern ACEO industries to manage building asset information over the whole life cycle. Nevertheless, the conventional LoD 100−500 is inadequate for the historic building, especially for building condition assessment (Brumana et al., 2018). Based on the complex nature and graphic shape of building defects, the LoDes, differing from LoD, aims to describe the information richness of building defects. Meanwhile, this chapter provides the LoDes guideline table (Table 14.1) for building defects modeling at both graphical and NGD. The GD are provided with illustration examples. Three levels of BD GD are proposed, and every project team should examine their limited resources to decide the optimal selection: GD LoDes 1, as low level, is a 3D bounding box of BD at a specific location; GD LoDes 2, as medium level, adds 3D wireframe with an approximate dimension of BD; GD LoDes 3, as high level, adds 3D geometry with a precise dimension of BD or local point cloud of BD. Three levels of GD have overlapped constraint relationships, where one single BD can be modeled as one or multiple representations. A crack on the wall is demonstrated as a GD LoDes example in Fig. 14.7. Generally, the higher level of GD would increase the modeling and processing time. NGD is categorized into five levels: NGD LoDes 1 contains basic information of BD type; NGD LoDes 2 adds external reference such as an image or local point cloud; NGD LoDes 3 adds sufficient internal materiality

Table 14.1 Building defect's Level of Description (LoDes) table.

BDs—LoDes					
LoDes	1	2	3	4	5
Graphical data	Bounding box at specific location	(+) 3D wireframe with approximate dimension	(+) 3D geometry with precise dimension or local point cloud		
Nongraphical data	Basic information of BD type	(+) Links external reference: image or point cloud	(+) Sufficient internal detail and materiality	(+) Condition assessment	(+) Restoration plan and detail

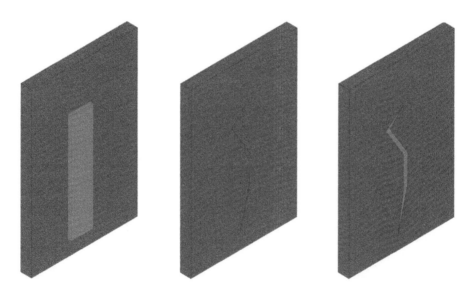

Figure 14.7 Crack LoDes as an example. Graphical data LoDes 1; LoDes 2; LoDes 3 (from left to right). *LoDes*, Lable of Description.

and detail information; NGD LoDes 4 adds condition assessment which engineers and experts can make; and NGD LoDes 5 adds restoration plan and detail to fix BD. Besides, NGD and GD can be mixed with different LoDes as the individual project needed.

14.3.5 Flight planning

Both mapping and waypoint methods in UAS flights should be adopted for building defects detection and monitoring. Initially, the mapping task is for a detailed examination to detect building defects, which usually operate with a lawnmower-pattern path. The mapping area could be set for building facades and roofs. After the detection and registration of building defects on the HBIM model, apply the waypoints flight path to track the potential damage progress of building defects. The central location of the building defect is essential for waypoint set-up, and the position and distance of the UAS sensor, which is represented by the waypoint vector, are vital for captured data quality (Fig. 14.8). When UAS arrive at the predefined waypoint, the subroute for the scan is an optional path designed for capturing building defects from diverse perspectives. The subroute for the scan, which enriches the completeness of information capture, should be created based on the scope and properties of a building defect. The waypoint-orientated UAS flight focuses on monitoring building defects, requiring identifying barriers that the flight path must detour. Current UAS waypoint software can take waypoints and barriers as separate inputs to generate subroute for obstacle avoidance automatically (Fig. 14.9).

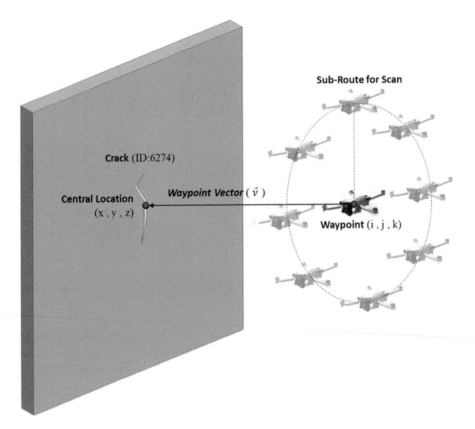

Figure 14.8 UAS subroute for scan. The UAS can capture general information of building defect from major waypoint. If needed, UAS can enter subroute to collect more comprehensive information of each building defect. *UAS,* Unmanned aerial systems.

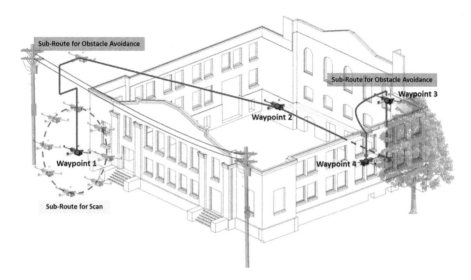

Figure 14.9 Waypoint-orientated UAS flight to monitor building defect. The waypoint-orientated flight path include major waypoints with subroute for scan and subroute for obstacle avoidance. *UAS*, Unmanned aerial systems.

14.4 Discussion and evaluation

This work has widely discussed entities' relationship and data model of heritage building defects in the BIM environment, which supplies two modification proposals to the current IFC data schema. To enrich and improve the data structure, more practical works are needed to validate its utilization. A well-defined data schema is critical for information consistency in long-term operations. As the study demonstrates, the BIM platform displays a promising future of supporting long-term documentation and building health monitoring for built heritage. Since HBIM has drawn research attention for preserving historic buildings, developing historic building health analysis and decision-making of restoration plans are also essential for built heritage. The future research of HBIM should intersect with diverse disciplines to enrich its methodology because historic buildings usually involve many stakeholders, and the availability of multiple perspectives can better preserve built heritage.

Besides, UAS have been proven as promising NDT tools for visual inspection, and their great mobility has demonstrated the automation potential to execute building defect detection and health monitoring. This research has demonstrated the feasibility of using the camera-based drone for information capture, and the bright future applications can be expanded as multidrone execution and multisensor UAS to scan and monitor historic buildings simultaneously. The development of autonomous UAS for structure inspection, such as Skydio 3D Scan, has dramatically progressed in processing information in real time, where a model is generated when UAS take flight (Skydio, 2023). However, the performance of captured data and model still needs further validation. Although the autonomous UAS and their software package are relatively expensive, they assist in obstacle avoidance and ensuring property safety. Future research can utilize this technique as an initial building defects detection method to automate flight paths and data capture rather than manually defining lawnmower-pattern flight paths. In addition, state-of-the-art literature is developing small unmanned ground vehicles (UGV) in the built environment, which can be applied for indoor and complex ground areas, such as the spot quadruped robot, from Boston Dynamics (2023). The combination of UAS and UGV can mitigate each drawback to capture more comprehensive information about built heritage. This chapter's framework focuses on heritage building defect management and HBIM development from a single building perspective. The future scope can be extended to historic districts, towns, and villages and the large area of built heritage sites, where UAS can carefully conduct sustainable conservation. The promising progress in UAS flight autonomy by leveraging AI and the enhanced regulations in flight beyond visual line of sight will benefit UAS adoption to preserve heritage buildings.

14.5 Conclusion

This research has proposed the framework of using UAS to monitor built heritage defects as the reflection of building health. Timely tracking and understanding of the built heritage condition can ensure the functionality and longevity of historic buildings, which require reality capture tools to regularly collect the data of as-built defects and well-defined data structures to store and update building information consistently. UAS as promising NDT tools and HBIM as favorable information documenting methodology can assist the automation of this use case. By analyzing entities' relationship among building defects with other essential features, the EER model is demonstrated to help modify data schema. Two ways of altering the current IFC data schema proposal with LoD guidelines on BD's information modeling can make the BIM platform efficient as a long-term building defects monitoring method, which serves as a contribution to support HBIM development. Besides, the data structure can also support UAS from data collecting to registering and advising the next flight planning. The goal of providing comprehensive data on heritage buildings is to better inform the decision-making process on the preservation and conservation of built heritage sites.

References

Alavi, H., Bortolini, R., & Forcada, N. (2022). BIM-based decision support for building condition assessment. *Automation in Construction*, *135*.

Albeaino, G., Eiris, R., Gheisari, M., & Issa, R. (2022). DroneSim: A VR-based flight training simulator for drone-mediated building inspections. *Construction Innovation*, *22*(4), 831–848.

Alizadeh, M., Sedaghat, A., & Kargar, E. (2014). Shape and orifice optimization of airbag systems for UAV parachute landing. *International Journal of Aeronautical and Space Sciences*, *15*(3), 335−343.

Banfi, F. (2019). HBIM generation: extending geometric primitives and bim modelling tools for heritage structures and complex vaulted systems. *Int. Arch. Photogramm. Remote Sens. Spatial Inf. Sci.*, *XLII-2/W15*, 139−148.

Banfi, F. (2020). *Drone meets Historic Building Information Modelling (HBIM): Unmanned aerial vehicle (UAV) photogrammetry for multi-resolution semantic models*. In *Drones - Systems of information on cultural heritage. For a spatial and social investigation*, (pp. 88−97). Pavia University Press.

Barazzetti, L., Banfi, F., Brumana, R., & Previtali, M. (2015). Creation of parametric BIM objects from point clouds using nurbs. *The Photogrammetric Record*, *30*(152), 339−362.

Barontini, A., Alarcon, C., Sousa, H., Oliveira, D., Masciotta, M., & Azenha, M. (2021). Development and demonstration of an HBIM framework for the preventive conservation of cultural heritage. *International Journal of Architectural Heritage*, *12*, 1451−1473.

Boston Dynamics. (2023). *Products* https://www.bostondynamics.com/products/spot#id_first.

Brumana, R., Torre, S., Previtali, M., Barazzetti, L., Cantini, L., Oreni, D., & Banfi, F. (2018). Generative HBIM modelling to embody complexity (LOD, LOG, LOA, LOI): surveying, preservation, site intervention—the Basilica di Collemaggio (L'Aquila). *Applied Geomatics*, *10*, 545−567.

Bruno, N., & Roncella, R. (2019). HBIM for Conservation: A new proposal for information modeling. *Remote Sensing*, *11*(15).

Croce, V., Caroti, G., Luca, L., Jacquot, K., Piemonte, A., & Véron, P. (2021). From the semantic point cloud to Heritage-Building Information Modeling: A semiautomatic approach exploiting machine learning. *Remote Sensing*, *13*(3).

Diara, F., & Rinaudo, F. (2020). IFC classification for FOSS HBIM: Open issues and a schema proposal for cultural heritage assets. *Applied Sciences*, *10*(23).

Dore, C., & Murphy, M. (2017). Current state of the art Historic Building Information Modelling. *The International Archives of the Photogrammetry. Remote Sensing and Spatial Information Sciences*, *XLII-2/W5*, 185−192.

Febro, J. (2020). 3D documentation of cultural heritage sites using drone and photogrammetry: a case study of philippine unesco-recognized baroque churches. *International Transaction Journal of Engineering, Management, & Applied Sciences & Technologies*, *11*.

Gheisari, M., Costa, D., & Irizarry, J. (2020). *Unmanned aerial system applications in construction. Construction 4.0* (pp. 264−288). Routledge.

Ibrahim, A., Golparvar-Fard, M., & El-Rayes, K. (2022). Multiobjective optimization of reality capture plans for computer vision−driven construction monitoring with camera-equipped UAVs. *Journal of Computing in Civil Engineering*, *36*(5).

Khodeir, L., Aly, D., & Tarek, S. (2016). Integrating HBIM (Heritage Building Information Modeling) tools in the application of sustainable retrofitting of heritage buildings in Egypt. *Procedia Environmental Sciences*, *34*, 258−270.

Li, B., Zeng, Q., Gentry, R., & Willkens, D. (2022). Establishing a decision-making support framework for optimizing Unmanned Aerial Systems (UAS) flight planning to monitor heritage building defects. *Transforming Construction with Reality Capture Technologies: The Digital Reality of Tomorrow*.

Lin, J., & Golparvar-Fard, M. (2020). Visual and virtual progress monitoring in Construction 4.0. In *Construction 4.0*, (pp. 240−263). Routledge.

Liu, J., Willkens, D., & Foreman, G. (2022). An introduction to technological tools and process of Heritage Building Information Modeling (HBIM). *EGE Revista de Expresión Gráfica en la Edificación* (pp. 50−65).

Liu, J., & Willkens, D. (2021). Reexamining the old depot museum in Selma, Alabama, USA. In *Building Information Modelling (BIM) in Design, Construction and Operations IV*, (pp. 171−186). WIT Press.

López, F., Lerones, P., Llamas, J., Gómez-García-Bermejo, J., & Zalama, E. (2018). A review of Heritage Building Information Modeling (H-BIM). *Multimodal Technologies and Interaction, 2*(2).

Masri, Y., & Rakha, T. (2020). A scoping review of non-destructive testing (NDT) techniques in building performance diagnostic inspection. *Construction and Building Materials, 265*, 120542.

Mishra, M. (2021). Machine learning techniques for structural health monitoring of heritage buildings: A state-of-the-art review and case studies. *Journal of Cultural Heritage, 47*, 227−245.

Namian, M., Khalid, M., Wang, G., & Kermanshachi, S. (2021). Ascending drones' safety risks in construction. *W099 & W123 Annual International Conference.*

Namian, M., Khalid, M., Wang, G., & Turkan, Y. (2021). Revealing safety risks of unmanned aerial vehicles in construction. *Transportation Research Record: Journal of the Transportation Research Board, 2675*(11), 334−347.

Rakha, T., & Gorodetsky, A. (2018). Review of unmanned aerial system (UAS) applications in the built environment: Towards automated building inspection procedures using drones. *Automation in Construction, 93*, 252−264.

Rocha, G., Mateus, L., Fernández, J., & Ferreira, V. (2020). A scan-to-bim methodology applied to heritage buildings. *Heritage, 3*(1), 47−67.

Rolin, R., Antaluca, E., Batoz, J.-L., Lamarque, F., & Lejeune, M. (2019). From point cloud data to structural analysis through a geometrical hBIM-oriented model. *Journal on Computing and Cultural Heritage, 12*(2), 1−26.

Ruiz, R., Lordsleem, A., Jr., Rocha, J., & Irizarry, J. (2021). Unmanned aerial vehicles (UAV) as a tool for visual inspection of building facades in AEC + FM industry. *Construction Innovation, 22*(4).

Sacks, R., Eastman, C., Lee, G., & Teicholz, P. (2018). *BIM Handbook: A Guide to Building Information Modeling for Owners, Designers, Engineers, Contractors, and Facility Managers.* John Wiley & Sons.

Schlenger, J., Yeung, T., Vilgertshofer, S., Martinez, J., Sacks, R., Borrmann, A. (2022). A comprehensive data schema for digital twin construction. *The 29th EG-ICE International Workshop on Intelligent Computing in Engineering.*

Skydio. (2023). *Skydio 3D scan - Adaptive scanning for autonomous inspection data capture* https://www.skydio.com/3d-scan?utm_source = google&utm_medium = cpc&utm_campaign = g_s_brand_ent-v2_brand&_bt = 541259346869&_bk = skydio%203d%20scan&_bm = e&_bn = g&gclid = Cj0KCQiA8t2eBhDeARIsAAVEga283HMzyQjoEfrQubGXtv-99t7r5FTp2CUiUpk5PkTXfEWSnWUhRKKQaArUWEALw_wcB.

Tejedor, B., Lucchi, E., Bienvenido-Huertas, D., & Nardi, I. (2022). Non-destructive techniques (NDT) for the diagnosis of heritage buildings: Traditional procedures and futures perspectives. *Energy and Buildings, 263.*

U.S. Department of the Interior. (1983). *Archeology and historic preservation*; Secretary of the Interior's Standards and https://www.nps.gov/subjects/historicpreservation/upload/standards-guidelines-archeology-historic-preservation.pdf.

U.S. Department of Transportation. (2019). U.S. Department of Transportation issues proposed rule on remote ID for drones https://www.transportation.gov/briefing-room/us-department-transportation-issues-proposed-rule-remote-id-drones.

Volk, R., Stengel, J., & Schultmann, F. (2014). Building Information Modeling (BIM) for existing buildings—Literature review and future needs. *Automation in Construction, 38,* 109−127.

Xu, X., Kargoll, B., Bureick, J., Yang, H., Alkhatib, H., & Neumann, I. (2018). TLS-based profile model analysis of major composite structures with robust B-spline method. *Composite Structures, 184,* 814−820.

Yasin, J., Mohamed, S., Haghbayan, M., Heikkonen, J., Tenhunen, H., & Plosila, J. (2020). Unmanned Aerial Vehicles (UAVs): Collision Avoidance Systems and Approaches. *IEEE Access, 8,* 105139−105155.

Zhang, Y., & Dong, W. (2021). Determining minimum intervention in the preservation of heritage buildings. *International Journal of Architectural Heritage, 15*(5), 698−712.

From 3D models to historic building information modeling (HBIM) and digital twins: A review

15

Susana Lagüela[1], Luis Javier Sánchez-Aparicio[2],
Enrique González-González[1], Alejandra Ospina-Bohórquez[1],
Miguel Ángel Maté-González[1], and Diego González-Aguilera[1]
[1]Department of Cartographic and Terrain Engineering, Universidad de Salamanca, Ávila, Spain, [2]Construction and Building Technology, Universidad Politécnica de Madrid, Madrid, Spain

15.1 Construction 4.0

Construction 4.0 (C4.0) is a subset of Industry 4.0 (I4.0) that specifically applies the principles and technologies of I4.0 to the construction industry. Like I4.0, C4.0 involves the integration of digital and physical systems to create a more efficient, flexible, and data-driven construction process (Statsenko et al., 2022). These technologies can help to optimize construction processes, reduce costs, improve safety, and increase the sustainability of construction projects. Some of the key trends and potential areas for development in C4.0, as identified in systematic literature reviews, include (1) data analytics and decision-making: the use of data analytics and machine learning to improve decision-making and optimize the performance of construction processes and assets (Mora et al., 2021). (2) Virtual and augmented reality: use of virtual and augmented reality technologies to improve design and planning, training, and maintenance (Sedano-Espejo et al., 2022). (3) Robotics and automation: implementation of robotics and automation to improve efficiency, reduce costs, and improve safety in construction (Gharbia et al., 2020). (4) BIM (Building Information Modeling): the application of BIM to improve collaboration, coordination, and information management in construction (Garcia-Gago et al., 2022). (5) Smart infrastructure: by using sensors, the Internet of Things (IoT), and other technologies to increase the performance and maintenance of infrastructure (Massafra et al., 2022). And (6), sustainability and resource efficiency: use of C4.0 technologies to improve sustainability and resource efficiency in the construction industry (Craveiro et al., 2019).

However, the successful implementation of C4.0 technologies will depend on the ability of the construction industry to adapt to and effectively utilize these technologies. It should be noted that the application of these technologies in the construction industry is still at the early stages compared to the manufacturing sector. One of the

Diagnosis of Heritage Buildings by Non-Destructive Techniques. DOI: https://doi.org/10.1016/B978-0-443-16001-1.00015-2

challenges to the adoption of I4.0 technologies in the construction industry is the complexity of construction projects (and even more in those where work is done to rehabilitate and conserve historic heritage buildings), which often involve multiple stakeholders, diverse materials and equipment, and numerous environmental conditions.

One of the most disruptive emerging technologies in I4.0 in recent years is the application of digital twin (DT) technology. A DT is the virtual/digital replica (avatar) of an element, object, product, service, or system that simulates the behavior of its physical counterpart with the aim of monitoring its behavior over time, analyzing its reaction to certain situations and anticipating possible problems, and improving its performance and efficiency (González-Aguilera et al., 2021). The concept of DT, although seems new, was coined at the end of the 20th century. At that time, technology did not allow for its practical application, but the concept did not stagnate and, progressively, thanks to the digital revolution, its development has been possible: its implementation has been successful thanks to new emerging technologies, such as artificial intelligence (AI), the IoT, big data, approaches and tools from the field of geomatics, blockchain, or virtual, augmented, or mixed reality techniques (Opoku et al., 2022). Although the field of I4.0 is the most prominent implementation sector, other sectors have found their own approaches, ideas, and requirements; the DT concept has been applied and extended to the point where different facets can be assumed depending on the application domain and intended use.

As is already known, BIM refers to a collaborative process among the stakeholders of a new construction project for the exchange of information through a digital object−based model. This methodology arose with the aim of improving the competitiveness of the construction sector, allowing the simulation of both the actual construction of the project and all the interactions among the different professionals/actors involved in the execution of the work. In this way, a significant reduction in the costs and construction time of the project is possible, in turn allowing for a more sustainable model that avoids problems and execution errors. In this sense, the application of BIM in historic buildings (HBIM, Historic Building Information Modeling) has been able to recognize the advantages of using these approaches by allowing the interoperability of information among all the agents involved in heritage conservation projects (through the open format IFC, Industry Foundation Classes) and the tools that allow the parametric modeling of the different parts that make up the heritage assets.

BIM and DT technology can be linked to provide a comprehensive view of the current state and future behavior of an architectural property, allowing (1) storage of historical data on the property or the assets located therein; (2) storage of information from periodic structural or auscultation analyses; (3) interoperability with other databases, such as those from monitoring sensors (IoT); and so on. In this way, the HBIM model is able to combine (1) geometric information (through the descriptive model or as-built model of the current state of the property), which includes data on the materials, techniques, and architectural styles of the asset under study; (2) historical data; (3) data from IoT sensors in real time; (4) data on the damage suffered by the asset; and (5) structural (determination of stresses, determination of strength and stiffness, material models, and so on) or auscultation analysis. All this information provides a true picture of the current state of conservation of the architectural asset. In addition, computer vision techniques through AI (machine learning and deep learning) and historical sensor data (IoT + big data) will help to better understand previous data, providing reliable classification techniques that will facilitate the diagnosis of damage and its state of evolution. These tools also provide information about the behavior of the architectural asset in the near future, allowing progress toward a predictive model of the object of study or the assets within it. Furthermore, if damage identification and quantification is added to the system, it is possible to achieve a prescriptive model that provides guidelines to advance in the preventive conservation processes.

The purpose of this chapter is to present, in a practical, direct and agile way, different success applications that highlight the concept of the DT in the field of conservation and safeguarding of architectural heritage.

15.2   From 3D models to HBIM

The BIM technique involves two main parts: the geometrical 3D modeling and the management of information such as materials and their thermophysical characteristics; and construction components and their year of construction and condition, along with general information about the building such as year of construction, use schedule, interventions (year, type, affected assets), and its history, and a high number of social, political, economic, and cultural issues relating to the external environment (Simeone et al., 2014). Thus the 3D model representation is the part of the project where the different elements of the building become construction elements with parametric intelligence, including its quantitative and qualitative description and relationship information with the other elements (Yang et al., 2020).

The application of heritage documentation has led to the appearance of the concept "HBIM," from Historic Building Information Modeling (Yang et al., 2020). Although its generation is increasingly common since 2009 (Diara, 2022), the generation of HBIM is still a difficult process because of both the great variety in the constructive complexity of heritage buildings (Malinverni et al., 2019) and the different types of information to be managed (Guzzetti et al., 2022).

Focusing on the 3D modeling, different techniques are applied for the geometric data acquisition, from photogrammetry to laser scanning, with the first being the most common (Yang et al., 2020). These techniques have in common that their main product is point clouds. Thus the generation of HBIM consists of the conversion from the 3D point cloud into solid building components. This remains mostly a manual process (Garagnani & Manferdini, 2013; Macher et al., 2017), called scan-to-BIM, although many commercial tools have appeared to enlighten the process, and many researchers have developed automation algorithms for the process (MacHer et al., 2019; Thomson & Boehm, 2015). The scan-to-BIM processes developed consist of different methodologies (computer graphics, AI, and ontologies) for the identification and separation of those points that correspond to different construction assets, as well as the automatic provision of semantics in terms of identification of the type of feature those points represent.

15.2.1 Conversion of the geometric representation

The interoperability and accessibility of the HBIM generated is a key aspect toward the extension of this technique for the management of historical assets. For this reason, the utilization of free and open-source software, or an OpenBIM approach, is the solution to guarantee data accessibility, transparency, and the possibility of adjusting each project to the characteristics of the heritage element (Diara, 2022), as well as to make possible the participation of professionals of different disciplines in the decision-making (Oostwegel et al., 2022).

The IFC open data exchange standard was developed by the buildingSMART International (buildingSMART International, 2021) organization and standardized by the International Organization for Standardization (ISO). Its main purpose is to allow data interchange between the different software solutions, avoiding problems of incompatibility associated with the existence of proprietary formats for each software. Thus buildingSMART International (2021) counts more than 200 software tools that can import or export IFC files.

However, the use of IFC presents some setbacks, such as the need for the users of all sectors to know the standard (Oostwegel et al., 2022), and the need to determine the resolution of representation for the adequate inclusion of all the types of data needed (Gabriele, 2021). In this sense, different concepts have been developed by the Canadian Architecture, Engineering, and Construction (AEC), the American Institute of Architects (AIA), and the NBS. These concepts define different levels of resolution possible for the different types of data: the geometry (spatial resolution and accuracy) and the information about the building (quantity and structure). In addition, as stated by Graham et al. (2018), the especifications of the BIM or HBIM should be defined according to the three categories, namely, level of detail (LoD), information (LoI), and accuracy (LoA), so all these aspects are closely related when applied to a specific model.

15.2.1.1 Level of detail

This concept, defined as Level of Development by AEC (CAN) and AIA, or as LoD by the NBS, refers to the progression of the graphic representation of the building. The lowest LoD would be the generic 2D representation of the building, increasing to various amounts of graphic or nongraphic information attached to the 3D modeled objects (Fai & Rafeiro, 2014). Thus, according to the definitions from AEC CAN (2014) and The American Institute of Architects (2015), the LoDs are as follows:

- LoD100: conceptual a-dimensional model
- LoD300: 3D model in the design phase

- LoD350−400: model implemented in the construction phase
- LoD500: as-built model, after the construction phase

With this classification, it is possible to manage of the information usable for the maintenance of the building (Brumana et al., 2018).

15.2.1.2 Level of information

The LoI is defined in parallel to the LoD, referring to the attributes expected (Bertin et al., 2020). LoI has four different levels, defined so that BIM objects with high LoI includes manufacturer-specific data sheets and maintenance instructions (Wang & Zhang, 2021). LoI is a critical issue when dealing with HBIM, since there could be heritage assets not geometrically parametrizable, but for which all the not-geometrical information related to its history and state of conservation and restoration could be available (Santagati et al., 2018). These would be the case of ruined historical buildings, for which the generation of an HBIM would need to require a nontraditional approach regarding its geometric representation.

15.2.1.3 Level of geometry

The LoG in a BIM refers to the accuracy and completeness of the geometrical representation of the building under study, which acts as a visual portal for embedded nongeometric data in a single information source (Volk et al., 2014). According to AIA (The American Institute of Architects, 2015), LoG has four different levels, where LoG 1 stands for a schematic or symbolic representation of a product, and LoG 4 for a detailed, manufacturer-specific representation (Wang & Zhang, 2021). However, in the case of HBIM, the definition of a high LoG can be complex due to the fragmentation and inconsistency of the information available (Charlton et al., 2021). In addition, in the case that the heritage buildings are complete and their geometry can be measured either with laser scanning or photogrammetric techniques, these can be complex in their form and the information they retain, in such a way that the processing of the geometric information toward an accurate representation is also an intricate procedure.

15.2.1.4 Level of accuracy

Modeling from point clouds, the LoA is closely related to the tools, techniques, and processes used to acquire and process the data to represent the objects (buildings) to be represented. In case the data is acquired with a laser scanning, the technical characteristics of the device used already establish a maximum accuracy to be met by the HBIM, while the photogrammetric technique is more subjected to the processing procedure. For all cases, the modeling process introduces error and inaccuracy.

Based on this, Rossi and Palmieri (2020) present a workflow for the modeling of historic buildings based on a certified LoA, which is defined according to UNI Ente Italiano di Normazione (2017) as a function of the geometric precision (LoG) and of the information reliability (LoI). Similarly, Santagati et al. (2018) present a modeling workflow for complex or ruined architectural heritage, where the determination of the LoA is critical due to the poor or lack of geometrical representation possibilities.

15.2.2   Examples of application

The current section presents some cases on which 3D point clouds have served to create the geometrical model for the HBIM methodology. Each one of these examples represents different stages of the conservation of a historic construction namely: (1) diagnostic, (2) intervention, and (3) preventive conservation.

With the aim of comparing each of these examples, we have considered appropriated to firstly introduce, in a brief way, the study case as well as the goal of each stage. Then, we present the surveying and modeling techniques. To conclude the presentation of the study case, we show the different families used as well as its different levels (LoD, LoI, LoG, and LoA) in accordance with the information available. It is worth mentioning that all the selected cases were modeled by using the well-known software Autodesk Revit. However, and due to the interoperability of the data, it is possible to extrapolate the approaches to other BIM software.

15.2.2.1   HBIM applied to the knowledge of the building: the Master Gate of San Francisco in Almeida, Portugal

The first example is a study case place at the village of Almeida (Portugal). More specifically, we show the case of the Master Gate of San Francisco. This construction is one of the main entrances of the village and it was erected in masonry. Nowadays, this building shows relevant conservation problems due to the high degradation of the stone which is promoting the reduction of the effective section of the barrel vault as well as the loss of detail in the ornamental parts, as shown in Fig. 15.1. For more

(A)   (B)

Figure 15.1 The Master Gate of San Francisco in Almeida: (A) general view of the main façade and (B) detail view of the conservation status of the vault that connect both façades. In this image it is possible to see the degradation of the stones as well as the presence of white crusts.

details about the construction systems the reader is referred to the study of Garcia-Gago et al. (2022).

This study case was selected because it represents a good example about the capacities of the HBIM approach for the diagnostics of historic constructions. The diagnosis is the first stage that we need to perform when we want to intervene in a historic construction and whose aim is to evaluate the current conservation status by using visual inspections, evaluation of available documentation (i.e., historic plans and previous interventions), as well as specific tests. The latter, which is commonly known as "previous tests," are tests that have the aim of understating specific questions of the building such as the physical properties of the construction material and the chemical composition of the damages. Under this basis, the authors decided to create an HBIM model that is able not only to organize the multisource information, but also to integrate the data obtained by evaluating the radiometric information contained in the point cloud (multispectral analysis, Garcia-Gago et al., 2022).

Due to the challenging illumination conditions, as well as the necessity of a high accuracy digitalization, the authors decided to digitalize the building by using a combination of laser scanning and panoramic imaging. More specifically, the authors used the Faro Focus 120 laser scanner as well as a Nikon D-5600 DLRS camera equipped with a NIKKOR 10.5 mm fisheye lens. These sensors allow to obtain the 3D point cloud of the building with not only geometric information but also radiometric data (intensity and real color) as well as 360-degree images that will be used as a support tool, all shown in Fig. 15.2. These data were complemented by several visual inspection campaigns as well as previous tests such as the ground-penetrating radar (GPR) and the boroscopic camera from previous experimental campaigns (Arce et al., 2018), as well as some new tests such as several X-ray power diffractions, scanning microscopy, and energy dispersive X-ray tests (Garcia-Gago et al., 2022).

All the information acquired was introduced within the HBIM framework by using three groups of families: (1) the families used for defining the construction elements on which the LoD is 300 and the LoI is 400; (2) the families used for defining the previous tests on which the LoD is 200 and the LoI is 500; and (3) the

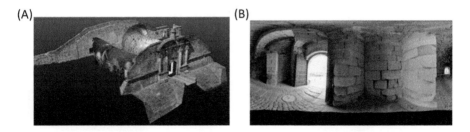

Figure 15.2 Images of the results obtained during the digitalization of the building: (A) general view of the points cloud without the masonry roof that covers the upper part of the building and (B) panoramic image on the inner part of the building.

families used for mapping the damages with an LoD of 200 and an LoI of 500. The authors do not define any data about the LoG and LoA used for the modeling. However, and according with the modeling techniques used, an LoG comprised between 1 and 8 can be expected.

For the first group of families, the authors use the 3D point cloud as well as the data provided by the GPR and the boroscopic camera (Fig. 15.3). These data allows to define the elements with the proper size, shape, location, and orientation by using parametric modeling strategies such as the extrusions or revolutions. Regarding the information contained in each element, the authors decided to define the type of construction technique in each element (as IfcText), the date of construction (introduced in the field IfcDate), a short description (also as IfcText), the date of inspection (as IfcDate), as well as the date associated with the properties of the stones by using different IfcText labels. The results of this modeling are shown in Fig. 15.3.

The second group of families encloses the previous tests carried out. In this case, the authors do not use the 3D point cloud for modeling, since the families are modeled by using path-objects that are placed on the constructive elements. These objects were of two types, as shown in Fig. 15.4: (1) spherical shapes for introducing the information related with the panoramic and 3D point cloud data and (2) rectangular or circular shapes for the rest of the previous tests. The information contained in the first type is just only a link for visualizing the data by using the FSPViewer (Free Viewer for Spherical Panoramic images) for the panoramic images and the Potree viewer for the 3D point cloud. For the second type, the authors decided to introduce a high level of information such as the type of test (IfcText), the name (IfcText), the date of acquisition (IfcDate), the date of processing (IfcDate), the equipment used (IfcText), a short description of the results obtained (IfcText), images and videos of the results (IfcURL), and metadata of the sample/s used (IfcText).

The third group is made up by all the families that are used to define the damages of the construction. In this case, the authors decide to use two groups shown in Fig. 15.4: (1) one used for integrating the damages detected by the radiometric and geometric analysis of the point cloud and (2) other used for the manual mapping of additional damages. Both damages were introduced into the HBIM

(A)   **(B)**

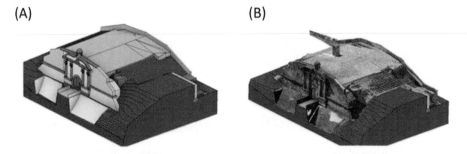

Figure 15.3 Views of the geometrical model used for the HBIM: (A) general view and (B) general view with the point cloud overimposed on the 3D model.

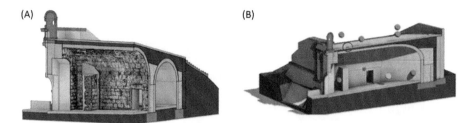

Figure 15.4 Result of the integration of damages and previous tests: (A) the automatic damage mapping based on the geometric and radiometric analysis carried out on previous campaigns and (B) use of path-based families to integrate the data, the spheres are used for the 3D point cloud and panoramic images and the rectangle based are used for the other previous tests. The boroscopic test attached to the roof is highlighted in pink color.

framework by using in-house Dynamo scripts. For the first type, the authors introduce each point of detection during the geometric and radiometric analysis of the point cloud as a type with spherical shape. Regarding the second type of damages, all those that are mapped in a manual way, the authors decided to use path-based object, similar to those used in the previous tests. These objects are attached to their corresponding constructive element. The algorithms used for detecting the damages by using point cloud processing strategies are described in detail in the work of Sánchez-Aparicio et al. (2018). Regarding the LoI, this type of family introduces the type of damage (IfcText), a three-level classification system (IfcText), a short description (IfcText), the direct and indirect causes (IfcText), associated damages (IfcText), proposed previous tests (IfcText), reparation proposal (Ifctext), and images (IfcURL), if they affect to a structural element (IfcText), its influence to the stability and urgency (IfcText).

15.2.2.2   HBIM applied during the intervention project: the Basilica di Collemaggio (L'Aquila)

The second example is the Basilica di Collemaggio within the city of L'Aquila (Italy). This building was erected by using masonry and timber elements and it was heavily affected by the earthquake that took place at the central region of Italy in 2009. Some of the damages are shown in Fig. 15.5. Due to this event, some parts of the Church collapsed requiring a large intervention to stabilize the structure as well as to recover its functionality. For more details about the construction systems, the reader is referred to the study of Brumana et al. (2018).

This study case was selected due to its role as representative of the current capacity of the technology for helping in the intervention of historic buildings and how the HBIM approach could be used as a framework of interoperability and for data storage. The intervention of a building requires to have a previous diagnosis of it and thus complete information about its current status (i.e., damage mapping, previous tests and historic information). With this information, it is possible to

Figure 15.5 Images of the building before and after its intervention: (A) general view of the building after the L'Aquila earthquake; (B) appearance of pillars placed at the central nave during the stabilization works; (C) aerial image of the building after its intervention; and (D) final appearance of the pillars placed at the main nave.

prescribe some intervention actions that will be used to recover the functionality of the building, including stabilization works or reconstruction of new parts.

During the process of intervention, a detailed view of the building in its previous state is required with the aim of having a robust image of the "original" elements as well as a second image after the intervention. In this sense, the authors decided to use several sensors. On the one hand, a total station is allowed to register, in a unique coordinate system, the 3D replicas of the building and on the other hand, the laser scanner and the digital camera (integrated in an aerial drone) to acquire the reality by using common laser scanning processing algorithms as well as the Structure from Motion approach. In the case of laser scanner, the Faro Focus 3D was the device used, and the photogrammetric survey was performed using the Falcon 8 drone equipped with a Sony NEX-5N camera, which was precalibrated. Complementary to this, several previous tests were carried out, involving material analysis, numerical simulations, damage mapping, and new construction elements, among others (Barazzetti et al., 2014).

According to Brumana et al. (2018), a total of three HBIM models were generated: (1) a simplified HBIM model, (2) a detailed HBIM model without the use of the LoG and its associated concept: the Grades of Generation (GOG), and (3) another detailed HBIM model which includes the LoG and GOG concepts.

Regarding the first one, the geometry was modeled with an LoD300 by generating simplified volumes for the provisional safety structures and other preliminary tasks (Brumana et al., 2018). In the second one, the authors used a higher LoD, with a value of 500 that allowed to define in depth the different constructive elements such as external and internal walls, openings, pillars (with the definition of ashlars), arches, vaulted systems, timber structures, and roof covering. These elements were modeled by using simple functionalities such as extrusion and substraction. These functionalities, as the authors declared, could lead to an LoG of about 1−8 (Brumana et al., 2018). The final model, which included the concept of LoG and GOG, was modeled with an LoD of 500 and an LoG of about 9−10 by using NURBs' (nonuniform rational basis-splines) surfaces for the vertical walls without plumb and nonplanarity shapes. This modeling strategy is able to reach an LoA of about 2−4 mm in damaged elements and could be carried out by using two modeling methods (1) the modeling based on slices and (2) the modeling based on control points obtained from the point cloud. A graphical explanation of these strategies could be consulted in Fig. 15.6.

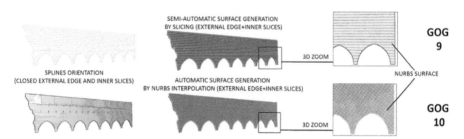

Figure 15.6 The different modeling strategies used for the detailed HBIM models.
Source: Adapted from Brumana, R., Della Torre, S., Previtali, M., Barazzetti, L., Cantini, L., Oreni, D., & Banfi, F. (2018). Generative HBIM modelling to embody complexity (LOD, LOG, LOA, LOI): surveying, preservation, site intervention—the Basilica di Collemaggio (L'Aquila). *Applied Geomatics. 10*(4), 545−567, http://www.springerlink.com/content/1866-- 9298/, https://doi.org/10.1007/s12518-018-0233-3.

If the readers want to evaluate more in depth the process, they can consult the work of Barazzetti et al. (2015).

With respect to the LoI, it is possible to observe a high level of information by including three groups of information: (1) the general information, (2) the materials and techniques, and (3) the damages/deterioration. For the first one, the authors proposed the use of different descriptors such as the dimension, technology, materials, construction technique, and stratigraphy. These descriptors could be filled by using IfcText and IfcReal fields. Visual information is also included in the form of images and graphical documentation. This information could be linked by using IfcURL labels. Regarding the second group, the authors propose a description of the material, processing, diagnostics, and information attached in the form of links by using, probably, the same labels like the previous group. For the damages, the authors propose to include a description (qualitative and quantitative) as well as an identification of the damages by using IfcText among other labels. The integration of this information, the damage, is carried out by using a similar procedure to the previous study case.

In Fig. 15.7, it is possible to observe the appearance of the HBIM model used for this study case.

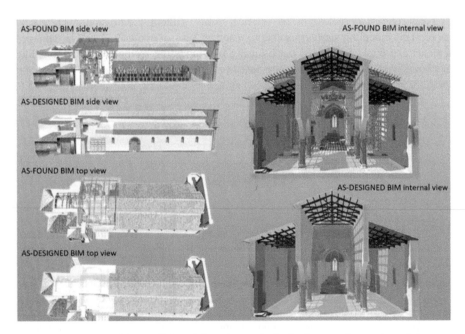

Figure 15.7 Images of the HBIM model used to assist in the intervention project of the building.

15.2.2.3   HBIM as a tool for supporting preventive conservation actions: The Ducal Palace of Bragança as study case

The last study case evaluated is the HBIM used for the preventive conservation of the Ducal Palace of Bragança within the locality of Guimaraes (Portugal) (Barontini et al., 2022), which state is shown in Fig. 15.8. The history of this

Figure 15.8 Images of the Ducal Palace in Guimaraes: (A) outdoors and (B) indoors.

building is intense since its construction, suffering a long period of decay, and partial dismantling. Nowadays, the Palace hosts a museum with dozens of assets (i.e., paintings and sculptures).

The huge amount of assets as well as the size of the building demands the design of a robust preventive conservation plan. Preventive conservation entails all those activities which are related with the maintenance of the building under the basis "prevention is better than cure." In this sense, it is necessary to evaluate the risk of degradation of each element (asset or not) to prioritize maintenance actions which need to be carried out by the manager of the building.

Coming back to the study case, the survey of the building was carried out by using historical documents, tradition survey methods as well as a laser scanner survey. The latter was performed with a laser scanner Leica P20, resulting in a huge point cloud made up by 1.164.664.694 points (Barontini et al., 2022). This 3D digitalization was used within a twofold BIM methodology which comprises the model generation as well as the nth inspection process, explained more in detail in Fig. 15.9. During these works, a total of three stakeholders were defined: (1) the asset manager, (2) the BIM modeler, and (3) the inspector, whose roles are defined in more detail in the work of Barontini et al. (2022).

With respect to the modeling strategy, the authors proposed the use of simple geometries on which the distortion from the ideal shape is not considered (i.e., out of plane movements) by using constant sections and extrusions, and being in line with an LoG comprised between $1-8$. On the contrary, the information related with these distortions are managed by using nongraphical information.

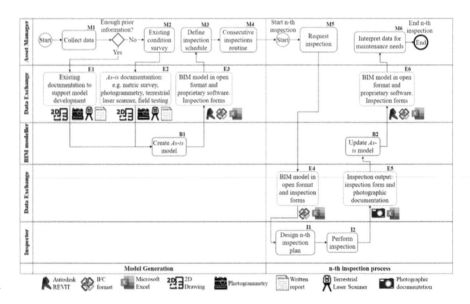

Figure 15.9 Inspection procedure proposed for the management of the Ducal Palace. *Source*: Adapted from (Barontini et al., 2022).

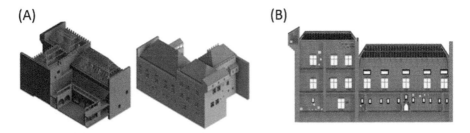

Figure 15.10 HBIM model of the Ducal palace: (A) general view of the construction elements and (B) section on which is possible to observe the damage families. The colors of the damage indicate different categories.

Regarding the families used, the authors included in this model the following (shown in Fig. 15.10): (1) families related with the physical elements and (2) families related with the damages observed.

The first type of families includes assets and construction elements which are modeled by using a low LoD (of about 300) and a high LoI (of about 500) (Fig. 15.10). This information is loaded within the model by using spreadsheets which are divided in different categories. For example, for the masonry wall, the authors proposed the following fields (Barontini et al., 2022): (1) general data, (2) construction data, (3) inspection data, (4) dimensional data, (5) structural data, (6) stone information, and (7) deformation, including the maintenance tasks. All this information could be defined by using Ifctext, Ifcreal, ifcboolean, ifcmassdensitymeasure, and ifclenghtmeasure.

With respect to the damage, the author declares the use of different types of strategies. The first one is the strategy which includes the information about the damage in the constructive element (i.e., tilt angle). On the contrary, the second group of damage (i.e., moisture, cracks, detachment, and corrosion) is modeled by using parallelepiped patches with fixed thickness (Fig. 15.10), obtaining an estimated LoD of about 200. These geometrical adoption allowed to update, in a fast way, its evolution along the time. The information contained in this families is structured as follows: (1) classification data, (2) inspection data (3) geometric data, (4) symptoms and diagnosis, and (5) evolution control data. This information is loaded by using IfcTex, Ifclenghtmeasure, and Ifccountmeasure. According to this, it can be concluded that the LoI is about 500.

15.3 From 3D models to Digital Twins

According to Grieves and Vickers (n.d.), a DT is "a model of the physical object or system, which connects digital and physical assets, transmits data in at least one direction, and monitors the physical system in real time." So, this section will show the current potentialities of the DT technology in the field of heritage by introducing several key concepts such as the geometric representation of the model, the connection with a sensor network as well as the data analytics. To finalize it, we will show several study cases which comprise from the generation of virtual visits and IoT sensors for real-time monitoring to numerical simulations able to predict the current and future behavior of a historical construction.

15.3.1   Conversion of the geometric representation

Regarding geometry conversion, different formats allow the exchange of 3D data between applications such as *COLLADA*, *FBX*, *X3D*, *VRML*, *gITF*, and *JSON*. This section will highlight some of the most used formats for creating DTs, indicating their characteristics, advantages, and disadvantages.

15.3.1.1   COLLADA Format

Collaborative Design Activity (*COLLADA*) is an open standard XML file format for storing 3D models. *COLLADA* files have a. *dae* extension and are compatible with an extensive selection of applications, which make it easy to share, view, and edit 3D files and animations. Also, it was designed by the Khronos Group, which allows free access to the format.

From its beginning, *COLLADA* was designed with the idea of compatibility between applications, which led ISO to adopt it as a standard (ISO/PAS 17506:2012,506:2012, 2022), for exchanging digital assets without losing information.

However, *COLLADA* has a complex file structure and a large file size, so in addition to taking up more storage space than other similar file types, exporting. *dae* files can be slow. On the other hand, this format is not usually updated and improved at same rate as other formats like *OBJ* or *FBX*.

15.3.1.2   OBJ Format

An *OBJ* file contains information about the geometry of 3D objects. The *OBJ* format is widely used for exchanging assets and nonanimated 3D models between applications. The OBJ format encodes the surface geometry of 3D models and their information on color and texture (materials). However, this format does not store other scene data, such as lighting or animations.

A considerable advantage of the. *obj* format is its simplicity and that it is open and widely supported for import and export between CAD software. Unlike other formats, such as COLLADA or FBX, opening and OBJ file in any CAD application is likely to interpret it correctly and consistently. In the case of COLLADA, the format is open but complex, which can lead to CAD software interpreting it differently and with inconsistencies. On the other hand, in the case of FBX, it is a closed format owned by Autodesk that offers a SDK to convert other formats to FBX. However, converting from FBX to another format is intricate, principally if Autodesk CAD software is not used. Also, due to the simplicity of the format and its native binary encoding, OBJ files are much more lightweight than FBX and COLLADA.

Despite its advantages, OBJ, in addition to not saving data on lighting and animations, does not support hierarchical information (parent−child relationship) of the parts of complex objects as DTs usually are.

15.3.1.3 FBX format

An FBX (. *fbx*) file is used for exchanging 3D geometry. Different programs can open, edit, and export 2D and 3D files through the FBX format. FBX is owned by Autodesk since 2006 and allows a high level of interoperability between Autodesk programs, that is, 3ds Max, AutoCAD, and Maya. It can also work with other non-Autodesk programs, although these may have limitations and inconsistencies, that is, exporting textures or lights may be problematic.

Despite being a widely used format, it has some limitations, that is,. *fbx* files can become quite complex depending on the amount of data to be transferred, therefore the import−export process can become slow. Then, it is recommended to export what is truly necessary. On the other hand, the preservation of attributes during the import and export is based on having sufficient knowledge about the source and destination programs and whether or not these programs are from Autodesk.

In any case, FBX is a helpful format that offers support for 2D and especially 3D graphics. It is a format that allows not only the exchange of 2D and 3D models but also animations, cameras, lighting effects, textures, colors, and even the hierarchical information of the models that can become fundamental in DTs of complex objects or processes.

15.3.1.4 JSON format

A JSON file stores simple data structures or objects in JavaScript Object Notation (JSON) format, a standard format for exchanging information. It is mainly used for data transmission between web applications and servers. JSON files are lightweight, text-based, and human-readable. These features have given rise to different formats for the exchange of 3D data based on JSON.

15.3.1.4.1 glTF

glTF (GL Transmission Format) is a format that saves information from a 3D model in JSON format. JSON minimizes the size of 3D objects and the execution time required for processing the assets. This format is intended to be an extensible publishing format for 3D content tools and services that simplifies workflows and enables interoperable use of content in different fields.

glTF files represent the 3D models through a complete description of the scene in JSON format that includes (1) the node hierarchy, materials, cameras, information about the meshes, and animations and (2) binary files (*.bin*) containing geometry and animation data; and (3) image files for the textures.

15.3.1.4.2 CityJSON

CityJSON is a JSON-based format for storing 3D models of cities and landscapes. This format defines ways to describe 3D features and objects that can commonly be found in cities (buildings, streets, rivers, bridges, and so on) and the relations that can exist between them. A CityJSON file defines the geometry and semantics of these assets, making it an appealing format for DT applications.

However, this format is considerably new, so it is still experimental in many cases, and it is not compatible with many 3D software applications and with those that is compatible, it is likely that the software does not have enough functionalities to work with it efficiently (i.e., view, generate, edit, and convert).

15.3.1.5 Datasmith format

Datasmith is a series of tools and plugins for bringing content to Unreal Engine (UE) from different 3D software applications. Datasmith's goal is to be able to bring complete prebuilt scenes and their complex relations to UE, no matter how big, dense, or heavy those scenes might be. To do this, Datasmith reuses the assets and layers built in other 3D design applications. This idea differs from other formats where the user is often forced to deconstruct scenes and divide them into fragments and then pass them to UE as separate parts (like FBX) and then reassemble them within UE.

Likewise, Datasmith supports a wide variety of applications (i.e., Autodesk 3ds Max, Cinema 4D, Revit, and Rhino 3D) and is constantly updating to increase this support. However, it only serves to transfer data from these applications to the UE.

15.3.2 Data in real time: the Internet of Things sensors

The aim of integrating IoT technology in heritage is to monitor physical and environmental parameters, providing real-time data for analyzing and supervising the conservation of cultural and historical legacy (Elabd et al., 2021; Lerario & Varasano, 2020).

There are several types of sensors with the purpose of heritage monitoring such as structural sensors, environmental sensors, or even cameras (Lynch & Loh, 2006). Depending on the pathology, the measurement device should be adapted to achieve the needs or the proposal of the case of study (Perles et al., 2018). Furthermore, there are other key aspects to consider in the device selection such as data transmission (Mahapatra et al., 2015), power source (Zhang et al., 2012), or cost (Lee & Lee, 2015).

Regarding the deployment of the IoT network, the determination of the technical aspects varies depending on the context of the heritage site (Kim et al., 2007). Thus, the technical configuration of the IoT devices is more complicated in locations with no energy neither data networks, where large areas of coverage are required to ensure the deployment of multiple IoT devices and that data are sent effectively from different locations through the study area. Due to the geometrical complexity of some heritage sites and their isolated location, the power demand of devices should be as low as possible (Callebaut et al., 2021) because the power source is normally a battery, simplifying the power requirements of IoT devices (Heble et al., 2018) deployed in heritage scenarios. These aspects can be supplied by a LPWAN (Ayele et al., 2018) (low-power wide area network) with a large area of coverage and low-power demand benefits, this being a compromise solution between the amount of data transmission and the duration of the battery.

The energy demand in IoT devices is directly affected by the type of sensor and the frequency and volume of data transmitted: for example, using a big resolution sensor or high amounts of sample data implies a huge demand of energy (Georgiou et al., n.d.). The selection of an adequate sensor and the establishment of a frequency of measurement to a specific proposal should be an important step for a successful IoT network deployment.

Depending on where the data are processed, there are two possibilities: edge data processing and server data processing (Tyagi et al., 2016). The edge data processing refers to the process of analyzing data at the edge of the network (Young et al., 2017), rather than in the server, which supposes a reduction in latency, the optimization of the network capacity, and the possibility to perform real-time actions. In the other hand, the complexity of the IoT device increases, as well as its energy demand due to the autonomous processing of data. According to the application, it should be recommended to process data on the edge, for example, in the case of detecting a crack or fracture by its characteristic sound applying IA models implemented in the device. When this process is not mandatory, it is recommended to simplify the IoT device as much as possible (Ammar et al., 2020), to maximize the energy autonomy, processing data on the server.

15.3.3   Data analytics

Data analytics studies the raw data to find trends and answer-specific questions. It consists of various techniques and processes that allow inferences from historical data. The main goal of data analytics is to transform data into optimization insights, and for this, there are two fundamental tasks: descriptive examination and advanced analysis.

During the descriptive examination, the trends are characterized based on historical data; this task does not focus on making predictions or decisions. Its objective is to summarize the data to make it descriptive and meaningful.

Then, in the advanced analysis, there are tools for extracting the data, making predictions and discovering new trends. These tools include machine learning technologies (i.e., neural networks), natural language processing, and sentiment analysis. Fig. 15.11 shows the steps that constitute the data analytics.

DTs often work together with data analytics to increase their potential. The data collected by DTs can be used to improve the accuracy of predictions made by forecasting software; or even for the training of predictive models with the same objective of improving the accuracy of predictions of the behavior of systems and processes. Likewise, the combined use of DTs and data analytics allows studies of how an operation would work under certain conditions, which makes it possible to identify potential problems and their solutions. So, it can be said that there are two fundamental data analytics areas for DTs: diagnosis and predictive analysis and prescriptive analysis.

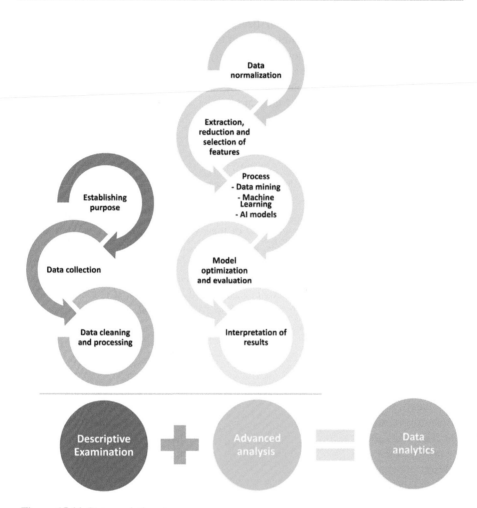

Figure 15.11 Data analytics steps.

In the diagnosis and predictive analysis, the DT must predict the future behavior of the system from some input data. Then, machine learning models based on IoT are usually applied. These allow the creation of intelligent processes, where the data from the sensors are analyzed in real time to diagnose and predict future behaviors that permit preventing problems and failures that may occur in the future.

On the other hand, the prescriptive analysis is made from the simulation of the entire system or process to identify its optimal operation. To do this, the prescriptive analysis starts with a series of variables and restrictions that must be taken into account. These optimization problems are widely used for planning and scheduling tasks. Then, machine learning models predict likely outcomes from a set of historical input data. Instead, optimization models allow deciding how to approach or take advantage of certain predicted outcomes that occurred.

15.3.4 Examples of application

15.3.4.1 Predictive potential

The example presented in this section highlights the potential that the joint use of virtual reality (VR) and IoT can offer for cultural heritage conservation. On the other hand, minimizing tourist visits to some historical buildings can be beneficial for their conservation. However, it is a pity that the public does not have access to historical heritage. Therefore this example intends to solve this situation, creating a DT of the General Historical Library of the University of Salamanca, which, in addition to enabling its monitoring, allows its virtual visit.

The preventive conservation approach for heritage buildings and the assets within them is considered the most efficient for their maintenance and protection (Kutasi & Vidovszky, 2010; Van Balan & Vandesande, n.d.). However, for the efficient implementation of this approach, it is necessary to start from a standardized workflow for documentation, registration, and information management, such as the one proposed in the European project HeritageCare. The HeritageCare system derives from the systematic inspection and monitoring process supported by digitization and the application of intelligent technologies (Masciotta et al., 2021).

A digitization of the General Historical Library of the University of Salamanca was performed, taking the HeritageCare system as reference. Once the 3D model was obtained, different approaches were considered to achieve both: monitoring and virtual tour objectives. After an arduous investigation, it was decided to apply this digitization to create a VR-IoT application since VR would allow realistic, interactive, and immersive virtual tours; and IoT, for its part, would allow the use of sensors for monitoring and eventually for damage prediction (when enough data is gathered).

VR is a simulation of a physical environment in which the user digitally practices their presence (Samadbeik et al., 2018). VR has come to be used in the construction field to visualize geographic data to support decision-making for maintenance work (Pirani, 2018). In the cultural heritage area, VR is a cognitive tool through which the user can interact with heritage and its information in a virtual environment (Bekele et al., 2018). Regarding heritage buildings, VR applications aim to preserve, document, investigate, educate, rebuild, and explore (Bruno et al., 2010).

In this line, from the 3D model, an interactive VR application of the General Historical Library of the University of Salamanca has been developed using UE. This application allows virtually visiting the library, as shown in Fig. 15.12.

IoT sensors have been used to achieve the monitoring and eventual prediction objectives. The sensors' values are shown in the VR application (Fig. 15.13) to monitor the environmental variables of the building. The control of these environmental variables is essential for the books' conservation, since some are very old and susceptible to deterioration.

15.3.4.2 Applicability of Digital Twins for solar potential simulation

Another application of DTs in historical constructions is to allow simulations regarding solar radiation. Knowing the solar radiation received by the surfaces is a

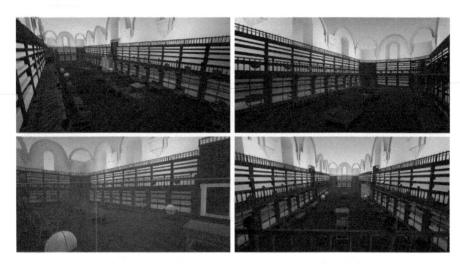

Figure 15.12 Virtual visit to the General Historical Library of the University of Salamanca.

Figure 15.13 Monitoring through IoT sensors. *IoT*, Internet of Things.

key issue for the conservation of heritage assets, due to the degradation of the materials produced during their exposure to the Sun. In addition, in a context of energy transition and sustainable conservation, the computation of the solar radiation may allow the design of adapted solar photovoltaic installations, using new technologies that do not perform any disturbance to the historic construction nor to its visual appearance, such as panels with colors similar to those of the roof materials, solar glasses for windows, flexible panels, and installation of the panels in the façades of the common spaces (Moschella et al., 2013).

An example of the capabilities of DTs of historical constructions for the evaluation of the degradation of materials due to sunlight is shown in the work of Tysiac et al. (2023). In this paper, the Autodesk Revit © tool is used for the modeling of the sunlight outside Saint Adalbert Church in Poland to determine the risk of degradation. The methodology developed is validated, consequently validating the use of DT and HBIM for sunlight evaluation toward architectural and conservation analysis.

Regarding the computation of solar radiation, the main requirement is that the DT is georeferenced, that is, its position and orientation in the reality is established. If that is the case, the first step for the computation of annual solar radiation consists of the determination of the shades that the buildings and constructions on the surroundings impose on the surface under study. This step requires the modeling of the historic construction and also of its surroundings, which is named as DEM (Digital Elevation Model). An example of shade modeling is shown in Fig. 15.14.

Then, the computation of the solar irradiation on the surface where the solar installation is planned is performed, based on solar radiation models that calculate the incoming solar radiation on a surface according to its position on the Earth and orientation (angle regarding the North and angle of inclination) toward the Sun. Several models can be used for the performance of the computation of annual solar radiation, such as PVGIS or PVWatts; for more information about the procedure and the solar radiation models, the readers are referred to the work of Sánchez-Aparicio et al. (2021).

The knowledge of the hours that the surface suffers from shading allows the subtraction of the lost incoming radiation from the total value. These values of solar irradiation on the surface where the installation is planned, together with the technical characteristics of the solar panel to be installed, allow for the determination of the production curve, both for the annual and monthly productions (Fig. 15.15).

Figure 15.14 Example of shade modeling.

15.3.4.3   Advanced numerical simulations potential: the study case of San Torcato (Guimaraes)

Apart from the applications previously shown, it is important to highlight the current potential of the computational methods for simulating the structural behavior of historical constructions. In this sense, we considered appropriated the work carried out by Funari et al. (2021), which could be considered an improvement to the work previously performed by Sánchez-Aparicio et al. (2014).

Both works were carried out on the Church of Saint Torcato within the municipality of Guimaraes (Portugal). Most of the parts of the building were erected on regular granite masonry with some elements (dome and apse) made of reinforced concrete (Funari et al., 2021). This Church has suffered important structural problems due to the weakness of the soil (Fig. 15.16). According to this, several in situ campaigns were carried out with the aim of diagnosing the building, being monitored from a static and dynamic point of view. This task started at 2009, with the integration of one tilt meter at the top of each tower as well as the

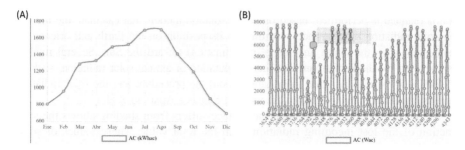

Figure 15.15 Example of solar production curves for the roofs of the Library of Salamanca: (A) annual production, per month and (B) production in the month of June, detailed per day.

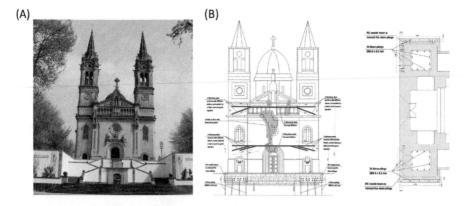

Figure 15.16 Saint Torcato Church: (A) appearance of the main façade and (B) intervention works carried out on the main façade to stabilize the structure.

dynamic identification by means of a total of 39 accelerometers. Regarding the dynamic network, it was decided to design a permanent network made up by just only four accelerometers, one per principal direction (longitudinal and transversal) of each tower. The data acquired by each accelerometer is automatically processed by using the methodology defined in the work of Ramos et al. (2013). Finally, it was decided to intervene the building by introducing several tied rods between the towers and some micropiles at the foundation, maintaining the dynamic monitoring system.

The data acquired by this system, before and after the intervention, has been used to generate and accurate numerical model in the work carried out by Funari et al. (2021). In this study, the geomatrical solid model was generated by using the laser scanner point cloud acquired by Leica ScanStation P20, requiring a total of 174 scans and generating a point cloud made up by 3 billion points (Funari et al., 2021). Then this 3D digitalization, as well as the data coming from other previous tests, was used to create the solid model by using a generative programming approach. This programming paradigm allow to modify 3D solid components since there are connected with specific variables (i.e., thickness of a wall). Under this basis, the authors propose a workflow which generates constructive elements by using different entities and subentities which are created by using Grasshopper, Rhino, and the Python programming language (Fig. 15.17). This approach allows to parametrize the model, reducing the amount of time invested with respect to previous modeling strategies based on the reverse engineering approaches carried out by Sánchez-Aparicio et al. (2014).

Finally, this solid model was introduced into a finite element environment, more specifically the Abaqus software, allowing the simulation of the structural behavior of the building by using the mechanic data provided by the different

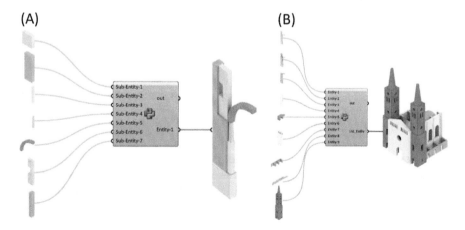

Figure 15.17 Examples of the generative modeling carried out in this study case: (A) the creation of structural units by using different subentities and (B) the creation of the complete 3D model.

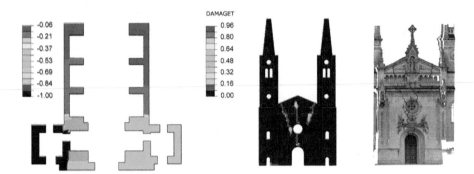

Figure 15.18 Results of the simulation carried out on the preintervention situation.
Source: Adapted from Funari, M. F., Hajjat, A. E., Masciotta, M. G., Oliveira, D. V., & Lourenço, P. B. (2021). A parametric scan-to-FEM framework for the digital twin generation of historic masonry structures. *Sustainability* (Switzerland). *13*(19), https://www.mdpi.com/2071-1050/13/19/11088/pdf, https://doi.org/10.3390/su131911088.

previous tests carried out on the Church. It is worth mentioning that the data coming from the permanent dynamic network was used with the aim of matching the real behavior (captured by the network) with the simulated one. This stage, named as finite element model updating, was carried within the tools previously mentioned (Grasshopper, Python, and Abaqus) by using an objective function that takes into consideration the four first frequencies of the Church (Funari et al., 2021).

This DT approach allowed to generate a computational model that represents properly the behavior of the Church before and after the intervention. Thanks to this, the authors were able to simulate different future scenarios by using a nonlinear finite element approach which considers the elastic and plastic behavior of the masonry (Funari et al., 2021). Among the different scenarios simulated by the authors, it is possible to highlight the foundation settlement produced before the intervention, obtaining results that match perfectly with respect to the real damage observed (Fig. 15.18).

15.4   Current challenges

The DT technology is rarely applied to heritage buildings, because the complexity that the generation of the DT entails implies its creation only for significant buildings which management includes a significant budget, or with complex needs of automation and domotics.

The first challenge to be overcome in the field of heritage buildings toward their consideration as candidates for DT generation is the geometric modeling. While modern buildings can be either complex or simple in design, heritage buildings usually present complex geometries, with nonstandard shapes and materials. This implies two challenges: first, the data acquisition of heritage buildings using terrestrial laser scanning is time-consuming and can present incremental error, due to the need to perform several acquisitions and the posterior registration of all point clouds acquired into the same coordinate system. The advances in 3D digitization using SLAM (simultaneous localization and mapping) will allow the increase on resolution and accuracy of the 3D representation. SLAM consists of the integration of laser scanning sensors on mobile platforms, with an inertial measurement unit and a control unit, in such a way that the geometrical 3D model is generated at the same time of the acquisition.

Second, the data processing is laborious and presents low level of automation. This step will benefit for the advances in the fields of AI and deep learning toward the segmentation of point clouds and their semantic recognition, although special adaptation will be required for the application of the developments on modern or residential buildings to historical constructions so that the AI models are able to recognize the particular assets present in heritage buildings and linked them with their particularities according to the year and type of construction.

The latter (information about construction, materials, and history of the buildings) implies a third challenge for the data compilation required for DT generation. Advances in AI and big data will allow the conversion of the 3D model into an HBIM (Historic Building Information Model).

As a fourth challenge, the inclusion of data from other sensors is required to enrich the HBIM toward the DT. Although the installation of IoT sensors is not completely generalized in the field of historical buildings, their presence in these scenarios is already not uncommon. The existence of low consumption sensors and networks has enlightened the process of digital and real-time monitoring, so that the provision of the character of the building into the DT toward the modeling of its structural and thermal behaviurs is the challenge closest to overcome.

15.5   Final remarks

This chapter presents the different technologies involved in the generation of DTs of buildings, and their particularities for application to historical constructions. The benefits of DT technology in the field of cultural heritage conservation come from the enlightening of the evaluation and decision-taking in the intervention phase and include the management of the building in all its phases, including possible intervention actions but also the organization of visits, the energy retrofitting of the buildings, and the control of the ambient conditions to minimize the degradation of the heritage assets present in the building.

Using point clouds as initial input for the DT generation, the main steps involved on the development of this new technology are (1) to develop a methodology for the generation of Building Information Models of historic sites (HBIM) from laser scanning data and imagery sensors (multispectral and thermal infrared can provide information about the state of the construction); (2) to incorporate information coming in real time from environmental sensors, using IoT communications, generating a DT of heritage sites for dynamic purposes; (3) to provide restorers with a digital support where simulations can be performed prior restoration actions to evaluate their impact on the heritage site and validate their implementation; and (4) to offer a digital avatar of the heritage sites that can help in the training of restorers, but also in the dissemination of the site to potential visitors, and to raise awareness of the effect of nonrespectful visiting activities on the maintenance of the site.

Thus the generation of DT of historical buildings still presents some challenging steps to be overcome toward their automation to make possible the spread of the use of this technology for the conservation of cultural heritage. However, the challenges are clearly identified, and the technologies required are all under active development, in such a way that their overcoming approaches at a promising speed.

References

AEC (CAN). (2014). *Protocol, implementing Canadian BIM standards for the architectural, engineering and construction industry based on international collaboration.*

Ammar, M., Crispo, B., & Tsudik, G. (2020) SIMPLE: A remote attestation approach for resource-constrained IoT devices. *ACM/IEEE 11th International Conference on Cyber-Physical Systems (ICCPS).* 2642–9500 247–258. Available from https://doi.org/10.1109/ICCPS48487.2020.00036.

Arce, A., Ramos, L. F., Fernandes, F. M., Sánchez-Aparicio, L. J., & Lourenço, P. B. (2018). Integrated structural safety analysis of San Francisco Master Gate in the Fortress of Almeida. *International Journal of Architectural Heritage, 12*(5), 761–778. Available from https://doi.org/10.1080/15583058.2017.1370507, http://www.tandf.co.uk/journals/titles/15583058.asp.

Ayele, E. D., Meratnia, N., & Havinga, P. J. M. (2018). Towards a new opportunistic IoT network architecture for wildlife monitoring system. *9th IFIP International Conference on New Technologies, Mobility and Security (NTMS).* 1–5. 2157–4960. https://doi.org/10.1109/NTMS.2018.8328721.

Barazzetti, L., Banfi, F., Brumana, R., & Previtali, M. (2015). Creation of parametric BIM objects from point clouds using nurbs. *Photogrammetric Record, 30*(152), 339–362. Available from https://doi.org/10.1111/phor.12122, http://www.blackwellpublishing.com/journals/PhotRec.

Barazzetti, L., Brumana, R., Oreni, D., Previtali, M., & Roncoroni, F. (2014). UAV-based orthophoto generation in urban area: The Basilica of Santa Maria di Collemaggio in L'Aquila. *Lecture Notes in Computer Science (including subseries Lecture Notes in Artificial Intelligence and Lecture Notes in Bioinformatics), 8582*(4), 1–13. Available from https://doi.org/10.1007/978-3-319-09147-1_1, 16113349. Springer Verlag Italy, http://springerlink.com/content/0302-9743/copyright/2005/.

Barontini, A., Alarcon, C., Sousa, H. S., Oliveira, D. V., Masciotta, M. G., & Azenha, M. (2022). Development and demonstration of an HBIM framework for the preventive conservation of cultural heritage. *International Journal of Architectural Heritage, 16*(10), 1451–1473. Available from https://doi.org/10.1080/15583058.2021.1894502.

Bekele, M. K., Pierdicca, R., Frontoni, E., Malinverni, E. S., & Gain, J. (2018). A survey of augmented, virtual, and mixed reality for cultural heritage. *Journal on Computing and Cultural Heritage, 11*(2), 1–36. Available from https://doi.org/10.1145/3145534.

Bertin, I., Mesnil, R., Jaeger, J. M., Feraille, A., & Le Roy, R. (2020). A BIM-based framework and databank for reusing load-bearing structural elements. *Sustainability*, *12*(8), 3147. Available from https://doi.org/10.3390/su12083147.

Brumana, R., Della Torre, S., Previtali, M., Barazzetti, L., Cantini, L., Oreni, D., & Banfi, F. (2018). Generative HBIM modelling to embody complexity (LOD, LOG, LOA, LOI): surveying, preservation, site intervention—the Basilica di Collemaggio (L'Aquila). *Applied Geomatics*, *10*(4), 545−567. Available from https://doi.org/10.1007/s12518-018-0233-3, http://www.springerlink.com/content/1866-9298/.

Bruno, F., Bruno, S., De Sensi, G., Luchi, M. L., Mancuso, S., & Muzzupappa, M. (2010). From 3D reconstruction to virtual reality: A complete methodology for digital archaeological exhibition. *Journal of Cultural Heritage*, *11*(1), 42−49. Available from https://doi.org/10.1016/j.culher.2009.02.006, https://www.sciencedirect.com/science/article/pii/S1296207409000958.

buildingSMART International. (2021). *Software Implementations—buildingSMART technical*.

Callebaut, G., Leenders, G., Van Mulders, J., Ottoy, G., De Strycker, L., & Van der Perre, L. (2021). The art of designing remote IoT devices—Technologies and strategies for a long battery life. *Sensors*, *21*(3), 1424−8220. Available from https://doi.org/10.3390/s21030913.

Charlton, J., Kelly, K., Greenwood, D., & Moreton, L. (2021). The complexities of managing historic buildings with BIM. *Engineering, Construction and Architectural Management*, *28*(2), 570−583. Available from https://doi.org/10.1108/ECAM-11-2019-0621.

Craveiro, F., Duarte, J. P., Bartolo, H., & Bartolo, P. J. (2019). Additive manufacturing as an enabling technology for digital construction: A perspective on Construction 4.0. *Automation in Construction*, *103*, 251−267. Available from https://doi.org/10.1016/j.autcon.2019.03.011, https://www.sciencedirect.com/science/article/pii/S0926580518310781.

Diara, F. (2022). HBIM open source: A review. *ISPRS International Journal of Geo-Information*, *11*(9). Available from https://doi.org/10.3390/ijgi11090472, http://www.mdpi.com/journal/ijgi.

Elabd, N. M., Mansour, Y. M., & Khodier, L. M. (2021). Utilizing innovative technologies to achieve resilience in heritage buildings preservation. *Developments in the Built Environment*, *8*, 100058. Available from https://doi.org/10.1016/j.dibe.2021.100058, https://www.sciencedirect.com/science/article/pii/S266616592100017X.

Fai, S., & Rafeiro, J. (2014). Establishing an appropriate level of detail (LOD) for a building information model (BIM) - West Block, Parliament Hill, Ottawa, Canada. *ISPRS Annals of the Photogrammetry, Remote Sensing and Spatial Information Sciences*, *5*(2), 123−130. Available from https://doi.org/10.5194/isprsannals-II-5-123-2014, 21949050 Copernicus GmbH Canada, http://www.isprs.org/publications/annals.aspx.

Funari, M. F., Hajjat, A. E., Masciotta, M. G., Oliveira, D. V., & Lourenço, P. B. (2021). A parametric scan-to-FEM framework for the digital twin generation of historic masonry structures. *Sustainability (Switzerland)*, *13*(19). Available from https://doi.org/10.3390/su131911088, https://www.mdpi.com/2071-1050/13/19/11088/pdf.

Gabriele, M. (2021). M.: HBIM-GIS integration with an IFC-to-shapefile approach: The previtali, palazzo, vimercate, case study, Ann ISPRS, Photogramm. *Remote Sens. Spatial Inf. Sci*. Available from https://doi.org/10.5194/isprs-annals-VIII-.

Garagnani, S., & Manferdini, A. M. (2013). Parametric accuracy: Building information modeling process applied to the cultural heritage preservation. *The International Archives of the Photogrammetry, Remote Sensing and Spatial Information Sciences*, *XL-5/W1*, 87−92. Available from https://doi.org/10.5194/isprsarchives-xl-5-w1-87-2013.

Garcia-Gago, J., Sánchez-Aparicio, L. J., Soilán, M., & González-Aguilera, D. (2022). HBIM for supporting the diagnosis of historical buildings: Case study of the Master Gate of San Francisco in Portugal. *Automation in Construction, 141*, 104453. Available from https://doi.org/10.1016/j.autcon.2022.104453, https://www.sciencedirect.com/science/article/pii/S0926580522003260.

Georgiou, K., Xavier-de-Souza, S., & Eder, K., (n.d.). The IoT energy challenge: A software perspective. *IEEE Embedded Systems Letters, 10*(3), 53−56. 1943-0671. Available from https://doi.org/10.1109/LES.2017.2741419.

Gharbia, M., Chang-Richards, A., Lu, Y., Zhong, R. Y., & Li, H. (2020). Robotic technologies for on-site building construction: A systematic review. *Journal of Building Engineering, 32*, 101584. Available from https://doi.org/10.1016/j.jobe.2020.101584, https://www.sciencedirect.com/science/article/pii/S2352710220313607.

González-Aguilera, D., Rodríguez-Gonzálvez, P., Rodríguez-Martín, M., & Sánchez-Aparicio, L. J. (2021). Inteligencia artificial y defensa: nuevos horizontesGemelo Digital: los modelos 3D y la inteligencia artificial como apoyo a la toma de decisiones en el ámbito de la seguridad. *Aranzadi Thomson Reuters*, 79−94.

Graham, K., Chow, L., & Fai, S. (2018). Level of detail, information and accuracy in building information modelling of existing and heritage buildings. *Journal of Cultural Heritage Management and Sustainable Development, 8*(4), 495−507. Available from https://doi.org/10.1108/JCHMSD-09-2018-0067.

Grieves, M., & Vickers, J., (n.d.) *Digital twin: Mitigating unpredictable, undesirable emergent behavior in complex systems.*

Guzzetti, F., Anyabolu, K. L. N., Biolo, F., & Dell'orto, R. (2022). A particular approach to digitization of the architectural and information heritage. International Archives of the Photogrammetry, Remote Sensing and Spatial Information Sciences - ISPRS Archives. *International Society for Photogrammetry and Remote Sensing Italy BIM AND CASTELLO SFORZESCO IN MILAN, 46*(5), 115−122. Available from https://doi.org/10.5194/isprs-archives-XLVI-5-W1-2022-115-2022, 16821750, http://www.isprs.org/proceedings/XXXVIII/4-W15/.

Heble, S., Kumar, A., Prasad, K. V. V. D., & Samirana, S., & Rajalakshmi, P., & Desai, U. B. (2018). A low power IoT network for smart agriculture. *IEEE 4th World Forum on Internet of Things (WF-IoT)* (pp. 609−614). Available from https://doi.org/10.1109/WF-IoT.2018.8355152.

ISO/PAS 17506:2012. (2022). ISO/PAS 17506:2012. Available from https://www.iso.org/standard/59902.html.

Kim, S., Pakzad, S., Culler, D., Demmel, J., Fenves, G., Glaser, S., Turon, M. (2007) Health monitoring of civil infrastructures using wireless sensor networks. *6th International Symposium on Information Processing in Sensor Networks* (pp. 254−263). Available from https://doi.org/10.1109/IPSN.2007.4379685.

Kutasi, D., & Vidovszky, I. (2010). The cost effectiveness of continuous maintenance for monuments and historic buildings. *Periodica Polytechnica Architecture, 41*(2), 57−61. Available from https://doi.org/10.3311/pp.ar.2010-2.03, https://pp.bme.hu/ar/article/view/13.

Lee, I., & Lee, K. (2015). The Internet of Things (IoT): Applications, investments, and challenges for enterprises. *Business Horizons, 58*(4), 431−440. Available from https://doi.org/10.1016/j.bushor.2015.03.008, https://www.sciencedirect.com/science/article/pii/S0007681315000373.

Lerario, A., & Varasano, A. (2020). An IoT smart infrastructure for s. domenico church in matera's "sassi": a multiscale perspective to built heritage conservation. *Sustainability, 12*(16), 6553. Available from https://doi.org/10.3390/su12166553.

Lynch, J., & Loh, K. (2006). A summary review of wireless sensors and sensor networks for structural health monitoring. *The Shock and Vibration Digest*, *38*(2), 91−128. Available from https://doi.org/10.1177/0583102406061499, http://svd.sagepub.com/cgi/doi/10.1177/0583102406061499.

MacHer, H., Chow, L., & Fai, S. (2019). Copernicus GmbH France automating the verification of heritage building information models created from point cloud data. *ISPRS Annals of the Photogrammetry, Remote Sensing and Spatial Information Sciences*, *42* (2), 455−460. Available from https://doi.org/10.5194/isprs-archives-XLII-2-W9-455-2019, 21949050, http://www.isprs.org/publications/annals.aspx.

Macher, H., Landes, T., & Grussenmeyer, P. (2017). From point clouds to building information models: 3D semi-automatic reconstruction of indoors of existing buildings. *Applied Sciences (Switzerland)*, *7*(10). Available from https://doi.org/10.3390/app7101030, http://www.mdpi.com/2076-3417/7/10/1030/pdf.

Mahapatra, C., Sheng, Z., Leung, V. C. M., & Stouraitis, T. (2015). A reliable and energy efficient IoT data transmission scheme for smart cities based on redundant residue based error correction coding. *12th Annual IEEE International Conference on Sensing, Communication, and Networking - Workshops (SECON Workshops)* (pp. 1−6). Available from https://doi.org/10.1109/SECONW.2015.7328141.

Malinverni, E. S., Pierdicca, R., Paolanti, M., Martini, M., Morbidoni, C., Matrone, F., & Lingua, A. (2019). Deep learning for semantic segmentation of 3D point cloud. *International Archives of the Photogrammetry, Remote Sensing and Spatial Information Sciences - ISPRS Archives*, *42*(2), 735−742. Available from https://doi.org/10.5194/isprs-archives-XLII-2-W15-735-2019, 16821750, http://www.isprs.org/proceedings/XXXVIII/4-W15/.

Masciotta, M. G., Morais, M. J., Ramos, L. F., Oliveira, D. V., Sánchez-Aparicio, L. J., & González-Aguilera, D. (2021). A digital-based integrated methodology for the preventive conservation of cultural heritage: The experience of heritagecare project. *International Journal of Architectural Heritage*, *15*(6), 844−863. Available from https://doi.org/10.1080/15583058.2019.1668985.

Massafra, A., Predari, G., & Gulli, R. (2022). Towards digital twin driven cultural heritage management: a hbim-based workflow for energy improvement of modern buildings. *Int. Arch. Photogramm. Remote Sens. Spatial Inf. Sci*, *XLVI-5/W1−2022*, 149−157. Available from https://doi.org/10.5194/isprs-archives-XLVI-5-W1-2022-149-2022, https://isprs-archives.copernicus.org/articles/XLVI-5-W1-2022/149/2022/.

Mora, R., Sánchez-Aparicio, L. J., Maté-González, M. Á., García-Álvarez, J., Sánchez-Aparicio, M., & González-Aguilera, D. (2021). An historical building information modelling approach for the preventive conservation of historical constructions: Application to the Historical Library of Salamanca. *Automation in Construction*, *121*, 103449. Available from https://doi.org/10.1016/j.autcon.2020.103449, https://www.sciencedirect.com/science/article/pii/S0926580520310293.

Moschella, A., Salemi, A., Lo Faro, A., Sanfilippo, G., Detommaso, M., & Privitera, A. (2013). Historic Buildings in Mediterranean Area and solar thermal technologies: architectural integration vs preservation criteria. *Mediterranean Green Energy Forum 2013: Proceedings of an International Conference MGEF-13*, *42*, 416−425. Available from https://doi.org/10.1016/j.egypro.2013.11.042, https://www.sciencedirect.com/science/article/pii/S187661021301744X.

Oostwegel, L. J. N., Jaud, Š., Muhič, S., & Malovrh Rebec, K. (2022). Digitalization of culturally significant buildings: ensuring high-quality data exchanges in the heritage domain using OpenBIM. *Heritage Science*, *10*(1). Available from https://doi.org/10.1186/s40494-021-00640-y, http://www.springer.com/materials/journal/40494.

Opoku, D. J., Perera, S., Osei-Kyei, R., Rashidi, M., Famakinwa, T., & Bamdad, K. (2022). Drivers for digital twin adoption in the construction industry: A systematic literature review. *Buildings, 12*(2), 2075−5309. Available from https://doi.org/10.3390/buildings12020113.

Perles, A., Pérez-Marín, E., Mercado, R., Segrelles, J. D., Blanquer, I., Zarzo, M., & Garcia-Diego, F. J. (2018). An energy-efficient internet of things (IoT) architecture for preventive conservation of cultural heritage. *Future Generation Computer Systems, 81*, 566−581. Available from https://doi.org/10.1016/j.future.2017.06.030, https://www.sciencedirect.com/science/article/pii/S0167739X17313663.

Pirani, M. (2018). International Association for Automation and Robotics in Construction (IAARC). Mixed reaiity approach for the management of building maintenance and operation. *Proceedings of the 35th International Symposium on Automation and Robotics in Construction (ISARC)* (pp. 2413−5844) 199−206. Available from https://doi.org/10.22260/ISARC2018/0028.

Ramos, L. F., Aguilar, R., Lourenço, P. B., & Moreira, S. (2013). Dynamic structural health monitoring of Saint Torcato Church. *Mechanical Systems and Signal Processing, 35*(1−2), 1−15. Available from https://doi.org/10.1016/j.ymssp.2012.09.007.

Rossi, A., & Palmieri, U. (2020). Modelling based on a certified level of accuracy: The case of the Solimene Façade. *Nexus Network Journal, 22*(3), 615−630. Available from https://doi.org/10.1007/s00004-019-00474-z.

Samadbeik, M., Yaaghobi, D., Bastani, P., Abhari, S., Rezaee, R., & Garavand, A. (2018). The applications of virtual reality technology in medical groups teaching. *Journal of advances in medical education & professionalism, 6*(3), 123−129.

Santagati, C., Lo Turco, M., & Garozzo, R. (2018). Reverse information modelling for historic artefacts: Towards the definition of a Level of Accuracy for ruined heritage. The International Archives of the Photogrammetry. *Remote Sensing and Spatial Information Sciences, XLII*(2), 1007−1014. Available from https://doi.org/10.5194/isprs-archives-XLII-2-1007-2018.

Sedano-Espejo, E., Méndez-Moreno, C., Sánchez-Aparicio, L. J., García-Morales, S., Aira, J. R., Moreno, E., Pinilla-Melo, J., Sanz-Arauz, D., Palma-Crespo, M., & González-Aguilera, D. (2022). *Use of a GIS-based solution for the design of preventive conservation plans in heritage constructions* (75−89). Cham: Springer International Publishing, 978-3-031-15676-2.

Simeone, D., Cursi, S., Toldo, I., & Carrara, G. (2014). ACADIA Italy Bim and knowledge management for building heritage. *ACADIA 2014 - Design Agency: Proceedings of the 34th Annual Conference of the Association for Computer Aided Design in Architecture.* 2014- 681−690. 9781926724515.

Statsenko, L., Samaraweera, A., Bakhshi, J., & Chileshe, N. (2022). Construction 4.0 technologies and applications: a systematic literature review of trends and potential areas for development. Construction Innovation. ahead-of-print (ahead-of-print). Available from https://doi.org/10.1108/CI-07-2021-0135.

Sánchez-Aparicio, L. J., Del Pozo, S., Ramos, L. F., Arce, A., & Fernandes, F. M. (2018). Heritage site preservation with combined radiometric and geometric analysis of TLS data. *Automation in Construction, 85*, 24−39. Available from https://doi.org/10.1016/j.autcon.2017.09.023.

Sánchez-Aparicio, M., Martín-Jiménez, J., Del Pozo, S., González-González, E., & Lagüela, S. (2021). Ener3DMap-SolarWeb roofs: A geospatial web-based platform to compute photovoltaic potential. *Renewable and Sustainable Energy Reviews, 135*, 110203. Available from https://doi.org/10.1016/j.rser.2020.110203, https://www.sciencedirect.com/science/article/pii/S1364032120304937.

Sánchez-Aparicio, L. J., Riveiro, B., González-Aguilera, D., & Ramos, L. F. (2014). The combination of geomatic approaches and operational modal analysis to improve calibration of finite element models: A case of study in Saint Torcato Church (Guimarães, Portugal). *Construction and Building Materials*, *70*, 118−129. Available from https://doi.org/10.1016/j.conbuildmat.2014.07.106.

The American Institute of Architects. BIM forum, Level of Development (LOD) specification. (2015).

Thomson, C., & Boehm, J. (2015). Automatic geometry generation from point clouds for BIM. *Remote Sensing*, *7*(9), 11753−11775. Available from https://doi.org/10.3390/rs70911753, http://www.mdpi.com/2072-4292/7/9/11753/pdf.

Tyagi, S., Agarwal, A., & Maheshwari, P. 2016 A conceptual framework for IoT-based healthcare system using cloud computing. *6th International Conference - Cloud System and Big Data Engineering (Confluence)*. 503−507. Available fromhttps://doi.org/10.1109/CONFLUENCE.2016.7508172.

Tysiac, P., Sieńska, A., Tarnowska, M., Kedziorski, P., & Jagoda, M. (2023). Combination of terrestrial laser scanning and UAV photogrammetry for 3D modelling and degradation assessment of heritage building based on a lighting analysis: case study—St. Adalbert Church in Gdansk, Poland. *Heritage Science*, *11*(1), 53. Available from https://doi.org/10.1186/s40494-023-00897-5.

UNI Ente Italiano di Normazione. (2017). UNI Ente Italiano di Normazione Unpublished content UNI 11337:2017. Edilizia e opere di ingegneria civile—Gestione digitale dei processi informativi delle costruzioni—Parte 1: Modelli, elaborati e oggetti informativi per prodotti e processi. Available from https://store.uni.com/uni-11337-1-2017.

Van Balan, K., & Vandesande, A. (n.d.) *Reflections on preventive conservation, maintenance and monitoring by the PRECOM³OS UNESCO chair*. 978-90-334-9342-3.

Volk, R., Stengel, J., & Schultmann, F. (2014). Building Information Modeling (BIM) for existing buildings — Literature review and future needs. *Automation in Construction*, *38*, 109−127. Available from https://doi.org/10.1016/j.autcon.2013.10.023, https://www.sciencedirect.com/science/article/pii/S092658051300191X.

Wang, G., & Zhang, Z. (2021). BIM implementation in handover management for underground rail transit project: A case study approach. *Tunnelling and Underground Space Technology*, *108*. Available from https://doi.org/10.1016/j.tust.2020.103684, http://www.elsevier.com/inca/publications/store/7/9/9/.

Yang, X., Grussenmeyer, P., Koehl, M., Macher, H., Murtiyoso, A., & Landes, T. (2020). Review of built heritage modelling: Integration of HBIM and other information techniques. *Journal of Cultural Heritage*, *46*, 350−360. Available from https://doi.org/10.1016/j.culher.2020.05.008, http://www.elsevier.com.

Young, R., Fallon, S., & Jacob, P. (2017). An architecture for intelligent data processing on IoT edge devices. *UKSim-AMSS 19th International Conference on Computer Modelling & Simulation (UKSim)*. 2473−3520 227−232. Available from https://doi.org/10.1109/UKSim.2017.19.

Zhang, D., Dong, D., & Peng, H. (2012). Research on development of embedded uninterruptable power supply system for IOT-based mobile service. *Computers & Electrical Engineering*, *38*(6), 1377−1387. Available from https://doi.org/10.1016/j.compeleceng.2012.04.001, https://www.sciencedirect.com/science/article/pii/S0045790612000730.

Section V

Other Techniques for Heritage Building Diagnosis

Monitoring of heritage buildings and dynamic simulation models

16

T. de Rubeis[1], L. Evangelisti[2], C. Guattari[3], G. Pasqualoni[4],
R. De Lieto Vollaro[2], D. Paoletti[4], F. Asdrubali[5], and D. Ambrosini[4]
[1]Department of Civil, Construction-Architectural and Environmental Engineering, University of L'Aquila, L'Aquila, Italy, [2]Department of Industrial, Electronic and Mechanical Engineering, Roma Tre University, Rome, Italy, [3]Department of Philosophy, Communication and Performing Arts, Roma TRE University, Rome, Italy, [4]Department of Industrial and Information Engineering and Economics, University of L'Aquila, L'Aquila, Italy, [5]Department of International Human and Social Sciences, Perugia Foreigners' University, Perugia, Italy

16.1 Introduction

It is well known that buildings are responsible for high energy consumption and energy efficiency within the building sector is a currently relevant issue. The reasons for this are strictly related to climate change, primarily triggered by human factors, the exhaustibility of fossil fuels, and, finally, the cultural enhancements of our societies involving sustainable choices and growing levels of comfort. Enhancing the building energy performance is crucial for achieving the objectives set by the European Community in terms of reduction of consumption and cutting emissions into the atmosphere. The environmental sustainability concept was used for the first time by the Bruntland Report in 1987. Since then, several protocols and directives have been issued to increasingly limit polluting emissions into the atmosphere, also by focusing on greater efficiency in the building sector. The Directive 2010/31/EU (European Commission, 2010) suggested, among the strategic objectives, the pursuit of energy saving of existing buildings. Moreover, this problem can be also observed from a complementary point of view. It is noteworthy that the built heritage can often have considerable aesthetic value, as is the case in historic buildings. Old European cities are characterized by old buildings with an architectural worth, thus requiring energy retrofit interventions capable of not compromising the aesthetic value of these structures. In these cases, there is a need to combine energy requalification within an optimal relationship between energy benefit and architectural sustainability (Alev et al., 2014; Farnaz et al., 2021; de Rubeis et al., 2021). This built heritage represents an important resource, constituting a public good and bearing witness to the identity of the different countries. The energy efficiency of the building stock is also aimed at the redevelopment of historic buildings.

Diagnosis of Heritage Buildings by Non-Destructive Techniques. DOI: https://doi.org/10.1016/B978-0-443-16001-1.00016-4

From a wider point of view, buildings are complex systems, characterized by several energy phenomena correlated to the characteristics of the building envelope (walls and windows), the systems (heating and cooling systems, electricity), the intended use (residences, offices, hotels, etc.), the presence of people or machinery that produce heat. For a proper design of retrofit interventions, it is essential to understand and reproduce realistically the energy behavior of buildings through tailored rating evaluations. In this context, the building energy simulation represents an important assessment tool. Simulation models are primarily used during the planning stage, allowing to verify the performances in advance and to optimize the energy efficiency of a project. One of the main objectives of the energy retrofitting of buildings is related to a better indoor climate (Azizi et al., 2019) and the reduction of energy management costs (Nair, 2012). Furthermore, an energy retrofit can increase the resale value of a building, thus providing an additional incentive to implement retrofit measures. However, historic buildings represent a subsector that cannot be subjected to the same renovations as conventional buildings (Mazzarella, 2015). These types of buildings consist of residential buildings, non-residential buildings, public monuments, and religious buildings. Heritage buildings are those characterized by artistic, historical, archaeological, and ethno-anthropological relevance (Filippi, 2015). Therefore retrofits in historic buildings need to be evaluated on a case-to-case basis (McCaig et al., 2016). Recently, several review papers related to energy efficiency in heritage buildings have been published, highlighting different issues related to operational energy use along with other aspects such as life cycle analysis, behavior, and cultural heritage value of buildings, criteria, analysis methods, and decision process for assessing energy retrofits, and examples of energy efficiency approaches and integration of renewable energy measures (Berg et al., 2017; Buda et al., 2021; Hao et al., 2020; Lidelöw et al., 2019; Martínez-Molina et al., 2016; Webb, 2017).

It is worthy to observe that the representativeness of the simulation models can be partially reduced by the uncertainties related to the input parameters. Some examples can be represented by climatic data, occupant behavior, materials' thermophysical properties, and operation of the HVAC (heating, ventilation, and air-conditioning) systems serving the buildings. The generation of accurate models is often complicated due to the large number of parameters involved in simulations. The lack of accurate data can introduce significant errors during the model definition, generating less reliable results (Evangelisti et al., 2018). Although the application of energy simulation models in modern buildings leads to important results, in historic buildings several uncertainties may occur, leading to inaccurate results. In particular, the approximations are related to materials, cold bridges, geometry, and construction techniques. Due to this, the energy simulation of historic buildings is challenging, with a frequent lack of data and a resulting difficult definition of their energy performance. However, energy efficiency improvement in historic buildings can lead to a reduction in terms of greenhouse gas emissions in the building sector (Nair et al., 2022).

From a general point of view, specific strategies need to be put in place to reduce the differences between actual and simulated buildings, and this issue appears even more evident in the case of simulation models of historic buildings.

The methodological approach is based on the models' calibration (de Rubeis et al., 2018; Mustafaraj et al., 2014). During the calibration of the models, the comparison between simulation results and specific parameters able to characterize the building—plant system allows for verifying the reliability of the calculations based on real data obtained from on-site monitoring or through the analysis of the energy consumption (Smarra et al., 2018). The calibration phase is a standardized procedure for reducing arbitrary factors during the acquisition and processing phases. Thus calibrated models allow for obtaining reliable energy-saving scenarios, encouraging investments in energy requalification projects for real savings prospects.

Measures to contain energy consumption can foresee two different actions, for which different tools (characterized by different degrees of complexity) can be applied: (1) optimized management of the existing building and (2) evaluation of energy retrofit interventions.

In the first case, it is necessary to identify more rational air-conditioning systems' working times, also considering optimal indoor air temperatures in the function of specific heat gains and a more rational operation of lighting systems. For this purpose, tailored rating dynamic simulations are needed.

In the second case, hypotheses and application of energy optimization interventions of the building—plant system are considered, taking into account the savings achieved about the initial investment and the economic payback times. Also in this case, for this purpose, tailored rating dynamic simulations are the best solution. However, even well-calibrated semistationary models can be a reliable answer in terms of reduced cost and time.

In addition, buildings' energy simulation tools are very useful in assessing whether indoor environmental conditions are suitable for the conservation of cultural heritage, often preserved in historic buildings. In fact, the heating strategies of historic buildings can have a twofold objective: on the one hand, cultural heritage conservation and, on the other hand, occupant thermal comfort optimization. However, these two goals may be conflicting (Camuffo & della Valle, 2007), because heating strategies often aim to provide thermal comfort, without considering that thermal stress could worsen cultural heritage conservation. Thus the search for optimal solutions requires the creation of accurate simulation models.

16.2 Materials and methods

16.2.1 Steady-state and dynamic simulation tool

The classification normally made on energy simulations concerns the simplifications that are adopted in the representation of the physics of the building and of the materials as well as in the hourly unit that is used to investigate the phenomenon. Essentially, the distinction between semistationary and dynamic simulations leads to a greater or lesser refinement of the results (Evangelisti, Battista, Guattari, Basilicata, De Lieto Vollaro, 2014a).

Semistationary simulation codes are characterized by a simplified calculation procedure based on Standard EN 13790. This analysis is commonly used for various reasons, among which is the simplicity of execution, due to a high level of simplification of the models. In general, the inputs of semistationary calculation codes are limited to poor geographical indications, simplified building geometry modeling, and an approximate indication of the systems and the building's use schedules. Due to the simplicity of this software, these simulations can be used during the initial stages of the design, to get preliminary information about the characteristics of the building and what its weaknesses may be or as a basis for urban building energy modeling approaches, with which to model the energy behavior of an entire building stock (de Rubeis et al., 2021). The climatic data used by semistationary software are monthly. However, the database characterized by monthly average weather data is considered unsuitable for the correct calibration of the models, as the data are not recent and may not be representative of the current reference climatic conditions. Taking into account historical buildings, often characterized by massive envelope components, semistationary simulation tools are not recommended (Evangelisti, et al., 2014b).

With dynamic simulations, the unit of time is the hour but often it is also possible to go down to the sub-hourly scale. This allows to provide a detailed description of the thermal behavior of the building, allowing to take into consideration the inertial behavior of the structure. These simulation tools use variable inputs, allowing to realistically simulate the effects of single change users make to the analyzed structure. Moreover, dynamic codes allow to simulate in detail the different air-conditioning systems in the building and to see the effects on consumption also on internal comfort. Greater modeling flexibility and accuracy means greater software complexity. In fact, this type of analysis requires in-depth knowledge of every single element of the project, both of the envelope and HVAC systems. Furthermore, the modeling must be extremely careful because a simple oversight can lead to different and unrealistic results.

16.2.2 Input data collection

Input data collection is a crucial and complicated step for the definition of building simulation models. Several information need to be acquired and often the lack of technical data or unreliable data represents an intricate issue for users. It is possible to affirm that the collection of data is a compromise between data availability and the accuracy of the model. In general, the necessary parameters can be obtained from direct and indirect sources. Direct sources are represented by long-term and short-term monitoring, punctual measurements, relief of environments, and interviews with building users. On the other hand, indirect sources are represented by project documentation (if available), technical data sheets of the materials, and operating manuals of the building systems. Finally, other sources can be technical standards, guidelines, and reference abacuses (Evangelisti et al., 2018).

16.2.2.1 Climatic data

Climatic data is a fundamental input for building energy simulations. The availability of accurate weather data can significantly affect the energy behavior of the models. Consequently, it is crucial to verify the reliability of the sources. Standard weather data and actual climatic data can be distinguished.

Standard weather data refer to a specific reference site, and they are characterized by statistical processing starting from data acquired for a time span equal to at least 10 years (ISO, 2005). Climatic data used to define the boundary conditions in buildings' simulations obviously have a huge influence on the results in terms of heating and cooling energy demands (Kočí et al., 2019). Weather data are also fundamental to predict the yearly performance of solar energy systems (Chicco et al., 2015; Muñoz & Perpiñán, 2016; Zang et al., 2012). Weather data usually used in building simulations are called typical meteorological years (TMYs): a series of 8760 hourly values (corresponding to one year) of weather parameters able to represent typical climatic conditions for a specific site (Janjai & Deeyai, 2009). Many efforts have been made to build TMYs for several locations. Findings in the literature demonstrated that climate change has considerable effects on energy use in buildings (de Rubeis, Falasca et al., 2020a; Lam et al., 2010; Wan et al., 2012) and, therefore, it must be considered in urban planning (Zhou et al., 2014). Due to global climate change and its crucial role in the use of energy, the time span useful for generating TMYs should be long enough to contain recent climatic data and to simulate climate trends properly (Chow et al., 2006). TMY needs a statistical analysis through which, after a long-term observation period, 12 typical months are selected to create the TMY (Zang et al., n.d.).

Compared to the past, the number of meteorological stations (intended as a set of measuring instruments that record data relating to the physical conditions of the atmosphere in each place, for an indefinite time, relative to its fundamental parameters, for meteorological and climatic purposes) has been increasing over time. Various weather stations of spread quality are available on national territories. However, not all stations collect climatic data that can be used for building physics applications. Only climatic data acquired by stations that apply the methods and prescriptions of the World Meteorological Organization (WMO) can be used (World Meteorological Organization). The weather stations installed following these prescriptions are generally of airport type and not always located in representative sites of urban areas. City airports are often located in the surroundings of cities, far from the more densely built neighborhoods. Consequently, climatic data monitored outside the city could be not always representative of the climatic conditions inside the city (or they may not be representative of all city areas). Fig. 16.1 shows several weather stations installed within and near the urban area of Rome (Italy), where FCO and CIA are Fiumicino and Ciampino Airports, respectively. The red points are the weather stations belonging to the Meteo Lazio meteorological network.

On the other hand, actual climatic data are logged by meteorological stations installed near the buildings under investigation, and data are registered for a time

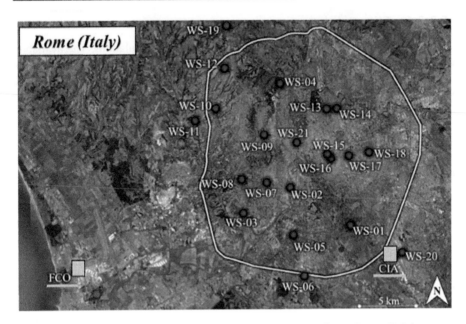

Figure 16.1 Locations of the weather stations installed within and near Rome (Italy) (Battista et al., 2023).
Source: An elaboration from Battista, G., Evangelisti, L., Guattari, C., Roncone, M., Balaras, C. A. (2023). Space-time estimation of the urban heat island in Rome (Italy): Overall assessment and effects on the energy performance of buildings. *Building and Environment* 228, 109878.

span useful for the subsequent model calibration. The use of real climatic data relating to the calibration period and collected through direct measurements represents the best solution during the model creation phase.

It is worthy to highlight that climatic data needs to be logged using the same weather station. It is not advisable to integrate any missing data with those of different weather stations. Significant correlations between the climatic variables have been observed in the literature (Bhandari et al., 2012; Guan et al., 2007).

16.2.2.2 Geometrical characteristics

A correct division of the building into thermal zones allows to define a representative model, as it allows to describe the building in detail in terms of boundary conditions and characteristics of the internal environments (De Lieto Vollaro, et al., 2015). It is worth mentioning that a thermal zone is a space or collection of spaces characterized by comparable space-conditioning needs. Moreover, the heating and cooling setpoint is the same, and it can be considered as the basic thermal unit (or zone) used in building modeling. A thermal zone can include one or more spaces. Thermal zones may be grouped together, but HVAC systems serving combined zones shall be subject to the efficiency and control requirements of the combined zones. To define the thermal zones of a building within a dynamic simulation model, the following parameters need to be considered: crowding index and equipment can determine internal gains; usage profiles of the premises of the structure can cause internal gains, influencing the working time of the systems; exposure is strictly related to solar gains; window surfaces can influence both heat losses and solar gains; and location of the premises within the building can be correlated to different solar gains depending on the floor of the building. All this is summarized in Table 16.1.

16.2.2.3 Building use profile (schedule)

Building use profiles (the so-called *schedules*) represent the conditions of use over time of the structure under investigation. In particular, they indicate the specific conditions of use of the HVAC system, thus representing an important viewpoint in the definition of simulation models in dynamic regime. When defining the schedules of a building, it is necessary to indicate the working time of the air-conditioning systems, electrical equipment, lighting systems, and the hours of the day during which the thermal zones are occupied. For a correct definition of the schedules, it is necessary to perform in situ investigations through which identifying the specific time intervals of the day. In addition to a daily operating regime, the building could also have a weekly operating regime, for example, varying the working week (full operation of the building plants and the maximum number of people) from weekends (building plants shutdown and lack of people; Battista et al., 2014).

16.2.2.4 Internal heat gains

Indoor environments are characterized by heat sources, represented by electrical equipment, lighting fixtures, and occupants. Internal gains cause a reduction in terms of heat loads and, on the contrary, an increase in terms of cooling loads. Taking into consideration the heating energy demand of a building, it is worthy to observe that internal gains are not constant during the winter. In the less cold months, it may happen that thermal gains are greater than heat losses and a consequent overheating of the rooms is generated. Dynamic simulation software allows to provide detailed evaluations of the internal gains over time.

Table 16.1 Parameters to be considered for thermal zones identification.

Factor to be considered	Affecting parameter
Crowding index	Internal heat gains
Occupancy profiles of the premises	• Internal heat gains • Working time of the heating, ventilation, and air-conditioning systems
Exposure of the premises	Solar gains
Window surfaces	Solar gains depending on the floor of the building

Indoor environments are characterized by the presence of people in the function of their intended use. Due to metabolic activity, people generate sensible and latent heat, making up a significant part of the total thermal load. Generally, internal gains related to people are estimated considering specific technical standards. ASHRAE Standard provides sensible and latent heat fluxes in the function of the metabolic activity that people carry out within environments (Table 16.2 lists some examples).

Devices and equipment inside rooms generate thermal contributions according to the working hours which is usually equal to the occupation time of the rooms. Also in this case, the ASHRAE Standard provides reference values in the function of the specific device (see Table 16.3).

16.2.3 In situ monitoring techniques

16.2.3.1 Infrared thermography

Very frequently, the envelopes of historic buildings have unknown stratigraphy, generally characterized by considerable heterogeneity (e.g., masonry walls), which can lead to highly variable thermal exchanges with the outside. In addition, the envelope of historic buildings may have undergone possible alterations over the years (e.g., window closures and elevations), whose knowledge allows for detailed energy assessments and interventions.

In this scenario, the use of non-destructive techniques (NDTs), such as infrared thermography (IRT), allows a thorough understanding of the envelope of historic buildings and the presence of any structural heterogeneities (Nardi et al., 2018).

Table 16.2 Internal gains in the function of the specific activity (ASHRAE, 2001).

Activity	Sensible heat flux [W]	Latent heat flux [W]
Sitting	65	30
Light work	70	45
Moderate activity	75	55
Walk	75	70
Sedentary work	80	80

Table 16.3 Internal gains due to equipment (ASHRAE, 2001).

Equipment	Internal gain [W]
Computer	55
Small monitor (330–380 mm)	55
Medium screen (400–460 mm)	70
Big monitor (480–510 mm)	80
Laser printer	70
Copy machine	300

In fact, IRT is the science of acquiring and analyzing information from noncontact thermal sensing devices. The great usefulness of IRT derives mainly from three of its unique features:

- the possibility of carrying out remote surveys, without any physical contact with the object being inspected (and therefore without altering or modifying it), operating even in darkness and in the presence of fumes and vapors (Daffara et al., 2020; Muttillo et al., 2020);
- the "two-dimensionality" of the inspection, given by the fact that the result obtained is a complete and detailed thermal image, capable of giving an overall view of the object, thus not the simple temperature measurement of a single point; and;
- the real-time observation of phenomena in motion.

IRT is a telemetric technique that can determine, using an infrared camera, the temperature of a surface with considerable spatial resolution and high thermal sensitivity (which depends on the characteristics of the thermal camera), by exploiting the infrared radiation that is emitted by each object proportionally to its temperature.

The thermal image is represented by means of a color scale that makes the temperature differences between contiguous areas more obvious.

The IRT applied to masonry surfaces (UNI EN 16714, 2016) makes it possible to highlight the presence of detachments between the layers of masonry, hidden architectural structures, areas affected by moisture, heat dissipation phenomena due to ineffective insulation, or leaks in ducts. In architectural diagnostics, the IRT is particularly useful for the inspection of frescoes and plasterwork, as it does not require material sampling, allowing complete and rapid mapping of extensive surfaces.

Moreover, the IRT applied to the masonry of historic buildings makes it possible to highlight the presence of thermal discontinuities (e.g., different materials), which cannot be detected visually, by exploiting the different temperatures caused by differences in conductivity and heat capacity between wall materials. Thermal stress is usually required for this type of investigation, which can be achieved by heating the structures under investigation or by exploiting changes in environmental parameters (e.g., solar radiation) (Nardi et al., 2016; Nardi et al., 2017; Sfarra et al., 2017). Fig. 16.2 shows some examples of IRT applications.

16.2.3.2 Heat flow meter method

The assessment of the thermal transmittance (U-value) of the building envelope is a very important step for evaluating the building's energy performance. According to the standard ISO 6946 (ISO, 2017), the U-value of a building component characterized by a known stratigraphy can be computed by means of the following formula:

$$U = \frac{1}{R_{tot}} = \frac{1}{R_{s,i} + \sum_i R_i + R_{s,e}} = \frac{1}{\frac{1}{h_i} + \sum_i \frac{s_i}{\lambda_i} + \frac{1}{h_e}}$$

where R_{tot} is the total thermal resistance of the wall; R_i is the thermal resistance of the ith layer; $R_{s,i}$ and $R_{s,e}$ are the internal and external surface thermal resistances (de Rubeis, Evengelisti et al., 2022a; 2022b). In turn, surface thermal resistances

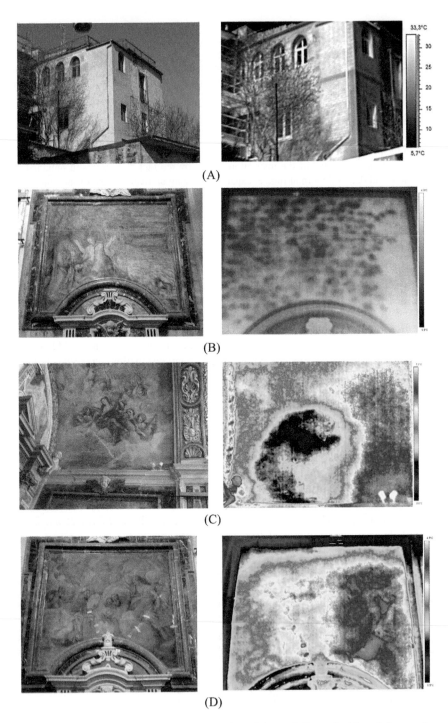

Figure 16.2 Examples of IRT tests. (A) Detection of thermal discontinuities of the envelope due to different materials. (B) Identification of masonry below a fresco. (C) Detection of moisture infiltration. (D) Detection of cracks and detachments in a fresco. *IRT*, Infrared thermography.

can be calculated through the internal and external heat transfer coefficients, called h_i and h_e, respectively, and the thermal resistance of each layer can be computed through the ratio between its thickness (s_i) and its thermal conductivity (λ_i). To identify the thermophysical properties of construction materials, when technical data are not available, specific reference standards can be used (see Table 16.4).

The main advantage of the theoretical method suggested by the standard ISO 6946 is related to a simple computation not including experimental tests (de Rubeis et al., 2019). However, a correct U-value calculation is challenging due to the potential lack of data associated with walls compositions and material thermophysical properties.

Due to this, it is advisable to carry out an in situ verification through experimental investigations which allow to determine in a realistic way the characteristics of the construction elements of buildings.

From an experimental point of view, the walls' U-value can be assessed through the heat flow meter method, which is a standardized technique. Measurements need to be performed according to the standard ISO 9869-1 (ISO, 2014), by recording heat fluxes and internal−external air temperature differences. Heat flux sensors for

Table 16.4 Reference standard for the thermophysical properties of construction materials.

Standard	Brief description
UNI 10351 (Italian)	The standard provides the method for finding the reference values for thermal conductivity, resistance to the passage of steam, and specific heat of building materials based on the time of installation. The standard integrates what is not present in UNI EN ISO 10456 with reference to insulating materials for building depending on whether or not they have the CE marking.
UNI 10355 (Italian)	This standard provides the thermal resistance values relating to the most common types of masonry and floors in Italy. It is based on the results achieved by laboratory tests and verifications by calculation, conducted over the last few years.
EN 1745	This standard provides guidelines for determining the values of thermal resistance and thermal conductivity of masonry and masonry products.
ISO 10456	This standard specifies the methods for determining the declared and design values for thermally homogeneous building materials and products. It also provides procedures for converting values obtained for one set of conditions into values valid for another set of conditions. These procedures are valid for design ambient temperatures between $-30°C$ and $+60°C$. The standard provides conversion coefficients for temperature and humidity. These coefficients are valid for average temperatures between $0°C$ and $30°C$. The standard also provides tabulated design values which can be used in heat and moisture transfer calculations for thermally homogeneous materials and products commonly used in building construction.

on-site measurements are usually characterized by plates of a few millimeters that must be installed on the inner surface of the wall. A preliminary thermographic survey should be performed for excluding cold bridges near the heat flux sensor (Baker, 2011; Ficco et al., 2015), thus causing deviations from the one-dimensional heat flow conditions, altering the results. Moreover, the sensors must not be influenced by thermal sources or air currents that modify the boundary conditions. The external surface of the wall needs to be protected from atmospheric agents such as rain, direct solar radiation, and snow (also through temporary protective screens). The measurement must be carried out continuously for at least 72 hours. However, longer measurement times may be required if the temperature at the heat flow meter has a variable trend over time.

If the thermal regime is almost stationary, the progressive average method can be used for data postprocessing. The U-value can be obtained by applying the following equation:

$$U = \frac{\sum\limits_{j=1}^{N} q_j}{\sum\limits_{j=1}^{N} \left(T_{a,ij} - T_{a,ej} \right)}$$

where j is the single measurement acquisition, N is the whole logged measurements, q is the heat flux density, $T_{a,i}$ is the indoor air temperature, and $T_{a,e}$ is the outdoor air temperature. This method is certainly easy to use, but stationary conditions over time are quite difficult to have. For walls characterized by a specific heat capacity of less than 20 kJ/m^{2}K, the standard ISO 9869-1 suggests using data acquired only during nights for the U-value calculation. The test can be halted when the results do not differ by more than $\pm 5\%$ after three consecutive nights. Otherwise, the data acquisition must be continued.

16.2.3.3 Thermo-hygrometric monitoring

Thermo-hygrometric monitoring allows to better comprehend the operating mode of HVAC systems or, more in general, the operating mode of an indoor environment. To describe the microclimate characteristics of a building, the following parameters need to be known: indoor and outdoor air temperatures, indoor and outdoor relative humidity, wind velocity, characteristic rainfall, and solar radiation. Moreover, additional information related to the working times of the building systems, the number of people within the environments, air change, and the type and arrangement of light sources can be acquired.

During an energy audit, it is necessary to employ measuring instruments able to simultaneously log the different thermo-hygrometric variables. They can be categorized into two categories: (1) digital detectors for instant monitoring, during energy audits, thermographic analysis, and blower door tests and (2) sensors connected to a data logger for long-term monitoring.

Long-term monitoring can have a duration ranging from 15 days up to 1 year. Fifteen days can be necessary to understand the thermo-hygrometric behavior of an indoor environment during the hours of the day, based on the different internal activities and the functioning of the building systems. For more detailed analyses, longer data acquisition times are needed to better comprehend the influence of seasonal climate variability on internal environmental conditions.

The degree of accuracy in measuring the temperature and relative humidity of the air must be 0.5°C and 0.5%, respectively. The measuring ranges are typically between $-35 \div 80$°C (for temperature sensors) and $0 \div 100$% (for relative humidity sensors).

16.2.4 Calibration procedure

Understanding how much a simulated building is representative requires a calibration phase. To confirm the effectiveness of a simulation model, control parameters able to represent the actual energy behavior of a building are required. Due to this, specific protocols (ASHRAE Guideline 14 (ASHRAE, 2002), IPMVP (IPMVP, 2007), FEMP (US DOE, 2008)) suggest making a comparison between the real building consumptions and those computed through the simulation. Nevertheless, disused buildings or buildings without air-conditioning systems represent examples where actual consumption cannot be traced back. In such situations, the calibration can be carried out through indoor air temperatures or wall surface temperatures. As mentioned before, indoor environment data can be acquired through thermo-hygrometric monitoring. Whatever the type of sensor used, the positioning of the instrumentation during the data acquisition phase is essential. Control environments should be chosen to be representative in terms of occupancy rate, internal inputs, and building envelope structures. Measuring instruments must not be installed near heat sources, or sources of direct solar radiation or air currents, just as in the case of acquisition of surface temperatures, the sensors must not be positioned over or near cold bridges.

In the following, only calibration via temperatures will be discussed. To calibrate simulation models, it is necessary to calculate error estimation coefficients able to highlight differences between simulated and actual values.

The mean bias error (MBE) is a nondimensional bias measure between measured and simulated data. It can be computed by applying the following equation:

$$MBE = \frac{\sum_{i=1}^{N}(m_i - s_i)}{\sum_{i=1}^{N}(m_i)}$$

where m_i is the measured value at time t_i, s_i is the simulated value for each time t_i, and N is the total number of samples. It is worthy to observe that positive bias compensates for negative bias (cancellation effect). For this reason, another index is required.

The coefficient of variation of root mean square error (CV(RMSE)) allows to define how well a model fits the data by using offsetting errors between measured and simulated data. This index can be calculated by:

$$CV(RMSE) = \frac{\sqrt{\sum_{i=1}^{N}(m_i - s_i)^2 / N}}{\overline{m}}$$

where $\overline{m}$ represents the mean value among the measured data.

The building energy model validation is based on the acceptance conditions listed in Table 16.5.

The calibration phase of simulation models has undergone considerable development in recent years. In addition to the statistical analysis techniques described previously, an approach based on the evaluation of the coefficient of determination (R^2) is often proposed (Coelho et al., 2018), computed with:

$$R^2 = \left(\frac{\sum_{i=1}^{N}(X_{i,meas} - \overline{X_{meas}}) \times \sum_{i=1}^{N}(X_{i,sim} - \overline{X_{sim}})}{\sqrt{\sum_{i=1}^{N}(X_{i,meas} - \overline{X_{meas}})^2 \times \sum_{i=1}^{N}(X_{i,sim} - \overline{X_{sim}})^2}} \right)^2$$

The calibration phase of a simulation model can be followed by an additional model validation phase. While the calibration phase involves an iterative variation of model parameters (e.g., temperature setpoints and air leakage) to improve model accuracy, the validation phase of a calibrated model is conducted to test the predictive ability of a model without changing any parameters.

16.2.5 Methodology

Starting from what has been said so far, the methodological approach for the creation and calibration of a building simulation model is shown in Fig. 16.3. The first step is represented by a complete in situ monitoring of the actual building to obtain

Table 16.5 Mean bias error (MBE) and coefficient of variation of root mean square error (CV(RMSE)) values for calibrated models.

	Monthly		Hourly	
	MBE [%]	CV(RMSE) [%]	MBE [%]	CV(RMSE) [%]
ASHRAE Guideline 14	± 5	± 15	± 10	± 30
IPMVP	± 20	–	± 5	± 20
FEMP	± 5	± 15	± 10	± 30

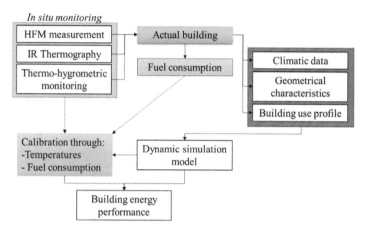

Figure 16.3 Flowchart of the methodological approach for generating calibrated building simulation models.

qualitative and quantitative data about the behavior of the structure and its conservation status. All the acquired data and information needs to be used during the generation of the dynamic model, both in terms of building geometry, use profile, weather data, and internal gains. As already mentioned, the subsequent calibration procedure needs a comparison between the real building consumptions and those computed through the simulation. Nevertheless, it is worth noting that disused buildings or buildings without air-conditioning systems represent examples where actual consumption cannot be traced back. In such situations, the calibration can be done using indoor air temperatures or wall surface temperatures. The calculation of MBE and the CV(RMSE) allows to understand and confirm the effectiveness of the simulation model, thus returning a correct picture of the building under investigation. Starting from this, two main paths can be undertaken: (1) optimized management of the existing building and (2) evaluation of energy retrofit interventions, with projects characterized by real savings prospects.

16.3   An applied example

16.3.1   Case study description

A methodology based on indoor microclimate monitoring, the creation of a dynamic simulation model, and the model calibration and validation steps was applied to the case study of the Church of Santa Maria Annunziata di Roio in L'Aquila, Italy (Fig. 16.4) (de Rubeis, Nardi et al., 2020b). The Romanesque-era church has a single nave, masonry structure, and wooden roof and was renovated following the earthquake that struck L'Aquila in 2009. The church was and is without HVAC systems. Inside the church, frescoes and a wooden choir are preserved.

Figure 16.4 The Church of Santa Maria Annunziata of Roio.
Source: Elaborated version from de Rubeis, T., Nardi, I., Muttillo, M., Paoletti, D. (2020b). The restoration of severely damaged churches — Implications and opportunities on cultural heritage conservation, thermal comfort and energy efficiency. *Journal of Cultural Heritage* 42, 186—203.

Table 16.6 Technical specifications of the measuring instruments.

Sensor	Model	Measurement range	Accuracy
Elitech	RC-4	−30.0°C to 60.0°C	± 0.5°C
Hobo	H08-003-02	−20.0°C to 70.0°C for temperature	± 0.7°C
		25%−95% for RH	± 5%

The envelope of the church is made of limestone mortar, and it is characterized by five different wall thicknesses. Based on the standard UNI 10351 (UNI, 2015), the thermophysical properties of the envelope were determined, while the total thermal resistances were determined according to UNI EN ISO 6946 (ISO, 2017).

16.3.2 Monitoring phase

The monitoring phase of the church was aimed at assessing the indoor microclimate (air temperature and relative humidity) and providing the necessary data for the calibration steps of the simulation model. The monitoring lasted 40 days, during which Hobo sensors, model H08-003-02, were used to measure air temperature and relative humidity, and Elitech sensors, model RC-4, were employed to measure air temperature. Table 16.6 shows the characteristics of the measuring instrumentation used.

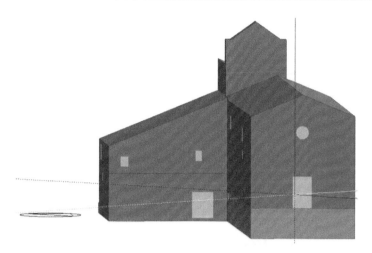

Figure 16.5 Simulation model of the Church of Santa Maria Annunziata.
Source: Elaborated version from de Rubeis, T., Nardi, I., Muttillo, M., Paoletti, D. (2020b).
The restoration of severely damaged churches − Implications and opportunities on cultural
heritage conservation, thermal comfort and energy efficiency. *Journal of Cultural Heritage*
42, 186−203.

16.3.3 Simulation modeling

The Church of Santa Maria dell'Annunziata was modeled using EnergyPlus software with DesignBuilder graphical interface. The model was built considering all
the thermal properties previously described. The input weather file was made from
weather data measured by a weather station located in L'Aquila and operated by
CETEMPS - Centre of Excellence of the University of L'Aquila. Fig. 16.5 shows
the simulation model of the church.

16.3.4 Model calibration and validation

The calibration phase was conducted by comparing the indoor temperature values
measured in the monitoring phase with the results obtained from modeling.
Fig. 16.6 shows a comparison between measured and simulated data, during the
period of September 28−October 13, for the model calibration.

Statistical analysis of the calibration phase was conducted by considering the
MBE, the CV(RMSE), and the coefficient of determination (R^2). Table 16.7 shows
the results of the calibration.

The calibrated model was then subjected to the validation phase by comparing
the simulated indoor temperature values with those measured during the period of
October 31−November 23. The simulation results used for the validation phase
were obtained without changing any parameters previously modified in the

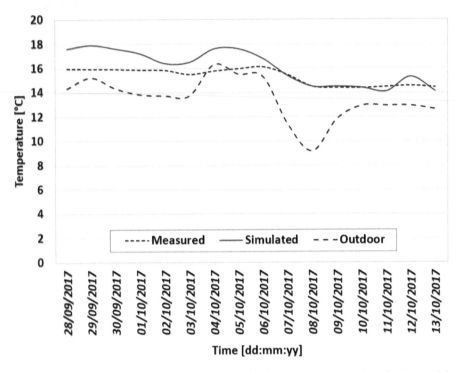

Figure 16.6 Comparison between measured and simulated temperature values for the model calibration.
Source: Elaborated version from de Rubeis, T., Nardi, I., Muttillo, M., Paoletti, D. (2020b). The restoration of severely damaged churches – Implications and opportunities on cultural heritage conservation, thermal comfort and energy efficiency. *Journal of Cultural Heritage* 42, 186–203.

Table 16.7 Model calibration results.

Description	Statistical parameters		
	MBE [%][a]	CV(RMSE) [%][a]	R^{2a}
Nave	4.91	15.00	0.87
Sacristy	4.73	14.56	0.92

CV(RMSE), Coefficient of variation of root mean square error; *MBE*, mean bias error.
[a]Limit values considered for the calibration: MBE < 5%, CV(RMSE) < 20% and R^2 > 0.75.

calibration phase. Fig. 16.7 shows the measured and simulated temperature data for the validation phase.

Numerically, the statistical analysis conducted during the validation phase led to the results shown in Table 16.8.

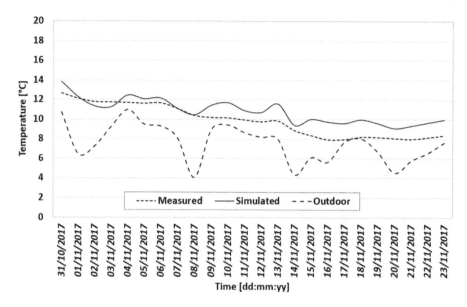

Figure 16.7 Comparison between measured and simulated temperature values for the model validation.
Source: Elaborated version from de Rubeis, T., Nardi, I., Muttillo, M., Paoletti, D. (2020b). The restoration of severely damaged churches — Implications and opportunities on cultural heritage conservation, thermal comfort and energy efficiency. *Journal of Cultural Heritage* 42, 186–203.

Table 16.8 Model validation results.

Description	Statistical parameters		
	MBE [%]	CV(RMSE) [%]	R^2
Nave	9.76	17.71	0.83
Sacristy	1.49	18.86	0.78

CV(RMSE), Coefficient of variation of root mean square error; *MBE*, mean bias error.

16.4 Conclusion

Buildings are responsible for high energy consumption and the building sector needs inclusive energy requalification. Today, we have more and more evident signs of climate change underway, primarily triggered by human factors, the exhaustibility of fossil fuels, and, finally, the cultural enhancements of our societies involving sustainable choices and growing levels of comfort. Improving the energy efficiency of the building stock is essential to reach the objectives set by the European Community in terms of reduction of consumption and cutting emissions. Ancient European cities are characterized by old buildings with architectural worth, thus requiring specific retrofit solutions able to not alter the aesthetic value of these structures. In such cases, there is a need to combine energy refurbishment within an optimal relationship between energy benefit and architectural sustainability. The energy efficiency of the building stock is also aimed at the renovation of historic buildings, which can be identified as complex systems. The application of indoor microclimate monitoring techniques, NDTs, and energy simulation of historic buildings represents a very useful approach for multiparametric and multidisciplinary evaluations. It is necessary, therefore, to make harmonious use of all the technologies in our possession to reach a thorough understanding of the historic building behavior, based on which a simulation model can be built. Simulation models, if accurate, allow outlining predictive scenarios with multidisciplinary purposes, ranging from energy and environmental assessment, indoor comfort, and detailed evaluation of microclimatic conditions of conservation of cultural heritage often preserved in historic buildings.

Nomenclature

h_e External heat transfer coefficient [W/m^{2}K]
h_i Internal heat transfer coefficient [W/m^{2}K]
q Heat flux density [W/m^2]
R_i Thermal resistance of the ith layer [m^{2}K/W]
$R_{s,e}$ External surface thermal resistance [m^{2}K/W]
$R_{s,i}$ Internal surface thermal resistance [m^{2}K/W]
R_{tot} Total thermal resistance [m^{2}K/W]
s_i Thickness of the ith layer [m]
$T_{a,e}$ Outdoor air temperature [°C]
$T_{a,i}$ Indoor air temperature [°C]
U Thermal transmittance [W/m^{2}K]
λ_i Thermal conductivity of the i-th layer [W/mK]

References

Alev, Ü., Eskola, L., Arumägi, E., Jokisalo, J., Donarelli, A., Siren, K., Broström, T., & Kalamees, T. (2014). Renovation alternatives to improve energy performance of historic rural houses in the Baltic Sea region. *Energy Build, 77*, 58−66.

ASHRAE. (2001). *ASHRAE fundamentals*. Atlanta, GA. e2001.

ASHRAE. (2002). Guideline 14−2002: Measurement of energy and demand savings, American Society of Heating, Refrigerating and Air-Conditioning Engineers, Atlanta, GA 30329.

Azizi, S., Nair, G., & Olofsson, T. (2019). Analysing the house-owners' perceptions on benefits and barriers of energy renovation in Swedish single-family houses. *Energy Build, 198*, 187−196.

Baker, P. (2011). *Technical Paper 10: U-values and traditional buildings-in situ measurements and their comparisons to calculated values*; Semantic Scholar: Seattle, WA, USA.

Battista, G., Evangelisti, L., Guattari, C., Basilicata, C., & De Lieto Vollaro, R. (2014). Buildings energy efficiency: interventions analysis under a smart cities approach. *Sustainability*, *6*(8), 4694−4705.

Battista, G., Evangelisti, L., Guattari, C., Roncone, M., & Balaras, C. A. (2023). Space-time estimation of the urban heat island in Rome (Italy): Overall assessment and effects on the energy performance of buildings. *Building and Environment*, *228*, 109878.

Berg, F., Flyen, A.-C., Godbolt, Å. L., & Broström, T. (2017). User-driven energy efficiency in historic buildings: A review. *J. Cult. Herit*, *28*, 188−195.

Bhandari, M., Shrestha, S., & New, J. (2012). Evaluation of weather datasets for building energy simulation. *Energy and Buildings*, *49*, 109−118.

Buda, A., de Place Hansen, E. J., Rieser, A., Giancola, E., Pracchi, V. N., Mauri, S., Marincioni, V., Gori, V., Fouseki, K., Polo López, C. S., et al. (2021). Conservation-compatible retrofit solutions in historic buildings: An integrated approach. *Sustainability*, *13*, 2927.

Camuffo, A., & della Valle, D. (2007). *Church heating: A balance between conservation and thermal comfort* Held in April 2007, in Tenerife *Contribution to the experts' roundtable on sustainable climate management strategies*. Spain.

Chicco, G., Cocina, V., Di Leo, P., Spertino, F., & Massi Pavan, A. (2015). Error assessment of solar irradiance forecasts and AC power from energy conversion model in grid-connected photovoltaic systems. *Energies*, *9*(1), 8.

Chow, T. T., Chan, A. L. S., Fong, K. F., & Lin, Z. (2006). Some perceptions on typical weather year—From the observations of Hong Kong and Macau. *Solar Energy*, *80*(4), 459−467.

Coelho, G. B. A., Silva, H. E., & Henriques, F. M. A. (2018). Calibrated hygrothermal simulation models for historical buildings. *Building and Environment*, *142*, 439−450.

de Rubeis, T., Evengelisti, L., Guattari, C., De Berardinis, P., Asdrubali, F., & Ambrosini, D. (2022a). On the influence of environmental boundary conditions on surface thermal resistance of walls: Experimental evaluation through a Guarded Hot Box. *Case Studies in Thermal Engineering*, *34*, 101915.

de Rubeis, T., Falasca, S., Curci, G., Paoletti, D., & Ambrosini, D. (2020a). Sensitivity of heating performance of an energy self-sufficient building to climate zone, climate change, and HVAC systems solutions. *Sustainable Cities and Society*, *61*, 102300.

de Rubeis, T., Nardi, I., Muttillo, M., & Paoletti, D. (2020b). The restoration of severely damaged churches − Implications and opportunities on cultural heritage conservation, thermal comfort and energy efficiency. *Journal of Cultural Heritage*, *42*, 186−203.

de Rubeis, T., Evangelisti, L., Guattari, C., Paoletti, D., Asdrubali, F., & Ambrosini, D. (2022b). How do temperature differences and stable thermal conditions affect the heat flux meter (HFM) measurements of walls? Laboratory experimental analysis. *Energies*, *15*, 4746.

de Rubeis, T., Giacchetti, L., Paoletti, D., & Ambrosini, D. (2021). Building energy performance analysis at urban scale: A supporting tool for energy strategies and urban building energy rating identification. *Sustainable Cities and Society*, *74*, 103220.

de Rubeis, T., Muttillo, M., Nardi, I., Pantoli, L., Stornelli, V., & Ambrosini, D. (2019). Integrated measuring and control system for thermal analysis of buildings components in hot box experiments. *Energies*, *12*, 2053.

de Rubeis, T., Nardi, I., Ambrosini, D., & Paoletti, D. (2018). Is a self-sufficient building energy efficient? Lesson learned from a case study in Mediterranean climate. *Applied Energy*, *218*, 131−145.

Daffara, C., Muradore, R., Piccinelli, N., Gaburro, N., de Rubeis, T., & Ambrosini, D. (2020). A cost-effective system for aerial 3D thermography of buildings. *Journal of Imaging*, *6*, 76.

De Lieto Vollaro, R., Guattari, C., Evangelisti, L., Battista, G., Carnielo, E., & Gori, P. (2015). Building energy performance analysis: a case study. *Energy and Buildings*, *87*, 87–94.

de Rubeis, T., De Vita, M., Capannolo, L., Laurini, E., Nardi, I., Ambrosini, D., Paoletti, D., & De Berardinis, P. (2021). *A multidisciplinary approach to retrofitting historic buildings: The case of the former San Salvatore Hospital, L'Aquila.*

European Commission. (2010). "Directive 2010/31/EU of the European Parliament and of the Council of 19 May 2010 on the energy performance of buildings (recast)", *Official Journal of the European Communities*.

Evangelisti, L., Battista, G., Guattari, C., Basilicata, C., & De Lieto Vollaro, R. (2014a). Analysis of two models for evaluating the energy performance of different buildings. *Sustainability*, *6*(8), 5311–5321.

Evangelisti, L., Battista, G., Guattari, C., Basilicata, C., & De Lieto Vollaro, R. (2014b). Influence of the thermal inertia in the European simplified procedures for the assessment of buildings' energy performance. *Sustainability*, *6*(7), 4514–4524.

Evangelisti, L., Guattari, C., Gori, P., & Asdrubali, F. (2018). Assessment of equivalent thermal properties of multilayer building walls coupling simulations and experimental measurements. *Building and Environment*, *127*, 77–85.

Farnaz, F., Ryser, J., Hopkins, A., & Mackee, J. (Eds.). (2021). Historic Cities in the Face of Disasters–Reconstruction, Recovery and Resilience of Societies. *The Urban Book Series* (pp. 623–637). Cham: Springer. Available from https://doi.org/10.1007/978-3-030-77356-4.

Ficco, G., Iannetta, F., Ianniello, E., Alfano, F. R. D., & Dell'Isola, M. (2015). U-value in situ measurement for energy diagnosis of existing buildings. *Energy Build*, *104*, 108–121.

Filippi, M. (2015). Remarks on the green retrofitting of historic buildings in Italy. *Energy Build*, *95*, 15–22.

Guan, L., Yang, V., & Bell, J. M. (2007). Cross correlation between weather variables in Australia. *Building and Environment*, *42*, 1054–1070.

Hao, L., Herrera-Avellanosa, D., Pero, C. D., & Troi, A. (2020). What are the implications of climate change for retrofitted historic buildings? A literature review. *Sustainability*, *12*, 7557.

IPMVP. (2007). *EVO, International Performance Measurement & Verification Protocol, Efficiency Valuation Organization,*.

ISO 15927–4:2005 - Hygrothermal performance of buildings—Calculation and presentation of climatic data—Part 4: Hourly data for assessing the annual energy use for heating and cooling.

ISO 6946. (2017). *Building components and building elements—Thermal resistance and thermal transmittance—calculation methods*. ISO: Geneva, Switzerland,

ISO 9869-1. (2014). *Thermal insulation—building elements—in-situ measurement of thermal resistance and thermal transmittance heat flow meter method*; ISO: Geneva, Switzerland.

Janjai, S., & Deeyai, P. (2009). Comparison of methods for generating typical meteorological year using meteorological data from a tropical environment. *Applied Energy*, *86*(4), 528–537.

Kočí, J., Kočí, V., Maděra, J., & Černý, R. (2019). Effect of applied weather data sets in simulation of building energy demands: Comparison of design years with recent weather data. *Renewable and Sustainable Energy Reviews*, *100*, 22–32.

Lam, T. N. T., Wan, K. K. W., Wong, S. L., & Lam, J. C. (2010). Impact of climate change on commercial sector air conditioning energy consumption in subtropical Hong Kong. *Applied Energy*, *87*(7), 2321–2327.

Lidelöw, S., Örn, T., Luciani, A., & Rizzo, A. (2019). Energy-efficiency measures for heritage buildings: A literature review. Sustain. *Cities Soc, 45*, 231−242.

Martínez-Molina, A., Tort-Ausina, I., Cho, S., & Vivancos, J.-L. (2016). Energy efficiency and thermal comfort in historic buildings: A review. *Renewable and Sustainable Energy Reviews, 61*, 70−85.

Mazzarella, L. (2015). Energy retrofit of historic and existing buildings. The legislative and regulatory point of view. *Energy Build, 95*, 23−31.

Mustafaraj, G., Marini, D., Costa, A., & Keane, M. (2014). Model calibration for building energy efficiency simulation. *Applied Energy, 130*, 72−85.

Muttillo, M., Nardi, I., Stornelli, V., de Rubeis, T., Pasqualoni, G., & Ambrosini, D. (2020). On field infrared thermography sensing for PV system efficiency assessment: Results and comparison with electrical models. *Sensors, 20*, 1055.

Muñoz, J., & Perpiñán, O. (2016). A simple model for the prediction of yearly energy yields for grid-connected PV systems starting from monthly meteorological data. *Renewable Energy, 97*, 680−688.

McCaig, I., Pender, R., & Pickles, D. (2016). *Energy efficiency and historic buildings: How to improve energy efficiency; Historic England: London, UK, 2018. — Irish Georgian Society. Energy Efficiency in Historic Houses. In Proceedings Produced from Ten Regional Seminars Held during 2009 & 2010; Irish Georgian Society: Dublin, Ireland, 2013. — Historic England. Energy Efficiency in Historic Buildings-Insulating Timber-Framed Walls.* Available online: https://historicengland.org.uk/images-books/publications/eehb-insulating-timber-framed-walls/heag071-insultatingtimber-framed-walls/ (Accessed 13.01.23).

Nair, G. (2012). *Implementation of energy efficiency measures in swedish single-family houses* (Ph.D. thesis). Mid Sweden University, Sundsvall, Sweden.

Nair, G., Verde, L., & Olofsson, T. (2022). A review on technical challenges and possibilities on energy efficient retrofit measures in heritage buildings. *Energies, 15*, 7472.

Nardi, I., de Rubeis, T., & Perilli, S. (2016). Ageing effects on the thermal performance of two different well-insulated buildings. *Energy Procedia, 101*, 1050−1057.

Nardi, I., de Rubeis, T., Taddei, M., Ambrosini, D., & Sfarra, S. (2017). The energy efficiency challenge for a historical building undergone to seismic and energy refurbishment. *Energy Procedia, 133*, 231−242.

Nardi, I., Lucchi, E., de Rubeis, T., & Ambrosini, D. (2018). Quantification of heat energy losses through the building envelope: A state-of-the-art analysis with critical and comprehensive review on infrared thermography. *Building and Environment, 146*, 190−205.

Sfarra, S., Perilli, S., Ambrosini, D., Paoletti, D., Nardi, I., de Rubeis, T., & Santulli, C. (2017). proposal of a new material for greenhouses on the basis of numerical, optical, thermal and mechanical approaches. *Construction and Building Materials, 155*, 332−347.

Smarra, F., Achin, J., de Rubeis, T., Ambrosini, D., D'Innocenzo, A., & Mangharam, R. (2018). Data-driven model predictive control using random forests for building energy optimization and climate control. *Applied Energy, 226*, 1252−1272.

UNI EN 16714. (2016). Non-destructive testing. Thermographic testing. General principles, Italian Standard, Milan.

UNI 10351. (2015). *Materials and product for construction − Thermo-hygrometricproperties − method for the selection of the design values*, Italian Standard, Milan.

US DOE, M& V Guidelines: Measurement and Verification for Federal Energy Projects Version 3.0, US Department Of Energy, 2008.

Wan, K. K. W., Li, D. H. W., Pan, W., & Lam, J. C. (2012). Impact of climate change on building energy use in different climate zones and mitigation and adaptation implications. *Applied Energy, 97,* 274–282.

World Meteorological Organization (n.d.). *Weather − climate − water.* (Accessed 03.03.21), from https://public.wmo.int/en

Zang, H., Wang, M., Huang, J., Wei, Z., &Sun, G. (n.d.). *A hybrid method for generation of typical meteorological years for different climates of China.* https://doi.org/10.3390/en9121094

Webb, A. L. (2017). Energy retrofits in historic and traditional buildings: A review of problems and methods. *Renewable and Sustainable Energy Reviews, 77,* 748–756.

Zang, H., Xu, Q., Du, P., & Ichiyanagi, K. (2012). A modified method to generate typical meteorological years from the long-term weather database. *International Journal of Photoenergy, 2012,* 1–9.

Zhou, Y., Clarke, L., Eom, J., Kyle, P., Patel, P., Kim, S. H., Dirks, J., Jensen, E., Liu, Y., Rice, J., Schmidt, L., & Seiple, T. (2014). Modeling the effect of climate change on U. S. state-level buildings energy demands in an integrated assessment framework. *Applied Energy, 113,* 1077–1088.

Understanding the invisible: Interpretation of results in ultrasonic tomography of gothic masonry using metrology and geometry

17

Laurent Debailleux and Morgane Palma Fanfone
Architectural and Urban Engineering Unit, Faculty of Engineering, University of Mons, Mons, Belgium

17.1 Introduction

Studies of historic structures include stability studies of buildings and archaeological investigations. As such, it is often an interdisciplinary approach requiring multiple skills. Regardless, the information gathered serves a common interest: a better understanding of the morphology and composition of structures. Ancient buildings and cultural heritage buildings require particular surveillance in structural assessment, because of the small amount of knowledge of the inner geometry, constructive materials, and state of conservation (Pérez-Gracia, 2014). Therefore the first step in assessing the state of the structures and designing prospective repairs is to comprehend their inner composition.

The scientific importance of gathering sufficient knowledge of a structure prior to any action that could affect its integrity is recalled in the Victoria Falls ICOMOS Charter (*Principles for the Analysis, Conservation and Structural Restoration of Architectural Heritage*). The charter also emphasizes the importance of a comprehensive knowledge of historical structures as being key to understanding and properly restoring ancient buildings (Binda, 2005; Martinho & Dionísio, 2014). Accurate studies of cultural heritage structures usually require the application of combined techniques, historic and structural knowledge also being necessary (Himi et al., 2016).

A broad range of analytical tests to examine and analyze the geometry of heritage buildings and physical characteristics of building materials are available to gain a knowledge of a structure. In the field of heritage studies, non-destructive tests must systematically take precedence over destructive ones that alter the artistic and historical values of the monument.

In this chapter, architectural surveys and structural analysis are combined in an integrated approach to study the inner composition of the 15th-century gothic column. This

Diagnosis of Heritage Buildings by Non-Destructive Techniques. DOI: https://doi.org/10.1016/B978-0-443-16001-1.00017-6

chapter highlights an innovative approach to provide a wide amount of complementary information at different levels of observation. Section 17.1 first presents a brief overview of the non-destructive techniques used to characterize the composition of inhomogeneous masonry structures in 2D and 3D. Section 17.2 focuses on methodological aspects and theory. The benefits of using metrology and geometry to recover the original construction plan of medieval buildings will be further discussed in Section 17.3, while Section 17.4 deals with the presentation, discussion, and interpretation of the results.

17.1.1   Non-destructive testing for studying the composition of built structures

In recent years, technological developments in non-destructive testing (NDT) have facilitated the data acquisition within ancient buildings (Fort et al., 2013). These include the evaluation of the integrity and composition of a structure, as well as physical properties of the materials (Yin et al., 2019). This chapter focuses on the constructive aspects of medieval constructions to characterize the composition of architectural elements, notably in terms of arrangement of stone blocks. Such study usually involves a metric survey to record the geometry of the accessible part of the structure. However, the visual assessment of a structure has a few limits within the brick work quality assessment (Cardani et al., 2012). Still, examining the surface of a masonry structure is not always enough to understand its interior composition. Therefore complementary tests are usually done to acquire and figure out the invisible parts of a masonry. In this field, NDTs are usually used to figure out the invisible parts of a construction and avoid deconstructing the masonry. In what follows, the most used techniques are presented, highlighting their possible applications.

- Infrared thermography (IRT) is an effective investigative tool used in the inspection and diagnosis of modern and historic buildings. The method uses electromagnetic radiations in the $10^{12}-1014$ Hz band of the electromagnetic spectrum. An infrared camera operating in the long wave band $(8-10\,\mu m)$ measures the emitted radiation of the object and converts the electromagnetic energy into the temperature distribution of its surface. Abnormal surface temperature distribution is often a strong indicator of possible subsurface problems, such as shallow subsurface voids and defects nearby the surface with a size-to-depth proportion of around 2 and higher (Moropoulou et al., 2018; Pitarma et al., 2019). The technique is generally used as an economical, efficient, non-destructive tool for qualitative detection. The thermal map of an object gives useful information about heat and moisture flux and thermophysical properties of its materials (Patrucco et al., 2020). Although the technique allows many applications from decay diagnosis to conservation interventions and monitoring actions.
- Radiographic testing (RT) is a NDT method which uses either X-rays or gamma rays to examine the internal structure. Waves are passing through the body and a portion of the radiation is absorbed or scattered by the internal structure, so that a 2D image may be recorded. The remaining X-ray photon patterns are transmitted to detectors for recording. By way of example, this technique was used to measure the penetration depth of consolidants within stones (Slavíková et al., 2012). However, 2D X-ray imaging is of little interest for the study of building walls, as it provides no information on the depth of the elements. To overcome this drawback, 3D tomography should be preferred for its ability to create the volumetric reconstruction of hidden objects. Both techniques were used to study painting layers and reinforcements in sculptures.

- Ground-penetrating radar (GPR) can be used as an alternative to RT. This geophysical method has been frequently used for investigating masonry structures (Binda et al., 1998; Masini et al., 2010; Santos-Assunçao et al., 2014). GPR deploys electromagnetic waves in the microwave band of the radio spectrum to investigate underground or internal structures to detect anomalies such as discontinuities, metal objects, or voids. Reflections are recorded in a radargram that indicates the subsurface location of reflections along the surveyed profile. The main restriction on the GPR's ability to penetrate deeply is the high rate of electromagnetic wave energy absorption in conductive materials (Nuzzo et al., 2010). Therefore the frequency of the electromagnetic waves affects the resolution of the data. Wave frequency of GPR ranges from 10 to 2000 MHz. Deep discontinuities can be recorded with low-frequency antennas but the resolution of the spatial feature will be limited, whereas high-frequency antennas provide a more detailed subsurface image, but with a shallow penetration depth (Elkarmoty et al., 2018). By way of example, a 1600 MHz antenna can reach about 1 m with a resolution of about 1 cm, while a 100 MHz antenna can investigate a medium down to a depth of about 15−20 m. High-frequency GPR can be used to survey masonry to figure out the layering of materials and localize degraded zones and anomalies (Ranalli et al., 2004), or metal reinforcements (Himi et al., 2016). On a smaller scale, GPR has been used to detect cracks of few millimeters into stone blocks (Zanzi et al., 2017), while Elkarmoty et al. (2018) used GPR data to model hidden voids and defects in 3D within the body of stone blocks. Nuzzo et al. (2010) highlighted the fundamental information provided by GPR for the study of defects and metal reinforcements within stones a medieval rose widow. Nonetheless, Pérez-Gracia (2014) pointed out the difficulty to interpret results when complex structures are concerned (). This may be due to irregular shapes, or arrangements of the various materials within a structure that generate numerous reflections and diffractions in all of the irregular target.
- Seismic tomography consists of the 2D or 3D reconstruction of an object based on the traveling wave velocity and/or amplitude of primary waves (P-waves) and secondary waves (S-waves). Data are collected with a high-frequency and high-sensitivity accelerometer to provide sections of an object which makes it easier to interpret the results. The applicability of seismic surveys in cultural heritage buildings has been proven as a successful tool in diverse works (Cardarelli & De Nardis, 2001; Pérez-Gracia, 2014). This technique has been used on stone pillars at the Cathedral of Mallorca and clearly shows the arrangement of stones and defects within the structure (Himi et al., 2016; Pérez-Gracia et al., 2013).
- Sonic and ultrasonic tests (UTs) are certainly the most NDTs used worldwide by engineers and archaeologists for assessing structures and subsoils. Both techniques are based on the transmission/reflection of acoustic waves inside a medium. Frequencies used range from 10^2 to 10^4 Hz for the sonic test and 10^4 to 10^6 Hz for the UT. Ultrasonic testing was developed in the early 1940s to assess the structural integrity of materials and structures (Hannachi & Guetteche, 2014; Ndagi et al., 2019). Since, ultrasonic techniques have been an important field of research in a variety of areas, including the conservation of cultural heritage. The ultrasonic technique allows the location of discontinuities within a structure and can work out the thickness of layers of materials. The results can be used to estimate their mechanical strengths (Martinho & Dionísio, 2014). In the last decade, acoustic tomography has benefited from technical advances, allowing 2D and 3D views of inner structures or subsoils (Misak et al., 2019). The ultrasonic reflection method uses piezo electric transducer above 20 kHz to generate high-frequency ultrasonic waves through a medium. Therefore it is widely used for flaw detection and dimensional measurements. However, the signals are much attenuated in heterogenous medium causing limited penetration depth into masonry structures (Nuzzo et al., 2010). While having been limited to transmission testing on samples, in situ measurements have now made their way into practical applications in the past 5 years and the construction

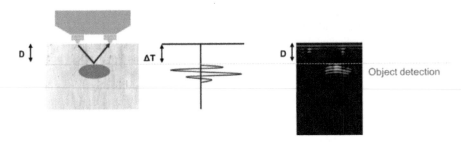

Figure 17.1 Ultrasonic pulse-echo technology.

industry has made a big step forward allowing two-dimensional and three-dimensional imaging helped by artificial intelligence (Niederleithinger et al., 2019). The working process of the technique is based on analysis waves that echo when passing through interfaces (Fig. 17.1). The velocity and attenuation of the propagating acoustic waves depend on the physical properties of the material (density and porosity) and the occurrence of discontinuities or defects within the object (Khairi et al., 2019; Mohd et al., 2019). Recent developments in ultrasonic tomography are based on new devices that work with low-frequency shear waves from 15 to 100 kHz. The commercially available MIRA and Proceq Pundit Live Array are both portable ultrasonic shear wave tomography devices developed for the analysis of concrete structures (Corbett et al., 2017; Dinh et al., 2023; Popovics et al., 2017). Both instruments operate on the same principle using a low-frequency ultrasonic pulse to reduce signal attenuation and optimize a 2D and 3D image reconstruction. Images are built of pixels representing finite, discrete, and small areas of the object and are associated with values of intensity reflection of the signal across the material.

Regardless of the technique used, the interpretation of results remains a central matter and results provided by different NDTs are often complemented with destructive evaluation that provides direct information about the internal shape of a structure and reduces interpretation ambiguities (Pérez-Gracia, 2014). By way of example, a multiscale integrated investigation of a medieval rose window was done by using GPR prospecting, sonic test and UT, as well as IRT (Nuzzo et al., 2010). The study highlighted the benefits of a multitechnique NDT approach to formulate a correct diagnosis and acquire accurate knowledge of the building internal structure.

In this context, this research explores the possible use of combined techniques that enhance architectural studies (metrology and geometry of sacred architecture) and NDT (ultrasonic imaging) as these methods usually provide independent conclusions when they are used individually.

17.2 Materials and methods

In this section, stereotomy, that is, the technique of cutting and assembling stones, is analyzed through architectural surveys and structural analysis of the gothic architecture. After a brief historical overview of the case study under scope, the metric survey of a gothic column is presented, and its geometry is studied by using rules in use for the design. Metrology, in which current measurements are expressed in terms of ancient metric units, is further presented as a tool for interpreting architectural drawings of gothic structures. A structural analysis combined to a visual inspection first examines the geometry of the drums to observe the division of the stone blocks. Finally, ultrasonic tomography is developed from a theoretical point of view to show how ultrasonic measurements are influenced by the composition of a composite masonry. The study involves a three-step procedure (Fig. 17.2):

1. building construction pattern hypothesis that explain the geometry of the crafts,
2. interpretation of ultrasonic signal reflections across the structure, and
3. confrontation between the hypothesis and ultrasonic measurements.

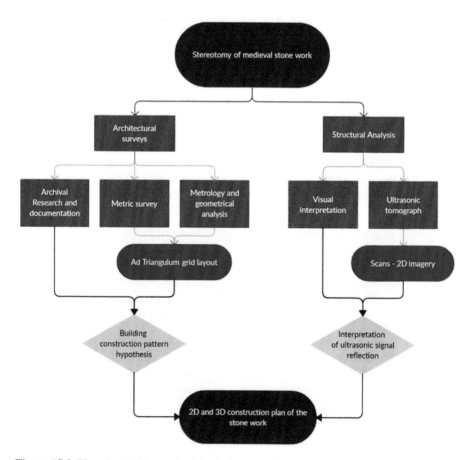

Figure 17.2 Flowchart of the methodological approach.

17.2.1 Architectural surveys

17.2.1.1 Archival research and documentation

Saint Waltrude Collegiate Church is a Catholic parish church, named in the honor of Saint Waltrude of Mons (Belgium). The building is protected by the heritage register of Wallonia (Fig. 17.3). It was built for the canonesses of the city between 1450 and 1691. The collegiate church of Mons is typical of the Brabantine Gothic style, a significant variant of Gothic architecture that is typical for the Low Countries (Belgium, the Netherlands, and Luxembourg). The main characteristics of this style are a large-scale cruciform floor plan with three-tier elevation along the nave and side aisles (pier arches, triforium, and clerestory). The Brabantine churches usually include an ambulatory with radiating chapels. The use of light-colored sandstone or limestone, which permitted rich detailing, is also what gives the Brabantine churches their distinctive appearance. In the nave, the pillars can be round or fasciculated, as in Mons. Such pillars are made up of at least five columns stuck together to form a whole. The columns continue often without interruption into the vault ribs. The triforium and the clerestory windows typically flow into one another, with the windows occupying the entire area of the pointed arch.

The collegiate church in Mons is no exception to these compositional rules. Its Latin cross shape formed by a nave of 115-m long is surrounded by 29 chapels. The vaults culminate at 24.5-m high. Visitors may be surprised by the architectural unity of the whole construction that took 236 years to be erected (Fig. 17.4).

Nevertheless, the Saint Waltrude Collegiate Church is not an exception of its own. Such monumental constructions usually took centuries to be built, involving successive contractors in charge of the construction. Over time, the original design wanted by the canonesses of Mons was followed without any adaptation. The masons were usually asked to continue the work "according to a form and mold" or

Figure 17.3 Saint Waltrude Collegiate Church.

Figure 17.4 View from the nave of the collegiate church of Mons.

"according to the portraiture" (Knoop & Jones, 1937). This obligation was clearly notified to contractors and stated in the building contract to guarantee the best quality in form and construction (Van Tussenbroek, 2017).

In 1448−49, the canonesses of Mons consulted three master masons named Michel de Rains, Jean Huelin, and Jean le Fèvre. They were asked for advice and plans of the collegiate. The three architects knew each other because of past collaborations in the north region of France. Michel de Rains was born in Valenciennes, in the north of France, about 40 km from Mons. He oversaw the construction of the collegiate Notre-Dame-La-Grande of Valenciennes and was in charge of several projects around this region, including Cambrai or Le Quesnoy. Jean Huelin, was born in Maubeuge, in the north of actual French Kingdom. He was sufficiently renowned at that time and recognized as master mason for Hainaut County by the Duke of Burgundy. Jean le Fèvre was a master mason from Mons who worked also in north of France, although he was the less experimented of the group.

The three master masons provided at least three preliminary drawings, also called patterns that can be consulted at the city archives of the city of Mons. The drawings show a core, a longitudinal half-section of a nave, and a detail of the vault. It is believed that Michel de Rains is the unique author of these drawings (Salingaros, 2010). However, the drawings do not correspond to the plans of the collegiate of Mons. These drawings were most probably used to contribute to the debate on the design of the future construction. According to Salamagne (2000), the first plan drawn by Michel de Rains was directly inspired by the cathedrals of Reims (1211−1345) and Orléans (1278−1568), which are both influenced by the early gothic regional style (Ph, 1988). The second drawing figures a nave with two aisles and a core with seven radiant chapels, with a deeper central aisle. The transept has also side aisles which are representative from French architecture. The third drawing is a map of a north-west chapel of Amiens cathedral, built between 1373 and 1375. All the drawings show similarities with the religious construction erected at that time in the north of France. Therefore there is little doubt that these drawings also inspired the architects.

In 1450, Philip the Good, Duke of Burgundy mandated the architect Jean Spiskin, native from Hainaut region, for the construction of the core of the collegiate church. Although there is no archive to confirm that Spiskin is the author of the design of the collegiate. He may have used the plans drawn by Michel de Rains. Nonetheless, the account books for the construction of the collegiate report that Spiskin made several trips in Belgium and France in 1450 and visited the cathedrals of Tournai, Grammont, Bruxelles, Leuven, Malines, and Lille. He was accompanied by Henri de Jauche, a master carpenter, and two master masons named Mathieu and Pierre de Layens from Leuven. The group was most probably looking for inspiration and technical solutions before starting the construction at Mons. The first cornerstone of the collegiate of Mons was led in 1451. Archives report that Mathieu de Layens was responsible for the control of the construction then replaced Spiskin who died in 1457. In 1465, Pierre de Layens replaced his father. No less than four master masons were in charge until 1527 when the construction of the transept ended. Contrary to their predecessors, de Layens and his son, as well as Pauwels, had no French origin but came from Brabant region. Compared to the core, it took twice as long to build the nave. In 1549, the erection of a bell tower of 190 m was started but this plan was never realized (Fig. 17.5).

17.2.2   Metric survey

A survey procedure requires the masonry qualification at a first level of investigation through visual observations and measurements. Today, visual inspection is mainly automated by using image processing on digital images. This allows the geometrical characterization of a masonry in 3D and helps to better understand the arrangement of the construction blocks of a wall (Oses & Dornaika, 2013).

A column located at the transept crossing was selected for the study. The column is built in Ecaussinnes blue stone with thin lime mortar joints. Each column shaft is made up of eight blocks of stone, the depth of which is unknown. Metric survey was preliminary done to acquire the outline of the structures and map the jointing planes of the drums. The 3D model was made by using photogrammetry. Horizontal cross sections of the drums were obtained with visual SFM (structure from motion) and Meshmixer. These two techniques provide a cross section with a

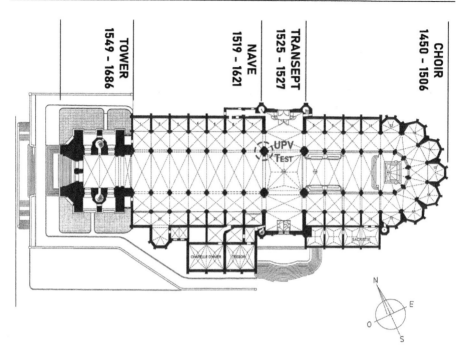

Figure 17.5 Ground floor of the collegiate church of Mons illustrating the four building phases. The red circle localizes the column under study.

measurement deviation of around 1.5%−3.6%. The column shafts are 263- and 265-cm wide. The drawing figures the vertical joints between stone blocks that constitute a pair of overlying drums, named type A and type B. Each pair of drums is constituted by two different assemblages of stone blocks (Fig. 17.6).

17.2.3   Metrology and geometry of sacred architecture

Ancient buildings and dimensioned drawings can be analyzed and better understood by the study and conversion of the units of measurement used at the time of construction. Units of measurement used in the past until the 19th century varied through ages and geo-cultural regions, reflecting social and economic influences. The *pied de Paris* for instance, also called *pied de roi* or king's foot, was used all over the kingdom of France in combination with other local foot values until 1831, when the meter was adopted as a unit of length in the International System. Although different units of measurement were used in a same place, this was not a source of confusion for the well-accustomed master masons who were able to convert any value into a different unit on a plan (Sebregondi & Schofield, 2016). This also highlights the probability that any ancient building that underwent different construction phases may have been designed according to different units of measurement (Masini et al., 2004).

Figure 17.6 Left and middle: stone pillar under study, right: 3D horizontal section.

Table 17.1 Units of measurement listed in the Dictionnaire universel des poids et mesures anciens et modernes, written by Horace Doursther in 1840.

Belgian cities	Foot conversion (cm)	French cities	Foot conversion (cm)
Antwerpen	28.7	Avignon	24.8
Bruges	27.4	Beauvais	29.8
Brussels	27.6	Bordeaux	35.7
Liège	29.5	Dijon	33.1
Mons	29.3	Nancy	28.6
Tournai	29.8	Paris	32.5

During the Middle Ages, units of measurement that referred to the proportions of the human body (the palm, the span, the foot, and the cubit) were used within construction. Today, one may observe that the conversion of these units into the International System gives values that vary from one place to another, although the numerical relation between the units always remains. The correspondence of a variety of foot values used in Belgian and French cities is shown in Table 17.1.

The rationality and the logic of a construction can be highlighted when the conversion of units of measurement is carried out. However, such kinds of observations should take into consideration that modern expectations of precision are different from the ancient, particularly as basic tools were used for construction (Rossi & Fiorillo, 2018). This should also temper any conclusions about construction laws based on the interpretation of overly precise computer-aided design (CAD) drawings.

The contribution of geometry for the design of gothic structures is a matter of interest and has been studied by researchers and architects for decades, highlighting its potential to enrich architectural practice in the 21st century, in much the same way as formal and archaeological analysis enriched architecture in the 19th and early 20th centuries (Bork, 2014). Nonetheless, few theories on the subject are strongly supported by evidence and often remain assumptions. Writings on the subject also abound, based on esoterism and scientific observations. One of the more famous being Ad Quadratum published by the controverted Norwegian historian Macody Lund in 1921 (Lund, 1921). Furthermore, CAD systems and 3D survey techniques render obsolete concerns about imprecision and approximation found in past research that also contributed to skepticism of geometrical research (Kidson, 1993; Murray & Addiss, 1990; Wu, 2002).

Besides the precision of a drawing, the design of its geometry remains a matter of study that relates firstly to construction rules rather than measurements, as stated by Shelby (1977), for whom rules based on geometry for the composition of sacred architecture materialize the Augustinian belief that the universe owes its stability to the perfect balance of its elements as instituted by the Creator (Shelby, 1977). The simplest geometric shapes, such as the circle, the square, or the equilateral triangle, were mostly used for design work (Fehér et al., 2019). Such vocabulary of forms is described by Salingaros (2010) as a pattern language. It was often symbolically linked to symbolic expressions of the human body too, as it can be observed in the treatise of Francesco di Giorio Martini (Fig. 17.7).

According to Ramzy (2015), Euclidean geometry was used during the medieval period as a visual tool for contemplating the mathematical nature of the universe under the authority of the divine (Ramzy, 2015). Although the relation between anthropomorphism and geometry does not uniquely refer to gothic architecture, since it was discussed from antiquity to the renaissance. De divina proportione written by Luca Pacioli at the end of the 15th century is one example, among others, that illustrates how a planning design can respect the proportions rules defined by the Fibonacci series or the Golden ratio. Although the overall state of the field remains strikingly primitive even today, the conception of gothic designs is undoubtedly based on an "unspeakable" process built on a geometrical logic that could only be achieved by a know-how training and orally transmitted from master masons to apprentices through explicit descriptions of the rules (Bork, 2011).

The drawings must be considered as the primary instrument of communication to transmit an architectural design from the master mason to the masons. Graphical methods for the design are explained in a few treatises written from the 13th to 15th centuries. Ramzy (2015) suggests that easy practical rules of Euclidian geometry were employed for the usually invisible, proportional, or working lines which are perceived only by the designer himself or by an analytical view of the drawings with a mesh of imaginary lines. The most repeated geometrical patterns produced according to the Euclidean theory are those designed with ad quadratum and ad triangulum patterns, from squares and equilateral triangles within a circle (Ramzy, 2015; Ržiha, 1883) (Fig. 17.8).

The portfolio of Villard De Honnecourt, dating from the 13th century, is probably the mostly known for its drawings that illustrate symbolic meanings with geometric accordance. The pensive face, for example, is drawn by using horizontal and vertical lines that intersect diagonals and form a square grid of four by four (Fig. 17.9).

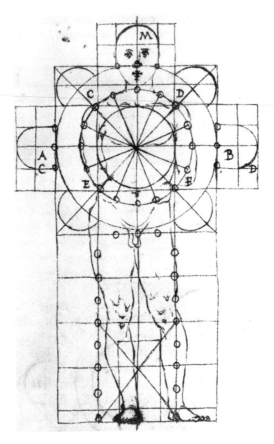

Figure 17.7 Church plan from the Trattato di Architettura of Francesco di Giorgio Martini, 1490.

Such a representation may be viewed as the perfection that each mason must look for to reach perfect harmony, where every part of the design is arranged to coincide with each other (Bechmann, 1993). Later on, in 1486, Matthäus Roriczer wrote *Geometria Deutch* and *Büchlein von der Fialen Gerechtigkeit*, two treatises that describe how a stone mason should use simple geometrical shapes to carve out stones. The booklet of Hans Schmuttermayer, from the early 16th century, also illustrates the manipulation of forms to compose the geometry of the crafts (Shelby, 1977). *Unterweisungen*, was written in 1516 by the mason Lorenz Lechler to teach his son the techniques for designing architecture of this time. However, Lechler specifies that his method is one among others to reach a good design. These examples illustrate the early need to theorize construction by means of rules to reproduce drawings. However, no unique theory to conceive a project existed. The construction, as an objective to achieve, appeared to be more important than the means of achieving it (Fig. 17.10).

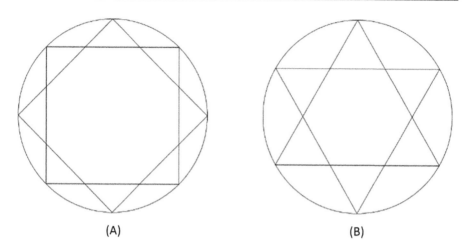

(A) (B)

Figure 17.8 Ad quadratum and ad triangulum patterns.

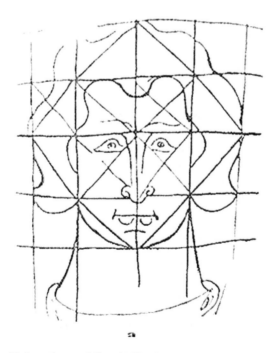

Figure 17.9 Folio 38 from the portfolio of Villard De Honnecourt (1230–1235).

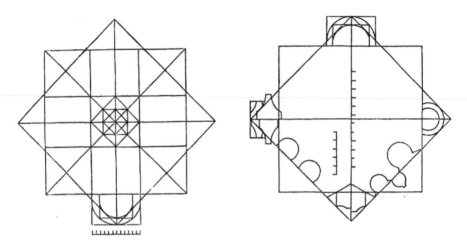

Figure 17.10 Examples of a quadrature diagrams with some mold profiles from the booklet of Lorenz Lechler (from Reichensperger 1856).

17.2.4   Structural analysis

17.2.4.1   Visual inspection

The materials used for the construction of the collegiate are local Bray sandstone for the solid sections and base walls, and a local blue stone ($CaCO_3$ 88%–99%), named Ecaussinnes blue stone (2687 kg/m^3, 0.31% vol.), used to build the columns of the nave and the upper sections of the walls. These stones usually bear mason's marks, which also makes it possible to date the different construction phases of the building (Fig. 17.11). Each overlying pair of drums is constituted by two different assemblages of stone. The vertical joints of the drums are set symmetrically on both sides of the horizontal and vertical axis, although some deviations also exist. In Fig. 17.12, the blue dots and red dots correspond to the vertical joints of drum types A and B, respectively.

17.2.4.2   Ultrasonic tomography

A portable ultrasonic tomography device is used for the research. The choice of the Pundit Live Array commercialized by Proceq is justified by the exploratory nature of the research, which aims to assess the instrument's ability to carry out surveys through thick masonry. The instrument consists of a linear array of 24 sending and receiving transducers arranged in 8 channels of 3 transducers with a central frequency of 50 kHz shear wave. In turn, one channel transmits an ultrasonic signal and its echoes are received by the seven other channels. The device uses the time-of-flight measurements between the different transducer pairs. The emitted energy propagates through the object and scattered energy is detected by the transducers (Fig. 17.13).

Figure 17.11 Mason's marks visible on the columns of the nave.

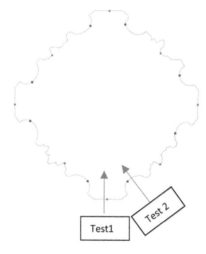

Figure 17.12 Cross section of the column under scope and localization of the vertical joints between stone blocks.

An ultrasonic signal sent from a transmitter transducer to a receiver transducer through a material provides an A-scan which represents the signal amplitude versus time of flight. A complete measurement will consist of 28 A-scans to compute and display in real time a single B-scan, also called tomogram (Marecos et al., 2016). Multiple scans in a grid pattern can be processed to additionally generate C- and D-scans. Portable ultrasonic tomography device proven their reliability for assessing the depth and composition of concrete structures and localize inside defects, such as poorly bonded interfaces, delamination zones, or honeycombing areas (Drobiec et al., 2020; Kwon et al., 2021; Luchin et al., 2020; Słoński et al., 2020). Recent

Figure 17.13 Pundit live array used for the tests.

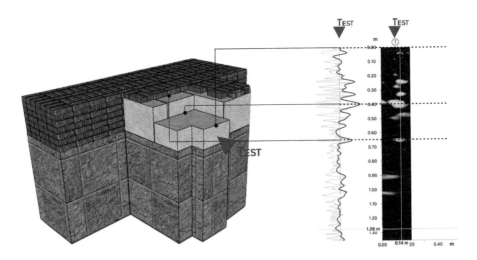

Figure 17.14 Ultrasonic test done on a composite masonry. Left: 3D representation of the stone and brick wall assemblage, right: B-scan illustrating the echoes between layers of building materials.

studies have also discussed the effectiveness of ultrasonic tomography for the assessment of masonry structures (di Prisco et al., 2019; Mohd et al., 2019; Zielińska & Rucka, 2018). However, little research has been dedicated to tomographic studies on stone or brick masonry with portative devices. These structures are much more heterogenous compared to concrete, with multiple interfaces of masonry units and mortar joints (Fig. 17.14). Therefore the capacity and accuracy of travel time tomography in assessing hidden defects, voids, or damages in such kinds of masonry is a matter under scope (Drobiec et al., 2020; Korgin et al., 2018; Luchin et al., 2020; Michaux & Grill, 2009; Wong, 2019).

The results can only be explained by understanding how waves propagate through the different materials that make up the complex structures. When an ultrasonic wave crosses an interface between two media (A and B) with different acoustic impedances Z_A and Z_B, the wave divides into two components: a transmitted and a reflected wave. Acoustic impedance (Z) represents the opposition to the flow of a sonic wave through a material, and it is dependent on the density of the material through which the wave propagates.

$$Z = \rho \times C$$

where ρ is the density of the medium, and C is the ultrasonic wave velocity.

By way of example, if we consider a lime base mortar with a matrix bulk density of 1570 kg/m³, a matrix porosity of 30%, and a propagation velocity of 2000 m/s, its acoustic impedance is 3.14 kg/m²s.

The amount of reflected and transmitted wave depends on the impedance of each one of the media, and it is quantified by the reflection coefficient (R) and transmission coefficient (T).

$$T = \frac{4Z_A \times Z_B}{(Z_A + Z_B)^2}$$

$$R = \frac{(Z_B - Z_A)^2}{(Z_A + Z_B)^2}$$

The R value varies between −1 and 1. An absolute value of R close to 1 indicates that the ultrasonic wave is completely reflected and no fraction of the incident wave will be propagated to the second medium.

Ancient masonry structures are never monolithic but made of multiple materials that constitute horizontal and vertical layers. The vertical joints cause signal attenuation of the penetrating mechanical wave, and reflections of the signal are much lower in amplitude due to difference of acoustic impedances between successive layers. These reflections may provide tomograms which are difficult to interpret if the composition of the masonry is unknown (Misak et al., 2019).

Blue stone has a compact structure that causes low attenuations of the waves unless the material is delaminated. Lime mortar can be considered as a two-phase media constituted by aggregate grains surrounded by a porous matrix (Arizzi et al., 2013). Therefore the surface of the aggregate constitutes an interface between the lime mortar and the aggregate itself, each having different acoustic impedances. In addition, it should be highlighted that voids and cavities within the structures notably reduce the transmitted waves through the structure. Values of densities (φ), ultrasonic velocity through the medium (C), and acoustic impedances are listed in Table 17.2. Transmission and reflection coefficients of materials that the masonry of the column under scope are shown in Table 17.3.

Table 17.2 Values of densities, ultrasonic velocity, and acoustic impedances for materials used for the construction.

	φ (kg/m^3)	C (m/s)	Z (kg/m^{2}s)
Stone	2687	5527	5.98 10^6
Lime	1570	2000	3.14 10^6
Aggregate	2660	6200	16.49 10^6
Air	1.22	340	414.80

Table 17.3 Values of wave transmissions and reflections through the interfaces of the column and reduce space between values.

	Stone	**Lime**	**Aggregate**	**Air**
Stone		$T = 0.90$ $R = 0.10$	$T = 0.71$ $R = 0.29$	$T = 0.06$ $R = 0.94$
Lime	$T = 0.90$ $R = 0.10$		$T = 0.47$ $R = 0.53$	$T = 0.03$ $R = 0.97$
Aggregate	$T = 0.71$ $R = 0.29$	$T = 0.47$ $R = 0.53$		$T = 0.17$ $R = 0.83$
Air	$T = 0.06$ $R = 0.94$	$T = 0.03$ $R = 0.97$	$T = 0.17$ $R = 0.83$	

The propagation velocity of waves and acoustic impedance are much higher in aggregates compared to mortar matrixes, causing ultrasonic wave attenuations from the aggregate grains to the matrix (Chekroun et al., 2009; Lafhaj & Goueygou, 2009). The transmission and reflection values at the interface between lime-based mortar and aggregate are 0.47 and 0.53, respectively, which means that considering an ultrasonic signal going from the aggregate to the matrix, 53% of the incidence wave is reflected, while 47% is propagated into the matrix.

One may observe that lime floated on a stone does not greatly interfere with the transmission of waves through the structure. However, waves are strongly attenuated when they cross air layers situated between two stone blocks and a lime layer. Higher reflections make it possible to locate such voids, but this clearly reduces the efficiency of the technique for deep inspection (Fig. 17.15).

17.3 Formal analysis and investigation, validation, calculation, and expression of results

In this section, we formulate hypotheses to clarify the stereotomy of the stones that make up the column's envelope. These hypotheses assume that stereotomy is dictated by geometric construction laws. The results obtained during the UT campaign are then presented.

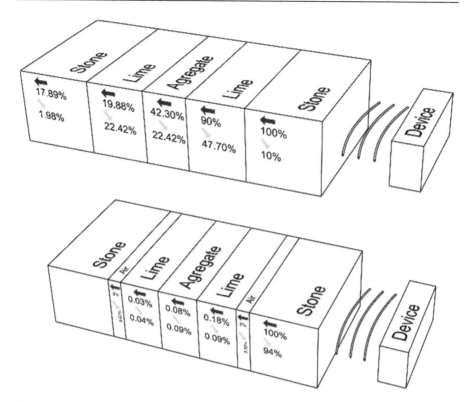

Figure 17.15 Transmission values of ultrasonic waves through walls that show a good cohesion (up) or air voids between layers (down).

17.3.1 Construction hypothesis for the architectural design

Knowing the value of the foot of Mons ($\cong 29.3$ cm) to interpret the metric values of the column, it can be noted that the whole section of the column ($\cong 263$ cm) is enclosed in a square of nine feet wide ($\cong 263.7$ cm) (Fig. 17.16A). An *ad triangulum* grid of 8 ft stood on its tip also encloses the structure. In this pattern, vertical joints are precisely located along the construction lines of the grid (Fig. 17.16B–D). Based on these observations, the study will further investigate if this scheme also rules the depth of the stones and the filling of a core. That is to say that following a logic of geometry, the stone blocks should be 1-ft deep with a core of 2-ft wide localized at the center of the column. Fig. 17.17 illustrates the plan and axiometry of the angle of the column with the dimensions and assembly of the stones.

17.3.2 Ultrasonic test campaigns

UTs were done on two opposite sides of the column. Measurements were calibrated according to the velocity of waves measured through the structure (global pulse

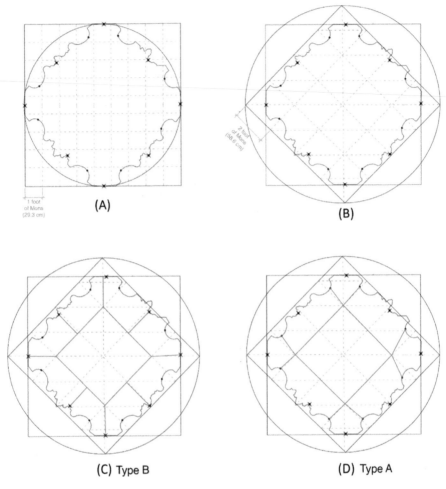

(A)

(B)

(C) Type B

(D) Type A

Figure 17.16 The section of the column is enclosed in a nine-foot ad quadratum square (A), but also in an 8-foot ad triangulum grid which locates the joints of the A and B drums along the construction lines (B–D).

velocity = 3228 m/s). Test one was done vertically on a flat corner from bottom to top across four overlying pairs of drums A and B. The 10 scans associated together by means of artificial intelligence are shown in Fig. 17.18. The dimensions of the instrument and the curvature of the stones made the second test only possible in a crook of drum A.

Test one gives a tomogram that clearly shows two repetitive echo patterns corresponding to the overlaying type A (drums 1 and 3) and B (drums 2 and 4) drums. Both types of drum echo differently depending on their depths. Considering type A drums, the tomograms show two homogeneous layers, each of approximatively 50-cm deep, separated by a zone of higher reflection. In type B drums, signals were reflected within three zones situated between: 0–30, 30–50, and 50–100 cm. Within the first 30 cm,

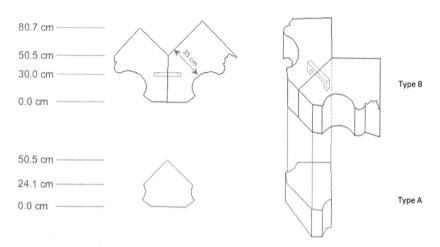

Figure 17.17 Plan and axonometry of stone blocks for drum types A and B.

no reflections were observed, while high reflections are observed between 30 and 50 cm. Beyond 50-cm deep, successive echoes of lower intensity were recorded. In no case, the backwall of the column, situated at $\cong 264$ cm, could be recorded, probably because of the total reflection of the ultrasonic waves through the deep and heterogeneous masonry. The second test done in the curvature of a type B drum indicated clearer reflections located at 30-, 60-, and 90-cm deep.

17.4 Discussion and evaluation

The differences in echo signals between drums A and B indicate a difference between the arrangement of stones of the overlaying drums. Tomograms of drums type A show two homogeneous zones of approximatively 50-cm deep with no signal reflection, while tomograms of drums type B drums reveal no reflection signal for the first 30 cm followed by high reflections around 30 cm. This area of strong reflection is interpreted as a metal reinforcement between two stones of the type B drum. This type of metal reinforcement has been seen on the outer buttresses of the collegiate church and strengthens this hypothesis (Fig. 17.19).

In addition, if we interpret the tomograms in the light of the construction plan, it can be argued that a variation in intensity at 50-cm deep could correspond to the rear face of the stones tested. However, as already mentioned, the appearance of the type A and B tomograms is very different and would also indicate that the stones are cut into a different shape at the back of the blocks. In fact, the ultrasound passing through drums type A generates very few echoes beyond 50 cm. Our hypothesis is that these type A drums have an outward angle that would reduce parasitic reflections through the stone. Conversely, a reentrant angle (type B drum) would cause

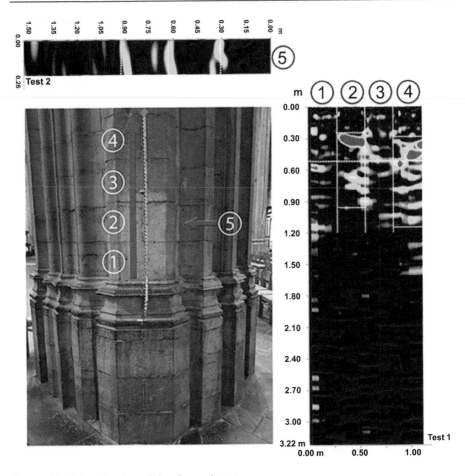

Figure 17.18 Localization of the ultrasonic tests.

total reflection of the signal into the blocks, which would induce multiple echoes through the stone (Fig. 17.20). This cut would result in a longer signal path between transmission and reception. This longer time between transmission and reception would be responsible for the false echoes visible beyond 50 in type B drums. This makes it very difficult to interpret the results for type B drums at a greater depth. However, the particular cut of the stones on the type A drums allows us to speculate further about the interior composition of the column. One may observe a reflection-free zone about 1-ft thick, between 50 and 75 cm ($\cong$ 2.5 ft), followed by another zone about 1.5 ft thick, between 75 and 120 cm (Fig. 17.18). This depth of 75 cm does not seem to be consistent since it represents 2.5 ft. In addition, these dimensions are consistent given that the center of the column is 132-cm deep. To sum up, the combined analysis indicates that the half-thickness of the column would be made up of four layers: 0−50/50−75/75−110/110−132 (Fig. 17.21).

Figure 17.19 Metal sample visible on the buttresses of the collegiate church.

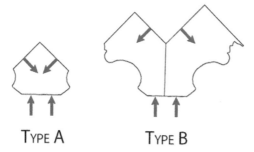

Figure 17.20 Reflection of the signal through a stone with an outward angle (type A drum) and a reentrant angle (type B drum).

Contrary to test one, the second test done in a curvature of a type A drum provides a very clear echo around 30 cm, followed by two false echoes around 60 and 90 cm deep. This scheme would indicate that incident signals are reflected by a surface most

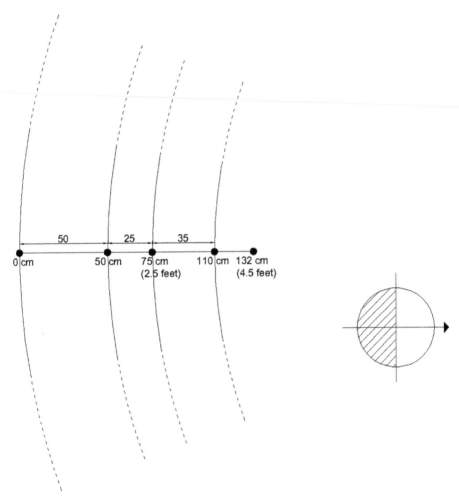

Figure 17.21 Stratigraphy of the column showing the different layers of its composition up to its center.

probably parallel to the instrument. The ultrasonic results are coherent with the construction hypothesis that locates the back side of the stone at 35-cm deep.

17.5   Conclusion

The research for geometrical theories that guide gothic architecture generally divides the scientific community between skepticism and belief. The aim of this research is to combine different methods of study into an integrated approach that can contribute to verifying the hypothesis made on NDT results.

This chapter sets out the results of an initial study of a gothic column built during the 15th century. A portable ultrasonic pulse-echo tomograph working with low-frequency shear waves was used to assess the depth of the different layers of building materials and the assemblage of stone blocks together. This instrument is originally designed for surveying reinforced concrete structures. One of the challenges of the study was therefore to test its performance in the analysis of masonry structures.

The instruments used for the tests provided reliable data at short range which does not exceed no more than 50-cm deep. As such, the study confirms the observations of Nuzzo et al. (2010) who reported that low-frequency signals are much attenuated in heterogenous medium causing limited penetration depth into masonry structures. This study emphasizes that the presence of air spaces between materials significantly reduces signal transmission through a heterogeneous structure. Here, the probable presence of metal staples used to bind the stones together also plays a role by causing a lot of short-range signal reflections in drum type B. Materials of different acoustic impedances and surfaces not parallel to the ultrasonic device also generate difficult results to interpret. The absence of echoes at depth made it impossible to specify the composition of the column core.

However, the study highlights the benefits of using combined approaches from different disciplines. Clearly, the metrological approach not only helps to convert the dimensions and interpret the gothic design, it also helps to clarify the hypothesis and assumptions made by the interpretation of the ultrasonic testing. Indeed, the interpretation of the results is not straightforward, as it may be unclear whether zones of high reflectance should be interpreted as a transition between two materials of different acoustic impedances or results from the geometry of the structure.

The study also confirms that a scheme of construction that uses the foot as a unit of measurement was used for the design of the gothic structure. The metrological analysis can be seen as a method to help better understand the logic hidden behind the drawing. The stereotomy of blocks and the assemblage of drums respect a canvas dictated by the *ad triangulum* scheme. This hypothesis was confirmed by the geometrical analysis and the ultrasonic testing results. The research on a possible logic of composition and assemblage helped to better understand the tomogram without resorting to a destructive observation through coring.

Future works will extend this study to several monuments and structures built during the Middle Age. In this way, the authors hope to verify that the *ad quadratum* and *ad triangulum* patterns were used in a recurring way for the constructions of this period in different regions. Further studies should also verify whether these schemes have been scrupulously observed everytimes and everywhere with the same rigor. In particular, future works should verify the impact of successive construction phases on compliance with these compositional rules. If the hypotheses are confirmed, the methodology could be used to date parts of a construction as well or to classify different parts of a building according to conception rules. Future investigations could combine other non-destructive techniques to compare the performance of each method. In that scope, past results obtained with GPR appear promising. This alternative technique could bring complementary data for the survey.

References

Arizzi, A., Martínez-Martínez, J., & Cultrone, G. (2013). Ultrasonic wave propagation through lime mortars: an alternative and non-destructive tool for textural characterization. *Materials and Structures*, *46*(8), 1321–1335. Available from https://doi.org/10.1617/s11527-012-9976-1.

Bechmann, A. (1993). *Villard De Honnecourt. La pensée technique au 13eme siècle et sa communication*. Picard.

Binda, L. (2005). *The importance of investigation for the diagnosis of historic buildings: application at different scales (centres and single buildings)*. Structural analysis of historical constructions (pp. 29–42). Taylor & Francis Group.

Binda, L., Lenzi, G., & Saisi, A. (1998). NDE of masonry structures: use of radar tests for the characterisation of stone masonries. *NDT & E International*, *31*(6), 411–419. Available from https://doi.org/10.1016/s0963-8695(98)00039-5.

Bork, R. (2011). *The unspeakable logic of gothic architecture*. Picard.

Bork, R. (2014). Dynamic unfolding and the conventions of procedure: geometric proportioning strategies in gothic architectural design. *Architectural Histories*, *2*(1), 14. Available from https://doi.org/10.5334/ah.bq.

Cardani, G., Cantini, L., Munda, S., Zanzi, L., & Binda, L. (2012). Non invasive measurements of moisture in full-scale stone and brick masonry models after simulated flooding: Effectiveness of GPR. *RILEM Bookseries*, *6*, 1143–1149. Available from https://doi.org/10.1007/978-94-007-0723-8_159.

Cardarelli, E., & De Nardis, R. (2001). Seismic refraction, isotropic anisotropic seismic tomography on an ancient monument (Antonino and Faustina temple ad 141). *Geophysical Prospecting*, *49*(2), 228–240. Available from https://doi.org/10.1046/j.1365-2478.2001.00251.x.

Chekroun, M., Le Marrec, L., Abraham, O., Durand, O., & Villain, G. (2009). Analysis of coherent surface wave dispersion and attenuation for non-destructive testing of concrete. *Ultrasonics*, *49*(8), 743–751. Available from https://doi.org/10.1016/j.ultras.2009.05.006, http://www.elsevier.com/wps/find/journaldescription.cws_home/525452/description#description.

Corbett, D., Gattiker, F., & Vonk, S. (2017). *Proceeding of the 15th Asia Pacific Conference for Non-Destructive Testing: A practice driven approach to the development of ultrasonic pulse echo technique applied to concrete structural assessment*.

Dinh, K., Tran, K., Gucunski, N., Ferraro, C. C., & Nguyen, T. (2023). Imaging concrete structures with ultrasonic shear waves—Technology development and demonstration of capabilities. *Infrastructures*, *8*(3). Available from https://doi.org/10.3390/infrastructures8030053, http://www.mdpi.com/journal/infrastructures.

Drobiec, Ł., Jasiński, R., & Mazur, W. (2020). The use of non-destructive testing (NDT) to detect bed joint reinforcement in AAC masonry. *Applied Sciences*, *10*(13), 4645. Available from https://doi.org/10.3390/app10134645.

Elkarmoty, M., Tinti, F., Kasmaeeyazdi, S., Bonduà, S., & Bruno, R. (2018). 3D modeling of discontinuities using GPR in a commercial size ornamental limestone block. *Construction and Building Materials*, *166*, 81–86. Available from https://doi.org/10.1016/j.conbuildmat.2018.01.091.

Fehér, K., Szilágyi, B., Bölcskei, A., & Halmos, B. (2019). Pentagons in medieval sources and architecture. *Nexus Network Journal*, *21*(3), 681–703. Available from https://doi.org/10.1007/s00004-019-00450-7, http://www.springerlink.com/content/1590-5896.

Fort, R., Alvarez de Buergo, M., & Perez-Monserrat, E. M. (2013). Non-destructive testing for the assessment of granite decay in heritage structures compared to quarry stone. *International Journal of Rock Mechanics and Mining Sciences, 61*, 296−305. Available from https://doi.org/10.1016/j.ijrmms.2012.12.048, http://www.elsevier.com/inca/publications/store/2/5/6/index.htt.

Hannachi, S., & Guetteche M. N. (2014). Review of the ultrasonic pulse velocity: Evaluating concrete compressive strength on site. *Proceeding of the Scientific Cooperations International Workshops on Engineering Branches* (pp. 103−112).

Himi, M., Pérez-Gracia, V., Casas, A., Caselles, O., Clapés, J., & Rivero, L. (2016). Nondestructive geophysical characterization of cultural heritage buildings: Applications at Spanish cathedrals. *First Break, 34*(8), 93−101. Available from http://fb.eage.org/index/home?p = 97.

Khairi, M. T. M., Ibrahim, S., Yunus, M. A. Md, Faramarzi, M., Sean, G. P., Pusppanathan, J., & Abid, A. (2019). Ultrasound computed tomography for material inspection: Principles, design and applications. *Measurement, 146*, 490−523.

Kidson, P. (1993) In Cocke, T., & Kidson, P. (Eds.), *Salisbury Cathedral: Perspectives on the architectural history. The historical circumstance and the principles of the design*. London: Royal Commission on Historical Monuments of England.

Knoop, D., & Jones, G. P. (1937). The decline of mason architect in England. *Journal of the Royal Institute of British Architects, 44*, 1004−1005.

Korgin, A., Kilani, L. Z., & Ermakov, V. (2018). Application of ultrasonic methods for determination location and sizes of granite blocks of the monument pedestal. *IOP Conference Series: Materials Science and Engineering, 365*, Institute of Physics Publishing Russian Federation. Available from https://doi.org/10.1088/1757-899X/365/5/052030, http://www.iop.org/EJ/journal/mse.

Kwon, H., Joh, C., & Chin, W. J. (2021). Pulse peak delay-total focusing method for ultrasonic tomography on concrete structure. *Applied Sciences, 11*(4), 1741. Available from https://doi.org/10.3390/app11041741.

Lafhaj, Z., & Goueygou, M. (2009). Experimental study on sound and damaged mortar: Variation of ultrasonic parameters with porosity. *Construction and Building Materials, 23*(2), 953−958. Available from https://doi.org/10.1016/j.conbuildmat.2008.05.012.

Luchin, G., Ramos, L. F., & D'Amato, M. (2020). Sonic tomography for masonry walls characterization. *International Journal of Architectural Heritage, 14*(4), 589−604. Available from https://doi.org/10.1080/15583058.2018.1554723, http://www.tandf.co.uk/journals/titles/15583058.asp.

Lund, M. (1921). *Ad quadratum; a study of the geometrical bases of classic & medieval religious architecture, with special reference to their application in the restoration of the Cathedral of Nidaros (Throndhjem) Norway*.

Marecos, V., Santos-Assunção, S., Fontul, S., & Pérez-Gracia, V. (2016). *Geophysics: Fundamentals and applications in structures and infrastructure. Non-destructive techniques for the evaluation of structures and infrastructure* (pp. 59−88). CRC Press.

Martinho, E., & Dionísio, A. (2014). Main geophysical techniques used for non-destructive evaluation in cultural built heritage: A review. *Journal of Geophysics and Engineering, 11*(5). Available from https://doi.org/10.1088/1742-2132/11/5/053001, http://iopscience.iop.org/1742-2140/11/5/053001/pdf/1742-2140_11_5_053001.pdf.

Masini, N., Fonseca, C. D., Geraldi, E., & Sabino, G. (2004). An algorithm for computing the original units of measure of medieval architecture. *Journal of Cultural Heritage, 5*(1), 7−15. Available from https://doi.org/10.1016/j.culher.2002.12.001, http://www.elsevier.com.

Masini, N., Persico, R., & Rizzo, E. (2010). Some examples of GPR prospecting for monitoring of the monumental heritage. *Journal of Geophysics and Engineering*, *7*(2), 190−199. Available from https://doi.org/10.1088/1742-2132/7/2/S05.

Michaux, C., & Grill, M. (2009). *NDT 3 D tomographic testing cases on concrete and national heritage buildings.*

Misak, L., Corbett, D., & Grantham, M. (2019). Comparison of 2D and 3D ultrasonic pulse echo imaging techniques for structural assessment. *MATEC Web of Conferences*. 289. EDP Sciences Switzerland.

Mohd, T., Sallehuddin, I., Mohd, A. Md. Y., Faramarzi, M., Pei, S. G., Pusppanathan, J., & Abid, A. (2019). Ultrasound computed tomography for material inspection: Principles, design and applications. *Measurement*, *146*, 490−523.

Moropoulou, A., Avdelidis, N. P., Karoglou, M., Delegou, E. T., Alexakis, E., & Keramidas, V. (2018). Multispectral applications of infrared thermography in the diagnosis and protection of built cultural heritage. *Applied Sciences (Switzerland)*, *8*(2). Available from https://doi.org/10.3390/app8020284, http://www.mdpi.com/2076-3417/8/2/284/pdf.

Murray, S., & Addiss, J. (1990). Plan and space at Amiens Cathedral: With a new plan drawn by James Addiss. *Journal of the Society of Architectural Historians*, *49*(1), 44−66. Available from https://doi.org/10.2307/990498.

Ndagi, A., Umar, A. A., Hejazi, F., & Jaafar, M. S. (2019). Non-destructive assessment of concrete deterioration by ultrasonic pulse velocity: A review. *IOP Conference Series: Earth and Environmental Science*, *357*(1), 012015. Available from https://doi.org/10.1088/1755-1315/357/1/012015.

Niederleithinger, E., Maack, S., Mielentz, F., Effner, U., & Strangfeld, Ch (2019). *Review of recent developments in ultrasonic echo testing of concrete.* Paper presented at the 5th conference on smart monitoring assessment and rehabilitation of civil structure.

Nuzzo, L., Calia, A., Liberatore, D., Masini, N., & Rizzo, E. (2010). Integration of ground-penetrating radar, ultrasonic tests and infrared thermography for the analysis of a precious medieval rose window. *Advances in Geosciences*, *24*, 69−82. Available from https://doi.org/10.5194/adgeo-24-69-2010, http://www.adv-geosci.net/volumes.html.

Oses, N., & Dornaika, F. (2013). Image-based delineation of built heritage masonry for automatic classification. *Lecture Notes in Computer Science*, *7950*, 782−789. Available from https://doi.org/10.1007/978-3-642-39094-4_90.

Patrucco, G., Cortese, G., Giulio Tonolo, F., & Spanò, A. (2020). Thermal and optical data fusion supporting built heritage analyses. *The International Archives of the Photogrammetry, Remote Sensing and Spatial Information Sciences*, *XLIII-B3-2020*(3), 619−626. Available from https://doi.org/10.5194/isprs-archives-xliii-b3-2020-619-2020.

Ph, K. J. (1988). Sainte-Waudru in Mons. Die Planungsgeschichte einer Stiftskirche 1449-1450. *Zeitschrift für Kunstgeschichte*, *51*, 372−413.

Pitarma, R., Crisóstomo, J., & Pereira, L. (2019). Portugal detection of wood damages using infrared thermography. *Procedia Computer Science*, *155*, 480−486, Elsevier B.V.

Popovics, J. S., Roesler, J., Bittner, J., Amirkhanian, A., Brand, A., Gupta, P., & Flowers, K. (2017). Ultrasonic imaging for concrete infrastructure condition assessment and quality assurance. A report of the findings of ICT-27−132. *Illinois Center for Transportation Series*.

di Prisco, M., Scola, M., & Zani, G. (2019). On site assessment of Azzone Visconti bridge in Lecco: Limits and reliability of current techniques. *Construction and Building Materials*, *209*, 269−282. Available from https://doi.org/10.1016/j.conbuildmat.2019.02.080.

Pérez-Gracia, V. (2014). Assessment of complex masonry structures with GPR compared to other non-destructive testing studies. *Remote sensing, 6*, 8220–8237. Available from https://doi.org/10.3390/rs6098220.

Pérez-Gracia, V., Caselles, J. O., Clapés, J., Martinez, G., & Osorio, R. (2013). Non-destructive analysis in cultural heritage buildings: Evaluating the Mallorca cathedral supporting structures. *NDT & E International, 59*, 40–47. Available from https://doi.org/10.1016/j.ndteint.2013.04.014.

Ramzy, N. (2015). The dual language of geometry in gothic architecture: The symbolic message of Euclidian geometry versus the visual dialogue of fractal geometry. *Peregrinations: Journal of Medieval Art and Architecture, 5*(2), 135–172.

Ranalli, D., Scozzafava, M., & Tallini, M. (2004). Ground penetrating radar investigations for the restoration of historic buildings: The case study of the Collemaggio Basilica (L'Aquila, Italy). *Journal of Cultural Heritage, 5*(1), 91–99. Available from https://doi.org/10.1016/j.culher.2003.05.001, http://www.elsevier.com.

Rossi, C., & Fiorillo, F. (2018). A metrological study of the Late Roman Fort of Umm al-Dabadib, Kharga Oasis (Egypt. *Nexus Network Journal, 20*(2), 373–391. Available from https://doi.org/10.1007/s00004-018-0388-6, http://www.springerlink.com/content/1590-5896.

Ržiha, F. (1883). *Studienüber Steinmetz-Zeichen. T. 3: Von dem graphischen Principe der Steinmetz-Zeichen.* Wien. Hof-und Staatsdruckerei.

Salingaros, N., (2010). *Twelve lectures on architecture: Algorithmic sustainable design.* Umbau Verlag.

Salamagne, A. 2000. Michel de Rains, Maître d'œuvre valenciennois, Sainte-Waudru de Mons et le problème de l'expertise au moyen-âge. Paper presented at the 6ème congrès de l'Association des Cercles Francophones d'Histoire et d'Archéologie de Belgique et 53ème congrès de la Fédération des Cercles d'Archéologie et d'Histoire de Belgique, Mons, Belgium, August 25.

Santos-Assunçao, S., Perez-Gracia, V., Caselles, O., Clapes, J., & Salinas, V. (2014). Assessment of complex masonry structures with GPR compared to other non-destructive testing studies. *Remote Sensing, 6*(9), 8220–8237. Available from https://doi.org/10.3390/rs6098220, http://www.mdpi.com/2072-4292/6/9/8220/pdf.

Sebregondi, G. C., & Schofield, R. (2016). First Principles: Gabriele Stornaloco and Milan Cathedral. *Architectural History, 59*, 63–122. Available from https://doi.org/10.1017/arh.2016.3, https://www.cambridge.org/core/journals/architectural-history/all-issues.

Shelby, L. (1977). *Gothic design techniques. The fifteenth-century design booklets of Mathes Roriczer and Hanns Schmuttermayer.* Southern Illinois University Press.

Slavíková, M., Krejčí, F., Žemlička, J., Pech, M., Kotlík, P., & Jakůbek, J. (2012). X-ray radiography and tomography for monitoring the penetration depth of consolidants in Opuka - the building stone of Prague monuments. *Journal of Cultural Heritage, 13*(4), 357–364. Available from https://doi.org/10.1016/j.culher.2012.01.010, http://www.elsevier.com.

Słoński, M., Schabowicz, K., & Krawczyk, E. (2020). Detection of flaws in concrete using ultrasonic tomography and convolutional neural networks. *Materials, 13*(7). Available from https://doi.org/10.3390/ma13071557, https://res.mdpi.com/d_attachment/materials/materials-13-01557/article_deploy/materials-13-01557-v2.pdf.

Van Tussenbroek, G. (2017). Building contracts in the Low Countries. Provisions concerning form and quality control in the construction industry (1350–1650). *Construction History, 32*(1), 1–20. Available from http://www.ciob.org.uk/home.

Wong, C. W. (2019). Applications of non-destructive tests for diagnosis of heritage buildings: Case studies from Singapore and Malaysia. *Built Heritage, 3*(1), 14–25. Available from https://doi.org/10.1186/BF03545732, https://built-heritage.springeropen.com/about.

Wu, N. Y. (2002). *Ad quadratum: The practical application of geometry in Medieval architecture*. Ashgate.

Yin, S., Cui, Z., Fu, J., & Kundu, T. (2019). Acoustic source localization in heterogeneous media. *Ultrasonics*, *99*, 105957. Available from https://doi.org/10.1016/j.ultras.2019.105957, http://www.elsevier.com/wps/find/journaldescription.cws_home/525452/description#description.

Zanzi, L., Hojat, A., Ranjbar, H., Karimi-Nasab, S., Azidi, A., & Arosio, D. (2017). GPR measurements to detect major discontinuities at Cheshmeh-Shirdoosh limestone quarry. *Bull. Eng. Geol. Environ.*

Zielińska, M., & Rucka, M. (2018). Non-destructive assessment of masonry pillars using ultrasonic tomography. *Materials*, *11*(12). Available from https://doi.org/10.3390/ma11122543, https://www.mdpi.com/1996-1944/11/12/2543/pdf.

Advances in artificial vision techniques applied to non-destructive tests in heritage buildings

David Marín-García[1], Juan Moyano[1], David Bienvenido-Huertas[2], and María Fernández-Alconchel[1]
[1]Department of Graphical Expression and Building Engineering, ETSIE, University of Seville, Seville, Spain, [2]Department of Building Construction, University of Granada, Spain

18.1 Introduction

18.1.1 Background

In the field of construction, especially in the rehabilitation or repair of buildings with elements protected for their historical-artistic value, there is a frequent need to carry out tests that do not affect or alter the properties of materials, elements, and structures subjected to said tests. These tests are called "non-destructive testing". Its beginnings go back to methodologies as simple, but it was not until the mid-19th century that they began to study and apply magnetic fields, ionizing radiation, and so on, being in 1941 when the American Society for Non-Destructive Testing (ASNT) is founded. Today, there are many similar societies around the world. According to the ASNT, "Non-destructive testing is any type of test performed on a material that does not permanently alter its physical, chemical, mechanical or dimensional properties."

At present, the methods and techniques are very varied, being able to distinguish between those that allow inspecting only surfaces, from those that can provide data on the interior of the elements. In the case of the former, there is VT (visual tests, visual inspection) and the aforementioned PT (penetrating liquids), although one could also consider here those that, in addition to the surface, can provide data from layers close to the surface, as is the case with MT (magnetic particles) and ET (electromagnetic tests, electromagnetism). Regarding those that allow the interior of the elements to be inspected, RT (radiographic tests, radiography), UT (ultrasonic tests, ultrasound), and AE (acoustic emission) stand out.

Leaving aside organoleptic tests, such as purely VT or VI, PT (penetrating liquids), and conventional photography, the main most advanced non-destructive tests are currently carried out using UT, infrared thermography (IRT), RT, ET, and AE.

If the question focuses on tests that produce images outside the visible spectrum, the technologies to carry them out have as their first precedent the discovery by

Diagnosis of Heritage Buildings by Non-Destructive Techniques. DOI: https://doi.org/10.1016/B978-0-443-16001-1.00018-8

Wilhelm Conrad Röntgen of X-rays (Nüsslin, 2020) and their application to medicine and, later, to other fields as in the case of the buildings.

As already mentioned, in addition to X-rays, over time other technologies have emerged, fundamentally based on acoustics (Schabowicz, 2019), electromagnetism (Huang & Wang, 2016), and the physics of materials, among others.

Leaving aside non-destructive physical-mechanical tests that are not the object of this study, such as the use of sclerometers or measurement of the rebound rate, today in the field of construction, in addition to the aforementioned X-rays, the most outstanding tests in this field are those based on ultrasonics or wave propagation velocity measurement, ground-penetrating radars based on the transmission of ultra-broadband electromagnetic, waves and thermography based on infrared (IR) radiation.

On the other hand, with the appearance and development of machine learning algorithms applied to the classification of images and the detection, identification, and location of multiclass objects in them, there has been an exponential advance in diagnostic tasks, assuming a great help quantitatively and qualitatively, for detecting and classification damages and applying the most appropriate treatments.

These algorithms are generally framed in the field of artificial vision and in recent years they are usually supported by the so-called deep learning (DL) (Lecun et al., 2015; Mu & Zeng, 2019; Sengupta et al., 2020) based on convolutional neural networks (CNNs) (Chen et al., 2021; Girshick et al., 2016; Long et al., 2015; Ciregan et al., 2012; He et al., 2016; Simonyan & Zisserman, 2015; Szegedy et al., 2013) that has been used in medicine, (Shen et al., 2017; Bakator & Radosav, 2018; Cai et al., 2020), industry (Sa et al., 2016), environment (Maggiori et al., 2017), among many other fields (Leiva et al., 2019).

In civil construction and buildings, work has also been carried out applying these algorithms (Akinosho et al., 2020; A.S. & Edayadiyil, 2022; Llamas et al., 2017; Abed et al., 2020) although usually from images of the visible spectrum, one of the applications being the identification of various pathologies based on symptoms reflected in these images.

Thus, based on images that reflect the visible spectrum, there are investigations related to the application of DL to identify fissures and cracks (Silva & Lucena, 2018), mold, stains, and paint deterioration (Perez et al., 2019), automatic detection of damage in historic buildings (Wang et al., 2019), efflorescence (Marín-García et al., 2023), and so on.

However, this research focuses on the detection and analysis of the most outstanding advances in the field of automated diagnosis with artificial vision through images that go beyond the visible spectrum.

18.1.2   Problem statement

In the field of diagnosis, the application of the so-called artificial vision in many cases is reaching accuracy close to 100%, in terms of detecting one or several objects within an image and classifying them, generally according to the training to which it has previously been subjected (supervised learning), although there is the possibility of using semisupervised, unsupervised, and reinforcement learning techniques.

On the other hand, in general, and in building, in particular, in the field on which this research is focused, in most cases, these algorithms perform the classification of pathologies (symptomatic elements detected) by visual analysis of the symptom or symptoms. There are few studies that cover other elements of interest such as the automatic assignment of the origin or cause, severity, the most appropriate way to repair or treat, and even, why not, the assessment of the application of said treatment or repair.

It should be noted that the uses of these algorithms have various problems and drawbacks, such as the need to have a graphic database that is quantitatively and qualitatively adequate for machine learning to be correct, the correct classification for training, problems related to the implementation and the limitations in its performance, and so on.

Therefore, when it comes to diagnosis through images and computer vision, it is necessary to indicate that this technology must be understood as an aid to the said diagnosis, since the supervision of an expert or experts who must always give the go-ahead is considered necessary final to the said diagnosis.

For all these reasons, the analysis of these technologies will be the central axis of this research.

18.1.3 Literature review

Non-destructive techniques (NDT) for building diagnosis are of great interest, especially in those cases in which it is not possible to alter the elements to be studied, as is the case of historic buildings (Tejedor et al., 2022).

On the other hand, artificial intelligence, machine learning, and smart technologies for non-destructive evaluation have already been studied by authors such as Taheri et al. (2022), which, in addition to "big data," mentions artificial vision to automate industrial and manufacturing processes, indicating that it helps to maintain and increase the quality of manufacturing, facilitates the analysis of information to detect material defects, and avoids human errors, revealing faults with high precision and at a faster rate faster than human inspectors can detect.

Artificial vision (machine vision [MV]) refers to technological methods aimed at machines' inspecting, recognizing, and analyzing images by providing data about them as a human being could.

When non-destructive tests are carried out on materials and these provide data in the form of images, humans with sufficient training can issue diagnoses in view of these images.

Therefore the MV applied to diagnosis through images provided by non-destructive tests is a methodology aimed at machines providing said diagnosis as said human beings could do.

This methodology usually focuses on the use of images to classify or detect objects or elements in them and, where appropriate, multiclassify them through techniques such as semantic segmentation.

Regarding the images generated by non-destructive tests, they need interpretation, since they do not offer information on the visible spectrum as a photograph does. For example, in 2021, Arbaoui et al. (2021) propose a new methodology for the detection and monitoring of cracks in concrete structures based on ultrasonic images processed by a wavelet and an automatic crack identification scheme supported by CNN, to detect cracks before they are detected visible, reaching precision close to 90%.

Another example is that of He et al. (2021) who propose images generated by IRT for diagnosis in various fields.

An et al. (2018) present a DL-based concrete crack detection technique that uses hybrid imaging of normal and vision thermographic and IR images to improve crack detection and minimize false alarms by detecting macro and micro cracks. automatically.

In short, there are currently various works on NDT for the diagnosis of existing buildings, even with a bibliometric and scientometric approach (Tejedor et al., 2022). There is also, as previously mentioned, research on computer vision applied to images of the results of said tests, although in the latter case, they are very specific applications, and no works have been detected that have compiled and studied the trends, growth, and evolution futures, comparatives, publications that lend greater interest to the topic and issues most discussed, and so on, this being precisely the central object of this work.

18.1.4 Safety considerations and standards

For standardization and normalization purposes, there are non-destructive testing societies such as the ASNT, the CHSNDT (Chinese Society for Non-destructive Testing), and the EFN-DT (European Federation for Non-destructive Testing), among others.

However, in terms of computer vision applied to non-destructive testing, it has not been detected that there are similar specific entities or that they deal with this issue.

Therefore the standards and norms established by said institutions are useful with respect to the methodology of carrying out the test itself, so that the data is reliable, but in the case of the application of computer vision, the specific aspects must be studied too, at least in the data production phase, and adjusted the methodology so that the resulting data, in addition to being reliable, is suitable for applying the appropriate computer vision technology.

This is also especially relevant when it comes to extracting patterns from nonvisible spectrum images (e.g., from IR and time series graphs that reflect data on nonvisible spectrum parameters) since in these cases, it is necessary to do a prior specific study before introducing the data in the algorithm and choose what type of algorithm, architecture, adjustments and tests it must be carried out so that the objectives pursued are achieved correctly and precisely.

18.2   Materials and methods

18.2.1   Method description

In this study, initially, a basic exploration and compilation of internationally recognized authors and publishers is carried out. This step is essential, since the approach used involves bibliometric and scientometric analysis, which requires prior knowledge of this information to select the most suitable databases and the literature pertinent to the desired bibliometric analysis, as well as to carry out the subsequent scientific analysis.

Bibliometric analysis plays a fundamental role in evaluating relevant scientific journals and exhaustively searching for appropriate literature. Its main objective is to guarantee the obtaining of reliable and precise results in the scientometric analysis.

Through the careful study of journals and the selection of the most pertinent literature, the aim is to build a solid base of information that supports subsequent analysis. This meticulous approach helps to ensure the quality and validity of the results obtained in scientometric analysis, thus providing a solid foundation for scientific research. In this study, a systematic bibliometric analysis was carried out, which has proven to be a valuable analytical tool for gaining a deeper understanding of research patterns and characteristics of different fields of study.

In addition, significant emphasis was placed on identifying and recognizing emerging issues, which will be further supported through the use of highly relevant databases spanning a broad spectrum of scientific journals globally. All scientific articles were registered and an exhaustive search for keyword matches was carried out, thus ensuring a complete and rigorous compilation of the relevant literature.

To carry out this research, a comprehensive query was carried out covering the relevant topics, titles, abstracts, and keywords. As a result, a wide variety of documents were collected, from which exclusions were made for the bibliometric analysis, discarding those that did not align with the study area, which allowed reducing the total number of documents processed. Among the selected documents, there were mainly scientific articles, works presented in congresses, and review articles. In addition, different bibliometric approaches were used to deepen the analysis.

Methodologies such as bibliographic coupling, historiography, the cooccurrence of terms, and the cocitation of documents, among others, were considered. These techniques made it possible to explore and analyze the relationships, connections, and trends present in the collected scientific literature, providing a more complete and enriching vision of the field of study in question.

Thanks to the exhaustive processing and analysis of the collected information, it was possible to visualize bibliometric networks that included journals, individual publications, and the affiliation of authors. To carry out this analysis, specialized software was used that focused on visualizing the statistical aspect of the investigation. Occasionally, additional tools such as BIBLIOMETRIX and other widely recognized open-source sources were used in scientometrics and bibliometrics research.

These tools provided a solid framework for the analysis and allowed to obtain valuable information on the interconnectedness and relevance of the publications, journals, and authors in the investigated field of study.

The procedure used for the bibliometric analysis, with the objective of evaluating both the quantity and the quality of scientific knowledge related to the keywords, was divided into three fundamental steps. First, the Scopus database was selected and the VOSviewer software was used to carry out the analysis and graphical representation of the scientific mapping. All relevant documents were exported in CSV format to be able to analyze them accurately in each corresponding program. Secondly, a comprehensive bibliometric analysis was applied to obtain detailed information on scientific production over time, by country, the cooccurrence of keywords, the most relevant sources of information, the most influential authors and affiliations, as well as the bibliographic grouping according to authors, documents, and countries.

In the third and last step, the results obtained in the previous step were reported and a detailed discussion was carried out, exploring the meaning and the correspondences found between the results. The implications were discussed and the appropriate context was established to properly interpret the findings.

This rigorous and structured approach in the bibliometric analysis allowed us to obtain a complete and significant vision of the scientific panorama related to the studied keywords, providing a solid base for the discussion and understanding of the obtained results. These techniques have been widely used in numerous fields; however, none of them has focused on technologies related to MV applied to nondestructive testing images on existing buildings. Therefore it is a qualitative study (assessment of the most appropriate publications) and quantitative in terms of scientometrics.

18.2.2 Materials and resources

For research and analysis, this document uses VOSviewer document viewing and analysis software developed by the Technical Research Centre of Leiden University in The Netherlands. Compared to other document measurement software, VOSviewer's main advantages are its powerful graphical display capabilities, which make it suitable for large amounts of data, as well as its versatility, which can be adapted to a variety of databases with source data.

On the other hand, as has been mentioned, in addition to the online tools of the databases consulted (SCOPUS, WOS, and so on), other tools have been used (BIBLIOMETRIX and other widely recognized open sources) but in a timely manner, since that with the aforementioned, practically all the needs were covered to meet the objectives.

18.2.3 Data sources

One of the most important data sources at the time of writing this paper is the Web of Science (WOS) database. However, according to Falagas et al. (2008), Cabeza et al. (2020), and Borri et al. (2021), the SCOPUS database offers a wide diversity of publications and can, in some cases, have 20% more coverage than WoS.

On the other hand, there are other well-known options such as Google Scholar, although with less consistency in terms of citation analysis (Cabeza et al., 2020; Borri et al., 2021), so they have not been used for this work.

Both SCOPUS and WOS cover the most relevant articles written by top scholars, have a high impact, have wide international reach, and include top academic journals. To guarantee the reliability and exhaustiveness of the data sources, this study mainly uses SCOPUS, although WOS has also been used occasionally for verification purposes.

For the search and selection of the literary, Boolean operators and synonyms of words were used, to increase the results and make better use of the search engines. These search engines allow you to select some of the fields to search for (title, Abstract, and so on). It also allows you to use logical operators such as AND, OR, or AND NOT.

In addition, some exclusion criteria were used: (1) methodologies that are not intended or are unfeasible for non-destructive testing in existing buildings; (2) equipment or technologies that incorporate the use of computer vision but are not intended for diagnosis through computational analysis of images resulting from non-destructive tests; (3) not related to building, even indirectly through civil constructions; and (4) documents with very similar content to each other or duplicates.

Through a comparative technique, the relevant information was synthesized, allowing the establishment of the necessary steps or guides that allowed the creation of the proposed conclusions.

Fundamentally, the extracted data is used for geographic analysis, the type and characteristics of the documents and authors, trends, importance, and weight of each specific theme, as well as a scientific one defined by Ren et al. (2020). Andersen & Swami (2021) as a quantitative approach that applies statistics and visualization techniques to classify and analyze bibliographic networks in a specific area.

As already indicated, for this scientific mapping, VOSviewer is used, to detect the relationship between different terms of publications (Cabeza et al., 2020; Borri et al., 2021; Shvindina, 2019) for evaluation purposes.

In this work, keywords are selected using the keyword retrieval method and a refinement of the search (inclusion/exclusion criteria). Therefore the keywords related to building, computer vision, and the technologies and algorithms for its realization were used, as well as the possible existing non-destructive tests in building, all in the period of the last 20 years (2002−22).

To further ensure the validity of the data sources, the language to English.

The keyword search results were included journals, conferences, and books to allow a more complete study of the subject area. The queries with these keywords and their results are shown in Table 18.1.

The process was as follows:

In the first place, keywords related to computer vision ("artificial vision" OR "computer vision" OR "machine vision" OR "artificial vision" OR "computer image") and with more frequent non-destructive tests in buildings that generate

Table 18.1 Categories, query formulas, and results.

Item	Category	Query formulas used	Number of publication (2002–22)
0	General	TITLE-ABS-KEY ("artificial vision" OR "computer vision" OR "machine vision" OR "artificial vision" OR "computer image" AND "infrared thermography" OR "IR technique" OR "IR thermography" OR "IRT" OR "ultrasonic" OR "ultrasound" OR "sonograms" OR "sonography" OR "multispectral images" OR "x-rays" OR "radiographic" OR "Gamma ray") AND PUBYEAR > 2001 AND PUBYEAR < 2023 AND PUBYEAR > 2001 AND PUBYEAR < 2023 AND (LIMIT-TO (LANGUAGE, "English"))	3101
1	General no unrelated subareas	TITLE-ABS-KEY ("artificial vision" OR "computer vision" OR "machine vision" OR "artificial vision" OR "computer image" AND "infrared thermography" OR "IR technique" OR "IR thermography" OR "IRT" OR "ultrasonic" OR "ultrasound" OR "sonograms" OR "sonography" OR "multispectral images" OR "x-rays" OR "radiographic" OR "Gamma ray") AND PUBYEAR > 2001 AND PUBYEAR < 2023 AND PUBYEAR > 2001 AND PUBYEAR < 2023 AND (EXCLUDE (SUBJAREA, "MEDI") OR EXCLUDE (SUBJAREA, "BIOC") OR EXCLUDE (SUBJAREA, "AGRI") OR EXCLUDE (SUBJAREA, "HEAL") OR EXCLUDE (SUBJAREA, "EART") OR EXCLUDE (SUBJAREA, "SOCI") OR EXCLUDE (SUBJAREA, "NEUR") OR EXCLUDE (SUBJAREA, "BUSI") OR EXCLUDE (SUBJAREA, "IMMU") OR EXCLUDE (SUBJAREA, "DENT") OR EXCLUDE (SUBJAREA, "PHAR") OR EXCLUDE (SUBJAREA, "ECON") OR EXCLUDE (SUBJAREA, "ARTS")	2124

(Continued)

Table 18.1 (Continued)

Item	Category	Query formulas used	Number of publication (2002−22)
		OR EXCLUDE (SUBJAREA, "NURS") OR EXCLUDE (SUBJAREA, "VETE")) AND (LIMIT-TO (LANGUAGE, "English"))	
2	"artificial intelligence"	TITLE-ABS-KEY ("artificial intelligence" AND Idem 1	161
3	"deep learning"	TITLE-ABS-KEY ("deep learning" AND Idem 1	391
4	"diagnosis"	TITLE-ABS-KEY ("diagnosis" AND Idem 1 or 2 or 3	355-59-136
5	"non destructive"	TITLE-ABS-KEY ("non destructive" AND Idem 1 or 2 or 3	86-3-13
6	"defect"	TITLE-ABS-KEY ("defect" AND Idem 1 or 2 or 3	149-7-31
7	"building"	TITLE-ABS-KEY ("building" AND Idem 1 or 2 or 3	68-7-11
8	"damage"	TITLE-ABS-KEY ("damage" AND Idem 1 or 2 or 3	59-3-7
9	"construction"	TITLE-ABS-KEY ("construction" AND Idem 1 or 2 or 3	39-5-3
10	"cracks"	TITLE-ABS-KEY ("crack" AND Idem 1 or 2 or 3	37-3-6
11	"concrete"	TITLE-ABS-KEY ("concrete" AND Idem 1 or 2 or 3	27-1-2
12	"facade or façade"	TITLE-ABS-KEY ("facade" AND Idem 1 or 2 or 3	3-0-1

images outside the visible spectrum (AND "infrared thermography" OR "IR technique" OR "IR thermography" OR "IRT" OR "ultrasonic" OR "ultrasound" OR "sonograms" OR "sonography" OR "multispectral images" OR "x- rays" OR "radiographic" OR "Gamma ray").

These words were tracked between 2002 and 2022, in the title, abstract, and in keywords of the publications, resulting in 3101 documents detected in English.

This figure was reduced to 2.124 documents, after eliminating the subarea of medicine, from biochemistry, genetics, and molecular Biology, from agricultural and biological sciences, health professions, from earth and planetary sciences; social science; neuroscience; business, management, and accounting, immunology, and microbiology; dentistry; pharmacology, toxicology and pharmaceutics; economics, econometrics, and finance; arts and humanities; veterinary; nursing.

On the other hand, the question was limited to documents in which the term "artificial intelligence" is used and, on the other, to those that use "deep learning," which is one of the most frequently used techniques for algorithms of image classification, detection, location and identification of multiclass objects in them, semantic segmentation, and so on (Guo et al., 2016; Minaee et al., 2022).

Based on this dataset, tests are carried out by introducing very specific categories of materials, properties, symptoms, or elements typical of buildings that are usually the subject of non-destructive testing. Although numerous and diverse categories were experimented with ("welding," "humidity," "cavities," deterioration, degradation, heritage, masonry, and so on), only those that after applying the previously established criteria gave some result fulfilling said criteria and a review and visual selection of the documents that are really related to the object of this investigation.

18.3   Results

It must be remembered that the scope of this study focuses on the application of artificial vision technology and artificial intelligence to non-destructive testing images in existing buildings.

Tables 18.1 and 18.2 show the categories used, query formulas, and results, as well as the documents found after applying preestablished criteria filters.

Finally, the appropriate checks were made with the WOS database.

In response to the purpose of the investigation, only the data from item 2 onward will be processed (161 artificial intelligence documents "AND" computer vision "AND" images of non-destructive tests such as thermography, x-rays, ultrasound, and so on).

Thus, with respect to item 2 in Fig. 18.1A, the growth in the number of publications in recent years can be observed; however, in Fig. 18.1B, it can also be observed with respect to keywords and their frequency, that with for this item 2, the fields of application are too varied and not very specific in most cases regarding those that are of interest, taking into account the objective set.

Regarding item 3 (391 documents of deep learning "AND" computer vision "AND" images of non-destructive tests such as thermography, X-rays, and ultrasound) in Fig. 18.2A, you can see the growth in the number of publications in recent years, especially from 2014 and 2015; however, in Fig. 18.2B, it can also be observed regarding the keywords and their frequency that the fields of application are still too varied and not very specific in the most cases, with respect to those that are of interest taking into account the objective set, since the most prominent terms are predominantly related to the field of medicine.Regarding item 4 "diagnosis and its combinations with items 1, 2 and 3," the predominance of the field of medicine is repeated.

For the following items, they will be grouped into ("non-destructive" OR "defect" OR "building" OR "damage" "construction" OR "cracks" OR "concrete" OR "façade" AND item 1), the search yields 54 results. If you add: AND "artificial

Table 18.2 Categories used and examples of documents detected after applying preestablished criteria.

Examples of some documents detected manually applying the established criteria

Category	Tittle	Authors	Year	Journals
"non destructive"	Infrared computer vision in non-destructive imaging: Sharp delineation of subsurface defect boundaries in enhanced truncated correlation photothermal coherence tomography images using K-means clustering (Risheh et al., 2022)	Risheh, A., Tavakolian, P., Melinkov, A., Mandelis, A.	2022	NDT and E International
"defect"	Automated Rust-Defect Detection of a Steel Bridge Using Aerial Multispectral Imagery (Li et al., 2019)	Li, Y., Kontsos, A., Bartoli, I.	2019	Journal of Infrastructure Systems
"building"	Automated subsurface defects' detection using point cloud reconstruction from infrared images (Montaggioli et al., 2021)	Puliti, M., Montaggioli, G., Sabato, A.	2021	Automation in Construction
	An Integrated Computational GIS Platform for UAV-based Building Façade Inspection (https://www.researchgate.net/profile/Kaiwen-Chen-10/publication/353785684_An_Integrated_Computational_GIS_Platform_for_UAV-based_Building_Facade_Inspection/links/62c483dfa306865ac92195bb/An-Integrated-Computational-GIS-Platform-for-UAV-based-Building-Facade-Inspection.pdf)	Chen, K., Xu, X., Reichard, G., Akanmu, A.	2021	EG-ICE 2021 Workshop on Intelligent Computing in Engineering, Proceedings
"damage"	Automatic defect detection for ultrasonic wave propagation imaging method using spatio-temporal convolution neural networks (Ye & Toyama, 2022)	Ye, J., Toyama, N.	2022	Structural Health Monitoring
	Attention-based generative adversarial network with internal damage segmentation using thermography (Ali & Cha, 2022)	Ali, R., Cha, Y.-J.	2022	Automation in Construction

(Continued)

Table 18.2 (Continued)

Examples of some documents detected manually applying the established criteria

Category	Tittle	Authors	Year	Journals
"construction"	Automatic defect recognition in x-ray testing using computer vision (Mery & Arteta, 2017)	Mery, D., Arteta, C.	2017	Proceedings - 2017 IEEE Winter Conference on Applications of Computer Vision, WACV
"cracks"	Automated damage detection of bridges sub-surface defects from infrared images using machine learning (Montaggioli et al., 2021)	Montaggioli, G., Puliti, M., Sabato, A.	2021	Proceedings of SPIE - The International Society for Optical Engineering
	Deep learning-based concrete crack detection using hybrid images (An et al., 2018)	An, Y.-K., Jang, K., Kim, B., Cho, S.	2018	Proceedings of SPIE - The International Society for Optical Engineering
	Deep learning–based autonomous concrete crack evaluation through hybrid image scanning (Jang et al., 2019)	Jang, K., Kim, N., An, Y.-K.	2019	Structural Health Monitoring
	Concrete cracks detection and monitoring using deep learning-based multiresolution analysis (Arbaoui et al., 2021)	Arbaoui, A., Ouahabi, A., Jacques, S., Hamiane, M.	2021	Electronics

"concrete"	Semantic Segmentation of Defects in Infrared Thermographic Images of Highly Damaged Concrete Structures (Pozzer et al., 2021)	Pozzer, S., Rezazadeh Azar, E., Dalla Rosa, F., Chamberlain Pravia, Z.M.	2021	Journal of Performance of Constructed Facilities
"facade"	Combined infrared imaging and structure from motion approach for building thermal energy efficiency and damage assessment (Sabato et al., 2020)	Sabato, A., Puliti, M., Niezrecki, C.	2020	Proceedings of SPIE - The International Society for Optical Engineering
"artificial intelligence" and others.	Infrared machine vision and infrared thermography with deep learning: A review (Y. He et al., 2021)	He, Y., Deng, B., Wang, H., (...), Cai, S., Ciampa, F.	2021	Infrared Physics and Technology
	The AIRES-CH Project: Artificial Intelligence for Digital REStoration of Cultural Heritages Using Nuclear Imaging and Multidimensional Adversarial Neural Networks (Bombini et al., 2022)	Bombini, A., Anderlini, L., dell'Agnello, L., (...), Ruberto, C., Taccetti, F.	2022	Lecture Notes in Computer Science (including subseries Lecture Notes in Artificial Intelligence and Lecture Notes in Bioinformatics)
	Materials Data Science for Microstructural Characterization of Archaeological Concrete (Ushizima et al., 2020)	Ushizima, D., Xu, K., Monteiro, P.J.M.	2020	MRS Advances

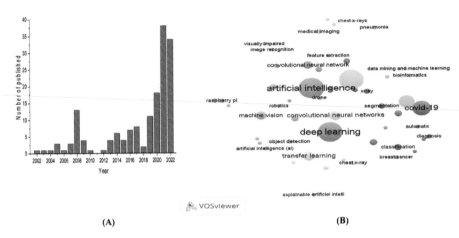

(A)   (B)

Figure 18.1 Item 2: (A) number of publications in the last years of item 2; (B) frequency of keywords. Considering the starting data of item 2, in the images of Fig. 18.1, it is possible to observe the growth in the number of publications in recent years, and a map of the keywords and their frequency of use.

Source: From Scopus (Elsevier) data, and image produced with the VOSviewer software with these data.

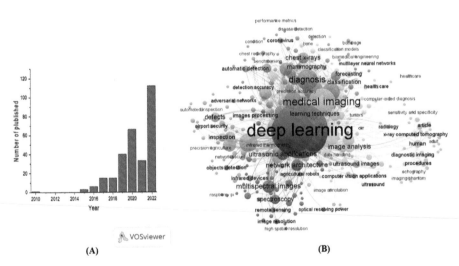

(A)   (B)

Figure 18.2 Item 3: (A) number of publications in the last years of item 3; (B) frequency of keywords . Considering the starting data of item 3, in the images of Fig. 18.2, it is possible to observe the growth in the number of publications in recent years, especially from 2014 and 2015, and a map of the keywords and their frequency of use.

Source: From Scopus (Elsevier) data, and image produced with the VOSviewer software with these data.

intelligence" OR "deep learning" OR "artificial vision" … the search yields 162 results that, when reviewed manually, largely reflect the majority of investigations related to the object of this investigation, so this combination will be the one used for the analysis and will be called the final item.

Fig. 18.3 shows the publications by year, highlighting that in the last 3 years, there has been a significant boom in research in this field.

Fig. 18.4 summarizes the number of journals that have published two or more relevant pieces of literature according to the bibliographic records resulting from the final combination described above. Of all of them, CiteScore and JIF are "Automation in Construction" and "Construction and Building Materials" and are also in the "CONSTRUCTION & BUILDING TECHNOLOGY" category, being magazines with a great impact on the research community.

This means that the scientific world is becoming increasingly interested in the recognition of the researchers involved.

Regarding the different countries, Fig. 18.5 generated by VOSviewer shows the spatial distribution of relevant articles starting from a minimum number of documents per country of 1, in a network of 36 items, 8 clusters, 57 links, and 66 total links. Strength in which the different interrelational links are shown. The size of the nodes indicates the total number of articles published in the country between 2002 and 2022.

The main countries with the most publications are China (36) and the United States (28), followed by others such as Canada (14), the United Kingdom (10),

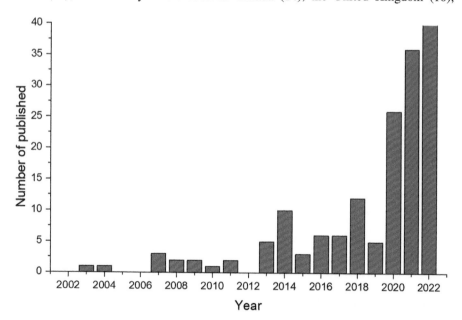

Figure 18.3 Final item: publications by years, highlighting that in the last 3 years, there has been a significant boom.

Source: From Scopus (Elsevier) data.

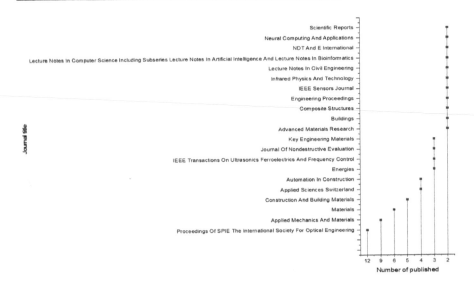

Figure 18.4 Number of journal publications. This figure summarizes the number of journals that have published two or more relevant works according to the bibliographic records resulting from the final combination.
Source: From Scopus (Elsevier) data.

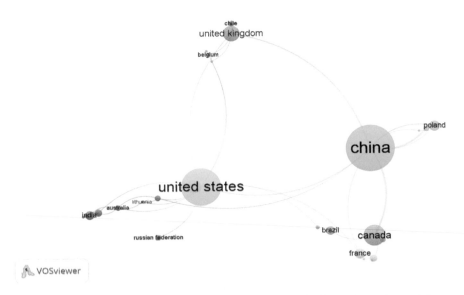

Figure 18.5 Documents published in different countries. The figure shows the spatial distribution of relevant articles starting from a minimum number of documents per country of 1, in a network of 36 items, 8 clusters, 57 links, and 66 total links.
Source: From Scopus (Elsevier) data, and image produced with the VOSviewer software with these data.

Italy (8), France (6), Germany (6), Poland (6), Taiwan (6), Brazil (5), and India (5). The United States was the first to carry out relevant investigations and is second in the number of articles after China, which in a few years has reached first place.

China, the United States, and several other nations have a high degree of intermediary centrality, which means that research institutions in these countries frequently collaborate with others.

Regarding the number of citations in Fig. 18.6, it can be seen that China, the United States, Canada, and the United Kingdom are the countries whose documents are most cited.

Regarding the most prominent authors in terms of citations, Fig. 18.7 shows the authors and the collaborative relationship between the authors. Central authors are the most cooperative. The colors indicate the dates.

Regarding cooccurrence keywords, it is of interest in terms of detecting development in the past and in which direction the issue tends to develop in the future in the study area. The results of keywords that have been cited at least 10 times are shown in Fig. 18.8. Twenty-one keywords are obtained.

The case study shows that this field began to emerge at the beginning of this century and reached its peak in 2017−18.

Node size indicates how often the keyword appears in the article.

The links between keywords indicate the frequency of occurrence in the cooccurrence of the document. This network can represent the trend of artificial vision

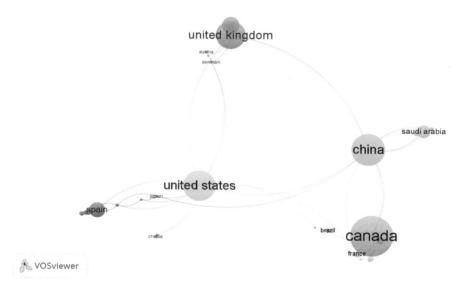

Figure 18.6 Citations in different countries. The figure shows that China, the non-destructive United States, Canada, and the United Kingdom are the countries whose documents are most cited.

Source: From Scopus (Elsevier) data, and image produced with the VOSviewer software with these data.

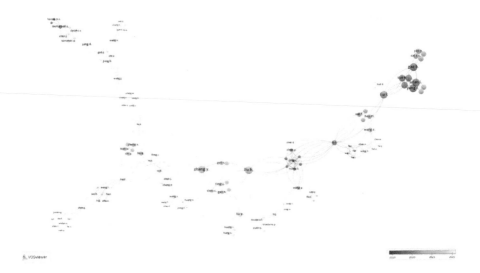

Figure 18.7 Author cocitation network. With colors, the figure shows that authors are the most cooperative.
Source: From Scopus (Elsevier) data, and image produced with the VOSviewer software with these data.

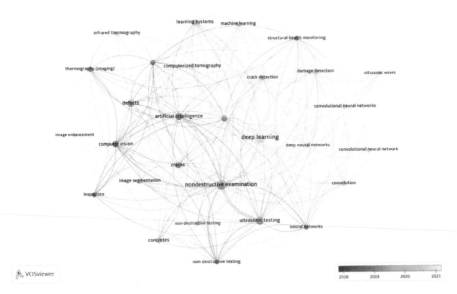

Figure 18.8 Keywords cooccurrence. The figure shows the cooccurrence of keywords to detect the evolution and trends taking into account those cited at least 10 times, obtaining 29 with these criteria.
Source: From Scopus (Elsevier) data, and image produced with the VOSviewer software with these data.

technology in the research field of non-destructive testing in buildings and related constructions.

According to Fig. 18.8, the five top keywords in frequency are Deep Learning (DL) 66; non-destructive examination (NDE) 51; ultrasound testing 34; defects 28; artificial intelligence (AI) 27; computer vision 24; crack 23; learning system 22; concrete 19; computerized tomography 19; damage detection 18; convolutional neural networks 18; thermography (imaging) 17; damage detection 17; crack detection 16; image processing 16; inspection 15; convolution 15; infrared Thermography 14; machine learning 14; deep neural networks 13; neural networks 13; structural health monitoring 13; ultrasonic waves 10; and non-destructive testing 10.

On the other hand, it must be said that in the published literature on the technology under study, algorithms that are trained and validated are frequently used, which involves the application of metrics whose frequent expressions correspond to "accuracy"; "precision"; recall; "F1-score"; "confusion matrix"; "true positive"; "true negative"; "false positive"; and "false negative".

These keywords, in a short time, have attracted a lot of interest from the scientific community. Words like "deep learning (DL)" and "non-destructive examination" have the highest burst intensity, so it can be seen that these keywords are the most popular research hotspots in this field of research.

Regarding the most recent and growing words, in addition to those related to the architecture and methodology of algorithms (DL; convolutional neuronal networks; convolution; deep neuronal networks; neuronal networks), of those related to non-destructive testing, ultrasounds stand out followed by those related to X-rays and thermography.

The higher the keyword strength, the more attention it will receive from researchers within the marked time window.

It is evident that the growth cycle of these words will continue, since, from the literature examined for the purposes of this study, they are generating a great deal of research interest. The word "computer vision" and its similar forms are frequently seen in bibliometric records.

Many studies explored how to use artificial vision in the classification of images in buildings, as well as the segmentation of objects in them for their location and identification, evaluating and comparing different algorithms using the aforementioned concepts of "accuracy," "precision," "recalls," and so on.

18.4   Discussion and evaluation

From the results obtained, it can be seen that the use of computer vision and artificial intelligence algorithms to analyze images generated by non-destructive testing is among the objectives of emerging studies, although currently, the data indicate that many could still be carried out more research in this regard compared to the potential that this field has.

Thus, although it is a growing research area, we identified two main barriers that hinder scientific development in this field: (1) the lack of real-world datasets and

(2) scarcity of the state of the art in terms of studies, especially comparative performance, that allow choosing or adequately developing the algorithms dedicated to this field.

Regarding temporal analysis, research in this field can be considered recent, noting that although various techniques included in the field of artificial intelligence have been developed to analyze images of all kinds, since the appearance of algorithms based on supervised learning, unsupervised, and reinforcement, most of the articles have focused on applying these algorithms and especially in the field of images those of supervised learning based on the so-called DL with CNNs, being perfectly reflected in the literature found in general and in particular in the field object of this investigation.

Therefore, taking into account the predominance of supervised learning and the application of DL with CNNs, to carry out research and advances regarding non-destructive testing images, it is necessary to have a dataset whose size is sometimes necessary to be considered with thousands of images for training and validation.

It has been observed that, in certain fields such as medicine, very valuable datasets are being developed and, therefore, numerous investigations are being carried out.

It is also true that in this scientific area, researchers in connection with health professionals, especially those related to diagnostic imaging and pathologists, daily perform and collect images, which means an accumulation of perfectly localizable data that is not available in other fields given, as is the case for the building.

Therefore, from the results of this research, it can be deduced that, at present, although progress is being made in the application of artificial vision and artificial intelligence technology to non-destructive testing images in existing buildings, it is still far from extensive application of these tools.

However, it is evident that the research and development of this technology has not diminished and more and more academics indicate the need to advance in this field, especially in terms of developing techniques that allow high precision as little data are available.

On the other hand, in addition to the fact that there are no well-defined standards in this field, the related image banks are very scarce, not to say nonexistent, with the researchers being the ones who in each case and in a timely and specific manner collect the data through fieldwork, in addition to normalization processing, segmentation where appropriate, labeling, and so on.

Regarding labeling, it is also problematic as it is a fundamentally manual procedure that must also be carried out by an expert, and even for greater reliability, it must be verified by other types of tests and trials, which means that the difficulty of building the image dataset is even greater.

To conclude this section and in summary, it should be noted that from the scientific literature detected and analyzed, it can be deduced the need to advance in the materialization of the training datasets since it can be seen that the work carried out in this regard in the different documents studied, sometimes it has not been enough to detect and, where appropriate, classify objects in images in the direction and with the desired precision since real precision is only achieved when training with sets of millions of labels and images.

On the other hand, although computers are becoming more powerful and faster, in much of the literature found, a long training time is still needed for errors to be tolerable.

Another matter of interest, as has been observed, is the evolution of the multiclass object detection method. Thus sliding window algorithms can sometimes produce high detection times, especially with large images and many objects to be detected in them. For this reason, R-CNN (Regions with Convolutional Neural Networks) systems are clearly being imposed, which only process the regions of the image that with a certain probability should contain the objects to be detected and classified.

However, there are many object detection algorithms, which currently stand out in addition to the aforementioned R-CNN, YOLO (you only look once), and SSD (single shot detector).

18.4.1 Limitations of the methodology

The research data in this chapter is based only on the SCOPUS core collection database, some WoS checks, and related keywords, so the research results may not fully cover the related literature the object of this study. However, future research in the field can overcome these limitations using other data sources and less relevant terms.

Regarding the limitations and technical problems detected in the development of the studies, it has already been mentioned that the difficulty in obtaining the appropriate dataset and reliably labeling it by experts is considered the most important problem to overcome in the future.

18.5 Conclusion

In the last 20 years and, especially in the last decade, there has been significant development of artificial intelligence techniques for image analysis and even significant changes in a very short period with respect to the said techniques.

With regard to the use of this technology for examination and diagnosis through images, the result of non-destructive tests, an important advance has also been experienced, especially in the field of health. These advances have also occurred in construction, although to a lesser extent, taking into account the study that combines scientometrics and bibliometrics.

Thus the United States and China are the countries that are at the forefront of these investigations in terms of global research in this field (China started late but has shown remarkable growth in recent years), with these countries having the teams with the most searches in this direction and number of journal publications. However, other countries also have relatively stable research teams in the field.

After the analysis of the network and cluster of keywords, using the "Keyword Highlight" method and based on the time factor, it is concluded that the most used terms are "Deep Learning" and "non-destructive examination" combined with words related to building defects and materials of its own, in addition to specific technologies such as ultrasound and thermography.

Regarding the evolution of keywords, the greatest growth has to do with DL.

In relation to influential authors and their alliances, there is a variety in this field of research and in the research teams, although sometimes the connections between them may not be very strong.

Regarding the frequency of publication, of all of them, the CiteScore and JIF are "Automation in Construction" and "Construction and Building Materials" and are also in the "CONSTRUCTION & BUILDING TECHNOLOGY" category. In order of the number of articles, the most prominent, with more than three documents, are Proceedings of SPIE the International Society for Optical Engineering, Applied Mechanics and Materials, Materials, Construction and Building Materials, Applied Sciences Switzerland, Automation in Construction, Energies, IEEE Transactions on Ultrasonics Ferroelectrics and Frequency Control, Journal of Non-destructive Evaluation, Key Engineering Materials. Regarding its publication frequency, no signs have been detected that it is a factor related to influence.

Regarding the evolution over time in terms of the number of publications and use of technology, it is detected that from 2002 to 2012 they are scarce and focus on issues of conception and development, although there are also some practical cases. It is from 2013 when the true growth and diffusion of experiences begins and from 2019 when the greatest growth and the explosion of results take place, indicating that MV technology in this field has a very broad research perspective in the area of use for diagnostics in buildings.

In relation to the problems detected in relation to these investigations and their future, this study offers data of interest for the purposes of said future investigations related to the application of artificial vision technology and artificial intelligence to images of non-destructive tests in existing buildings, highlighting that it is necessary to advance in the resolution of the problem, especially in terms of the need to have image datasets whose size and content are adequate for training and corresponding validation of the algorithms.

The related image banks are very scarce, not to say nonexistent, being the researchers who in each case and in a timely and specific manner carry out the data collection through fieldwork, in addition to normalization processing, segmentation, where appropriate, labeling, and so on.

In this sense, it is also necessary to create accessible databases using information and communication technologies that allow it to be fed massively, in addition to being able to be shared with the scientific community.

However, there is a need to advance in developing techniques that allow for high precision, as little data is available.

It should also be noted that, although progress is being made in the application of artificial vision and artificial intelligence technology to non-destructive testing images in existing buildings, the extensive application of these tools in real practice is still far from being achieved.

On the other hand, there are no well-defined standards in this field.

Regarding the labeling of the images and the objects or elements within them, it is also problematic as it is a fundamentally manual procedure that must also be carried out by an expert and even for greater reliability be contrasted by other types of tests and tests, which makes the difficulty of building the image dataset even greater.

Acknowledgments

The authors would like to appreciate the reviewers for all their helpful comments.

References

Abed, M.H., Al-Asfoor, M., & Hussain, Z.M. (2020). *CEUR Workshop Proceedings 16130073 1-12 CEUR-WS Iraq Architectural heritage images classification using deep learning with CNN*, 2602. http://ceur-ws.org/.

Akinosho, T. D., Oyedele, L. O., Bilal, M., Ajayi, A. O., Delgado, M. D., Akinade, O. O., & Ahmed, A. A. (2020). Deep learning in the construction industry: A review of present status and future innovations. *Journal of Building Engineering* (32). Available from https://doi.org/10.1016/j.jobe.2020.101827, http://www.journals.elsevier.com/journal-of-building-engineering/.

Ali, Rahmat, & Cha, Young-Jin (2022). Attention-based generative adversarial network with internal damage segmentation using thermography. *Automation in Construction, 141*, 104412. Available from https://doi.org/10.1016/j.autcon.2022.104412.

An, Y.K., Jang, K., Kim, B., & Cho, S. (2018). *Proceedings of SPIE - The International Society for Optical Engineering* 10.1117/12.2294959 1996756X SPIE South Korea Deep learning-based concrete crack detection using hybrid images, 10598 http://spie.org/x1848.xml.

Andersen, N., & Swami, V. (2021). Science mapping research on body image: A bibliometric review of publications in Body Image, 2004−2020. *Body Image, 38*, 106−119. Available from https://doi.org/10.1016/j.bodyim.2021.03.015, http://www.elsevier.com/inca/publications/store/6/7/2/9/3/2/index.htt.

Arbaoui, A., Ouahabi, A., Jacques, S., & Hamiane, M. (2021). Concrete cracks detection and monitoring using deep learning-based multiresolution analysis. *Electronics, 10*(15), 1772. Available from https://doi.org/10.3390/electronics10151772.

A.S., G., & Edayadiyil, J. B. (2022). Automated progress monitoring of construction projects using machine learning and image processing approach. *Materials Today: Proceedings, 65*, 554−563. Available from https://doi.org/10.1016/j.matpr.2022.03.137.

Bakator, M., & Radosav, D. (2018). Deep learning and medical diagnosis: A review of literature. *Multimodal Technologies and Interaction, 2*(3), 47. Available from https://doi.org/10.3390/mti2030047.

Bombini, A., Anderlini, L., dell' Agnello, L., Giaocmini, F.., Ruberto, C., & Taccetti, F. (2022). *Lecture notes in computer science (including subseries lecture notes in artificial intelligence and lecture notes in bioinformatics)*. The AIRES-CH Project: Artificial Intelligence for Digital REStoration of Cultural Heritages Using Nuclear Imaging and Multidimensional Adversarial Neural Networks. Springer Science and Business Media Deutschland GmbH, Italy, 13231. 10.1007/978-3-031-06427-2_57 16113349 685-700 https://www.springer.com/series/558.

Borri, E., Zsembinszki, G., & Cabeza, L. F. (2021). Recent developments of thermal energy storage applications in the built environment: A bibliometric analysis and systematic review. *Applied Thermal Engineering*, *189*, 116666. Available from https://doi.org/10.1016/j.applthermaleng.2021.116666.

Cabeza, L. F., Chàfer, M., & Mata, É. (2020). Comparative analysis of web of science and scopus on the energy efficiency and climate impact of buildings. *Energies*, *13*(2). Available from https://doi.org/10.3390/en13020409, https://www.mdpi.com/1996-1073/13/2/409.

Cai, L., Gao, J., & Zhao, D. (2020). A review of the application of deep learning in medical image classification and segmentation. *Annals of Translational Medicine*, *8*(11), 713. Available from https://doi.org/10.21037/atm.2020.02.44, −713.

Chen, L., Li, S., Bai, Q., Yang, J., Jiang, S., & Miao, Y. (2021). Review of image classification algorithms based on convolutional neural networks. *Remote Sensing*, *13*(22), 4712. Available from https://doi.org/10.3390/rs13224712.

Ciregan, D., Meier, U., & Schmidhuber, J. (2012). Multi-column deep neural networks for image classification. *Proceedings of the IEEE Computer Society Conference on Computer Vision and Pattern Recognition. Switzerland.*

Falagas, M. E., Pitsouni, E. I., Malietzis, G. A., & Pappas, G. (2008). Comparison of PubMed, Scopus, Web of Science, and Google Scholar: Strengths and weaknesses. *FASEB Journal*, *22*(2), 338−342. Available from https://doi.org/10.1096/fj.07-9492LSFGreece, http://www.fasebj.org/cgi/reprint/22/2/338.

Girshick, R., Donahue, J., Darrell, T., & Malik, J. (2016). Region-based convolutional networks for accurate object detection and segmentation. *IEEE Transactions on Pattern Analysis and Machine Intelligence*, *38*(1), 142−158. Available from https://doi.org/10.1109/TPAMI.2015.2437384.

Guo, Y., Liu, Y., Oerlemans, A., Lao, S., Wu, S., & Lew, M. S. (2016). Deep learning for visual understanding: A review. *Neurocomputing*, *187*, 27−48. Available from https://doi.org/10.1016/j.neucom.2015.09.116, http://www.elsevier.com/locate/neucom.

He, K., Zhang, X., Ren, S., & Sun, J. (2016). *IEEE Computer Society United States Deep residual learning for image recognition*. Proceedings of the IEEE Computer Society Conference on Computer Vision and Pattern Recognition (pp. 770−778).

He, Y., Deng, B., Wang, H., Cheng, L., Zhou, K., Cai, S., & Ciampa, F. (2021). Infrared machine vision and infrared thermography with deep learning: A review. *Infrared Physics & Technology*, *116*, 103754. Available from https://doi.org/10.1016/j.infrared.2021.103754.

Huang, S., & Wang, S. (2016). *New technologies in electromagnetic non-destructive testing*. Springer.

Jang, K., Kim, N., & An, Y. K. (2019). Deep learning−based autonomous concrete crack evaluation through hybrid image scanning. *Structural Health Monitoring*, *18*(5-6), 1722−1737. Available from https://doi.org/10.1177/1475921718821719, http://shm.sagepub.com/.

Lecun, Y., Bengio, Y., & Hinton, G. (2015). Deep learning. *Nature*, *521*(7553), 436−444. Available from https://doi.org/10.1038/nature14539, http://www.nature.com/nature/index.html.

Leiva, G., Ortuño, & Muñoz, J.V. (2019). *Técnicas y usos en la clasificación automática de imágenes/Techniques and uses in the automatic classification of images Actas Del IV Congreso ISKO España y Portugal* (pp. 11−26), Available from https://doi.org/10.5281/zenodo.3733409.

Li, Y., Kontsos, A., & Bartoli, I. (2019). Automated rust-defect detection of a steel bridge using aerial multispectral imagery. *Journal of Infrastructure Systems*, *25*(2). Available from https://doi.org/10.1061/(ASCE)IS.1943-555X.0000488.

Llamas, J., Lerones, P. M., Medina, R., Zalama, E., & Gómez-García-Bermejo, J. (2017). Classification of architectural heritage images using deep learning techniques. *Applied Sciences, 7*(10), 992. Available from https://doi.org/10.3390/app7100992.

Long, J., Shelhamer, E., & Darrell, T. (2015). *Proceedings of the IEEE Computer Society Conference on Computer Vision and Pattern Recognition. IEEE Computer Society United States Fully convolutional networks for semantic segmentation.*

Maggiori, E., Tarabalka, Y., Charpiat, G., & Alliez, P. (2017). Convolutional neural networks for large-scale remote-sensing image classification. *IEEE Transactions on Geoscience and Remote Sensing, 55*(2), 645–657. Available from https://doi.org/10.1109/TGRS.2016.2612821.

Marín-García, D., Bienvenido-Huertas, D., Carretero-Ayuso, M. J., & Torre, S. D. (2023). Deep learning model for automated detection of efflorescence and its possible treatment in images of brick facades. *Automation in Construction, 145.* Available from https://doi.org/10.1016/j.autcon.2022.104658, https://www.journals.elsevier.com/automation-in-construction.

Mery, D., & Arteta, C. (2017) Chile automatic defect recognition in x-ray testing using computer vision. *Proceedings - 2017 IEEE Winter Conference on Applications of Computer Vision, WACV 2017* 1026-1035. Institute of Electrical and Electronics Engineers Inc. 10.1109/WACV.2017.119 9781509048229.

Minaee, S., Boykov, Y., Porikli, F., Plaza, A., Kehtarnavaz, N., & Terzopoulos, D. (2022). Image segmentation using deep learning: A survey. *IEEE Transactions on Pattern Analysis and Machine Intelligence, 44*(7), 3523–3542. Available from https://ieeexplore.ieee.org/servlet/opac?punumber = 34.

Montaggioli, G., Puliti, M., & Sabato, A. (2021). Automated damage detection of bridges sub-surface defects from infrared images using machine learning. *Proceedings of SPIE - The International Society for Optical Engineering.* SPIE Italy, 11593. 10.1117/12.2581783 1996756X http://spie.org/x1848.xml.

Mu, R., & Zeng, X. (2019). A review of deep learning research. *KSII Transactions on Internet and Information Systems, 13*(4), 1738–1764. Available from https://doi.org/10.3837/tiis.2019.04.001, http://www.itiis.org/digital-library/manuscript/file/2310/TIIS + Vol + 13, + No + 4-1.pdf.

Nüsslin, F. (2020). Wilhelm Conrad Röntgen: The scientist and his discovery. *Physica Medica, 79*, 65–68. Available from https://doi.org/10.1016/j.ejmp.2020.10.010, http://www.fisicamedica.org.

Perez, H., H.M., J., & Amir Mosavi, T. (2019). Deep learning for detecting building defects using convolutional neural networks. *Sensors, 19*(16), 3556. Available from https://doi.org/10.3390/s19163556.

Pozzer, S., Rezazadeh Azar, E., Dalla Rosa, F., & Chamberlain Pravia, Z. M. (2021). Semantic segmentation of defects in infrared thermographic images of highly damaged concrete structures. *Journal of Performance of Constructed Facilities, 35*(1). Available from https://doi.org/10.1061/(ASCE)CF.1943-5509.0001541, https://ascelibrary.org/journal/jpcfev.

Ren, R., Hu, W., Dong, J., Sun, B., Chen, Y., & Chen, Z. (2020). A systematic literature review of green and sustainable logistics: Bibliometric analysis, research trend and knowledge taxonomy. *International Journal of Environmental Research and Public Health, 17*(1). Available from https://doi.org/10.3390/ijerph17010261, https://www.mdpi.com/1660-4601/17/1/261/pdf.

Risheh, A., Tavakolian, P., Melinkov, A., & Mandelis, A. (2022). Infrared computer vision in non-destructive imaging: Sharp delineation of subsurface defect boundaries in enhanced truncated correlation photothermal coherence tomography images using K-means clustering.

NDT & E International, *125*, 102568. Available from https://doi.org/10.1016/j.ndteint.2021.102568.

Sa, I., Ge, Z., Dayoub, F., Upcroft, B., Perez, T., & McCool, C. (2016). DeepFruits: A fruit detection system using deep neural networks. *Sensors*, *16*(8), 1222. Available from https://doi.org/10.3390/s16081222.

Sabato, A., Puliti, M., & Niezrecki, C. (2020). Combined infrared imaging and structure from motion approach for building thermal energy efficiency and damage assessment. Proceedings of SPIE - The International Society for Optical Engineering SPIE United States, 11381. http://spie.org/x1848.xml.

Schabowicz, K. (2019). Non-destructive testing of materials in civil engineering. *Materials*, *12*(19), 1–13. Available from https://doi.org/10.3390/ma12193237, https://res.mdpi.com/d_attachment/materials/materials-12-03237/article_deploy/materials-12-03237.pdf.

Sengupta, S., Basak, S., Saikia, P., Paul, S., Tsalavoutis, V., Atiah, F., Ravi, V., & Peters, A. (2020). A review of deep learning with special emphasis on architectures, applications and recent trends. *Knowledge-Based Systems*, *194*, 105596. Available from https://doi.org/10.1016/j.knosys.2020.105596.

Shen, D., Wu, G., & Suk, H. I. (2017). Deep learning in medical image analysis. *Annual Review of Biomedical Engineering*, *19*, 221–248. Available from https://doi.org/10.1146/annurev-bioeng-071516-044442, http://arjournals.annualreviews.org/loi/bioeng.

Shvindina, H. (2019). Coopetition as an emerging trend in research: Perspectives for safety & security. *Safety*, *5*(3). Available from https://doi.org/10.3390/safety5030061, https://www.mdpi.com/2313-576X/5/3/61.

Silva, W. R. L. da, & Lucena, D. S. de (2018). Concrete cracks detection based on deep learning image classification. *Proceedings*, *2*(8), 489. Available from https://doi.org/10.3390/ICEM18-05387.

Simonyan, K., & Zisserman, A. (2015). Very deep convolutional networks for large-scale image recognition. *Conference Track Proceedings International Conference on Learning Representations*, ICLR United Kingdom. 3rd International Conference on Learning Representations, ICLR 2015 - https://dblp.org/db/conf/iclr/iclr2015.html.

Szegedy, C., Toshev, A., & Erhan, D. (2013). *Deep Neural Networks for object detection. Advances in neural information processing systems.* Neural Information Processing Systems Foundation, United States.

Taheri, H., Bocanegra, M. G., & Taheri, M. (2022). Artificial intelligence, machine learning and smart technologies for non-destructive evaluation. *Sensors*, *22*(11), 4055. Available from https://doi.org/10.3390/s22114055.

Tejedor, B., Lucchi, E., Bienvenido-Huertas, D., & Nardi, I. (2022). Non-destructive techniques (NDT) for the diagnosis of heritage buildings: Traditional procedures and futures perspectives. *Energy and Buildings*, *263*, 112029. Available from https://doi.org/10.1016/j.enbuild.2022.112029.

Ushizima, D., Xu, K., & Monteiro, P. J. M. (2020). Materials data science for microstructural characterization of archaeological concrete. *MRS Advances*, *5*(7), 305–318. Available from https://doi.org/10.1557/adv.2020.131, https://www.cambridge.org/core/journals/mrs-advances.

Wang, N., Zhao, X., Zhao, P., Zhang, Y., Zou, Z., & Ou, J. (2019). Automatic damage detection of historic masonry buildings based on mobile deep learning. *Automation in Construction*, *103*, 53–66. Available from https://doi.org/10.1016/j.autcon.2019.03.003.

Ye, J., & Toyama, N. (2022). Automatic defect detection for ultrasonic wave propagation imaging method using spatio-temporal convolution neural networks. *Structural Health Monitoring*, *21*(6), 2750–2767. Available from https://doi.org/10.1177/14759217211073503, https://journals.sagepub.com/home/SHM.

Non-destructive approach for the study of decorative revetments: Implementation of spectroscopic techniques

19

María José Ayora-Cañada and Ana Domínguez-Vidal
Department of Physical and Analytical Chemistry, Universidad de Jaén, Jaén, Spain

19.1 Introduction: Spectroscopic non-destructive technique in heritage studies

Analytical science plays a significant role in the preservation of cultural heritage as they can provide crucial information about the conservation state of heritage assets. A large array of analytical technologies are available nowadays; however, the principle of minimal impact must always be considered when dealing with cultural heritage investigations. This is why concerns associated with the accessibility and the need for irreversible destructive sampling on monuments prevent the generalized use of many of the available analytical resources. To address this problem, in the last two decades, non-destructive techniques (NDTs) have played an increasingly important role in the field of built cultural heritage conservation, particularly those that can be applied in situ and do not require destructive sampling (Moropoulou et al., 2013; Tejedor et al., 2022). In this sense, the implementation of non-destructive spectroscopic methods that provide a deep understanding of the material composition of historic buildings while respecting their integrity is of great interest. This chapter will focus on the study of ornaments and decorative motifs, elements added to an otherwise merely structural form that blends into the building as a whole and beautify it. The broad spectrum of ornamentation is a distinguishing characteristic of buildings and its style refers to the specific culture from which it originates. Ornamental structures are particularly delicate and, in general, more prone to deterioration than structural elements. Thus a deep knowledge of the materials employed either originally or in different past interventions, of the execution techniques, and of the decay products and mechanisms is essential to guarantee their preservation. The feasibility of the in situ characterization of different decorative revetments using portable spectroscopic techniques will be illustrated here for the study case of the Alhambra and Generalife Monumental Ensemble.

Diagnosis of Heritage Buildings by Non-Destructive Techniques. DOI: https://doi.org/10.1016/B978-0-443-16001-1.00019-X

19.2   Case of study: The Alhambra in Granada, Spain

The Alhambra monument was included in the UNESCO world heritage list in 1884 and it is nowadays one of the most visited sites in the world. Located in Granada (Spain), it is a unique example of medieval Islamic architecture. It was founded in the 9th century as a fortress and got its maximum splendor during the Nasrid period (1232−1492) when it became a palatine city with plenty of halls, baths, gardens, and mosques for all the people living or working there. After the conquest of Granada, the Alhambra became a Christian court in 1492 and thus a significant remodeling of the space was performed and more palaces and chapels were added. These interventions were followed by several adaptations with different purposes, such as restoration or conservation needs, and have continued to the present day (Mileto & Vegas, 2007).

For the conservation of this unique space, it is imperative to develop new approaches and strategies to get good knowledge about the transformations experienced by the decorations of the Alhambra throughout its history and to make a proper diagnosis of the conservation state. Most previous studies on the ornamental structures in the Alhambra deal with a visual inspection, centered on stylistic aspects, whereas only more recently, scientific investigations have been devoted to the characterization of materials and execution techniques. These initial scientific-technical studies have focused on fragments or isolated samples of plasterwork (Bueno & Flórez, 2004; Cardell-Fernández & Navarrete-Aguilera, 2006; Romero-Pastor et al., 2011) or polychrome wood (Cardell et al., 2009) and there is a need for more complete studies of the different decorative ornamental structures using non-destructive techniques like those presented in this chapter. Furthermore, the formation of multidisciplinary teams of scientists from several areas and conservators working in close cooperation is mandatory. This chapter describes an extensive study of several decorative revetments along different spaces of the monument and is mainly based on the use of noninvasive analytical techniques. The study of some restored spaces was also included as they contribute to understanding the complex history of the Alhambra building.

19.2.1   Decorative revetments

In the Alhambra, the fundamental elements of Islamic architecture found their highest expression evolving into a highly ornamented art based on geometric, natural, and calligraphic elements profusely used to decorate all the interior spaces. A description of the different decorative structures under study is presented next and some examples can be seen in Fig. 19.1.

The key architectural features in Islamic architecture are plasterwork and carved wooden ceilings. Two expressions of plasterwork can be found in the Alhambra. First, stalactite vaults with ornaments known as *mocarabes* are one of the most creative features of Hispano-Islamic architecture (Ettinghausen et al., 2003) and decorate many rooms in the Alhambra. *Mocarabes* are made up of prismatic units

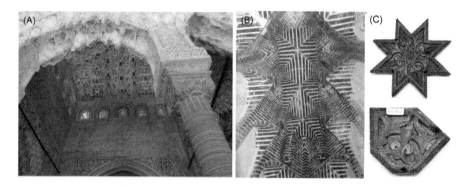

Figure 19.1 Examples of the studied decorative revetments in the Alhambra.
(A) Mocarabes (plasterwork) and marble capital, (B) faux bricks mural painting, and
(C) pieces of demounted ceiling.

joined to build a self-supporting structure. The final result of the decoration gives
the impression of the stalactites of a cave (see Fig. 19.1A). Furthermore, in the
Alhambra, it is very usual to find the *mocarabes* colorfully decorated with geometric and vegetable motifs. This study focuses on the seven stalactite vaults of the
Hall of the Kings (Sala de los Reyes), in the Lions Palace. In addition to *mocarabes*
vaults, polychrome plasterwork decorations are a constant feature of wall decoration in the Alhambra with a wide range of ornamental designs and a vibrant palette
of colors. In this case, our study focused on the restored polychrome plasterwork
decorations of the Room of the Beds (Sala de las Camas) in the Royal Bath of
Comares. This controversial redecoration, which aimed to imitate the original
appearance of Nasrid plasterwork, was carried out by Rafael Contreras in the 19th
century (Pérez, 2017) and now also belongs to the Alhambra's history.

Wooden ceilings have also a long tradition in the Hispano-Islamic area being
particularly important during the Nasrid period. In this chapter, the polychrome carpentry of the ceilings of two rooms in the Alhambra complex is described, namely,
the Hall of the Abencerrages (Sala de Abencerrajes) and the Hall of the Two Sisters
(Sala de Dos Hermanas). Both date to the reign of Muhammad V in the 14th century. These ceilings are called *alfarjeataujerado* and are formed by several pieces
of wood nailed to a supporting structure forming geometric patterns. These paneled
ceilings were disassembled during the conservation works carried out by the
Council of the Alhambra and the Generalife (*Wood materials restoration workshop*), allowing a noninvasive investigation of the pieces without scaffoldings. The
pieces (Fig. 19.1C) are carved and decorated mainly in white, red, blue, green, and
black colors.

Carved marble capitals (Fig. 19.1A) constitute also a characteristic decoration
along the Alhambra complex. In this study, six columns of five different locations
were considered. Three of them were located inside halls (in the Mexuar and the
Hall of the Abencerrages), two were in courtyards (the Court of the Myrtles/Patio

de los Arrayanes and the Court of the Main Canal/Patio de la Acequia), which is completely outdoors, and one was located in the limit between the hall and the courtyard (in the Hall of the Kings). All of them were profusely decorated with vivid colors and gildings. The state of preservation of the polychromy is very different, being those of the Mexuar the best preserved (with testimonies of different interventions, since the early redecoration of this space shortly after the Christian conquest to recent and documented interventions at the end of the 20th century).

In contrast with all the above-described decorative elements, mural paintings in the Alhambra complex are not well known by most visitors. The simplest manifestation of wall painting in the Alhambra monument simulates red brick walls (Fig. 19.1B). They are found both on exterior walls and on some other architectural elements such as vaults. These paintings are quite common in Nasrid architecture although it is not clear if their objective was only aesthetical or intended to cause the impression of a more robust building. Scientific investigations about this kind of decorative work are scarce. More elaborated mural paintings in the Alhambra, showing geometrical and vegetable motifs in several colors, are located on the wall socles of certain small rooms reserved for domestic activities and private places, such as the paintings of the Partal house and the Harem Court, the latter considered in this chapter.

19.3 Methods

As commented earlier, the preservation of the objects is one of the first issues to consider in the study of heritage. Thus the use of analytical techniques that avoid damaging and reduce sampling is mandatory. Guidelines about the investigation protocols on artworks and buildings preserved as cultural heritage, given by the UNESCO-ICOMOS (International Council of Monuments and Sites) (ICOMOS, 2011), recommend the use of NDTs as far as possible. However, it is necessary to clarify the usual terminology employed when talking about these NDT analytical techniques (Vandenabeele & Donais, 2016). The following classification can be considered:

- Noninvasive techniques: This term refers to portable techniques that can be applied in situ and do not require samples (even of micrometer size) to be removed from the object. Most of the noninvasive techniques are spectroscopic techniques, which allow performing direct measurements on the area of interest using different radiation beams. Strictly speaking, the technique should also leave the object in essentially the same state before and after analysis (Edwards & Vandenabeele, 2012; Thickett et al., 2017). In this regard, there are techniques that do not require sampling but cannot be considered noninvasive because they cause small damage to the object on a micrometer scale. This is the case of laser ablation inductively coupled plasma mass spectrometry (LA-ICP-MS) (Giussani et al., 2009; Kyriakou et al., 2019; Schreiner & Grasserbauer, 1985). They could be referred to as microdestructive techniques.
- NDTs that require sampling: Here the term "non-destructive" refers to the possibility of preserving the sample for further examination or analysis (with another technique).

However, since they require the removal of samples, they cannot be considered non-destructive with regard to the artwork. For this reason, the distinction between these techniques and those not requiring any sample extraction is of particular importance in the conservation field. Nevertheless, many research scientists generally use the term "non-destructive" for any of the abovementioned analysis methods (Adriaens et al., 2005).

This chapter will focus on two true noninvasive spectroscopic techniques implemented for in situ studies in the Alhambra monumental ensemble: portable X-ray fluorescence (XRF) and portable Raman spectroscopy. In addition, complementary studies on microsamples using non-destructive benchtop spectroscopic techniques (Raman and infrared microscopy) will be also considered. This approach will be limited to those issues for which the information gathered in situ was insufficient. In this sense, during the studies, minimization of impact on the artwork is aimed trying to maximizing the information obtained.

19.3.1 Implementation of spectroscopic techniques

19.3.1.1 Portable X-ray fluorescence

The principle of operation of this technique is that X-rays from an excitation source interact with the test material causing the emission of secondary fluorescence X-rays. This happens because of the ionization of the materials when excited with high-energy radiation (e.g., X-rays). If this energy is sufficient to remove an inner shell electron, an outer electron will replace the ejected one to stabilize the atom. When this happens, energy is released in the form of X-ray radiation of lower energy than the primary incident X-rays. The so-emitted secondary radiation is called fluorescent radiation. Each chemical element emits fluorescent X-radiation of characteristic energy with an intensity related to its concentration that can be useful to detect the different chemical elements present in the sample. Technology breakthroughs in the last decades permitted the miniaturization of the XRF instrumentation (Drake & MacDonald, 2022), and many handheld instruments are now frequently used for in situ analysis (Barone et al., 2018; Bonizzoni et al., 2010; Madariaga, 2021).

The principal advantages of XRF are that it is non-destructive and fast and detects major and minor elements. Portable devices equipped with state-of-the-art sensitive detectors can detect chemical elements with atomic numbers (Z) higher than 12, which is from Mg. Lighter elements cannot be detected by XRF fluorescence, which makes impossible the characterization of organic materials by means of this technique. Furthermore, it must be taken into account that relatively light elements (with $Z < 16$) such as P, Si, Al, and Mg emit low-energy photons that are partially absorbed by the matrix of the sample or the atmospheric constituents. Thus the sensitivity for these elements is limited.

The studies in the Alhambra were performed with a handheld NitonXL3t GOLDD + XRF analyzer with a rhodium anode (50 kV, 200 µA) (Thermo Fisher Scientific, Waltham MA, United States) (Fig. 19.2A). It holds a camera that allowed seeing the measurement area, which can be changed from 8 to 3 mm

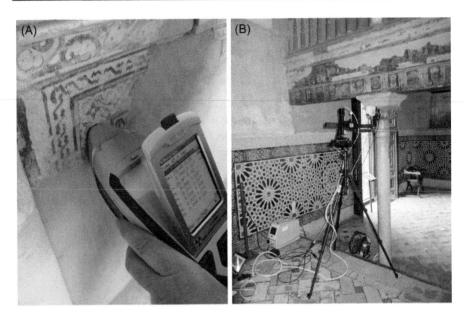

Figure 19.2 Portable instruments working on the site. (A) XRF analyzer and (B) Raman spectrometer with the XYZ accesory mounted on the tripod. *XRF*, X-ray fluorescence.

diameter by collimation of the primary X-ray beam. The Niton XL3t GOLDD + spectrometer's sensitivity for the different elements is enhanced by appropriate excitation filters. Therefore *Main, high, and low* filters provide good sensitivity from Mn to Bi, from Ba to Ag, and from Ti to Cr, respectively. Furthermore, the *Light range* filter is useful for light elements (down to Mg). The measuring mode "mining" with measurement times of 20 seconds for each filter was used. In this way, each spot analysis took 80 seconds. NITON Data Transfer (NDT) software was used for data transfer and management.

19.3.1.2   Portable Raman spectroscopy

Raman spectroscopy is based on the interaction of the light of a laser with the material under study. When this monochromatic light strikes on the matter, a small fraction of the incident radiation is scattered at different wavelengths, due to interaction with molecular vibrations. By analyzing this wavelength shift, it is possible to obtain information about the chemical composition of the material, including both organic and inorganic compounds. The Raman spectrum of a chemical compound is like a fingerprint, being even possible to distinguish polymorphic substances since in the low-frequency region Raman spectroscopy is sensitive to crystal structure.

Since the Raman effect is a very weak physical phenomenon, the presence of background light must be avoided and the presence of fluorescence from the sample constituents is a major drawback. The excitation laser and the optical properties of the investigated materials are the most important aspects for gathering good-quality Raman spectra. Nowadays, most state-of-the-art portable Raman spectroscopic devices are equipped with a diode laser emitting at 785 nm as an excitation source as a good choice to reduce fluorescence backgrounds in most materials although double laser analyzers are now commercially available (Rousaki & Vandenabeele, 2021).

For the on-site studies in the Alhambra monument, an innoRam spectrometer (B&W TEK Inc., Newark, United States) with Charge Coupled Device (CCD) detector, thermoelectrically cooled to $-20°C$, was used. It was equipped with a fiber-optic probe with a 785 nm laser for excitation that was coupled to a video microscope with a long focal distance objective (20 ×) and an integrated camera for visualization of the measurement area. The video microscope head was attached to a motorized XYZ accessory with remote control for positioning and focusing (see Fig. 19.2B). Optimization of the measurement parameters included the laser power, the exposition time, and the number of accumulations to avoid any damage to the materials and gain a good signal-to-noise ratio. The spectra were recorded from 65 to 2565 cm^{-1} Raman shift with 2 cm^{-1} of spectral resolution.

19.3.1.3 Benchtop spectroscopic techniques for complementary non-destructive analysis on microsamples

Raman measurements in the laboratory were carried out with a Renishaw inVia Raman confocal microspectrometer equipped with a CCD detector camera (working with a Peltier system at $-70°C$). Two lasers were available as excitation sources, namely, 514 (green) and 785 nm (red), and four objectives (5 ×, 20 ×, 50 ×, 100 ×) for enhanced spatial resolution focusing on small sample areas.

Infrared spectra were obtained by using an FTIR 6300 spectrometer coupled to an FTIR IRT-7000 microscopy (objective lens of 16 × magnification) with a mercury cadmium telluride detector from JASCO (Tokyo, Japan). Spectra were registered in transmission mode using a Spectra-Tech microcompression cell with diamond windows (Thermo Scientific, United States) in the spectral range 750–4000 cm^{-1}.

19.4 The role of a non-destructive technique in the characterization of decorative revetments

19.4.1 Challenges of using portable instruments

This section considers the main obstacles that can be found when working in the field using portable instruments, and particularly, in the Alhambra monumental site. Different challenges were encountered during the analyses, as the experimental conditions were far from those found in a laboratory. Raman measurements are particularly challenging since in laboratories they are usually made in the dark to avoid sunlight perturbations. Thus previous studies with portable instruments have reported experiencing difficulties in this regard since the Raman effect is very weak and background radiation from sunlight interfered with the spectra (Aramendia et al., 2012; Vandenabeele et al., 2004). In our case, when working in indoor locations, no strong influence due to sunlight was experienced probably because the characteristics of the *halls*, with their small lattice windows and the cave-like spatial disposition of the *mocarabes* prevented direct sunlight from reaching the measurement areas. In contrast, more difficulties were found when working on the exterior. There, daylight was a problem that was partially solved by either using the Raman fiber-optic probe without the microscope or by covering the objective with a dark foam to prevent the sunlight from reaching the aperture. Nevertheless, the Raman spectra obtained in exterior locations were always of less quality than those recorded in the darker interior halls. Furthermore, the mandatory use of the extensible bar on the tripod when reaching high motifs (sometimes around 3 m), like the capitals of the columns and certain remains of wall paintings, was an issue in any wind, even light, as it moved the equipment bringing the optics out of focus. A special situation was the wide study performed on the *mocarabes* of the Hall of the Kings. To reach the painted areas situated on the vaults of the hall, the Alhambra Council installed scaffoldings for each vault at a height of c. 12 m on ground level to carry out conservation works. The instruments, as well as, their accessories, and the positioning equipment were installed on top of these platforms. Vibrations of the scaffolding sometimes made not possible the recording of Raman spectra with long acquisition time, which hindered the identification of weak Raman scatterers.

Other problems experienced with both XRF and Raman measurements were related to the intricate forms of the Islamic decorations that made it difficult to position the measuring heads of the instruments. In this regard, sometimes it was not possible to acquire spectra in certain motifs. Despite all these difficulties, the possibility of registering both XRF and Raman spectra of reasonably good quality in situ allows for a wide exploration of the spaces as it is not limited to sampling. Thus a representative identification of the different pigments employed and their degradation products is provided.

19.4.2   Identification of pigments used along the history of the monument

As described earlier, Hispano-Muslim architecture is characterized by a profuse decoration of the inside locations. Our studies using noninvasive spectroscopic techniques focused on the characterization of ornaments with polychromy. Polychromy is usually composed of a series of layers with distinct roles. Preparation layers are used not only to provide a suitable substrate to apply the color but sometimes also to hide the underlying substrate or previous colors. Then, layers of color can be applied, usually employing mineral pigments mixed with some binder. The pigments are the visible part of the polychromy and the most prone to suffer alterations with an important visual impact on the aspect of the artwork. The identification of the pigments employed in the different types of decorative revetments in the Alhambra monument is the main topic of this section.

The investigation of the materiality of a work of art always starts with an inspection with the naked eye. When someone visits the Alhambra site, two main colors come to the eyes: red and blue. Furthermore, black color is also observed, mainly in the contours of the drawings and in some geometrical decorations with circles or zig-zag lines. Here, the results of the study will be presented and organized by the color of the pigments, summarized in Table 19.1. However, since the use of color strongly depends on the type of ornamentation, considerations about the type of decorative revetment and its characteristic are also relevant for the discussion.

The most simple decorative element that can be found in many locations throughout the Alhambra monumental complex, covering many vaults and walls are mural paintings with a pattern that could be described as "faux red brick." Curiously, the supporting materials of these mural paintings were, in most locations, brick walls covered with a mixture of sand and lime plaster (De La Torre López et al., 1996). Finally, these walls were painted with a pattern that imitates red bricks (see Fig. 19.1B). Due to its stylistic simplicity and the poor state of conservation of some of these motifs, mainly those located outdoors, these decorations have received much less attention than plasterwork, carved marble, or wooden ceilings, much popular in the Alhambra. The inspection of these motifs in the different locations revealed that the red rectangles simulating bricks were painted over a continuous white base of plaster. The results obtained in situ (Arjonilla et al., 2021) using XRF suggested the use of natural iron oxide pigments to paint the red areas with the identification of iron, calcium, sulfur, and silicon. The presence of hematite (α-Fe_2O_3), the compound responsible for the red tonality of red ocher pigments, was confirmed by Raman spectroscopy, thanks to the characteristic bands at 224, 292, 411, and 614 cm^{-1}(Zoppi et al., 2008). Typical XRF and Raman spectra obtained in situ from these decorations are shown in Fig. 19.3. In general, the white base layer was made of lime mortar, as revealed by the spectral features of calcite ($CaCO_3$) in the Raman spectra (see Fig. 19.3). Only in one of the indoor locations, the vaults of the Infant's Tower, gypsum ($CaSO_4 \cdot 2H_2O$) plaster was used for the white base of the red brick motifs (see spectra 2 and 7 in Fig. 19.3).

Gypsum plaster is, however, the most commonly used substrate for most indoor decorations in the Alhambra, both in the form of *mocarabes* vaults and in other panels covering the inner walls. These plasterwork motifs were richly painted, mainly in red and blue, being green and black colors also present together with gilding remains (see spectra in Fig. 19.4). The conservation state of these polychrome works is very different depending on the location. Two red pigments were clearly identified in the stalactite vaults of the Hall of the Kings, namely, cinnabar (HgS) and red lead or minium (Pb_3O_4). It is remarkable the high quality of the Raman spectra recorded for these two pigments despite the already discussed challenges of working in the field (Dominguez-Vidal et al., 2012). The location of cinnabar in the decorations of the vaults suggested its original use by Nasrid artists, although it was probably also employed in later redecorations. On the contrary, according to the judgment of the conservator of our team, the red lead did not seem

Table 19.1 Summary of the in situ identified pigments in the decorative revetments in several spaces in the Alhambra.

Type of ornament	Location	Blue	Red	Black	Green	Yellow
Plasterwork	Hall of the Kings (Mocarabes)	Lapis lazuli	Cinnabar/red lead	Carbon black	Altered Azurita	NF
	Room of the Beds (Reliefs)	Ultramar	Cinnabar/hematite	Carbon black	Copper arsenate	NF
Marble	Hall of the Boat (Carved Relief)	Lapis lazuli/Azurite	Cinnabar	Carbon black	NF	NF
	Hall of the Abencerrages	Lapislazuli	Cinnabar	Carbon black	NF	NF
	Mexuar	Lapis lazuli/Azurite Christian	Cinnabar/red lead	Carbon black	NF	NF
	Hall of the Kings	Lapis lazuli	NF	Carbon black	NF	NF
	Court of the Myrtles	Lapis lazuli	NF	Carbon black	NF	NF
	Court of the Main Canal	NF	NF	Carbon black	NF	NF
Wooden Ceilings	Hall of the Abencerrages	Lapis lazuli	Cinnabar/red lead	NF	Altered Azurita	NF
	Hall of the Two sisters	NF	Cinnabar/red lead	Carbon black	NF	Massicot/ Orpiment
Mural painting	Court of Harem	Lapis lazuli	Hematite/cinnabar very altered	Carbon black	NF	NF
	Infant's Tower	NF	Hematite	NF	NF	NF
	Partal Dwellings/Tower of the Ladies	NF	Hematite	NF	NF	NF
	Diverse outdoor locations - Faux bricks pattern decoration*	NF	Hematite	NF	NF	NF

NF, Not found.

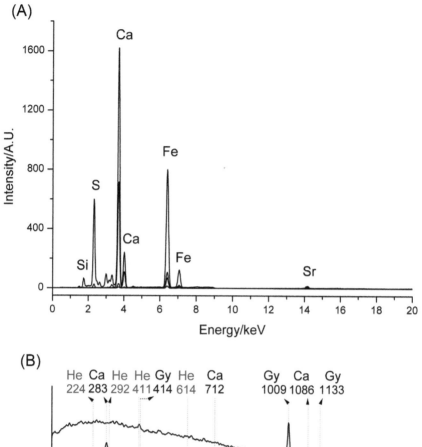

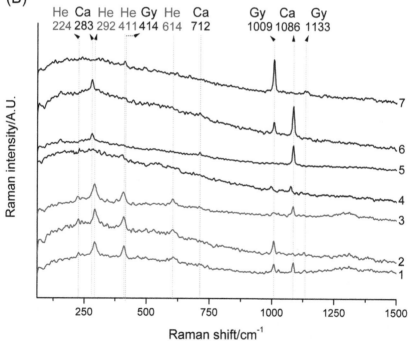

Figure 19.3 Typical XRF and Raman spectra of faux bricks decorations. (A) XRF spectra on red areas and (B) Raman spectra of (1−3) red and (4−7) white areas. He: hematite, Ca: calcite, Gy: gypsum. Raman spectra have been stacked for clarity. *XRF*, X-ray fluorescence. From Arjonilla, P., Ayora-Cañada, M. J., de la Torre-López, M. J., Gómez, E. C., Domene, R. R., Domínguez-Vidal, A. (2021). Spectroscopic investigation of wall paintings in the Alhambra monumental ensemble: Decorations with red bricks. *Crystals. 11*(4). https://www.mdpi.com/2073−4352/11/4/423/pdf, https://doi.org/10.3390/cryst11040423.

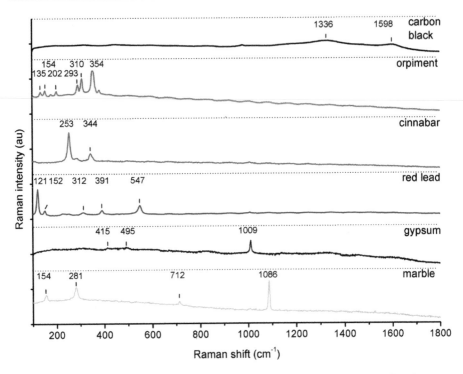

Figure 19.4 Typical in situ Raman spectra obtained in the revetments in the Alhambra Spectra of different pigments together with those of gypsum and marble substrates.

to correspond to the oldest decorations. At the naked eye, clear signs of alteration of both red pigments were observed. The main decay products identified and the alteration causes hypothesized will be discussed in detail in Section 19.4.2. The same two red pigments were identified in the polychrome carved marble capitals of the columns of the Mexuar (Arjonilla et al., 2016) and in the two wooden ceilings studied. The differential use of the red pigments made by Nasrid artists, depending on the type of ornamentation, is remarkable. In particular, the cheapest pigment red ocher (hematite) was used only in the simplest decorations, employed also in out-door walls, whereas more expensive pigments such as cinnabar were reserved for the more elaborated inner decorations. Interestingly, in the redecoration of the 19th century, in the Room of the Beds, both cinnabar and hematite were identified. In this case, cinnabar was used to achieve red coloring, whereas hematite was employed mixed with carbon black, to obtain a brown color (Arjonilla, Ayora-Cañada, et al., 2019). In the case of polychrome carpentry, cinnabar was the pigment employed for the red color, whereas red lead was not really employed as a pigment. It was used in a priming layer, probably with the intention of protecting wood against insects, fungi, and mold. Massicot, another lead oxide, was also identified in this layer. Its presence is probably not intentional but accidental, as an impurity formed during red lead synthesis (Arjonilla, Domínguez-Vidal, et al., 2019).

Furthermore, a light yellow pigment with a golden glow, orpiment (As_2S_3) was found in these ceilings but not in the rest of the studied revetments.

Black color is mainly used for lines and contours in the studied revetments. To obtain Raman spectra from these areas was challenging as the black pigments absorbed most of the radiation and the laser power had to be maintained very low to avoid burning. Consequently, the spectra obtained were, in general, of poor quality. Despite this, the two broad bands were always found (Tomasini et al., 2012) around 1334 and 1594 cm^{-1}, typical of carbon black. Furthermore, in some cases, spectra presented a weak feature around 960 cm^{-1}, which could be attributed to phosphate (Edwards, 2018). With this information, the use of both black pigments, carbon black, obtained from the partial burning of natural wood or vegetables, or ivory black (or bone black), obtained by burning bones can be suggested. However, the Raman spectra were always weak and noisy, which made it difficult to assess the origin of the carbon pigment employed.

As exposed previously, blue is one of the most abundant colors in the studied decorations and it was very interesting to obtain information about the use of this color in the history of the monument. Most of the registered Raman spectra showed a band at 548 cm^{-1} that can be attributed to lazurite, a blue feldspathoidmineral, together with a series of bands in the region 1250–1900 cm^{-1} (see Fig. 19.5). Natural lazurite is the mineral that provides its characteristic blue tonality to the lapis lazuli rock and is one of the most precious pigments employed since Antiquity, genuine ultramarine. The pigment is purified via a floatation process after crushing and grinding the lapis lazuli rock.

The rest of the bands only appear when using 785 nm excitation (González-Cabrera et al., 2020; Schmidt et al., 2009) and have been attributed to impurities present in the lapis lazuli rock (González-Cabrera et al., 2022; Schmidt et al., 2009). Thus these bands are only present in natural ultramarine pigments and can be used to distinguish them from synthetic ultramarine blue, first obtained in the 19th century. This synthetic pigment was also found in the Alhambra as can be seen in Fig. 19.5. In some cases, its presence can be related to documented interventions, as in the case of the Room of the Beds (Contreras redecoration) and the marble capitals of the Mexuar (documented 20th-century intervention with color reintegration). However, in other cases, there is no information about the intervention when it was applied, for example, in one of the vaults of the Hall of the Kings. In any case, natural ultramarine blue (genuine lapis lazuli) was the most commonly blue pigment identified in the different spaces investigated, including all the studied revetments. It is worth noting that the relative intensity of the lazurite/impurities bands in the spectra of the natural ultramarine blue could also be useful to inform about the quality of the pigments employed. For example, the pigment employed in the marble capitals of the Hall of the Mexuar showed a high value for the ratio of lazurite/impurities bands revealing the use of a better quality pigment. In contrast, in the case of the wooden ceilings, the impurity bands were very intense, while the lazurite band at 548 cm^{-1} was almost imperceptible, suggesting the use of a poorer quality pigment.

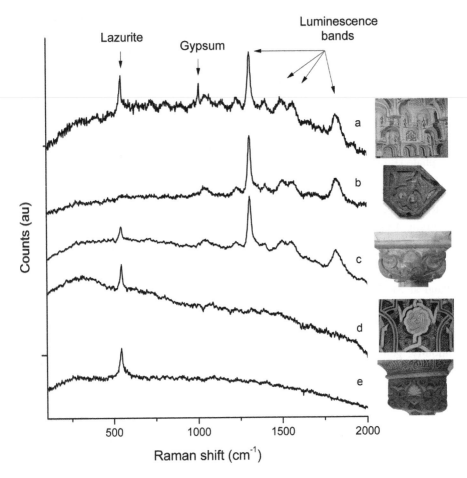

Figure 19.5 In situ Raman spectra registered by using 785 nm laser excitation on blue decorations Different halls and supports were considered being plasterwork (a, Hall of the Kings, and d, Room of the Beds), wooden ceilings (b, Hall of the Abencerrages), and marble capitals (c and e, Hall of the Mexuar). Spectra were stacked for better visualization and linked to the images of the decorative motifs.

Source: From González-Cabrera. M., Arjonilla, P., Domínguez-Vidal, A., Ayora-Cañada, M. J. (2020). Natural or synthetic? Simultaneous Raman/luminescence hyperspectral microimaging for the fast distinction of ultramarine pigments. *Dyes and Pigments. 178.* https://www.scopus.com/ inward/record.uri?eid = 2-s2.0-85082402930&doi = 10.1016%2fj.dyepig.2020.108349 &partnerID = 40&md5 = 440a8cd69f2c34a6750c3d3cb5b04a64, https://doi.org/10.1016/j. dyepig.2020.108349.

Green color was scarcely used in the decorations studied although some pale blue-greenish decorations were found. In all the cases, they did not provide any useful Raman spectra during the in situ study. This has been reported to occur due to the strong absorption of the near-infrared light by green pigments (Vandenabeele & Moens, 2004).

Furthermore, in many cases, the probable presence of an organic binder induced high fluorescence backgrounds limiting the identification of the pigments. Nevertheless, the portable ED-XRF instrument was extremely useful for the detection of the elements. In these pale blue and green motifs, copper was identified, and in one case (in the redecoration of the Room of the Beds plasterwork) also arsenic. These results suggested the use of copper-based pigments, but they could not be unambiguously identified in the in situ study. For these reasons, microsamples were collected for their analysis in the laboratory. Taking into account that these pigments presented a high degree of degradation, the detailed discussion belongs to the following section.

19.4.3   Investigation of pigment alterations and other degradation products

After the identification of the main pigments, the conservation stage and the origin of the different chromatic alterations observed will be discussed. Lapis lazuli pigment always conserved its intense and bright blue color, although in many cases it was partially detached or covered with dirt and dust. This is consistent with the already-known chemical stability of this pigment (Eastaugh et al., 2004). The most intense color alterations were observed in the red decorations on gypsum plasterwork (Fig. 19.6). In the case of the motifs painted with red lead, they appeared often darkened, showing a brownish tonality. In these areas, different alteration products were identified as can be seen in Fig. 19.6 (spectra 1 and, 2). Although most of the spectra registered showed the characteristic bands of red lead, additional features were identified. For example, the band at 978 cm^{-1}(spectra 1b) was attributed to anglesite, $PbSO_4$. In previous studies about read lead alterations on model samples, the disproportionation of read lead to form both lead dioxide (in the form of plattnerite $ß\text{-}PbO_2$) and Pb(II) ions in acidic conditions has been reported (Aze et al., 2007). In a medium rich in sulfate, such as gypsum, Pb (II) ions can form anglesite. Since the anglesite is white, the darkened color observed should be attributed to the black plattnerite formed. However, the typical Raman features of this compound were not detected because plattnerite is a very weak Raman scatterer. Furthermore, it easily degrades under laser exposure, even using very low power. The presence of plattnerite is, in fact, the cause of the band observed at 141 cm^{-1} (see spectra 1a), which corresponds to the strong scatterer lead monoxide (PbO) formed from plattnerite under laser irradiation (Burgio et al., 2001). In addition to those dark locations of altered red lead, other areas where the color was totally lost were observed. The only Pb species identified in these areas was anglesite which can be explained by the slow transformation of the plattnerite into the more stable anglesite. Thus this alteration of red lead, although initially causing darkening, ends with a total loss of the red color. This effect was mainly found in areas with evidence of water leaking through the cracks of the structure (Dominguez-Vidal et al., 2014).

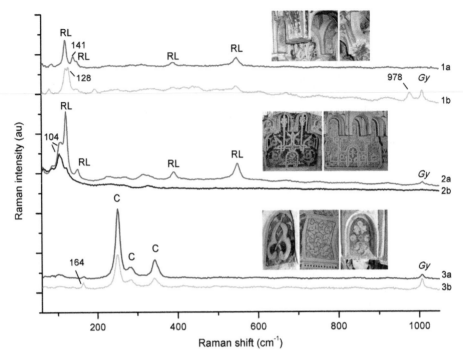

Figure 19.6 Degradation of red pigments. Raman spectra registered in situ in red areas where the pigments, red lead (1 and 2) and cinnabar (3), are in different stages of degradation, showing bands of different decay products. Bands assigned to the original pigments are labeled: C: cinnabar, RL: red lead. Gypsum substrate band is also marked, Gy. Images of the motifs show the typical chromatic alterations observed.

Curiously, in other motifs more exposed to sunlight penetrating from the windows, a different alteration of red lead was detected although the color change observed was also the darkening of the pigment. In this case, the Raman spectra in the altered areas (see spectra 2a) showed a general decrease of the bands of red lead and a new Raman band developing at 104 cm^{-1}. In the most darkened areas of the top, the bands of minium were totally lost (see spectra 2b). This decay product could not be unambiguously identified but the broadband at 104 cm^{-1} could be assigned to mixed crystals of lead oxides showing Pb−Pb stretching vibrations shifted from that occurring in red lead (Pb$_3$O$_4$) at 121 cm^{-1}.

The other red pigment, cinnabar, showed also important chromatic alterations. In well-preserved motifs, it showed its characteristic intense red tonality but often it appeared very darkened. Cinnabar blackening is a common phenomenon, which has been observed since Antiquity. Already Vitruvius described the change of the bright red of cinnabar when exposed to light (Colombo, 2003) but its causes and mechanism are still not fully understood (Pérez-Diez et al., 2022).

Studies on different artworks together with laboratory artificial aging tests have shed some light on the problem (Cotte et al., 2011; Elert et al., 2021; Pérez-Diez et al., 2022). These studies have ruled out the classical hypothesis of transformation into black metacinnabar, describing the photosensitization of cinnabar induced by chlorides with the formation of different Cl−Hg and Hg−S−Cl compounds, as well as the reduction of Hg(2) to black Hg(0). However, there is no total agreement about the identity of the species responsible for the black color. In the *mocarabes* vaults of the Hall of the Kings, an additional band at $164 \, \text{cm}^{-1}$ was identified in some of the altered cinnabar spectra obtained (Fig. 19.6: spectra 3a and 3b). This band can be attributed to calomel (Hg_2Cl_2). In our case, as happened with the above-described degradation of red lead, the initial darkening of the surface evolves to a gray color and, finally, to the total loss of the pigment. Nevertheless, even in the gray areas, the characteristic Raman spectrum of cinnabar could be clearly registered, although with low intensity, when focusing on small red particles thanks to the use of the microscope. Cinnabar appeared particularly degraded in the wall paintings of the Harem Court with scarce remainings of the pigment. In that case, the quality of the spectra recorded in outdoor locations is not optimal to detect the weak Raman signal of calomel. Nevertheless, again the use of XRF allowed clarifying the degradation process as the presence of low amounts of chlorine was systematically detected in the paintings of the Court. Natural impurities of cinnabar have been reported to induce the photodegradation of the pigment (McCormack, 2000). However, this source cannot explain the amount of chloride found and the fact that it is not always associated with Hg. Environmental pollution, especially in geographical locations close to the sea, is a plausible explanation for chlorine-induced cinnabar degradation (Cotte et al., 2006; Keune & Boon, 2005; Spring & Grout, 2002). Other possibilities include the pigment preparation or the binders employed but further studies on microsamples would be necessary to clarify this point.

Chlorine ions have also been found to play a crucial role in the degradation of copper-based pigments with pale blue or green color. This study had to be performed on microsamples, as even using benchtop instruments, the Raman spectra from the green areas were extremely difficult to gather. In these cases, the 514 nm laser was employed with low laser power and long acquisition times. In these samples, the blue pigment azurite, ($Cu_3(CO_3)_2(OH)_2$), severely degraded to copper hydroxychlorides, was found. Copper hydroxychlorides were characterized by Raman features such as a characteristic peak at $511 \, \text{cm}^{-1}$ (attributed to O-Cu-O symmetric stretching) accompanied by bands of Cu-O-H bending in the region from 800 to $1000 \, \text{cm}^{-1}$ and those typical of OH stretching vibrations ($3300-3450 \, \text{cm}^{-1}$).

A different type of green color corresponds to the redecoration of the Room of the Beds. In these motifs, copper and arsenic were detected by XRF (see Fig. 19.7) and therefore the use of pigments such as Scheele's green or Emerald green, widely employed during the 19th century (Eastaugh et al., 2004) was hypothesized.

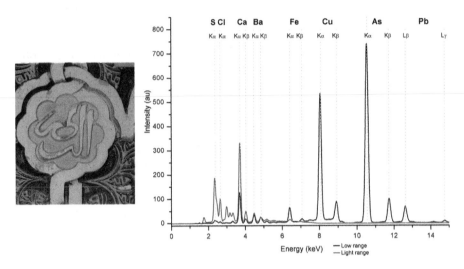

Figure 19.7 Green decoration in the Room of the Beds. Image of a green motif together with the corresponding XRF spectrum. *XRF*, X-ray fluorescence.

However, the Raman spectra obtained from the green microsamples did not match the reported Raman spectra of either of these two pigments. In fact, they were dominated by a strong band at 860 cm^{-1} that can be attributed to the symmetric stretching vibration of the arsenate anion (AsO_4^{3-}), probably resulting from the degradation of a copper arsenite pigment like Scheele's green ($CuHAsO_3$).

The case of gilding decorations is also interesting as they suffer from severe decay. Only remnants with brilliant rests and blackened areas could be found, mainly in protected and hidden areas. Tin monoxide (SnO) with Raman bands at 110 and 207 cm^{-1} (Fig. 19.8) was found in the black areas. It is worth mentioning that the identification of this compound was not straightforward. First, it was not considered in the most usual databases of artistic materials (Caggiani et al., 2016; Castro et al., 2005) and, second, a portable spectrometer able to record Raman spectra from 65 cm^{-1} was used, whereas in many databases the spectral region below 200 cm^{-1} is not available due to the limitations of the filters used to eliminate the Rayleigh radiation. The presence of SnO is due to the particular execution technique of these gildings (De La Torre-López et al., 2014). The gilding process involved the use of two different metal layers, one over another. Thus a thick tin layer was employed to make easier the application of very thin gold foils. The oxidation of tin led to an expansion in volume causing the detaching of the gold foil and leaving exposed the dark color of SnO.

Furthermore, and particularly in plasterwork decorations, other typical degradation products such as oxalates (Rampazzi, 2019), in the form of whewellite ($CaC_2O_4 \cdot H_2O$) and weddellite ($CaC_2O_4 \cdot 2H_2O$), were identified. Their origin is not completely clear but they could be generated due to the microbial degradation

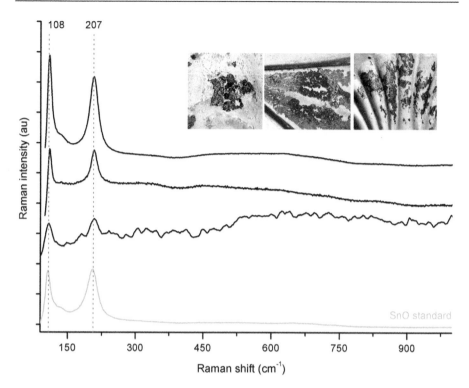

Figure 19.8 Tin monoxide as decay product in gildings. Raman spectra of SnO registered in the black areas of different gilded motifs (in the images). Spectrum of SnO standard was added for comparison.

of the organic binders used to fix the polychromies in a calcium-rich ambient provided by the porous support of gypsum plasterwork.

Finally, gypsum ($CaSO_4 \cdot 2H_2O$) was also identified as a decay product on the surface of the wall paintings, particularly in outdoor decorations. The sulfation of the carbonatic matrix of the lime plaster due to the reaction with environmental pollutants is a well-described phenomenon (Charola et al., 2007). Nevertheless, in some locations, the presence of gypsum on the surface of the paintings was due to the incomplete removal of plaster coverings applied in ancient interventions to hide the decorations (Arjonilla et al., 2021, 2022).

19.4.4   Study on the execution techniques

During the noninvasive studies, the execution methodologies of the different decorative revetments were also investigated. Thus the combination of the information obtained from XRF and Raman spectra about the materials together with the visual and microscopic inspection of the decorations was attempted. Especially interesting for this purpose were those areas where the deterioration of the most external painting layers allowed for the investigation of the underlying layers. In the simplest form of wall paintings, with false brick motifs painted in red color, lime mortar was applied over the supporting structure. The presence of calcite in the red pictorial layer and the absence of organic binders or its degradation products are typical of lime-based painting techniques, such as *fresco*. An example of a different execution technique was also found in these motifs. In one of the indoor locations, gypsum ($CaSO_4 \cdot 2H_2O$) plaster was used instead of lime plaster to receive the polychromy. Here, the identification of weddellite ($CaC_2O_4 \cdot 2H_2O$) can be related to the degradation of an organic binder, suggesting a *secco* technique. The *secco* technique was also employed in the Harem Court wall paintings, showing much more elaborated geometrical and vegetal motifs in different colors, but in this case over a lime plaster substrate.

For the *mocarabes* decorations, Nasrid artist applied the pigments over a white finishing layer constituted by very pure gypsum, which contrasted with the underlying substrate, made also of gypsum plaster but richer in Si, Al, and K, typical of aluminosilicate impurities. The ubiquitous identification of calcium oxalates in these decorations suggests the use of organic additives to settle the gypsum plaster and to fix the pigments. The execution technique employed by the Nasrid artist in these decorations is completely different from that used by Contreras in the redecoration of the Room of the Beds, who applied a thick preparation layer of white lead containing barium and lead sulfates over the gypsum plaster.

In the case of marble capitals, the pigments were applied directly above the marble substrate. The thinness of the pictorial layer in these ornaments did not allow us to gain any information about the nature of the organic binder by NDTs, even working with microsamples. Using gas chromatography coupled to mass spectrometry (GC-MS), the binder was identified as animal glue in most of these decorations. Furthermore, in the Mexuar marble capitals, evidence of old interventions during the early Christian period, when this space was completely transformed, was encountered. In some places, remains of a redecoration were found. Only very fragmented rests were preserved and it was less delicate in its execution than the Nasrid one, employing a very thick layer of gypsum to hide the original paintings. In this Christian redecoration, azurite was used for the blue color instead of lapis lazuli.

Finally, in the polychrome wood pieces of the ceilings, an orange preparation layer of red lead was always found on the wooden support. In addition, a white layer was detected sometimes under the pigments. The complex stratigraphy of these decorations had to be further investigated on microsamples. Using infrared spectroscopy as a complementary non-destructive analytical technique and working on selected samples, it was possible to identify a proteinaceous binder (possibly animal glue) employed to fix the pigments, a natural resin probably used as a mordant for false gilding, and an external layer of beeswax, presumably applied as a protective treatment in modern restorations Fig. 19.9.

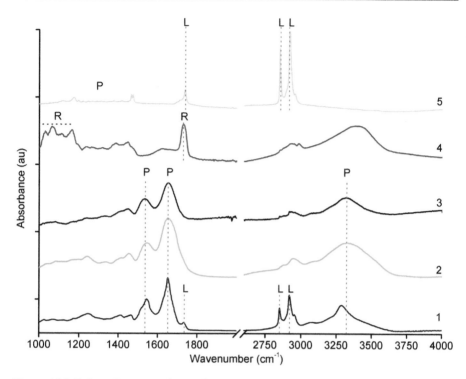

Figure 19.9 Infrared spectra of organic material identified in the decorative revetments. Spectra registered in the plasterwork of Room of the Beds (1 and 2) and in the wooden ceilings (3–5). Characteristic bands used for the identification of the organic material are labeled: P, proteinaceous material; L: lipids; R: resin.

19.5 Conclusions and challenges

In this chapter, the implementation of two portable spectroscopic techniques, namely, XRF and Raman spectroscopy, for the noninvasive characterization of decorative revetments in the Alhambra monumental ensemble has been presented. These two techniques provide complementary information at different levels, that is, elemental and molecular, respectively. This approach allows for in situ analysis of buildings' decorations with a maximization of the information-to-risk ratio on damage to the heritage asset. However, working out of the laboratory is also rather troublesome. In particular, when using Raman spectroscopy daylight causes certain interference problems. Furthermore, working on scaffoldings, elevated platforms, or with extension accessories involves stability issues for long measurements. Despite these difficulties, the advantages of the noninvasive approach have been illustrated here with the identification of all the pigments employed in different spaces of the monument, as well as its more common alteration mechanisms. Avoiding the need for the physical removal of samples has allowed much more representative investigations, with a high number of measurement points at each location. In this way, only a few samples from selected locations were taken to provide information on those issues where the portable techniques did not achieve satisfactory results (as in the case of green pigments). The most important limitation of these noninvasive spectroscopic techniques is that they provide information only about the most external layers. In this sense, the recent developments in the field of microspatially offset Raman spectroscopy (micro-SORS) could offer new possibilities to explore the subsurface of materials in a noninvasive way (Conti et al., 2020). However, even this approach cannot provide information about the thickness of the layers or the depth of the sublayer from which the Raman signal originated. Thus information about the execution techniques with noninvasive techniques is rather limited, and detailed knowledge of the stratigraphy must still be gathered from cross sections.

References

Adriaens, A., Cassar, J., & Degrigny, C. (2005). *Benefits of non-destructive analytical techniques for conservation. European cooperation in science and technology.* Publications Office.

Aramendia, J., Gomez-Nubla, L., Castro, K., Martinez-Arkarazo, I., Vega, D., Sanz López De Heredia, A., García Ibáñez De Opakua, A., & Madariaga, J. M. (2012). Portable Raman study on the conservation state of four CorTen steel-based sculptures by Eduardo Chillida impacted by urban atmospheres. *Journal of Raman Spectroscopy*, *43*(8), 1111−1117. Available from https://doi.org/10.1002/jrs.3158.

Arjonilla, P., Ayora-Cañada, M. J., de la Torre-López, M. J., Gómez, E. C., Domene, R. R., & Domínguez-Vidal, A. (2021). Spectroscopic investigation of wall paintings in the Alhambra monumental ensemble: Decorations with red bricks. *Crystals*, *11*(4). Available from https://doi.org/10.3390/cryst11040423, https://www.mdpi.com/2073-4352/11/4/423/pdf.

Arjonilla, P., Ayora-Cañada, M. J., Rubio Domene, R., Correa Gómez, E., de la Torre-López, M. J., & Domínguez-Vidal, A. (2019). Romantic restorations in the Alhambra monument: Spectroscopic characterization of decorative plasterwork in the Royal Baths of Comares. *Journal of Raman Spectroscopy*, *50*(2), 184−192. Available from https://doi.org/10.1002/jrs.5422, http://onlinelibrary.wiley.com/journal/10.1002. (ISSN)1097−4555.

Arjonilla, P., Domínguez-Vidal, A., Correa-Gómez, E., Domene-Ruiz, M. J., & Ayora-Cañada, M. J. (2019). Raman and Fourier transform infrared microspectroscopies reveal medieval Hispano−Muslim wood painting techniques and provide new insights into red lead production technology. *Journal of Raman Spectroscopy*, *50*(10), 1537−1545. Available from https://doi.org/10.1002/jrs.5660, http://onlinelibrary.wiley.com/journal/10.1002/ (ISSN)1097−4555.

Arjonilla, P., Domínguez-Vidal, A., de la Torre López, M. J., Rubio-Domene, R., & Ayora-Cañada, M. J. (2016). In situ Raman spectroscopic study of marble capitals in the Alhambra monumental ensemble. *Applied Physics A: Materials Science and Processing*, *122*(12). Available from https://doi.org/10.1007/s00339-016-0537-2, http://www.springer.com/materials/journal/339.

Arjonilla, P., Domínguez-Vidal, A., Rubio Domene, R., Correa Gómez, E., de la Torre-López, M. J., & Ayora-Cañada, M. J. (2022). Characterization of Wall Paintings of the Harem Court in the Alhambra monumental ensemble: Advantages and Limitations of In Situ Analysis. *Molecules (Basel, Switzerland)*, *27*(5). Available from https://doi.org/10.3390/molecules27051490, https://www.mdpi.com/1420-3049/27/5/1490/pdf.

Aze, S., Vallet, J. M., Pomey, M., Baronnet, A., & Grauby, O. (2007). Red lead darkening in wall paintings: Natural ageing of experimental wall paintings versus artificial ageing tests. *European Journal of Mineralogy*, *19*(6), 883–890. Available from https://doi.org/10.1127/0935-1221/2007/0019-1771, https://www.european-journal-of-mineralogy.net/index.html.

Barone, G., Mazzoleni, P., Cecchini, A., & Russo, A. (2018). In situ Raman and pXRF spectroscopic study on the wall paintings of Etruscan Tarquinia tombs. *Dyes and Pigments*, *150*, 390–403. Available from https://doi.org/10.1016/j.dyepig.2017.12.008, http://www.journals.elsevier.com/dyes-and-pigments/.

Bonizzoni, L., Caglio, S., Galli, A., & Poldi, G. (2010). Comparison of three portable EDXRF spectrometers for pigment characterization. *X-Ray Spectrometry*, *39*(3), 233–242. Available from https://doi.org/10.1002/xrs.1253Italy, http://www3.interscience.wiley.com/cgi-bin/fulltext/123326585/PDFSTART.

Bueno, A. G., & Flórez, V. J. M. (2004). The Nasrid plasterwork at "qubba Dar al-Manjara l-kubra" in Granada: Characterisation of materials and techniques. *Journal of Cultural Heritage*, *5*(1), 75–89. Available from https://doi.org/10.1016/j.culher.2003.02.002, https://www.sciencedirect.com/science/article/pii/S1296207403001171.

Burgio, L., Clark, R. J. H., & Firth, S. (2001). Raman spectroscopy as a means for the identification of plattnerite (PbO2), of lead pigments and of their degradation products. *Analyst*, *126*(2), 222–227. Available from https://doi.org/10.1039/b008302j, http://pubs.rsc.org/en/journals/journal/an.

Caggiani, M. C., Cosentino, A., & Mangone, A. (2016). Pigments Checker version 3.0, a handy set for conservation scientists: A free online Raman spectra database. *Microchemical Journal*, *129*, 123–132. Available from https://doi.org/10.1016/j.microc.2016.06.020, http://www.elsevier.com/inca/publications/store/6/2/0/3/9/1.

Cardell, C., Rodriguez-Simon, L., Guerra, I., & Sanchez-Navas, A. (2009). Analysis of Nasrid polychrome carpentry at the Hall of the Mexuar Palace, Alhambra complex (Granada, Spain), combining microscopic, chromatographic and spectroscopic methods. *Archaeometry*, *51*(4), 637–657. Available from https://doi.org/10.1111/j.1475-4754.2008.00438.x.

Cardell-Fernández, C., & Navarrete-Aguilera, C. (2006). Pigment and plasterwork analyses of Nasrid polychromed lacework stucco in the Alhambra (Granada, Spain). *Studies in Conservation*, *51*(3), 161–176. Available from https://doi.org/10.1179/sic.2006.51.3.161, https://doi.org/10.1179/sic.2006.51.3.161.

Castro, K., Pérez-Alonso, M., Rodríguez-Laso, M. D., Fernández, L. A., & Madariaga, J. M. (2005). On-line FT-Raman and dispersive Raman spectra database of artists' materials (e-VISART database). *Analytical and Bioanalytical Chemistry*, *382*(2), 248–258. Available from https://doi.org/10.1007/s00216-005-3072-0.

Charola, A. E., Pühringer, J., & Steiger, M. (2007). Gypsum: A review of its role in the deterioration of building materials. *Environmental Geology*, *52*(2), 207–220. Available from https://doi.org/10.1007/s00254-006-0566-9.

Colombo, L. (2003). *I Colori Degli Antichi*. Nardini Editore, Rome (Italy).

Conti, C., Botteon, A., Colombo, C., Pinna, D., Realini, M., & Matousek, P. (2020). Advances in Raman spectroscopy for the non-destructive subsurface analysis of artworks: Micro-SORS. *Journal of Cultural Heritage*, *43*, 319–328. Available from https://doi.org/10.1016/j.culher.2019.12.003, http://www.elsevier.com.

Cotte, M., Susini, J., Metrich, N., Moscato, A., Gratziu, C., Bertagnini, A., & Pagano, M. (2006). Blackening of Pompeian cinnabar paintings: X-ray microspectroscopy analysis. *Analytical Chemistry, 78*(21), 7484−7492. Available from https://doi.org/10.1021/ac0612224.

Cotte, M., Szlachetko, J., Lahlil, S., Salomé, M., Solé, V. A., Biron, I., & Susini, J. (2011). Coupling a wavelength dispersive spectrometer with a synchrotron-based X-ray microscope: A winning combination for micro-X-ray fluorescence and micro-XANES analyses of complex artistic materials. *Journal of Analytical Atomic Spectrometry, 26*(5), 1051−1059. Available from https://doi.org/10.1039/c0ja00217h.

De La Torre-López, M. J., Dominguez-Vidal, A., Campos-Suñol, M. J., Rubio-Domene, R., Schade, U., & Ayora-Cañada, M. J. (2014). Gold in the Alhambra: Study of materials, technologies, and decay processes on decorative gilded plasterwork. *Journal of Raman Spectroscopy., 45*(11-12), 1052−1058. Available from https://doi.org/10.1002/jrs.4454, http://www3.interscience.wiley.com/journal/3420/home.

De La Torre López, M. J., Sebastián, P. E., & Rodríguez, G. J. (1996). A study of the wall material in the Alhambra (Granada, Spain). *Cement and Concrete Research, 26*(6), 825−839. Available from https://doi.org/10.1016/0008-8846(96)00075-0, https://www.scopus.com/inward/record.uri?eid = 2-s2.0-0030166290&doi = 10.1016%2f0008-8846%2896%2900075-0&partnerID = 40&md5 = 9ecb10adc785f38b728e8cb6cf12dcc1.

Dominguez-Vidal, A., De La Torre-López, M. J., Campos-Suñol, M. J., Rubio-Domene, R., & Ayora-Cañada, M. J. (2014). Decorated plasterwork in the Alhambra investigated by Raman spectroscopy: Comparative field and laboratory study. *Journal of Raman Spectroscopy., 45*(11−12), 1006−1012. Available from https://doi.org/10.1002/jrs.4439, http://www3.interscience.wiley.com/journal/3420/home.

Dominguez-Vidal, A., Jose De La Torre-Lopez, M., Rubio-Domene, R., & Ayora-Cañada, M. J. (2012). In situ noninvasive Raman microspectroscopic investigation of polychrome plasterworks in the Alhambra. *Analyst, 137*(24), 5763−5769. Available from https://doi.org/10.1039/c2an36027f, http://pubs.rsc.org/en/journals/journal/an.

Drake, B. L., & MacDonald, B. L. (2022). Advances in portable x-ray fluorescence spectrometry: Instrumentation. In B. L. Drake, & B. L. MacDonald (Eds.), *Application and Interpretation*. The Royal Society of Chemistry.

Eastaugh, N., Walsh, V., Chaplin, T., & Siddall, R. (2004). *Pigment compendium: A dictionary of historical pigments*. Elsevier Butterworth-Heinemann. Available from 10.4324/9780080473765.

Edwards, H., & Vandenabeele, P. (2012). In H. Edwards, & P. Vandenabeele (Eds.), *Analytical archaeometry -selected topics*. The Royal Society of Chemistry, 978-1-84973-162-1 978-1-78262-624-4.

Edwards, H. G. M. (2018). Volume 2: Analytical Raman spectroscopy of inks. In P. Vandenabeele, & H. Edwards (Eds.), *Raman spectroscopy in archaeology and art history*. The Royal Society of Chemistry. 978-1-78801-138-9 Available from https://doi.org/10.1039/9781788013475-00001.

Elert, K., Pérez Mendoza, M., & Cardell, C. (2021). Direct evidence for metallic mercury causing photo-induced darkening of red cinnabar tempera paints. *Communications Chemistry, 4*(1), 174. Available from https://doi.org/10.1038/s42004-021-00610-2, http://nature.com/commschem/.

Ettinghausen, R., Grabar, O., & Jenkins-Madina, M. (2003). *Islamic art and architecture, 650−1250 The Yale University Press Pelican History of Art Series*. The Yale University Press. Available from https://yalebooks.yale.edu/book/9780300088694/islamic-art-and-architecture-6501250/.

Giussani, B., Monticelli, D., & Rampazzi, L. (2009). Role of laser ablation−inductively coupled plasma−mass spectrometry in cultural heritage research: A review. *Analytica*

Chimica Acta, 635(1), 6−21. Available from https://doi.org/10.1016/j.aca.2008.12.040, https://www.sciencedirect.com/science/article/pii/S0003267008021946.

González-Cabrera, M., Arjonilla, P., Domínguez-Vidal, A., & Ayora-Cañada, M. J. (2020). Natural or synthetic? Simultaneous Raman/luminescence hyperspectral microimaging for the fast distinction of ultramarine pigments. *Dyes and Pigments, 178.* Available from https://doi.org/10.1016/j.dyepig.2020.108349, https://www.scopus.com/inward/record. uri?eid = 2-s2.0-85082402930&doi = 10.1016%2fj. dyepig.2020.108349&partnerID = 40&md5 = 440a8cd69f2c34a6750c3d3cb5b04a64.

González-Cabrera, M., Wieland, K., Eitenberger, E., Bleier, A., Brunnbauer, L., Limbeck, A., Hutter, H., Haisch, C., Lendl, B., Domínguez-Vidal, A., & Ayora-Cañada, M. J. (2022). Multisensor hyperspectral imaging approach for the microchemical analysis of ultramarine blue pigments. *Scientific Reports, 12*(1). Available from https://doi.org/ 10.1038/s41598-021-04597-7, https://www.scopus.com/inward/record.uri?eid = 2-s2.0-85123124896&doi = 10.1038%2fs41598-021-04597-7&partnerID = 40&md5 = cd3b21b865e6b2cb5a22e07168bce944.

ICOMOS. (2011). *ICOMOS charter principles for the analysis, conservation and structural restoration of architectural heritage.* https://www.icomos.org/en/about-the-centre/179-articles-en-francais/ressources/charters-and-standards/165-icomos-charter-principles-for-the-analysis-conservation-and-structural-restoration-of-architectural-heritage.

Keune, K., & Boon, J. J. (2005). Analytical imaging studies clarifying the process of the darkening of vermilion in paintings. *Analytical Chemistry, 77*(15), 4742−4750. Available from https://doi.org/10.1021/ac048158f.

Kyriakou, L., Theodoridou, M., & Ioannou, I. (2019) *International Conference on Sustainable Materials, Systems and Structures (SMSS 2019), novel methods for characterization of materials and structures.* Standardised experimental techniques and novel micro-destructive methods for the assessment of lime mortar properties (pp. 20−22) Rovinj, Croatia. Unpublished content.

Madariaga, J. M. (2021). Analytical strategies for cultural heritage materials and their degradation. In J. M. Madariaga (Ed.), *X-ray fluorescence (XRF) techniques* (pp. 2052−3076). Royal Society of Chemistry (RSC). Available from http://doi.org/ 10.1039/9781788015974-00023.

McCormack, J. K. (2000). The darkening of cinnabar in sunlight. *Mineralium Deposita, 35* (8), 796−798. Available from https://doi.org/10.1007/s001260050281.

Mileto, C., & Vegas, F. (2007). Understanding architectural change at the Alhambra: stratigraphic analysis of the western gallery, court of the myrtles. In: Revisiting al-Andalus. 193−207. Available from https://doi.org/10.1163/ej.9789004162273.I-304.37.

Moropoulou, A., Labropoulos, K. C., Delegou, E. T., Karoglou, M., & Bakolas, A. (2013). Non-destructive techniques as a tool for the protection of built cultural heritage. *Construction and Building Materials, 48,* 1222−1239. Available from https://doi.org/ 10.1016/j.conbuildmat.2013.03.044, https://www.sciencedirect.com/science/article/pii/ S0950061813002547.

Pérez, A. G. (2017). Reconstructing the Alhambra: Rafael Contreras and architectural models of the Alhambra in the nineteenth century. *Art in Translation, 9*(1), 29−49. Available from https://doi.org/10.1080/17561310.2017.1297041, http://www.tandfonline.com/doi/ pdf/10.1080/17561310.2017.1297041.

Pérez-Diez, S., Larrañaga, A., Madariaga, J. M., & Maguregui, M. (2022). Unraveling the role of the thermal and laser impacts on the blackening of cinnabar in the mural paintings of Pompeii. *European Physical Journal Plus, 137*(10). Available from https://doi.org/10.1140/ epjp/s13360-022-03392-1, https://www.springer.com/journal/13360.

Rampazzi, L. (2019). Calcium oxalate films on works of art: A review. *Journal of Cultural Heritage*, *40*, 195–214. Available from https://doi.org/10.1016/j.culher.2019.03.002, http://www.elsevier.com.

Romero-Pastor, J., Duran, A., Rodríguez-Navarro, A. B., Grieken, R. V., & Cardell, C. (2011). Compositional and quantitative microtextural characterization of historic paintings by micro-X-ray diffraction and Raman microscopy. *Analytical Chemistry*, *83*(22), 8420–8428. Available from https://doi.org/10.1021/ac201159e, https://doi.org/10.1021/ac201159e.

Rousaki, A., & Vandenabeele, P. (2021). Analytical strategies for cultural heritage materials and their degradation Raman spectroscopy. In M. Madariaga Juan (Ed.), *Analytical strategies for cultural heritage materials and their degradation*. The Royal Society of Chemistry, 978-1-78801-524-0 doi: 10.1039/9781788015974-00124.

Schmidt, C. M., Walton, M. S., & Trentelman, K. (2009). Characterization of lapis lazuli pigments using a multitechnique analytical approach: Implications for identification and geological provenancing. *Analytical Chemistry*, *81*(20), 8513–8518. Available from https://doi.org/10.1021/ac901436g, https://www.scopus.com/inward/record.uri?eid = 2-s2.0-70449914835&doi = 10.1021%2fac901436g&partnerID = 40&md5 = 7ab5e3c 0556ebc60f34c5cc0fe64c70.

Schreiner, M., & Grasserbauer, M. (1985). Microanalysis of art objects: Objectives, methods and results. *Fresenius' Zeitschrift für analytische Chemie*, *322*(2), 181–193. Available from https://doi.org/10.1007/BF00517657, https://doi.org/10.1007/BF00517657.

Spring, M., & Grout, R. (2002). The Blackening of Vermilion: An analytical study of the process in paintings. *National Galley Technical Bulletin*, *23*, 50–60. Available from http://www.nationalgallery.org.uk/technical-bulletin/spring_grout2002.

Tejedor, B., Lucchi, E., Bienvenido-Huertas, D., & Nardi, I. (2022). Non-destructive techniques (NDT) for the diagnosis of heritage buildings: Traditional procedures and futures perspectives. *Energy and Buildings*, *263*, 112029. Available from https://doi.org/10.1016/j.enbuild.2022. 112029, https://www.sciencedirect.com/science/article/pii/S0378778822002006.

Thickett, D., Cheung, C. S., Liang, H., Twydle, J., Gr Maev, R., & Gavrilov, D. (2017). Using non-invasive non-destructive techniques to monitor cultural heritage objects. *Insight - Non-Destructive Testing and Condition Monitoring.*, *59*(5), 230–234. Available from https://doi.org/10.1784/insi.2017.59.5.230.

Tomasini, E. P., Halac, E. B., Reinoso, M., Di Liscia, E. J., & Maier, M. S. (2012). Micro-Raman spectroscopy of carbon-based black pigments. *Journal of Raman Spectroscopy*, *43*(11), 1671–1675. Available from https://doi.org/10.1002/jrs.4159 10974555.

Vandenabeele, P., & Donais, M. K. (2016). Mobile spectroscopic instrumentation in archaeometry research. *Applied Spectroscopy*, *70*(1), 27–41. Available from https://doi.org/ 10.1177/0003702815611063, https://www.osapublishing.org/as/browse.cfm.

Vandenabeele, P., & Moens, L. (2004). Pigment identification in illuminated manuscripts. *Comprehensive Analytical Chemistry*, *42*, 635–662. Available from https://doi.org/ 10.1016/S0166-526X(04)80018-7.

Vandenabeele, P., Weis, T. L., Grant, E. R., & Moens, L. J. (2004). A new instrument adapted to in situ Raman analysis of objects of art. *Analytical and Bioanalytical Chemistry*, *379*(1), 137–142. Available from https://doi.org/10.1007/s00216-004-2551-z.

Zoppi, A., Lofrumento, C., Castellucci, E. M., & Sciau, Ph (2008). Al-for-Fe substitution in hematite: The effect of low Al concentrations in the Raman spectrum of Fe2O3. *Journal of Raman Spectroscopy*, *39*(1), 40–46. Available from https://doi.org/10.1002/jrs.1811, https://www.scopus.com/inward/record.uri?eid = 2-s2.0-38549133125&doi = 10.1002% 2fjrs.1811&partnerID = 40&md5 = 221af6597ca915fcbea4b94565cb2b05.

Evaluation of heritage stone deterioration through non-destructive techniques (ultrasonic pulse velocity, rebound hammer test, SEM, and X-ray diffraction)

20

Supriya Patil[1], A.K. Kasthurba[2] and Mahesh Patil[3]
[1]School of Architecture, D.Y. Patil College of Engineering and Technology, Kolhapur, Maharashtra, India, [2]National Institute of Technology, Calicut, Kerala, India, [3]Sardar Vallabhbhai National Institute of Technology, Surat, Gujarat, India

20.1 Introduction

Since ancient times, stone has been used as a building material to construct significant monuments such as temples, forts, tombs, and palaces. These monuments carry evidences of history that needs to be preserved for their architectural, historical, and cultural values (Padilla-ceniceros & Pacheco-martínez, 2017). Rich architectural heritage of India is also manifested in stone monuments. Amongst those monuments, temples of exceptional architecture throughout the country assume significance as they were built by the regional rulers employing locally available materials and technology. Various types of heritage monuments such as early rock-cut architecture from the 6th century up to fully developed temple complexes from the 17th century are found in Maharashtra. Many of these temples are still in use with high religious and architectural values. Archaeological Survey of India (ASI) has currently started the restoration process at Markandeshwar temple from Vidharbha. Dharashiv caves at Osmanabad and Elephanta caves at Mumbai are some other examples of the built heritage which has experienced the conservation and restoration processes in recent times. At present, there are 285 heritage monuments recognized by ASI in Maharashtra. Many of these were predominantly built using local basalt stone in Dravidian style (Patil & Kasthurba, 2020). The preservation and conservation of these heritage stone structures are of paramount importance to maintain our cultural heritage. These structures, such as historical monuments, archaeological sites, and traditional buildings, often face the challenges of deterioration caused by natural

Diagnosis of Heritage Buildings by Non-Destructive Techniques. DOI: https://doi.org/10.1016/B978-0-443-16001-1.00020-6

processes, environmental factors, and human activities. Effective evaluation of the condition of heritage stones is crucial for developing appropriate conservation strategies (Patil et al., 2021). The Kopeshwar temple and Panhala Fort (located in Kolhapur district of Maharashtra, India), selected for the study area, are of national importance. The construction of these monuments depicts the architecture, culture, and history of that era. The Kopeshwar temple is built in Dravidian style with black basalt. Panhala Fort is built in Bijapur style with black basalt and laterite. Documentation of these monuments is essential for conservation purposes and policy-making. Scientific investigation and understanding of weathering mechanisms are highly essential for sustainable preservation of these monuments which can be achieved through various non-destructive tests (NDTs). These non-destructive techniques offer valuable tools to assess the deterioration without causing further harm to these precious structures. This chapter aims to explore the evaluation of heritage stone deterioration through non-destructive techniques, including ultrasonic pulse velocity (UPV), rebound hammer test, scanning electron microscopy (SEM), and X-ray diffraction (XRD). By understanding the principles, methodologies, and interpretations of these techniques, conservation experts can gain insights into the condition of heritage stones and make informed decisions for their preservation.

20.1.1 Background and significance

Heritage stone structures embody historical, architectural, and cultural values. They serve as tangible links to our past and contribute to the identity of communities and nations. However, these structures are susceptible to various forms of deterioration, such as weathering, erosion, chemical decay, biological growth, and mechanical damage (Bozda et al., 2019). Factors such as climate change, pollution, urbanization, and inadequate maintenance further accelerate the degradation process. The preservation of heritage stone structures is essential to safeguard our cultural heritage for future generations (Fitzner, 1997). Maharashtra, located in western India, is renowned for its rich architectural heritage, characterized by an array of stone structures such as temples, forts, caves, and ancient buildings. These structures exhibit a diverse range of stone types, including basalt, sandstone, limestone, and granite, which have been used for construction over centuries. However, the heritage stones in Maharashtra face various deterioration challenges due to environmental, climatic, anthropogenic, and geological factors. The deterioration of heritage stones in Maharashtra carries immense significance due to the cultural, historical, and architectural value of these structures. Many stone structures in Maharashtra represent significant architectural styles and techniques prevalent during different periods. The intricate carvings, sculptures, and ornamentation on these stones showcase the artistic skills and craftsmanship of past civilizations. Protecting heritage stones ensures the preservation of architectural history and serves as a valuable resource for architectural research and education. Studying heritage stone deterioration in Maharashtra provides valuable insights into the interaction between geological, environmental, and anthropogenic factors. This knowledge aids in developing effective conservation strategies, materials, and techniques not only for Maharashtra but also for similar heritage sites worldwide. Research in this field promotes interdisciplinary collaboration and contributes to the advancement of heritage conservation science. Traditional destructive testing methods, which involve taking samples from the stones, can cause irreversible damage. Hence, the need for non-destructive techniques arises. non-destructive techniques provide a means to assess the condition of heritage stones without altering or harming their integrity. These techniques enable conservation professionals to identify the areas of deterioration, understand the underlying causes, and develop appropriate conservation strategies. By using non-destructive techniques, the conservation process becomes more sustainable and preserves the authenticity and historical integrity of the structures.

20.1.2 Preservation challenges for heritage stone structures

Preserving heritage stone structures presents several challenges due to their historical significance, unique characteristics, and the complexity of deterioration processes. Some of the key challenges are discussed in the following subsections.

20.1.2.1 Material vulnerability

Heritage stones are often composed of natural materials, such as limestone, sandstone, marble, or granite, which are susceptible to various forms of degradation. The properties of these materials, including porosity, mineral composition, and structural integrity, influence their susceptibility to weathering and decay processes.

20.1.2.2 Environmental factors

Heritage stone structures are exposed to diverse environmental conditions, including temperature variations, moisture, humidity, air pollution, and salt crystallization. These factors contribute to physical, chemical, and biological deterioration mechanisms.

20.1.2.3 Historical significance

Heritage stone structures possess historical and cultural significance, making it essential to preserve their authenticity and original fabric. Interventions should be minimal and respectful of the original materials and construction techniques.

20.1.2.4 Structural stability

Deterioration of heritage stones can affect the structural stability of the entire building or monument. Understanding the condition of the stones is crucial for ensuring the safety and longevity of the structures.

20.1.3 Role of non-destructive techniques in heritage conservation

Many archaeological sites exhibit different types of weathering patterns due to variation in climate, geography, building materials, construction techniques, level of air pollution, internal structure of materials, and so on. The stone weathering which shows signs on the facade starts from core of the stone initially. Therefore it is necessary to study the internal structure behavior of the stone for detail diagnosis of damage. NDT methods such as XRD analysis, scanning electron microscope, and energy-dispersive X-ray spectroscopy (EDS) are used for this. Thin section with high quality and fluorescent epoxy impregnation are mainly suggested for finding the origin and provenance of the building materials in the historical monuments. When the stones possess fine graining or if stones have weathered significantly, thin section technology becomes extensively useful. Sometimes, thin sections become the only way to trace the originality of the stones. Durability of any building material depends on differences in open and closed porosities, pore size distribution, and interconnectivity between minerals (Umar et al., 2020). Microstructure and texture of any stone are basically different from each other and thus the durability also varies accordingly. Therefore the characterization of impact of stone microstructure on decay phenomenon is very important. When changes in micro structural, physical, and chemical properties are observed into stones, it leads to the complications in use of such stones for engineering purposes (Patil et al., 2022; Tenconi et al., 2018). Thus non-destructive techniques play a vital role in heritage conservation by offering valuable insights into the condition and deterioration processes of heritage stones. These techniques have several advantages that are defined in the following subsections.

20.1.3.1 Preservation of integrity

Non-destructive techniques allow for the evaluation of heritage stones without causing additional damage or altering the original fabric. These techniques are employed to evaluate the structural integrity of heritage buildings, monuments, and archaeological sites. Ground-penetrating radar (GPR) and seismic tomography are used to detect subsurface features, soil conditions, and potential structural weaknesses. UPV and rebound hammer tests help determine the soundness of masonry and concrete structures. These assessments aid in identifying areas requiring reinforcement or stabilization to prevent collapse or further deterioration.

20.1.3.2 Early detection of deterioration

Non-destructive techniques enable the identification and assessment of early signs of deterioration, allowing for timely intervention and conservation measures. By detecting problems at an early stage, the extent of damage and associated restoration costs can be minimized. non-destructive techniques allow experts to assess the condition of heritage structures without altering their original form. Techniques such as visual inspection, photography, and thermography help identify the visible signs of deterioration, cracks, weathering, and structural instability. These assessments provide crucial information for planning conservation strategies and prioritizing restoration efforts.

20.1.3.3 Comprehensive evaluation

Non-destructive techniques provide a holistic approach to assess different aspects of deterioration, including physical, mechanical, chemical, and mineralogical properties of the stones. By combining multiple techniques, a comprehensive evaluation of heritage stone deterioration can be achieved. non-destructive techniques are used to analyze the materials used in heritage structures, which helps in understanding their composition, properties, and deterioration mechanisms. Techniques such as X-ray fluorescence (XRF), XRD, and SEM enable experts to identify the presence of specific elements, minerals, or compounds. This information guides conservationists in selecting appropriate conservation materials and methods.

20.1.3.4 Conservation planning and decision-making

The data obtained from non-destructive techniques assist conservation professionals in making informed decisions regarding appropriate conservation strategies, materials selection, and maintenance practices. These techniques provide a scientific basis for prioritizing interventions and monitoring the effectiveness of conservation measures. non-destructive techniques enable conservationists to evaluate the effectiveness of conservation treatments without causing harm. Techniques such as infrared thermography and UV fluorescence imaging help assess the penetration, distribution, and effectiveness of protective coatings and cleaning methods. This evaluation ensures that conservation interventions are properly executed and have the desired impact. non-destructive techniques have become indispensable tools in heritage conservation, enabling a thorough understanding of the condition and deterioration mechanisms of heritage stone structures. By utilizing these techniques, conservation experts can ensure the long-term preservation of these valuable assets and maintain our cultural heritage for future generations.

20.2 Ultrasonic pulse velocity

UPV is a non-destructive technique used to assess the condition of materials, including heritage stones. The principle behind UPV is based on the measurement of the time taken for an ultrasonic pulse to travel through a material. The velocity of the pulse is influenced by the material's elastic properties, density, and presence of defects such as cracks or voids. When the ultrasonic pulse is passed through the building stone, its velocity is measured and the strength and quality is checked accordingly. The UPV is measured by electroacoustic transducer with frequency range of 20−150 kHz. It contains electrical pulse generator, one pair of transducer, an amplifier, and electronic timing device. One transducer is kept in contact with stone surface on one side and, on other side, another transducer is attached to the surface. When ultrasonic pulse is generated from transducer on one side, it passes through the stone surface through known path (L) in the form of wave and receiving transducer receives these waves in the form of electronic pulse. Electronic timing circuit records the timing (T) of the wave. Then the UPV is measured as per following formula:

$$V(\text{ultrasonic pulse velocity}) = L(\text{length of sample})/T(\text{travel time})$$

Good quality and continuity of stones is expressed through higher UPV value. On the other hand, when the stone contains voids and cracks, its UPV value decreases. Amongst many of the influencing factors for UPV, average particle size is the significant one (Salvatici et al., 2020). The traveling distance of pulse should be greater than 10 times of average particle size of the stone. The fresh basalt values are between 4400 and 5600 m/s. UPV values were taken from 10 different places in both the monuments for basalt and laterite. Fig. 20.1 shows the results of UPV of basalt samples. UPV test for fresh laterite ranges around 2700–2800 m/s. Fig. 20.2 illustrates that only few samples of laterite were within the limit of the value of fresh sample. Most of the samples showed lower values comparatively. It can be concluded that the samples which had lower velocities might contain voids or cracks or discontinuities inside. These voids or cracks or discontinuities affect the strength of the material. Thus these samples were the representative of the building stones which had low strength.

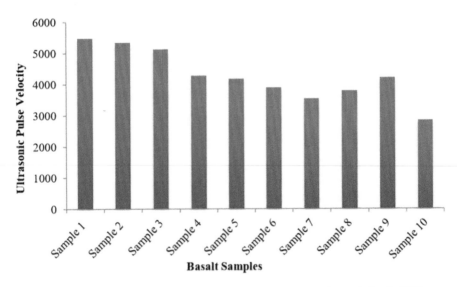

Figure 20.1 UPV graph for basalt samples 1–10. Graph shows the variation in UPV values of different Basalt samples. *UPV*, Ultrasonic pulse velocity.

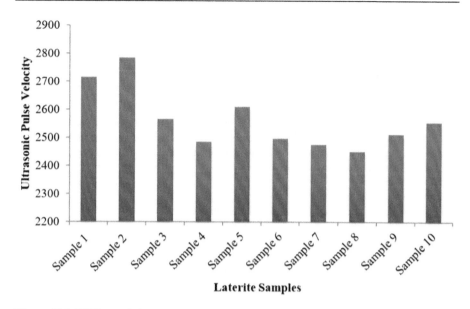

Figure 20.2 UPV graph for laterite samples 1−10. Graph shows the variation in UPV values of different Laterite samples. *UPV*, Ultrasonic pulse velocity.

The microstructure analysis provides insights about the factors that affect the physical properties of the stone (De la Fuente et al., 2013). Therefore, by understanding the microstructural characteristics, the behavior of a stone can be predicted. This is also significant in the prediction of intrinsic stone behavior which can cause failures. This process involves the determination of mineral compositions, stress behavior, and strength of the stone with aid of some laboratory analyses such as XRD and SEM.

20.2.1 Various applications of ultrasonic pulse velocity in the evaluation of heritage stone deterioration

- Structural integrity assessment: UPV can be used to assess the structural integrity of heritage stone structures by detecting areas with reduced velocity values, indicating potential structural weaknesses or deterioration.
- Detection of defects and deterioration: UPV is effective in detecting defects such as cracks, voids, and areas of decay within the stone. It helps in identifying areas that require targeted conservation interventions.
- Quality control in stone conservation: UPV can be used as a quality control tool during stone conservation interventions. It helps evaluate the effectiveness of repair or consolidation techniques by comparing pre- and posttreatment velocity values.
- Selection of stone for restoration: UPV measurements can aid in the selection of replacement stones during restoration projects. By comparing the velocity values of the existing stone with potential replacement stones, a suitable match in mechanical properties can be ensured.

20.2.2 Limitations and considerations of non-destructive techniques in heritage conservation

While non-destructive techniques such as UPV offer valuable insights into heritage stone deterioration, they have certain limitations and considerations that should be taken into account:

- Surface accessibility: non-destructive techniques often require direct access to the stone surface, which can be challenging for complex or delicate structures. The presence of coatings, vegetation, or surface irregularities may hinder accurate measurements.
- Limited depth of assessment: non-destructive techniques generally provide information about the surface or near-surface conditions of the stone. Deeper internal defects or deterioration may not be detectable using these techniques alone.
- Influence of environmental factors: Environmental factors, such as temperature, humidity, and presence of moisture, can affect the accuracy of non-destructive measurements. These factors should be considered and controlled during data collection to ensure reliable results.
- Operator skill and interpretation: Proper training and expertise are required for accurate data collection, analysis, and interpretation of non-destructive measurements. Skilled professionals with knowledge of both the techniques and heritage conservation principles should be involved in the evaluation process.

Despite these limitations, non-destructive techniques, including UPV, offer significant advantages in the evaluation of heritage stone deterioration. When used in combination with other techniques, they contribute to a comprehensive understanding of the stone's condition, aiding in the development of effective conservation strategies.

20.3 Rebound hammer test

The rebound hammer test is a widely used non-destructive technique for assessing the surface hardness and strength of materials, including heritage stones. The principle behind the rebound hammer test is based on the rebound of a hammer mass after it strikes the surface of the material under investigation. The rebound value is indicative of the surface hardness and, to some extent, the compressive strength of the material. The rebound value obtained from the test is used to estimate the surface hardness and, to some extent, the compressive strength of the heritage stone. The interpretation and analysis of rebound hammer test data involve the following aspects:

- Comparison with reference values: The rebound value is compared with reference values or ranges for the specific stone type to assess its relative hardness and strength. Deviations from the reference values indicate variations in stone condition and potential areas of deterioration.
- Trend analysis: Repeated measurements over time at the same test points can reveal trends in surface hardness and provide insight into the progression of stone deterioration.

Significant changes in rebound values may indicate the need for further investigation or conservation interventions.

- Correlation with other parameters: The rebound value can be correlated with other material properties, such as density, porosity, or moisture content, to gain a more comprehensive understanding of the stone's condition and deterioration mechanisms.

20.3.1 Applications in heritage stone evaluation

The rebound hammer test has several applications in the evaluation of heritage stone deterioration:

- Surface hardness assessment: The rebound value serves as an indicator of the surface hardness and relative strength of the stone. It helps identify areas of varying hardness and potential deterioration.
- Detection of deterioration: Significant deviations in rebound values compared to reference values may indicate areas of deteriorated or weakened stone surfaces. It assists in identifying localized deterioration and planning targeted conservation interventions.
- Quality control in stone conservation: The rebound hammer test can be used as a quality control tool during conservation interventions. It helps assess the effectiveness of cleaning, consolidation, or repair measures by comparing pre- and posttreatment rebound values.
- Site mapping and prioritization: The rebound hammer test can be used to create maps of surface hardness across heritage stone structures. These maps aid in prioritizing conservation efforts and identifying areas requiring immediate attention.

20.3.2 Limitations and considerations

The rebound hammer test, like other non-destructive techniques, has certain limitations and considerations:

- Surface sensitivity: The rebound hammer test primarily assesses the surface hardness of the stone. It may not provide information about internal defects, structural stability, or deeper deterioration.
- Surface preparation: The accuracy of rebound hammer test results relies on proper surface preparation. Irregularities, coatings, or surface treatments can affect the rebound values and require careful consideration during testing.
- Influence of surface texture: The rebound value can be influenced by the surface texture of the stone. Rough or porous surfaces may yield lower rebound values compared to smooth surfaces.
- Calibration and reference values: The rebound hammer requires calibration using reference materials specific to the stone type being evaluated. Accurate calibration and reference values are essential for meaningful interpretation of results.
- Operator skill and interpretation: The rebound hammer test requires skilled operators who are trained in proper usage, calibration, and interpretation of results. Operator proficiency plays a crucial role in obtaining reliable data.
- Complementary techniques: The rebound hammer test should be used in conjunction with other non-destructive techniques to obtain a comprehensive evaluation of heritage stone

deterioration. Different techniques provide complementary information about different aspects of stone condition.

20.4   X-ray diffraction analysis

XRD is a non-destructive technique used to analyze the crystallographic structure of materials, including heritage stones. XRD provides information about the mineral composition, crystalline phases, and crystallographic orientation present in the stone (Mohamed Aly Abdelhamid et al., 2020). The principle behind XRD is based on the diffraction of X-rays by the crystal lattice of the sample. Similar to XRF analysis, in this method, an X-ray is emitted and allowed to pass through the crystalline sample. When the X-ray goes through each element of the sample, it is deflected. Through this deflection, elements are identified from the sample. Due to frequent changes in environmental conditions and increasing pollution levels at microlevel, it becomes very important to recognize the weathering processes and measures to control them in minerals and rocks (Emara & Korany, 2016). And it is important to illustrate microscopically the weathering reactions which include changes in mineral and chemical structure which are further responsible for mineral breakdown. For detail information of structure of rock and minerals and chemical composition, XRD analysis was done.

20.4.1   Interpretation and analysis

XRD analysis provides information about the mineral composition, crystalline phases, and crystallographic properties of heritage stones. The interpretation and analysis of XRD data involve the following aspects:

- Mineral identification: XRD analysis allows for the identification of mineral phases present in the stone. By comparing the diffraction pattern of the sample with reference patterns, mineral phases can be identified and quantified.
- Crystalline structure: XRD provides information about the crystallographic structure of the minerals. It reveals details about the lattice parameters, crystal symmetry, and crystal orientation within the sample.
- Quantitative analysis: XRD data can be used for quantitative analysis, such as determining the relative abundance of different mineral phases or estimating the degree of amorphous content in the stone.
- Phase transformation: XRD analysis can detect phase transformations or mineralogical changes occurring in the stone due to weathering, aging, or conservation treatments. It helps in understanding the stability of the stone over time.

20.4.2   Applications in heritage stone evaluation

XRD has several applications in the evaluation of heritage stone deterioration:

- Mineralogical analysis: XRD analysis helps identify the mineralogical composition of heritage stones. It provides valuable information about the primary constituents, trace minerals, and potential alteration products.

- Weathering assessment: XRD analysis assists in assessing the degree of weathering or alteration in the stone. Changes in mineral phases or the formation of new minerals can indicate the progression of deterioration processes.
- Compatibility of repair materials: XRD analysis aids in assessing the compatibility of repair materials with the original stone. It helps identify whether the new materials have similar mineralogical characteristics to the existing stone, ensuring long-term compatibility and stability.
- Conservation monitoring: XRD analysis can be used to monitor the effectiveness of conservation treatments. It helps evaluate the impact of treatments on the mineralogical composition and the stability of the stone.

Fig. 20.3 shows the XRD analysis of the basalt samples 1−5 and Fig. 20.4 shows the XRD analysis for laterite samples. The XRD analysis revealed that the minerals present in the samples were pyroxene, plagioclase, olivine, chlorite, zeolite, hematite, and kaolinite. When CO_2 reacts with water, it produces weak carbonic acid. Zeolite minerals present in rock are dissolved by carbonic acid. And thus it reduces the stiffness and strength of the rock (Rodriguez-Navarro et al., 2003). This carbonic acid when it reacts with plagioclase, it forms kaolinite and carbonate ions. From XRD graphs, the presence of kaolinite is seen, which is the result of weathering. When olivine undergoes oxidation, it forms free ions as well as silicic acid. When pyroxene weathers, it produces some clay minerals as well as some Fe oxide minerals. Weathering of pyroxene forms cleavage cracks and formation of noncrystalline materials (Wilson, 2004). In the hydrothermal environment, pyroxene is altered to chlorite. In the weathering process, hematite is formed due to weathering and dehydration reactions of free iron oxides or oxyhydroxides.

P—Pyroxene, L—plagioclase, O—olivine , C—chlorite, K—kaolinite, Z—zeolite, H—hematite, and Q—quartz

Kaolinite was ubiquitous in all laterite samples. The presence of Fe-rich minerals, namely, goethite and hematite, in major quantities was confirmed in laterite. Goethite and hematite are necessary minerals in laterite. The rocks are formed at high temperature and pressure under the Earth's surface. When the rocks are exposed on the Earth's surface at comparatively low temperature and pressure, they become unstable. The presence of CO_2, water, and oxygen accelerates the changes in basic structure of the stone, and this is how the actual weathering process begins. This weathering process starts with the reaction of weathering agents with the minerals in the stone. This may be the resultant of the chemical weathering. Minerals which are formed during this process are found on the rock surface and they may be reacting with other minerals and leading to the morphological changes of the stone. More the surface of the stone is uneven, more are the chances of the biological weathering at such places. These surfaces are the attracted by lichen, algae, and other plants also. The acid generated by plants reacts with the stone surfaces and finally leads to the chemical weathering.

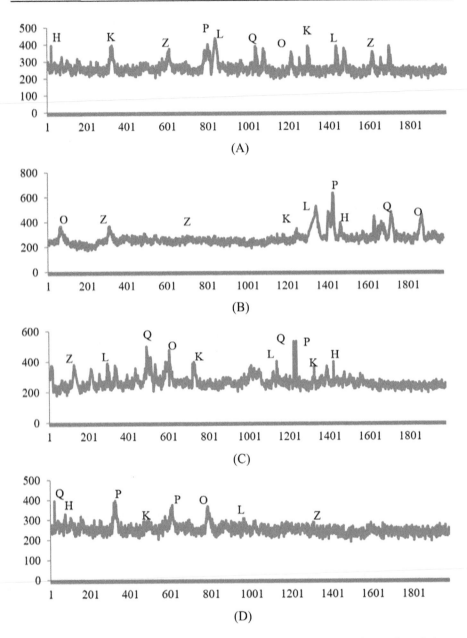

Figure 20.3 XRD graphs for Basalt samples 1-4. Graphs showing various minerals in Basalt samples. (A) XRD graph for basalt sample 1, (B) XRD graph for basalt sample 2, (C) XRD graph for basalt sample 3, (D) XRD graph for basalt sample 4. *XRD*, X-ray diffraction.

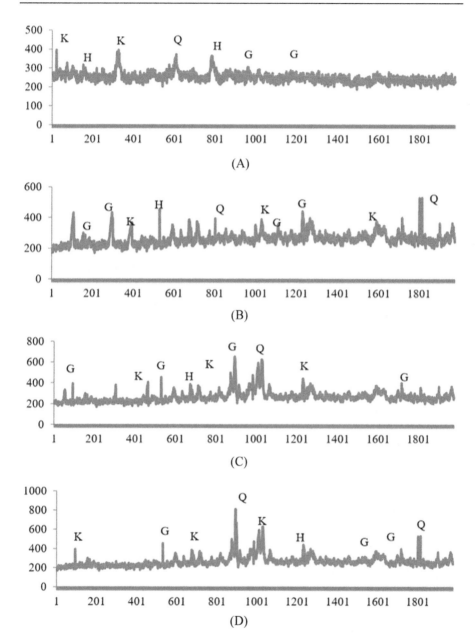

Figure 20.4 XRD graphs for laterite samples 1-4 . Graphs showing various minerals in laterite samples. (A) XRD graph for laterite sample 1, (B) XRD graph for laterite sample 2, (C) XRD graph for laterite sample 3, (D) XRD graph for laterite sample 4. *XRD*, X-ray diffraction.

20.4.3 Limitations and considerations

XRD analysis, like other non-destructive techniques, has certain limitations and considerations:

- Sample preparation: XRD analysis requires powdered samples, which may alter the original state of the stone. The grinding process may affect the texture and introduce potential sampling biases.
- Surface sensitivity: XRD analysis provides information about the bulk mineralogical composition and may not capture variations in mineralogy at the surface or near-surface layers.
- Detection limits: XRD analysis may have limitations in detecting minor phases or low-level crystalline components present in the stone. The sensitivity of the technique depends on factors such as sample quality, instrument capability, and data analysis methods.
- Complementary techniques: XRD should be used in conjunction with other non-destructive techniques to obtain a comprehensive evaluation of heritage stone deterioration. Different techniques provide complementary information about different aspects of stone condition.
- Conservation ethics: XRD analysis may require sample collection, which should be done following ethical considerations, with minimal impact on the integrity of the heritage structure.

Despite these limitations, XRD analysis is a valuable tool in the evaluation of heritage stone deterioration. It provides crucial information about the mineralogical composition and crystallographic properties of the stone, aiding in understanding its condition, deterioration mechanisms, and conservation requirements.

20.5   SEM analysis

SEM is a powerful imaging technique used in the evaluation of heritage stone deterioration. SEM allows for high-resolution imaging of the stone's surface and provides detailed information about its microstructure. The principle behind SEM is based on the interaction between an electron beam and the sample's surface. For identification of surface topography, chemical analysis and identification of minerals, and microanalysis in stone samples, scanning electron microscope analysis was done. In SEM analysis, a focused beam of electrons is passed through the sample; these electrons are interacted with the atoms from sample. And various signals are produced; these signals are then converted into high magnification images containing surface topography and sample composition.

20.5.1   Interpretation and analysis

SEM analysis provides detailed information about the microstructure, composition, and surface morphology of heritage stones. The interpretation and analysis of SEM data involve the following aspects:

- Microstructural analysis: SEM images reveal the microstructural features of the stone, such as grain boundaries, mineral phases, porosity, and the presence of cracks or defects. It helps in understanding the stone's formation, fabric, and potential degradation mechanisms.

- Surface morphology: SEM images provide high-resolution views of the stone's surface, allowing the examination of surface textures, weathering patterns, roughness, and alterations. It aids in identifying specific weathering features, such as dissolution, exfoliation, or biological colonization.
- Elemental analysis: SEM coupled with EDS systems allows for elemental analysis. It provides information about the chemical composition and distribution of elements within the stone. Elemental mapping helps identify specific minerals, contaminants, or alteration products.

20.5.2   Applications in heritage stone evaluation

SEM has various applications in the evaluation of heritage stone deterioration:

- Identification of deterioration mechanisms: SEM analysis helps identify the microstructural features associated with deterioration processes, such as salt crystallization, freeze-thaw cycles, or chemical weathering. It aids in understanding the underlying mechanisms of stone decay.
- Characterization of alteration products: SEM analysis allows for the identification and characterization of alteration products formed on the stone's surface. It helps in evaluating the effectiveness of conservation treatments and monitoring the progression of deterioration.
- Assessment of surface weathering: SEM images reveal the surface morphology and texture of weathered stones. It aids in assessing the extent and severity of surface degradation, such as erosion or disintegration.
- Conservation planning: SEM analysis provides valuable information for conservation planning and decision-making. It assists in selecting appropriate conservation treatments, understanding the compatibility of repair materials, and monitoring the effectiveness of conservation interventions.

SEM images of basalt samples are presented in Fig. 20.5. SEM analysis reveals morphological changes in the basalt stone. Basically, basalt is very compact and fine-grained rock. But, fine particles with irregular shapes are found in the basalt structure, which indicates that the particles have already separated out from the parent rock. It was clear from SEM images that the particles were irregular in shape and were dissimilar in size; fine particles of clay were in scattered form and could be seen in Fig. 20.5A and B. Fig. 20.5A shows the presence of albite on the stone surface, which is sodium-rich mineral from feldspar group and occurs in silicates only. This indicated that sodium and silica were reacting with the weathering agents and due to this formation of albite was found on the rock surface. As this albite was very unstable, it started reacting with other minerals and thus the separation of minerals started. Successive layers were seen around the mineral grains with small silica particles (Fig. 20.5C). Fig. 20.5D shows rough surface and angular shape particles with small amount of illite. This illite was probably the effect of plagioclases weathering. Fig. 20.5F and H shows the formation of minerals such as hematite and illite on stone surfaces. Hematite was formed due to free iron oxides, which in turn changed the color of the stone (Zhao et al., 2015). Illite was formed in the stone

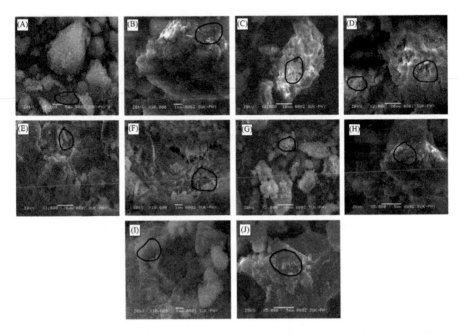

Figure 20.5 SEM images of basalt samples 1 − 10. SEM analysis reveals morphological changes and presence of various minerals.

when the weathering occurred in hygrothermal environments. Fig. 20.5E−G shows some elongated crystals. These were mainly calcite crystals and were rich in Fe and Ca. when the basalt reacted with CO_2 in atmosphere; it formed carbonic acid as well as released Fe and Ca (Guha Roy et al., 2016). Particles of kaolinite forming the matrix were found in Fig. 20.5H. Formation of detritus minerals was observed through Fig. 20.5I and J. This formed due to weathering on the surface of parent rock. Thus, through these images, it can be concluded that there were changes in shape and forms of minerals due to weathering. As more organisms are adhered on surface of the stone, bioreceptivity in future can be related to these types of morphological differences. The organic acid which is found mainly in plants, when it reacts with the major constituents of pyroxene and olivine from basalt rock, crystals of smaller size are formed and parts of original grains become amorphous. Fe, Mg, and other major constituents such as Ca and Al of parent rock are then released in large amount. Fig. 20.5G and I shows the formation of minerals such as albite, hematite, and plagioclase. Here also the cleavage planes were seen very clearly. Thus it indicated that the stone was having compact structure as compared to other stones. The cleavage planes were seen clearly in Fig. 20.5J. There were detritus minerals present on the cleavage surfaces, which were formed due to the weathering of the stone due to weathering agents (Dobiszewska & Beycioğlu, 2020). Fig. 20.6 shows the SEM analysis of fresh basalt adapted from literature. It was observed that fresh sample had fine, angular, and compact grains. Cleavage planes

Figure 20.6 SEM image of fresh basalt sample (Parthasarathy et al., 2019). Image shows fine, angular, and compact grains of basalt.

were observed clearly without any deposition. Detritus minerals were absent in the image. Under the Earth's surface, formation of plagioclase takes place at both high temperatures and at low temperatures. These two types of plagioclase have different characteristics. Plagioclases formed at high temperatures are more disordered and are considerably less stable as compared to others and are readily impacted by moisture and weathering. Plagioclases which are formed at lower temperatures exhibit ordered crystal lattice. Sericite and kaolinite are generated due to thermal changes and under the metamorphic conditions. Some other minerals such as albite, zoisite, epidote, quartz, and actinolite are formed. During hygrothermal processes, the high-temperature plagioclases are replaced by plagioclases formed under low temperature. Fig. 20.7 shows the SEM images for laterite samples. Presence of quartz grains in Fe-rich matrix of laterite is clear from SEM micrographs. The matrix observed here is fully porous in nature and predominated by tortuous cracks. It was observed that there was formation of white layers on the stone surfaces. Also the formation of sodium aluminosilicate was visible. The white layer formation indicated the separation of minerals from parent stone. The sodium aluminosilicate hydrate is responsible for the strength gain properties (Latifi et al., 2014). As this was separating out from the stone, it caused a decrease in strength properties of stone. It was found that iron exists in various forms in the matrix. Presence of heavy minerals such as hematite and goethite are primarily responsible for hardness and strength. Fibrous structure of clayey portions (Fig. 20.7J) is the probable cause of high porosity and water absorption. Fig. 20.8 shows the SEM image of fresh laterite (Kasthurba et al., 2007). It was observed that the fresh sample was free from

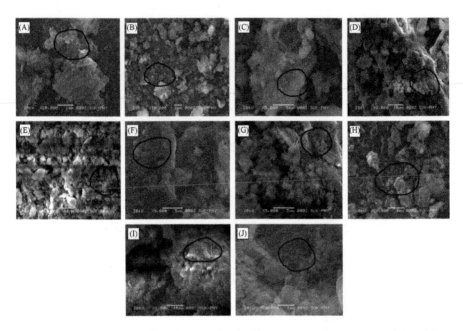

Figure 20.7 SEM images of laterite samples 1–10. Presence of quartz grains in Fe-rich matrix of laterite is clear from SEM micrographs.

Figure 20.8 SEM image of fresh laterite sample (Kasthurba et al., 2007). Image is free from deposition.

any deposition of minerals. The structure of the rock was observed sharp and clear. It was the indication of good strength and low porosity. The laterite used in monument was in the less polluted area, therefore there were no evidence of black crust formation on the stone surfaces, but there were some depositions seen on the surfaces, which may have occurred due to high velocity wind. Such active component tends to deposit between the grains of the substrate forming noncontinuous films of irregular morphology (Karatasios et al., 2017).

20.5.3 Limitations and considerations

While SEM is a powerful tool in heritage stone evaluation, it has certain limitations and considerations:

- Sample preparation: SEM analysis requires carefully prepared samples, including cutting, sectioning, and coating. Sample preparation techniques can potentially alter the original state of the stone and may introduce artifacts.
- Localized analysis: SEM provides information about the analyzed area only and may not represent the overall condition of the entire stone structure. Multiple samples or locations may need to be analyzed for comprehensive assessment.
- Sample size: The size of the sample that can be analyzed in the SEM chamber is limited. Large or bulky stone samples may require specialized sample preparation techniques or alternative imaging methods.
- Time and cost: SEM analysis is a time-consuming and costly technique. It requires specialized equipment, skilled operators, and significant data analysis and interpretation.
- Destructive sampling: SEM analysis often requires the extraction of samples from the stone, which can be considered a destructive method. Careful consideration should be given to the sampling strategy, ensuring minimal impact on the integrity of the heritage structure.
- Complementary techniques: SEM should be used in conjunction with other non-destructive techniques to obtain a comprehensive evaluation of heritage stone deterioration. Different techniques provide complementary information about different aspects of stone condition.

Despite these limitations, SEM analysis plays a crucial role in understanding the microstructural features and deterioration mechanisms of heritage stones. It assists in formulating effective conservation strategies and aids in the preservation of these valuable cultural assets.

20.6 Conclusion

Non-destructive techniques play a crucial role in the evaluation and conservation of heritage stone structures. This comprehensive approach allows for a detailed assessment of stone deterioration, identification of underlying mechanisms, and informed decision-making regarding conservation strategies. The integration of multiple techniques, such as UPV, rebound hammer test, SEM, XRD, and emerging technologies, provides a holistic understanding of the stone's condition, facilitating effective preservation efforts.

20.6.1   Summary of key findings

Throughout this chapter, various non-destructive techniques and their application in heritage stone evaluation have been explored. Following are the key findings:

1. Non-destructive techniques, such as UPV, rebound hammer test, SEM, and XRD, provide valuable insights into different aspects of heritage stone deterioration, including mechanical properties, surface hardness, microstructural features, mineralogical composition, and weathering patterns.
2. The combination of multiple non-destructive techniques allows for a more comprehensive evaluation, as each technique provides unique information that complements the others. This integrated approach enhances the accuracy and reliability of the assessment.
3. Advanced technologies and emerging techniques, such as remote sensing, hyperspectral imaging, GPR, and terahertz imaging, offer new opportunities for evaluating heritage stone structures with improved efficiency and precision.

20.6.2   Recommendations for future research

Establishing standardized protocols, calibration procedures, and data interpretation guidelines for non-destructive techniques in heritage conservation is essential. This ensures consistency and comparability of results across different studies and allows for better data integration and analysis. Further exploration of AI and machine learning algorithms in non-destructive techniques can enhance data analysis, interpretation, and decision support systems for heritage conservation. This includes developing predictive models, automated data processing, and real-time monitoring algorithms. By addressing these research areas, the field of non-destructive techniques in heritage stone evaluation can continue to evolve, providing more accurate, efficient, and reliable methods for the conservation and preservation of our valuable cultural heritage.

References

Bozda, A., İ, İ., Bozda, A., Ergün Hat, M., Bahad, M., & Korkanç, M. (2019). An assessment of deterioration in cultural heritage: The unique case of Eflatunpınar Hittite Water Monument in Konya, Turkey. *Bulletin of Engineering Geology and the Environment. (Darga 1992).*

De la Fuente, D., Vega, J. M., Viejo, F., Díaz, I., & Morcillo, M. (2013). Mapping air pollution effects on atmospheric degradation of cultural heritage. *Journal of Cultural Heritage, 14*(2), 138−145. Available from https://doi.org/10.1016/j.culher.2012.05.002.

Dobiszewska, M., & Beycioğlu, A. (2020). Physical properties and microstructure of concrete with waste basalt powder addition. *Materials, 13*(16). Available from https://doi.org/10.3390/MA13163503.

Emara, A.-A. S., & Korany, M. S. (2016). An analytical study of building materials and deterioration factors of Farasan Heritage Houses, and the recommendations of conservation and rehabilitation (German House Case Study. *Procedia - Social and Behavioral Sciences, 216*(October 2015), 561−569. Available from https://doi.org/10.1016/j.sbspro.2015.12.021.

Fitzner, B.(1997). *Diagnosis of weathering damage on stone monuments* (pp. 21–28)

Guha Roy, D., Vishal, V., & Nath Singh, T. (2016). Effect of carbon dioxide sequestration on the mechanical properties of Deccan basalt. *Environmental Earth Sciences, 75*(9). Available from https://doi.org/10.1007/s12665-016-5587-4.

Karatasios, I., Michalopoulou, A., Amenta, M., & Kilikoglou, V. (2017). Modification of water transport properties of porous building stones caused by polymerization of silicon-based consolidation products. *Pure and Applied Chemistry, 89*(11), 1673–1684. Available from https://doi.org/10.1515/pac-2016-1104.

Kasthurba, A. K., Santhanam, M., & Mathews, M. S. (2007). Investigation of laterite stones for building purpose from Malabar region, Kerala state, SW India - Part 1: Field studies and profile characterisation. *Construction and Building Materials, 21*(1), 73–82. Available from https://doi.org/10.1016/j.conbuildmat.2005.07.006.

Latifi, N., Eisazadeh, A., & Marto, A. (2014). Strength behavior and microstructural characteristics of tropical laterite soil treated with sodium silicate-based liquid stabilizer. *Environmental Earth Sciences, 72*(1), 91–98. Available from https://doi.org/10.1007/s12665-013-2939-1.

Mohamed Aly Abdelhamid, M., Li, D., Ren, G., & Zhang, C. (2020). Estimating deterioration rate of some carbonate rocks used as building materials under repeated frost damage process, China. *Advances in Materials Science and Engineering, 2020.* Available from https://doi.org/10.1155/2020/3826128.

Padilla-ceniceros, R., & Pacheco-martínez, J. (2017). Rock deterioration in the masonry walls of the Cathedral Basilica of Aguascalientes, Mexico. *Revista Mexicana De Ciencias Geológicas, 34*(2), 138–149.

Parthasarathy, G., Pandey, O. P., Sreedhar, B., Sharma, M., Priyanka, T., & Nimisha, V. (2019). First observation of microspherule from the infratrappean Gondwana sediments below Killari region of Deccan LIP, Maharashtra (India) and possible implications. *Geoscience Frontiers, 10*(6), 2281–2285. Available from https://doi.org/10.1016/j.gsf.2019.04.005.

Patil, S. M., & Kasthurba, A. K. (2020). Weathering of stone monuments: Damage assessment of basalt and laterite. *Materials Today: Proceedings, 43*(part 2), 1647–1658. Available from https://doi.org/10.1016/j.matpr.2020.10.022.

Patil, S. M., Kasthurba, A. K., & Patil, M. V. (2022). Damage assessment through petrographic and microscopic studies of stone monuments. *Journal of Building Pathology and Rehabilitation, 7*(1). Available from https://doi.org/10.1007/s41024-022-00223-9.

Patil, S. M., Kasthurba, A. K., & Patil, M. V. (2021). Characterization and assessment of stone deterioration on Heritage Buildings. *Case Studies in Construction Materials, 15*, e00696. Available from https://doi.org/10.1016/j.cscm.2021.e00696.

Rodriguez-Navarro, C., Rodriguez-Gallego, M., Ben Chekroun, K., & Teresa Gonzalez-Muñoz, M. (2003). Conservation of ornamental stone by *Myxococcus xanthus*-induced carbonate biomineralization. *Applied and Environmental Microbiology, 69*(4), 2182–2193. Available from https://doi.org/10.1128/AEM.69.4.2182-2193.2003.

Salvatici, T., Calandra, S., Centauro, I., Pecchioni, E., Intrieri, E., & Alberto Garzonio, C. (2020). monitoring and evaluation of sandstone decay adopting non-destructive techniques: On-site application on building stones. *Heritage, 3*(4), 1287–1301.

Tenconi, M., Karatasios, I., Bala'awi, F., & Kilikoglou, V. (2018). Technological and microstructural characterization of mortars and plasters from the Roman site of Qasr Azraq, in Jordan. *Journal of Cultural Heritage, 33*(2017), 100–116. Available from https://doi.org/10.1016/j.culher.2018.03.005.

Umar, M. U., Hanafi, M. H., & Abdul Latip, N. (2020). Analysis of non-destructive testing of historic building structures analysis of non-destructive testing of historic building structures. *Australian Journal of Basic and Applied Sciences*, *9*(February 2015), 326–330.

Wilson, M. J. (2004). Weathering of the primary rock-forming minerals: Processes, products and rates. *Clay Minerals*, *39*(3), 233–266. Available from https://doi.org/10.1180/0009855043930133.

Zhao, D. F., Liao, X. W., & Yin, D. D. (2015). An experimental study for the effect of CO2-brine-rock interaction on reservoir physical properties. *Journal of the Energy Institute*, *88*(1), 27–35. Available from https://doi.org/10.1016/j.joei.2014.05.001.

Evaluation of the mechanical characteristics of marble using non-destructive techniques: Ultrasound versus Schmidt hammer rebound tests

21

Sáez-Pérez Maria Paz[1], Durán-Suárez Jorge A.[2], Rodríguez-Gordillo José[3], Castro-Gomes João[4], and Di Benedetto Giacomo[5]

[1]Building Constructions Department, University of Granada, Granada, Spain, [2]Sculpture Department, University of Granada, Granada, Spain, [3]Mineralogy and Petrology Department, University of Granada, Granada, Spain, [4]Civil Engineering and Architecture Department, University of Beira Interior, Covilha, Portugal, [5]Enginlife Engineering Solutions, Rome, Italy

21.1 Introduction

As part of the overall cultural heritage, architectural heritage plays a key role in generating new spaces and consolidating the development of our cities. It is essential that the built heritage is well preserved, both for its use and enjoyment, and to generate tourist attractions. This will entail greater development and exert a great influence on society in general as it will lead to an improvement in the life of its inhabitants, as well as economic and cultural aspects (Borri & Corradi, 2019; Du et al., 2016). Currently, one of the main challenges in the protection of our heritage, inevitably linked to sustainable development, demands a new interpretation in which protection and conservation actions come together (Huynh et al., 2020; Rodrigues et al., 2018; Ruggiero et al., 2021), to recognize the different heritage values attributed to buildings (Chaves Moreno et al., 2017; Mardones, 2019; Rey et al., 2018; Sampaio et al., 2021), helping to protect the environment and generate the sustainability of our cities (Onecha et al., 2021; Salman & Hmood, 2019). Within this context, it is implied that the conservation of cultural heritage and its enhancement immediately contributes to the fulfilment of Sustainable Development Goals, a transversal generator of objectives related to the safety and sustainability of cities. This slows down environmental degradation, generates sustainable and nonpolluting energy, and encourages alliances to achieve the objectives (UNESCO, 2022); all involved in the promotion of peaceful and inclusive societies (Hosagrahar, 2022). In this respect, the European Union

Diagnosis of Heritage Buildings by Non-Destructive Techniques. DOI: https://doi.org/10.1016/B978-0-443-16001-1.00021-8

proposed the sustainable renovation of architectural heritage buildings as one of its main strategic options (Europe EU, 2022) for the 21st century.

In recent decades, the issue of property conservation in a heritage context has aroused greater interest, due to the existence and increase in the number of buildings with a marked historical-artistic character (Theocharis Katrakazis et al., 2018). Its recovery and intervention requires adapting the buildings to new uses and conditions in compliance with new regulations, to which is added action on structures and construction systems of the first order, to avoid damage and environmental actions that may limit or make the assets disappear. Without being able to forget the need to maintain the image and achieve conservation without alterations. In this line of action, interventions for conservation and maintenance are also part of the life cycle of buildings in general, but even more so for heritage buildings (Balayssac & Garnier, 2017; Ramesh et al., 2021; Yi et al., 2013), thus preventing the advance of deterioration and the control of aging processes (Drobiec et al., 2021; Nowogońska, 2020).

Intervention in the heritage field requires specific knowledge about the buildings and the necessary study into evaluation techniques. In this sense, historical studies include those related to the characterization of materials (knowing their nature and properties), the state and level of degradation and environmental conditions, the state of stress to which they are subjected, and their foreseeable evolution. To achieve this, it is necessary to apply experimental techniques, including sampling. However, due to the cultural value, the number of samples is considerably reduced, which requires the use of non-destructive techniques (NDTs) and may be the only way of analyzing the building in many cases. As Menéndez (2016) confirms, the need to preserve the most valuable objects in the most intact way possible has made the use of this type of technique mandatory, even during the construction phase, which allows us to know the suitability of the improvements implemented, as well as compliance with construction and regulatory requirements (Deepak et al., 2021). There are limitations to destructive testing and the less representative results need to be contrasted with a greater number of tests, to allow an adequate level of knowledge and evaluation in the shortest possible time (Diaferio, 2022). Over the years, one of the main concerns in the application of these techniques has been trying to improve their use for diagnostic and follow-up purposes, through quantitative assessment, to try to achieve complete reliability, for which it is necessary to use more techniques (Tavukçuoğlu et al., 2018). Highlighting the importance of non-destructive tests as relevant techniques in the study of built heritage has been investigated (El Masri & Rakha, 2020; Grzyb et al., 2022; Lourenço, 2013; Ruggiero et al., 2021; Tejedor et al., 2022; Umar, Hanafi, Latip 2015; Umar, Hanafi, Latip 2015).

With respect to natural materials, the many studies that have been carried out confirm that their behavior derives from the formation conditions, of which their composition, mineralogical and petrological disposition fundamentally refer, and to which their anisotropy is added (Fort et al., 2011). From the extraction conditions that affect initial exposure and are implemented with the environmental conditions that may occur during its use, the alteration that may occur is derived, to a greater or lesser degree, throughout its useful life. Therefore, knowing the evolution and how it develops once put into service are key aspects for guaranteeing the capacity of the material, its behavior, and durability. The fundamental issues to consider when establishing the intervention with different considerations are structural, constructive, energetic, and aesthetic.

For verification, accelerated aging tests are carried out, which allow us to know and evaluate the incidence that the change of conditions causes in the material (Andriani & Germinario, 2014; Boudani et al., 2015; Rodríguez-Gordillo & Sáez-Pérez, 2006); this type of test is used to improve the design of the regulations (Bellopede et al., 2016).

In relation to the damage most commonly evaluated through NDTs, those that stand out are more important to the capacity and functionality of buildings. These include those that affect the structural behavior (discontinuities, defects, fissures, and cracks) and hygrothermal behavior (humidity conditions, temperature, thermal inertia, and so on), which are recognized through the modification of the results of their physical properties (Bramanti & Bozzi, 2001).

In line with these comments, the correlation of techniques is the most suitable solution for the evaluation required in heritage buildings through NDT with multiple (existing) investigations, in various materials and case studies (Kouddane et al., 2022; Samson & Moses, 2014; Theodoridou et al., 2012). At present, other properties are correlated, see the studies of Forestieri et al. (2017) and Fort et al. (2022) for examples.

From a review of the existing bibliography, the ultrasound and Hammer sclerometer techniques stand out as the most common NDTs in the study and analysis of structural behavior, due to their ease of use and low cost. The variation of the properties of the original material is the main evidence for the presence of deterioration or defects.

The importance of these studies means that they are carried out through projects of international importance. This is the case of the WARMEST project (WARMEST PROJECT, 2020), in which this research is included. Its main purpose is integrating different detection, analysis, and evaluation methodologies, which allow the development of evolutionary models. These are currently recognized as being a great challenge to the research community and reveal the validity and capacity of NDT at any stage. There is a global need to try to find the best solution to the problem of studying and conserving heritage in the most effective way, in the shortest possible time and at a reasonable cost. This supposes the implication of different techniques and disciplines that in a rigorous way allow to predict the behavior of the materials to establish the measures that in order of priority must be applied. The project provides a necessary approach for the use and integration of different methods and NDTs, with algorithms and numerical developments. A prototype is used to provide the monitoring and evaluation of heritage sites, helping managers and intervention teams to carry out the most suitable proposals in each case, favoring the improvement of cultural and tourist destinations (Prados-Peña et al., 2021).

Based on the earlier discussion, the present investigation was carried out in one of the most emblematic heritage places worldwide, The Alhambra in Granada, and with one of the most known material, Macael marble. The objective focused on determining the correlation that exists between the destructive (compression strength) and non-destructive mechanical resistance determination techniques (through the parameters of ultrasonic waves and Schmidt hammer rebound hardness) in the diagnosis of deterioration of the Macael marble. This will help to evaluate the conservation interventions that can be proposed as part of the strategic planning at the monument scale, allowing evaluation, management, protection, and sustainable development actions to be carried out on built cultural heritage.

The methodological process followed during this research has been carried out according to the scheme shown in Fig. 21.1. In the first stage, previous studies were carried out which, based on previous research, revealed the existing problems and, consequently, the need to study in detail the behavior of the material in the heritage environment under study. Then, the fieldwork was carried out. The data relating to

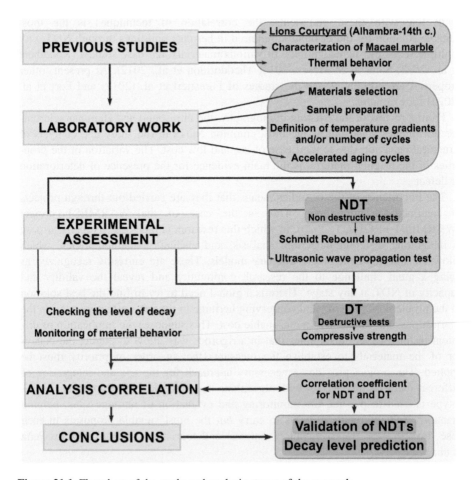

Figure 21.1 Flowchart of the study and analysis stages of the research.

the number of specimens, thermal gradients to be applied in the accelerated ageing tests, and the number of cycles to be addressed in this study were decided. The experimental evaluation was carried out in two groups, the first for non-destructive tests (Schmidt rebound hammer test, ultrasonic wave propagation test [UPVs]) and the second for destructive tests (compressive strength test). Once the tests were completed, the results were evaluated to establish the correlation coefficient between the three tests performed. Finally, in the conclusions, the techniques' suitability and the correlation's validity were established.

21.2   Materials and methods

21.2.1   Materials

The present investigation, as commented, focused on so-called Macael white marble, which is considered to be one of the most outstanding stone materials of cultural heritage worldwide and particularly in Spain (Navarro et al., 2017). It was first recognized between 2700 and 2000 years BCE (Siret & Siret, 2006) in decorative objects and jewelry, the period of its greatest expansion being the Muslim period (Carretero-Gómez, 1995). It was used in the construction of widely known buildings, such as the Alcazaba de Almería (11th century), the Medina Azahara Palace in Córdoba (10th century), and the Monastery of San Lorenzo de El Escorial (16th century). Numerous cathedrals have also incorporated this material. In the 20th century, it was used in the Metropolitan Museum of New York (1913) and the Monument to the Spanish Constitution (1978). Currently, at an international level, White Macael marble is present on five continents, although its penetration into the United States of America (USA), Asia, and the United Arab Emirates stands out. It is being considered as a possible world heritage stone resource (Navarro et al., 2019).

The most representative site is probably the Alhambra in Granada (12th−14th centuries), which recognizes the use of this material in various locations in its Nasrid palaces, notably the colonnade and fountain of the Patio de Los Leones (14th century) and some of its rooms. This is in addition to the Palace of Comares (14th century) and its later use in the Palace of Carlos V (S. XVI), whilst under Christian rule (Navarro et al., 2013).

From a geological point of view, Macael marble is part of the so-called Nevado-Filábride complex, which is located in the lowest part of the Internal Zones of the Bética mountain range (Egeler, 1964). According to Navarro et al. (2017), the mountainous complex is subdivided into different units by tectonic criteria, which Martín-Algarra et al. (2004) designate as being upper and lower. The Macael quarries are located in the upper unit. The formation processes were metamorphic with a low thermal gradient and high pressure, followed by others with higher thermal gradients (Navarro et al., 2019). The formation period of the materials that make up this complex extends from the Precambrian to the Jurassic. Regarding the age of the metamorphism, it places fundamentally during the Alpine Orogeny (Upper Cretaceous-Miocene). Therefore the resulting material is carbonate rocks of Upper

Triassic age, which comprise alternating types of mica schists (calcareous, quartz-itic, quartzitic with garnet, and quartzitic mica schists with amphibole).

Focusing on the Macael marble used in the present investigation, mineralogical characterization testing, obtained from previous publications (Sáez-Pérez, 2004), confirm that it is mainly composed of calcite with small amounts of quartz and iso-lated muscovite and feldspar crystals. It presents a medium-type granulometry between 0.16 and 3.20 mm. Visually, a homogeneous appearance with a mosaic texture predominates; it is of the granoblastic type that varies from equigranular to heterogranular, presenting grains of different sizes, varying from large to medium-fine. White and gray tones alternate to form bands, with preferential directions in the extracted blocks. Fig. 21.2 shows an image of one of the quarries in Macael, where the study material can be identified as having white and greyish banding. Of the physical properties, porosity stands out: it is of the open type, between 0.1% and 0.6%, and apparent density values oscillate between 2.50 and 2.75 g/cm³.

Regarding its morphology, Macael marble is composed mainly of calcite crystals that show an anisotropic behavior from a thermal point of view. As confirmed by vari-ous investigations, (Rodríguez-Gordillo & Sáez-Pérez, 2006, 2010; Sáez-Pérez & Rodríguez-Gordillo, 2009), the crystals expand parallel to the direction of the c axis with the increase in temperature and contract in the perpendicular direction, which gen-erates an opening of the contact limits between grains. If the situation continues over time and temperatures increase, the effects of expansion increase, which causes greater sensitivity to environmental conditions. This behavior is considered the main responsible for the deterioration of the marble, which requires studies that allow knowing the level of affection, and the reduction of its qualities to carry out the decision-making that allows its intervention.

With respect to the categorization of White Macael, different commercial varie-ties are recognized, depending on the area from which they are extracted and the marketing company. Some of the best known varieties are: Blanco Macael Río, Blanco Macael Puntilla, Blanco Macael Polonia, Blanco Macael Cañaílla, and Blanco Macael Tranco (Navarro et al., 2019).

Figure 21.2 Quarry image showing white marble fronts and greyish banded fronts.

To investigate the behavior of the purest material, for this study the quarry material considered to be white (with practically zero greyish banding) was analyzed; Table 21.1 shows the chemical composition.

As previously indicated, the main component is CaO, followed by MgO, SiO_2, and Al_2O_3.

Prior to carrying out the tests, and to adapt the material to the applicable regulations, the chosen marble was cut into $15 \times 5 \times 5$ cm prisms.

As indicated in the objectives proposed in this research based on previous research (Boudani et al., 2015; Rodríguez-Gordillo & Sáez-Pérez, 2006), it is essential to carry out a study of the material both in its natural state and subjected to accelerated aging cycles, which specifically allow for detecting the incidence of these in the behavior of the material, as well as in the correlation of the properties of the object of study, that is, different cycles and aging conditions. Moreover, previous studies (Rodríguez-Gordillo & Sáez-Pérez, 2006, 2010; Sáez-Pérez & Rodríguez-Gordillo, 2009) revealed that, to study the effects of the meteorological conditions and thermal gradients, prior to carrying out the different tests, the material was subjected to the thermal cycles. In this research, the design of the cycles, based on previous research (Rodríguez-Gordillo & Sáez-Pérez, 2006; Sáez-Pérez & Rodríguez-Gordillo, 2009), was proposed to emulate the daily temporal cycle of the material when it is exposed to outdoor atmospheric conditions. In this case, both the number of cycles and the maximum and minimum temperatures were increased, to achieve a more immediate effect. The first phase of the study of gradients, addressed in this publication, focused on temperature ranges between $-20°C$ and $100°C$ and for a maximum of 50 cycles analyzed in bundles of 10 cycles. What it means to work with four groups of material under different conditions. The data related to the cycles and temperatures of each group are shown in Table 21.2.

In all cases, including the material in its natural state, the number of specimens was 25 for each group, with a total of 100 specimens tested. The measurements of ultrasonic propagation speed, rebound value, and mechanical resistance until breakage were made according to the direction of greatest entity of the specimen (z direction or height), coinciding with the direction of maximum load in which the material was available.

The heating/cooling cycles were carried out by keeping the specimens in a convection heated oven and a freezer, respectively. For the freezing phase, a GGM Gastro International GmbH Freezing Chamber (model SKU TC2126) was used. For the heat aging process, a natural HD-E804 convection oven (from HAIDA EQUIPMENT) was used.

Table 21.1 Chemical composition of the Macael marble by XRF analysis (wt.%). Data are normalized to 100% (LOI-free).

CaO	SiO$_2$	Al$_2$O$_3$	MgO	Na$_2$O	K$_2$O	Fe$_2$O$_3$	TiO$_2$	P$_2$O$_3$	LOI
56.10	0.27	0.08	0.75	0.02	0.02	0.06	<0.01	<0.01	42.1

Table 21.2 Name of the groups of test pieces, number of cycles and environmental conditions of the accelerated aging tests (heating/freezing) applied.

Sample groups	Number of daily cycles	Tª MÁX/hours	Tª MÍN/hours
Group-G1	0	Laboratory conditions. $T = 22°C \pm 1°C$ and RH $= 37\% \pm 5\%$	Laboratory conditions
Group-G2	50	50°C/10 H	−20°C/10 H
Group-G3	50	75°C/10 H	−20°C/10 H
Group-G4	50	100°C/10 H	−20°C/10 H

In the other 4 hours, to complete the 24 daily hours, it was spent 2 hours changing the cycle from maximum temperature to minimum temperature and the remaining 2 hours for the opposite change.

The specimens were tested at the same time so that they were carried out under the same laboratory conditions, at a temperature of $22°C \pm 1°C$ and relative humidity of $37\% \pm 5\%$.

21.3   Methods

21.3.1   Ultrasound propagation velocity test

The measurement of the ultrasonic wave velocity test was carried out by the direct transmission method. For this, two sensors were placed at each end of the sample, at a constant pressure; one of them was the emitter and the other, the receiver. The material used to achieve the proper coupling between the probe and the transducers was a viscous crystalline Quick Eco ecogel.

The ultrasound equipment used was an STEINKAMP BP V with a low-power wave generator and high-power emission and reception transducers, capable of generating 50- and 100-KHz waves. Given the material and the size of the blocks, 100-KHz frequency waves were used. During its implementation, the recommendations of (ISO 16823:2012, 2012) were followed.

The measurements were made on untreated material, as well as after each of the heating/cooling cycles was described. Each measurement was repeated three times, thus guaranteeing the reproducibility of the test, as well as the results obtained.

V_p was calculated according to the following formula:

$$V_p = L/t$$

where L is the length of the sample (distance between the transducers), in meters, and t is the average transit time of the ultrasound signal, in seconds, for each of the three measurements made at the same location.

21.3.1.1   Schmidt rebound hammer tests

To measure the rebound index test, a Rock Schmidt RS8000 Schmidt rebound hammer was used, which is recommended for rock testing applications. Prior to carrying out the test, it was verified that the surface of the specimen was smooth, clean, and dry. Six rebounds were made for each specimen, in a direction perpendicular to the measurement plane, at a distance from the edge >10 mm. The data corresponding to each tube tested were obtained from the arithmetic mean of all the measurements carried out on each specimen, eliminating anomalous data. During its performance, the recommendations of the standard (ASTM Stand. International: West, 2001) were followed.

As in the previous section, the test was carried out on untreated material, as well as after each of the heating/cooling cycles described.

21.3.1.2 Compression strength

To carry out the compression strength test, an Ibertest Eurotest MD2 universal testing machine was used. This equipment has charge characteristics that can be varied from 2 to 2000 kN and offers a continuous record of stress (MPa) and deformation (mm/m), until the sample block breaks. The speed applied during the test was 5 mm/min. The results obtained were determined for each specimen tested in the different groups. During its implementation, the recommendations of the standard were followed (UNE-EN, 2007).

21.4   Results and discussion

21.4.1   Results

21.4.1.1   Non-destructive techniques

The results obtained by carrying out the ultrasonic wave velocity measurement and rebound value for the different groups of specimens are presented in Tables 21.3 and 21.4 and Figs. 21.3 and 21.4.

To facilitate clarity of reading, in the tables of results and the comments, the percentages of loss only with whole numbers have been reported.

21.4.1.2   Ultrasonic wave propagation tests

In the results obtained during the ultrasonic wave measurement test, it is observed that, in all of the groups, there is a variable decrease, depending on the number of cycles and conditions to which they were subjected. For the first 10 cycles, it is verified that the reduction oscillates between 5% for the specimens subjected to 50/ − 20°C (G2) cycles, 30% in the specimens subjected to 100/ − 20°C (G4) cycles (an intermediate value), and 19% for Group 3 (G3).

In the rest of the cycles (20, 30, 40, and 50 cycles), the reduction is different. In the group G2, the cycles of 10 and 20 lost 5% and 9%, respectively, expanding to 15% in 30 cycles and 23% in 40 cycles. In this group, a final reduction in speed of 33% was observed in the maximum number of cycles applied.

Table 21.3 Average velocity of the ultrasonic wave velocity (m/s) and standard deviation for the different batches of 10 cycles in the three groups of specimens subjected to the aging tests. Percentage loss occurred between the batch of "0 cycles" and each of the following batches ("20 cycles," "30 cycles," "40 cycles," and "50 cycles").

Number of cycles	G1 without cycles	G2 (50/ − 20°C)		G3 (75/ − 20°C)		G4 (100/ − 20°C)	
	V_p (m/s) + σ	V_p (m/s) + σ	Loss of value (%)	V_p (m/s) + σ	Loss of value (%)	V_p (m/s) + σ	Loss of value (%)
0	5660 ± 18.32	5666 ± 20		5635 ± 19		5675 ± 15	
10		5360 ± 61	5%	4580 ± 159	19%	3972 ± 166	30%
20		5154 ± 81	9%	4334 ± 98	23%	3650 ± 93.71	36%
30		4837 ± 120	15%	3986 ± 109	29%	3331 ± 103	41%
40		4342 ± 194	23%	3570 ± 178	37%	2871 ± 111	49%
50		3770 ± 178	33%	3014 ± 145	47%	2531 ± 96	55%

Table 21.4 Rebound value and standard deviation for the different batches of 10 cycles in the three groups of specimens subjected to aging tests; percentage loss occurred between the batch of "0 cycles" and each of the following batches: "20 cycles," "30 cycles," "40 cycles," and "50 cycles").

Number of cycles	G1 without cycles	G2 (50/ − 20°C)		G3 (75/ − 20°C)		G4 (100/ − 20°C)	
	Rebound + σ	Rebound + σ	Loss of value (%)	Rebound + σ	Loss of value (%)	Rebound + σ	Loss of value (%)
0	36.05 ± 0.05	36.10 ± 0.06		36.00 ± 0.07		36.00 ± 0.07	
10		35.06 ± 0.61	3%	34.41 ± 1.25	4%	32.20 ± 2.80	11%
20		33.62 ± 0.42	7%	30.74 ± 0.79	15%	26.92 ± 0.92	25%
30		31.31 ± 0.80	13%	25.92 ± 1.22	28%	23.92 ± 0.38	34%
40		27.87 ± 0.99	23%	22.56 ± 0.95	37%	21.43 ± 0.80	40%
50		24.34 ± 1.13	24%	19.81 ± 1.07	45%	18.77 ± 1.04	48%

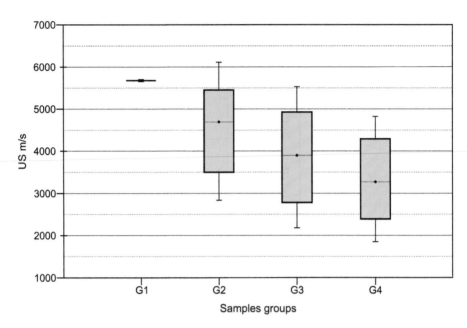

Figure 21.3 Velocity of ultrasonic waves for the four groups of specimens, classified according to the conditions of the aging test applied (G1: "0 cycles"; G2: 50 cycles' − 20/50°C; G3: 50 cycles' − 20/75°C; and G4: "50 cycles" − 20/100°C). The maximum, mean and minimum values, as well as the standard deviations, are presented for each group.

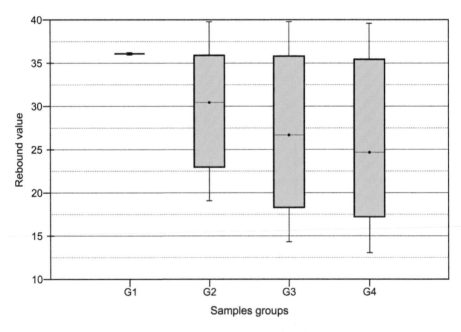

Figure 21.4 Rebound value of the Schmidt hammer rebound test for the four groups of test pieces, classified according to the aging test conditions applied (G1: "0 cycles"; G2: 50 cycles " − 20/50°C; G3: 50 cycles " − 20/75°C; and G4: 50 cycles " − 20/100°C). The maximum, average and minimum values, as well as the resulting standard deviations, are represented for each group.

In Group 3 (G3), a higher reduction was observed from the first round of cycles, the difference being 19% when only 10 cycles were applied, progressively increasing the difference, as in the previous group, but with a greater percentage of incidence. The differences reach a maximum of 47% reduction between the material subjected to 0 cycles and the material subjected to 50 cycles.

Finally, in Group 4 (G4), after the high reduction verified in the first round of cycles (30%), it was observed that the variation was less in the following cycles; it increased by 6% between 10 and 20 cycles and similar percentages between the following consecutive batches. Regarding the material without cycles, which was subjected to the maximum gradient applied in the study, the ultrasonic wave velocity values decreased by 55% after 50 cycles.

Fig. 21.3 shows the results graphically, in which the loss of velocity occurred as the number of cycles in each group of test pieces was observed. In this case, the speed reduction between groups subjected to cycles is similar, the variation being between 12% and 14%.

The evolution of the measured values shows that the number of cycles is relevant to the results, observing important differences, especially in the G2 and G3 groups (21% and 33%, respectively).

The values are in line with those verified for similar studies (Rodríguez-Gordillo & Sáez-Pérez, 2006, 2010), in which the reduction of the ultrasound speed is confirmed according to the condition of the material which, in this case, referred to different batches and temperature gradients.

21.4.1.3   Schmidt hammer rebound

Regarding the results obtained in the Schmidt hammer rebound test, it should be noted that a reduction in values is observed, as in the previous test, although in this case it is more evident because the intensity of the aging test progresses (groups) and depends on the number of cycles to which the specimens have been subjected.

It is confirmed that, in Group 2 (G2), the reduction at the end of the test was 24%; the reduction was 45% in Group 3 (G3), and 48% in Group 4 (G4), which implies a condition by type of aging conditions applied.

If the behavior for the different batches is analyzed, the results for the first 10 cycles show a reduction of 3% in G2, 4% in G3, and 11% in G4. The evolution of the different batches of cycles in each group shows the decrease in the rebound value as the number of cycles progresses.

Fig. 21.4 graphically shows the results, in which the loss of rebound value can be observed, as the number of cycles in each group of test tubes increases. The graph also shows a greater difference in the resulting values between the G2 and G3 groups than the difference observed between the G3 and G4 groups. This further reinforces the impact that aging conditions have on the properties of the material.

Considering the average values of ultrasound and rebound index, it can be seen that the values obtained in the ultrasound testing range from 5660 m/s in G1, with 0 cycles, and the minimum value obtained was 2531 m/s. G4, after being subjected to 50 accelerated aging cycles, assumed a total decrease of approximately 55%. In the case of the rebound value test, starting from a maximum value of 36.05 for G1 with 0 cycles, a minimum value of 18.77 is reached for G4 in 50 cycles. This represents a maximum difference of 48%, which is lower than the percentage obtained for the ultrasound test.

Comparing the results obtained in the non-destructive tests for the different groups, a similar evolution is confirmed in all cases. There is a reduction as a consequence of the number of cycles and the intensity of the aging tests determined by each group. The percentage reduction of the values in both tests does not occur in the same way, being more intense in the ultrasonic wave velocity test; however, in both cases, a reduction in the results is evident.

21.4.1.4 Compressive strength test

The results of the destructive testing, through compression strength of the specimens, are shown in Table 21.5 and Fig. 21.5.

In all cases, the same trend as that for non-destructive testing is confirmed, that is, there was a reduction of the mechanical resistance of the specimens in all the groups subjected to aging cycles.

The batches of cycles to which the test pieces were subjected indicate a reduction in compressive strength of 19%, in the case of Group 2 (G2), 46% in the case of Group 3 (G3) and 51% for Group 4 (G4).

By analyzing the data for each batch of cycles, it was found that (for the first 10 cycles) the reduction in compressive strength in Group 2 (G2) is exclusively 4%, but 9% in Group 3 (G3) and 35% in Group 4 (G4). These differences are considerably reduced in the following batches: 1% at 20 cycles, 8% at 30 cycles, 13% at 40 cycles, and 19% at 50 cycles. In the case of the G3 group, the differences between 0 cycles and each of the subsequent batches increased, being 22% at 20 cycles, 29% at 30 cycles, 36% at 40 cycles, and 66% at 50 cycles. Finally, for the G4 group, the reduction obtained confirms that, between 0 and 20 cycles, compressive strength decreased by 35%−43% at 30 cycles, 48% at 40 cycles, and 51% at 50 cycles.

The incidence was observed to be much more pronounced between the groups, G1−G2 and G2−G3 than between the two final groups G3 and G4, where the reduction observed is lower.

From all of the above, the same trend can be deduced in the results obtained for each group and number of cycles, as in the non-destructive tests. However, the percentages vary, with the difference/reduction obtained in the non-destructive tests being higher than in the destructive test.

Finally, if the results obtained in the different tests are compared, it is verified that they all maintain the same trend, obtaining lower speeds, rebound values, and compressive strength as the number of cycles increases. These results, and their trends, are in accordance with the findings of previous research (Rodríguez-Gordillo & Sáez-Pérez, 2010; Sáez-Pérez, 2004), observing differences in the absolute values which could be explained by the difference in the conditions of the accelerated aging tests.

Table 21.5 Compressive strength and standard deviation for the different batches of 10 cycles in the three groups of specimens subjected to the aging tests, percentage loss occurred between the batch of "0 cycles," and each of the following batches ("20 cycles," "30 cycles," "40 cycles," and "50 cycles").

Number of cycles	G1 Without cycles	G2 (50/ − 20°C)		G3 (75/ − 20°C)		G4 (100/ − 20°C)	
	MPa + σ	MPa + σ	Loss of value (%)	MPa + σ	Loss of value (%)	MPa + σ	Loss of value (%)
0	68.67 ± 0.04	69.00 ± 0.09		69.00 ± 0.16		69.00 ± 0.16	
10		67.13 ± 0.98	3%	66.27 ± 2.06	4%	62.65 ± 5.35	9%
20		65.86 ± 0.07	5%	60.11 ± 2.78	13%	50.45 ± 3.47	27%
30		64.46 ± 0.73	7%	51.95 ± 1.20	25%	42.80 ± 1.77	38%
40		61.37 ± 1.10	11%	46.90 ± 1.45	32%	37.29 ± 0.95	46%
50		57.21 ± 1.36	17%	41.02 ± 2.34	41%	34.86 ± 0.74	49%

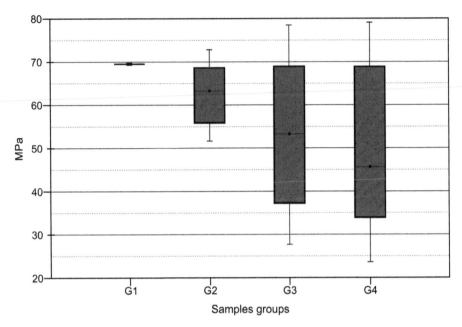

Figure 21.5 Compressive strength for the four groups of test pieces, classified according to the aging test conditions applied (G1: "0 cycles"; G2: 50 cycles " − 20/50°C; G3: 50 cycles " − 20/75°C; and G4: 50 cycles " − 20/100°C). The maximum, average and minimum values, as well as the resulting deviations, were represented for each group.

21.4.1.5 Correlation results

The correlation between each of the non-destructive tests and the destructive tests is shown in the following figures. Firstly, the correlation obtained between destructive and non-destructive tests is shown, followed by the correlation obtained between the non-destructive tests.

21.4.1.6 Correlation of velocity of ultrasonic waves and compressive strength

The correlation of the non-destructive testing, (speed of ultrasonic waves) and the destructive testing, (compressive strength) of the material under study, offering value of R^2 equal to 0.8747. In Fig. 21.6, the results are showed. This indicates a good correlation between both assays, as confirmed by previous research (Ciornei et al., 2004; Fort et al., 2013). This emphasizes that, in the case of the present investigation, this correlation is also valid for the material in which deterioration is recognized after being affected by heating-cooling cycles (50 cycles in this study). Less dispersion is observed, even at the lowest values.

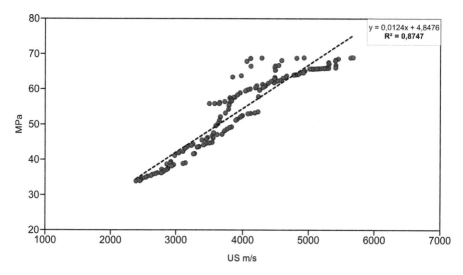

Figure 21.6 Ultrasonic velocity (V_p) versus compressive strength of the studied material. Results for all groups (G1, G2, G3, and G4).

21.4.1.7   Correlation of Schmidt hammer rebound and compressive strength

The second correlation made corresponds to the rebound values and compressive strength (see Fig. 21.7). This results in an R^2 value of 0.8819, which is 1% higher than the previous correlation, which offers greater reliability between the results of these two tests.

The graph shows the direct linear relationship that exists between the results obtained in both tests and confirms that the higher the rebound value, the greater the compressive strength of the marble studied. Likewise, the alteration offered by the material subjected to cycles is reflected in both tests in a similar way, the values obtained being reduced.

21.4.1.8   Correlation of ultrasonic velocity and Schmidt hammer rebound

Finally, the correlation between the two non-destructive tests is performed. The correlation established between these two tests is shown in Fig. 21.8. The correlation coefficient (R^2) offers a value of 0.8531, which, as in the previous correlations, indicates a strong relationship between the values obtained from the speed of ultrasonic waves (V_p) and the results of the rebound value (rebound number).

After analyzing the results, all the relationships presented R^2 values ≥ 0.85, indicating that the properties of the material are strongly interdependent; this coincides with the findings in the work of Aldeeky et al. (2020).

In this study, therefore, the correlations show high reliability in all cases, with R^2 being between the minimum value of 0.8531 (rebound value-ultrasonics velocity

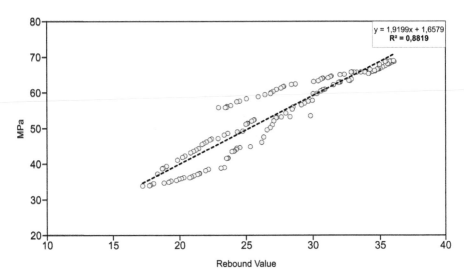

Figure 21.7 Rebound values (*R*) versus compressive strength of the studied material. Results for all groups (G1, G2, G3, and G4).

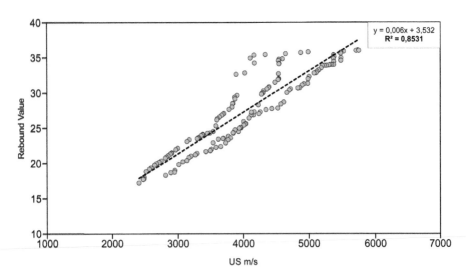

Figure 21.8 Ultrasonic velocity (V_p) versus rebound values (*R*). Results for all groups (G1, G2, G3, and G4).

correlation) and maximum value of 0.8819 (compressive strength-rebound value correlation). In all cases, it is observed that the material affected by the cycles had reduced values, compared to the material without cycles.

21.5   Discussion

From the test results and the correlations carried out, it is confirmed that these characteristics are related to the existence of alterations and discontinuities present in the material, even if they are not visible. In this regard, this material is especially sensitive to thermal changes, as confirmed in previous publications and studies (Rodríguez-Gordillo & Sáez-Pérez, 2006, 2010; Sáez-Pérez, 2004), when the expansion−contraction of the calcite crystals occurs, which are affected to a greater extent, once the material is subjected to continuous and high thermal stress. In this case, observation of the continuous reduction of loss of resistance, rebound value, and speed of ultrasonic waves as the number of cycles progressed showed the loss or decrease of these values being greater as the conditions of the aging test became more intense.

The same trend in the results, observed in the three tests carried out in the study, confirms the usefulness of NDTs in estimating defects or imperfections in the investigated material. The correlations obtained were high between the destructive and non-destructive tests, as well as between the different non-destructive tests.

In line with Benavente et al. (2022) and Fort et al. (2022), the results obtained confirm that the combination of the two ND tests carried out allows improvement of the evaluation of compressive strength when, as in the present study, each of the parameters (ultrasonic wave speed and rebound value) presents a similar correlation with said characteristics.

21.6   Conclusion

The research carried out deals with the analysis of the behavior of the Macael marble material after the application of 50 cycles of different conditions of thermal gradient. This type of study is decisive for the conservation of built heritage, by providing knowledge about its resistant capacity and offering information on its compressive strength, the state of deterioration of the material, and its evolution after the application of aging cycles.

The main conclusions are as follows:

1. Temperature cycling (with different temperatures and the number of repeats) has shown that there is a direct relationship between the temperature gradient and the loss of mechanical properties. Macael marble is recognized by the modification of the intergranular state, producing the separation of its limits because of thermal expansion.

2. The increase in the thermal gradient generates a greater incidence in the values of the tests carried out, observing the progressive reduction of the qualities, with respect to the material that has not been subjected to any gradient.

3. The ultrasonic pulse velocity (V_p) is reduced in all groups subjected to temperature gradients. This reduction increases as the number of cycles and the gradient to which the material is subjected increases. The reduction is not percentage equivalent to the number

of cycles or the temperature gradient, producing a similar progressive increase in the groups and gradients applied.

4. The rebound index repeats a behavior pattern similar to that commented for the ultrasonic pulse velocity, the values obtained being lower as the number of cycle's progresses and the temperature gradient increases. Only a percentage reduction in the quality of lesser entity is recognized than that observed in the case of the ultrasonic pulse velocity. This may be due to the fact that the rebound index analysis recognizes the behavior of the material for more superficial thicknesses.

5. The ultrasonic pulse velocity and the rebound index are considered an effective measure of the reduction in the resistant capacity of Macael marble when it is subjected to thermal cycles, which leads to an increase in the microcracking state of this material, as described has confirmed in previous research.

6. The reduction of the applied NDT values recognizes its equivalence in the loss of mechanical capacities as the temperature gradient increases, observing the same trend in the compressive strength test.

7. The application of thermal cycles in the studied material confirms once again the incidence of temperature gradients in the intergranular behavior as a consequence of thermal expansion (intergranular thermal expansion) even from the first 10 cycles.

8. The high correlation coefficient (R^2) obtained confirms the quality of the results between the techniques and, therefore, the usefulness of the NDT to evaluate the behavior of Macael marble.

9. The evaluation of P waves and rebound values have been chosen as properties associated with the microstructural characteristics of the stone material which, in turn, are directly related to mechanical resistance. This is confirmed by the results of mechanical resistance that were determined by NDTs and the high correlation value obtained >0.85.

10. When focusing on carrying out tests on the existing material in our heritage, it is important to take into account both the costs and the means for its realization. The suitability of NDT, which is guaranteed by this study, justifies the great interest shown in its application for carrying out tests "in situ."

11. Conclusive results have been obtained in the application of NDTs and the compression resistance test. The use of a large group of samples guarantees the results.

12. Assessing the conditions of deterioration and its foreseeable behavior using the NDT for the investigated marble offers relevant information to those who will have to manage the heritage and establish the need for actions on the specific site, known as the use at an international level that recognizes this material.

Acknowledgments

This research was supported by the REMINE Project Programme for Research and Innovation Horizon 2020 Marie Skłodowska-Curie Actions, Horizon 2020, WARMEST Project Research and Innovation Staff Exchange (RISE) H2020-MSCA-RISE-2017, RRRMaker project Marie Skłodowska-Curie Research and Innovation Staff Exchange, Precompetitive Research Projects Program, belonging to the Research and Transfer Plan of the University of Granada (PP2022.PP.27), and Scientific Unit of excellence "Ciencia en la Alhambra," ref. UCE-PP2018−01 (University of Granada) and was carried out under the auspices of Research Group RNM 0179 and HUM 629 of the Junta de Andalucía.

References

Aldeeky, H., Al Hattamleh, O., & Rababah, S. (2020). Assessing the uniaxial compressive strength and tangent Young's modulus of basalt rock using the Leeb rebound hardness test. *Materiales de Construcción, 70*(340), 230. Available from https://doi.org/10.3989/mc.2020.15119.

Andriani, G. F., & Germinario, L. (2014). Thermal decay of carbonate dimension stones: fabric, physical and mechanical changes. *Environmental Earth Sciences, 72*(7), 2523−2539. Available from https://doi.org/10.1007/s12665-014-3160-6, https://link.springer.com/journal/12665.

ASTM Stand. International: West. (2001). *ASTM D 5873−00 Standard test method for determination of rock hardness by rebound hammer method.*

Balayssac, J. P., & Garnier, V. (2017). *Non-destructive testing and evaluation of civil engineering structures* (pp. 1−356). France: Elsevier Inc. Available from http://www.sciencedirect.com/science/book/9781785482298.

Bellopede, R., Castelletto, E., Schouenborg, B., & Marini, P. (2016). Assessment of the European Standard for the determination of resistance of marble to thermal and moisture cycles: Recommendations for improvements. *Environmental Earth Sciences, 75*(11). Available from https://doi.org/10.1007/s12665-016-5748-5, http://www.springerlink.com/content/121380/.

Benavente, D., Martinez-Martinez, J., Galiana-Merino, J. J., Pla, C., de Jongh, M., & Garcia-Martinez, N. (2022). Estimation of uniaxial compressive strength and intrinsic permeability from ultrasounds in sedimentary stones used as heritage building materials. *Journal of Cultural Heritage, 55*, 346−355. Available from https://doi.org/10.1016/j.culher.2022.04.010, http://www.elsevier.com.

Borri, A., & Corradi, M. (2019). Architectural heritage: A discussion on conservation and safety. *Heritage, 2*(1), 631−647. Available from https://doi.org/10.3390/heritage2010041, https://www.mdpi.com/2571-9408/2/1/41/pdf.

Boudani, M. E., Wilkie-Chancellier, N., Martinez, L., Hébert, R., Rolland, O., Forst, S., Vergès-Belmin, V., & Serfaty, S. (2015). Marble characterization by ultrasonic methods. *Procedia Earth and Planetary Science, 15*, 249−256. Available from https://doi.org/10.1016/j.proeps.2015.08.061.

Bramanti, M., & Bozzi, E., (2001). *An ultrasound based technique for the detection and classification of flaws inside large sizes marble or stone structural elements.* International Cultural Heritage Informatics Meeting, (pp. 283−287).

Carretero-Gómez, A. (1995). *La industria del mármol en Almería. Servicio de Publicaciones de la.* Universidad de Almería, Almería.

Chaves Moreno, E.A., Pachon Garcia, P., Camara Perez, M., Compan, & Cardiel, V. (2017). *V Congreso Internacional sobre documentación, conservación y reutilización del patrimonio arquitectónico y paisajístico,* Granada Caracterización de propiedades dinámicas de edificios patrimoniales mediante análisis modal operacional.

Ciornei, N., Facaoaru, I., & Cetean, V. (2004). Non-destructive method for rapid "in situ" characterization of rocks. *Bulletin of the Geological Society of Greece, 36*(4), 1912. Available from https://doi.org/10.12681/bgsg.16674.

Deepak, J. R., Bupesh Raja, V. K., Srikanth, D., Surendran, H., & Nickolas, M. M. (2021). Non-destructive testing (NDT) techniques for low carbon steel welded joints: A review and experimental study. *Materials Today: Proceedings, 44*, 3732−3737, Elsevier Ltd. India. Available from https://doi.org/10.1016/j.matpr.2020.11.578, https://www.sciencedirect.com/journal/materials-today-proceedings.

Diaferio, M. (2022). Correlation curves for concrete strength assessment through non-destructive tests. *SSRN Electronic Journal*. Available from https://doi.org/10.2139/ssrn.4177623.

Drobiec, Ł., Grzyb, K., & Zając, J. (2021). Analysis of reasons for the structural collapse of historic buildings. *Sustainability*, *13*(18), 10058. Available from https://doi.org/10.3390/su131810058.

Du, F., Okazaki, K., Ochiai, C., & Kobayashi, H. (2016). Post-disaster building repair and retrofit in a disaster-prone historical village in China: A case study in Shangli, Sichuan. *International Journal of Disaster Risk Reduction*, *16*, 142−157. Available from https://doi.org/10.1016/j.ijdrr.2016.02.007, http://www.journals.elsevier.com/international-journal-of-disaster-risk-reduction/.

Egeler, C. G. (1964). On the tectonics of the eastern Betic Cordilleras (SE Spain. *Geologische Rundschau*, *53*(1), 260−269. Available from https://doi.org/10.1007/BF02040750.

El Masri, Y., & Rakha, T. (2020). A scoping review of non-destructive testing (NDT) techniques in building performance diagnostic inspections. *Construction and Building Materials*, *265*, 120542. Available from https://doi.org/10.1016/j.conbuildmat.2020.120542.

EUROPE EU (2022). https://eur-lex.europa.eu/legal-content/ES/TXT/HTML/?uri = CELEX: 52019DC0022&from = PT. 2022 11 3.

Forestieri, G., Tedesco, A., & Ponte, M. (2017). The stone in a monumental masonry building of the Tyrrhenian coast (Italy): New data on the relationship between stone properties and structural analysis. *Ge-Conservacion*, *1*(11), 102−109. Available from http://www.ge-iic.com/ojs/index.php/revista/article/download/459/798/.

Fort, R., Varas, M. J., Alvarez de Buergo, M., & Martin-Freire, D. (2011). Determination of anisotropy to enhance the durability of natural stone. *Journal of Geophysics and Engineering*, *8*(3), S132−S144. Available from https://doi.org/10.1088/1742-2132/8/3/s13.

Fort, R., Alvarez de Buergo, M., & Perez-Monserrat, E. M. (2013). Non-destructive testing for the assessment of granite decay in heritage structures compared to quarry stone. *International Journal of Rock Mechanics and Mining Sciences*, *61*, 296−305. Available from https://doi.org/10.1016/j.ijrmms.2012.12.048, http://www.elsevier.com/inca/publications/store/2/5/6/index.htt.

Fort, R., Feijoo, J., Varas−Muriel, M. J., Navacerrada, M. A., Barbero-Barrera, M. M., & De la Prida, D. (2022). Appraisal of non-destructive in situ techniques to determine moisture- and salt crystallization-induced damage in dolostones. *Journal of Building Engineering*, *53*, 104525. Available from https://doi.org/10.1016/j.jobe.2022.104525.

Grzyb, K., Drobiec, Ł., Blazy, J., & Zając, J. (2022). The use of NDT diagnostic methods and calculations in assessing the masonry tower crowned with the steel dome. *Materials*, *15*(20), 7196. Available from https://doi.org/10.3390/ma15207196.

Hosagrahar, J. (2022). *Culture at the heart of the SDGs*. The UNESCO Courier https://es.unesco.org/courier/april-june-2017/cultura-elemento-central-ods.

Huynh, T.-C., Nguyen, B.-P., Man Singh Pradhan, A., & Pham, Q.-Q. (2020). Vision-based inspection of bolted joints: Field evaluation on a historical truss bridge in Vietnam. *Vietnam Journal of Mechanics*. Available from https://doi.org/10.15625/0866-7136/15073.

ISO 16823:2012. (2012). *Non-destructive testing-Ultrasonic testing-Transmission technique*. 16823.

Kouddane, B., Sbartaï, Z. M., Alwash, M., Ali-Benyahia, K., Elachachi, S. M., Lamdouar, N., & Kenai, S. (2022). Assessment of concrete strength using the combination of NDT —Review and performance analysis. *Applied Sciences (Switzerland)*, *12*(23). Available from https://doi.org/10.3390/app122312190, http://www.mdpi.com/journal/applsci/.

Lourenço, P. B. (2013). Conservation of cultural heritage buildings: Methodology and application to case studies. *Revista ALCONPAT*, *3*(2), 98−110. Available from https://doi.org/10.21041/ra.v3i2.46.

Mardones, G. (2019). La Defensa del patrimonio arquitectónico. *Arquitextos*, *33*, 73−86. Available from https://doi.org/10.31381/arquitextos.v0i33.1861.

Martín-Algarra, A. C., Alonso-Chaves, F. M., Andreo, B., Azañón, J. M., Balanyá, J. C., Booth-Rea, G., Crespo-Blanc, A., Delgado, F., Federico., Estévez, A., Galindo-Zaldívar, J., García-Casco, A., García-Dueñas, V., Garrido, C. J., Gervilla, F., González-Lodeiro, F., Jabaloy, A., López-Garrido, A. C., Martín-Algarra, A., ... Vera, J. A. (2004). Geología de España. *SGE-IGME*, 395−444.

Menéndez, B. (2016). *Non-destructive techniques applied to monumental stone conservation*. InTech. Available from https://doi.org/10.5772/62408.

Navarro, R., Cruz, A. S., Arriaga, L., & Baltuille, J. M. (2013). White Macael marble: A key element in the architectonic heritage of Andalusia for over 25 centuries. *Geophysical Research Abstracts*, *15*, 2013−4252.

Navarro, R., Cruz, A. S., Arriaga, L., & Baltuille, J. M. (2017). Caracterización de los principales tipos de mármol extraídos en la comarca de Macael (Almería, sureste de España) y su importancia a lo largo de la historia. *Boletín geológico y minero*, *128*(2). Available from https://doi.org/10.21701/bolgeomin.128.2.005.

Navarro, R., Pereira, D., Cruz, A. S., & Carrillo, G. (2019). The significance of "white Macael" marble since ancient times: Characteristics of a candidate as global heritage stone resource. *Geoheritage*, *11*(1), 113−123. Available from https://doi.org/10.1007/s12371-017-0264-x, http://www.springer.com/earth + sciences/geology/journal/12371.

Nowogońska, B. (2020). Consequences of abandoning renovation: Case study—neglected industrial heritage building. *Sustainability*, *12*(16), 6441. Available from https://doi.org/10.3390/su12166441.

Onecha, B., Dotor, A., & Marmolejo-Duarte, C. (2021). Beyond cultural and historic values, sustainability as a new kind of value for historic buildings. *Sustainability*, *13*(15), 8248. Available from https://doi.org/10.3390/su13158248.

Prados-Peña, M. B., Sáez-Pérez, M. P., & Piernikowska, A. (2021). Heritage destination: Weaknesses and strengths based on the opinions of its visitors—Case Study: The Alhambra and Generalife. *Journal of Tourism and Hospitality Management*, *9*, 327−342.

Ramesh, G., Srinath, D., Ramya, D., & Vamshi Krishna, B. (2021). Repair, rehabilitation and retrofitting of reinforced concrete structures by using non-destructive testing methods. *Materials Today: Proceedings*. Available from https://doi.org/10.1016/j.matpr.2021.02.778.

Rey Rey, J., Vegas González, P., & Ruiz Carmona, J. (2018). Structural refurbishment strategies on industrial heritage buildings in Madrid: Recent examples. *Hormigón y Acero*, *69*(285), e27−e35. Available from https://doi.org/10.1016/j.hya.2018.05.001.

Rodrigues, F., Matos, R., Di Prizio, M., & Costa, A. (2018). Conservation level of residential buildings: Methodology evolution. *Construction and Building Materials*, *172*, 781−786. Available from https://doi.org/10.1016/j.conbuildmat.2018.03.129.

Rodríguez-Gordillo, J., & Sáez-Pérez, M. P. (2006). Effects of thermal changes on Macael marble: Experimental study. *Construction and Building Materials*, *20*(6), 355−365. Available from https://doi.org/10.1016/j.conbuildmat.2005.01.061.

Rodríguez-Gordillo, J., & Sáez-Pérez, M. P. (2010). Performance of Spanish white Macael marble exposed to narrow- and medium-range temperature cycling. *Materiales de Construcción*, *60*(297), 127−141. Available from https://doi.org/10.3989/mc.2010.44107.

Ruggiero, G., Marmo, R., & Nicolella, M. (2021). A methodological approach for assessing the safety of historic buildings' façades. *Sustainability*, *13*(5), 2812. Available from https://doi.org/10.3390/su13052812.

Sáez-Pérez, M. P. (2004). *Estudio de elementos arquitectónicos y composición de materiales del Patio de los Leones*. Interacciones en sus causas de deterioro. Servicio de publicaciones de la Universidad de Granada, Granada. Available from: http://hdl.handle.net/10481/4586.

Sáez-Pérez, M. P., & Rodríguez-Gordillo, J. (2009). Structural and compositional anisotropy in Macael marble (Spain) by ultrasonic, XRD and optical microscopy methods. *Construction and Building Materials*, *23*(6), 2121–2126. Available from https://doi.org/10.1016/j.conbuildmat.2008.10.013.

Salman, M., & Hmood, K. (2019). *"Challenges and paradigms of the contemporary city": CPSV conservation of historic districts: Challenges of integration in modern cities within sustainable perspective*. XIII CTV 2019 Proceedings: XIII International Conference on Virtual City and Territory, UPC. https://doi.org/10.5821/ctv.8530.

Sampaio, A. Z., Pinto, A. M., Gomes, A. M., & Sanchez-lite, A. (2021). Generation of an hbim library regarding a palace of the 19th century in Lisbon. *Applied Sciences (Switzerland)*, *11*(15). Available from https://doi.org/10.3390/app11157020, https://www.mdpi.com/2076-3417/11/15/7020/pdf.

Samson, D., & Moses, O. (2014). Correlation between Non-Destructive Testing (NDT) and Destructive Testing (DT) of Compressive Strength of Concrete. *International Journal of Engineering Science Invention*, *3*, 2319–6726.

Siret, E., & Siret, L. (2006) *Las primeras edades del metal en el sudeste de España. Álbum*. Edición Facsimilar. Dirección General de Cultura. Museo Arqueológico de Murcia. 182,

Tavukçuoğlu, A., Diouri, A., Boukhari, A., Ait Brahim, L., Bahi, L., Khachani, N., Saadi, M., Aride, J., & Nounah, A. (2018). Non-destructive testing for building diagnostics and monitoring: Experience achieved with case studies. *MATEC Web of Conferences*, *149*, 01015. Available from https://doi.org/10.1051/matecconf/201814901015.

Tejedor, B., Lucchi, E., Bienvenido-Huertas, D., & Nardi, I. (2022). Non-destructive techniques (NDT) for the diagnosis of heritage buildings: Traditional procedures and futures perspectives. *Energy and Buildings*, *263*, 112029. Available from https://doi.org/10.1016/j.enbuild.2022.112029.

Theocharis Katrakazis, T., Heritage, A., Dillon, C., Juvan, P., & Golfomitsou, S. (2018). Enhancing research impact in heritage conservation. *Studies in Conservation*, *63*(8), 450–465.

Theodoridou, M., Dagrain, F., & Ioannou, I. (2012). Correlation of stone properties using standardized methodologies and non-standardized micro-destructive techniques. *12th International Congress on the Deterioration and Conservation of Stone Columbia University*.

UNE-EN 2007. (2007). *1926 stone test methods - Determination of uniaxial compressive strength*.

Umar, U., Hanafi., & Latip. (2015). Analysis of non-destructive testing of historic building structures. *Australian Journal of Basic and Applied Sciences*, *9*(7), 326–330.

Umar, U., Hanafi., & Latip, A. A. (2015). Strengthening of historic buildings through structural repair works: review of the methods and process. *Australian Journal of Basic and Applied Sciences*, *9*(7), 358–362.

UNESCO (2022). https://es.unesco.org/sdgs. 2022 12 8.

WARMEST PROJECT (2020). https://warmestproject.eu/. 2023 1 10.

Yi, J. H., Park, W. S., Lee, S. Y., Huynh, T. C., Kim, J. T., & Seo, C. K. (2013). Evaluation of vibration characteristics of an existing harbor caisson structure using tugboat impact tests and modal analysis. *International Journal of Distributed Sensor Networks* (2013). Available from https://doi.org/10.1155/2013/806482.

Applications of the ground-penetrating radar technique to heritage buildings: Case studies and combination with other non-destructive testing

22

Mercedes Solla[1], Vega Pérez-Gracia[2], Susana Lagüela[3], and Simona Fontul[4]

[1]CINTECX, GeoTECH Research Group, Universidade de Vigo, Vigo, Spain, [2]Dpt Resistencia de Materials i Estructures a l'Enginyeria, RMEE, Universitat Politècnica de Catalunya UPC-Barcelona Tech, Barcelona, Spain, [3]Department of Cartographic and Terrain Engineering, Universidad de Salamanca, Ávila, Spain, [4]Transportation Department, LNEC - National Laboratory for Civil Engineering, Lisbon, Portugal

22.1 Introduction

The determination of the conservation state of heritage buildings requires the performance of a high number of measurements, to determine both the state of the structures and materials at their surface, as well as at their interior. The characteristics of interest can be measured in a direct way in the case of modern buildings, but the performance of these measurements in a heritage building presents a series of setbacks that complicate the process. As such, heritage buildings are unique elements, so that their diagnosis must be made without any level of destruction. In many cases, the condition and the materials of the heritage asset cannot even support contact, so measurements should be made from a distance. What is more, many heritage buildings have complex geometries, with difficult access areas, where the entrance of heavy or large equipment is not possible.

For these reasons, non-destructive testing (NDT) techniques are currently the solution for the diagnosis of heritage buildings, since their application allows to measure the geometry, and also the thermophysical and electrical conditions of the materials in the interior of the structure, in such a way that it is possible to infer the different layers of materials present in the structure, as well as their conservation state, or the existence and type of interior and surface defects, such as cracks, voids, corrosion, and presence of moisture.

Among all NDT methods, geophysical methods are those applied for the study of the interior of the heritage buildings, due to their capacity to measure through

Diagnosis of Heritage Buildings by Non-Destructive Techniques. DOI: https://doi.org/10.1016/B978-0-443-16001-1.00022-X

the materials without incurring any damage. There are different geophysical techniques, based on the measurement of different physical properties (e.g., electromagnetic waves, electric potential, and elastic waves) on the surface of the medium, which are affected by the distribution of different physical parameters (e.g., conductivity, resistivity, elastic parameters, and density) inside the medium. It is therefore highly recommended to combine different geophysical methods, based on different parameters, to achieve more reliable and accurate interpretations.

Regarding other NDT techniques, the geospatial techniques have also wide application in heritage buildings, due to their capacity to measure the 3D geometry without contact and from a distance, in an accurate and agile manner. The knowledge of the geometry is important for the description of the construction, but it also allows for the detection and characterization of pathologies that affect the structure of the building, such as cracks and voids, variations in thickness, and level of deformation. Thus, provided that geospatial techniques can measure the surface, their combination with geophysical techniques is key for the integral characterization of heritage buildings, including the state of the interior and the exterior of the construction components.

However, in some cases, the performance of direct measurements is needed, especially in heritage buildings in a complex state of deterioration, where several pathologies appear and their interpretation using geophysical methods is not straightforward. In these cases, the application of mechanical NDT techniques, such as gravimetric methods, is required, for the complete understanding of the situation in the interior of the materials. In the case of heritage buildings where the physical integrity is not challenged, another set of mechanical NDT techniques, such as small drills for the introduction of sensors such as penetration resistance sensors, can be applied for the direct measurement of damages. These can be used as validation of the interpretation of the geophysical results, as well as for their calibration for further analysis.

Thus this chapter shows how NDT techniques in general, and geophysical methods in particular, focusing on different combinations of ground-penetrating radar (GPR) with other techniques, are applied for the auscultation and monitoring of heritage buildings through time, assisting decision-making toward intervention actions or toward possibilities of use.

22.2 Review of ground-penetrating radar applications to heritage buildings

The GPR technique has the capability of producing both 2D and 3D high-resolution imaging of the subsurface. Typically, the GPR systems used on the field of archaeology prospection and cultural heritage management consist of ground-coupled antennas, which work in contact with the surface or as close as possible to the surface (distance less than 5 cm). For this application, the ground-coupled antennas are preferable to air-coupled antennas due to their higher range of penetration and image resolution. Moreover, the development of multiantenna systems and multichannel antenna arrays has opened up new possibilities for dense 3D data collection and 3D visualization of the underground space, which allows for a better knowledge of the original appearance of heritage sites and buildings, as well as the location and extent of damages.

Ground-coupled antennas have frequencies ranging between 10 MHz and 4 GHz. The frequency and depth of penetration are related, with higher frequency pulses achieving lower penetration, but better resolution. Low-frequency antennas (from 10 MHz to 400 MHz) are generally used for deeper investigations such as analysis of soil systems and stratigraphy, while medium-frequency antennas (from 500 to 900 MHz) are most commonly used for the detection of shallow structures (e.g., hidden crypts), and high-frequency antennas (up to 1 GHz) are highly recommended for more detailed studies such as thickness estimation and detection of small defects such as cracking and fissures.

Table 22.1 presents some relevant published works using the GPR in the documentation and evaluation of heritage buildings. Through the literature, the GPR technique has demonstrated to be useful to document and characterize structural elements (thicknesses, material composition, and so on), to detect hidden artifacts and structures (crypts, tombs, and so on), as well as to identify anomalies or damages (cracks, moisture, subsidence, and so on).

22.3 Review of non-destructive techniques most commonly combined with ground-penetrating radar

Despite being a technique with high applicability for the detection of defects in heritage buildings, GPR can be also combined with other NDT techniques with the aim at performing a more comprehensive inspection in terms of variety of defects detected and completeness in the depth of the materials under study.

Regarding the techniques that can be combined with GPR, many different types can be used, as a function of the objective: if the aim is at completing the information of the interior of the building materials, geophysical techniques are mostly used; while if the objective is acquiring data from different depths and the surface, geospatial techniques are applied. When direct measurements are needed with high resolution, mechanical techniques are used.

Geophysical techniques measure the physical properties from the subsurface of a material or element, either naturally (like elastic properties measured with the seismographic technique) or as a response to an induced event (like electrical resistivity [ERT], measured as the response of the materials to the introduction of an electrical field in ERT technique—electric tomography). Examples of these properties include elastic movement (seismic tomography), electrical resistivity (electrical resistivity tomography), changes in density (gravimetric technique), magnetic field (magnetometry), and induced vibrations (ultrasounds).

Table 22.1 Review of ground-penetrating radar applications to heritage buildings.

Applications	References	Frequency antennas used
To estimate thicknesses and quality of walls	Cintra et al. (2020), Pérez-Gracia et al. (2013), Solla et al. (2020), Yalçıner et al. (2019)	2.3 GHz, 1.5 GHz, 1.2 GHz, 900 MHz, 800 MHz
To define roof and columns composition	Perez-Gracia et al. (2019)	900 MHz, 400 MHz
To detect moisture in facades, walls, columns, and floors	Cataldo et al. (2005), Diz-Mellado et al. (2021), Garrido et al. (2020), Işık et al. (2022), Martínez-Garrido et al. (2018), and Yalçıner et al. (2019)	2.3 GHz, 1 GHz, 500 MHz, 100 MHz
To detect cracks and fractures in facades, walls, and columns	Catapano et al. (2018), Pérez-Gracia et al. (2013), Santos-Assunçao et al. (2014), and Yalçıner et al. (2019)	2.3 GHz, 2 GHz, 1.6 GHz, 1.5 GHz
To detect hidden artifacts/structures (crypts, sarcophagus, and galleries)	Cataldo et al. (2005), Leucci et al. (2021), Novo et al. (2010), and Núñez-Nieto et al. (2014)	500 MHz, 600 MHz, 270 MHz, 250 MHz, 200 MHz
To investigate subsidence and settlement phenomena, soil system analysis	Diz-Mellado et al. (2021), Işık et al. (2022), and Solla et al. (2022)	500 MHz, 100 MHz
To inspect basements, foundations and floor systems	Dabas et al. (2000), Evangelista et al. (2017), González-Drigo et al. (2008), Guadagnuolo et al. (2014), Perez-Gracia et al. (2019), Rucka et al. (2020), and Yalçıner et al. (2019)	2 GHz, 1.6 GHz, 600 MHz, 500 MHz, 450 MHz, 400 MHz, 270 MHz, 200 MHz
To detect corrosion in walls and ceiling/floor slabs	Işık et al. (2022) and Solla et al. (2019)	2.3 GHz, 100 MHz
To map moisture and detachment in frescoes and mural paintings	Danese et al. (2018) and Garrido et al. (2022)	2 GHz, 500 MHz
To define stratigraphy and moisture content in mosaic Pavements	Caldeira et al. (2019) and Calia et al. (2013)	2 GHz, 1.6 GHz, 400 MHz
To inspect timber floors/beams/ceilings	Fontul et al. (2018) and Perez-Gracia et al. (2022)	2.3 GHz, 900 MHz

Geospatial techniques acquire information about the geometry of the elements on their surface (photogrammetry or laser scanning techniques), which can possibly be measured together with data related to the condition of the surface regarding aspect (visible image), composition (infrared image), or temperature (infrared thermography [IRT]). Their classification as geospatial techniques comes from the fact that these studies perform the georeferencing of the data, in such a way that information about the coordinates of the element, either in absolute or relative reference systems, is always available. Although infrared thermographic inspections can be performed with independence of the coordinates, this technique is here classified as geospatial because a reference system is applied in combination technique studies to allow data comparison.

Regarding the mechanical techniques mostly combined in GPR studies, there can be destructive and non-destructive ones. The non-destructive mechanical techniques consist of the measurement of the mechanical conditions of the element under study, using sensors such as accelerometers and inertial sensors, which measure the acceleration of the element in the three axes that define the 3D movement. The destructive techniques are only used when data from the element is needed to verify the non-destructive measurements through the comparison with a ground truth, and they mostly consist of drilling, sampling, and measurement of penetration resistance.

Table 22.2 shows a summary of applications of different non-destructive techniques applied in combination with GPR, the objective of the combination and the contribution of each technique.

22.4 Selected case studies

22.4.1 Monastery of Batalha (Leiria, Portugal)

The Monastery of Batalha, which mainly has Gothic and Manueline styles, is inscribed on the UNESCO World Heritage List. The main part of the monastery, in which the NDT surveys were conducted, was built between 1388 and 1402. Different restoration works were carried out throughout its history, which included the demolition of some parts, rebuilding damaged elements, and also introducing some modifications. The monastery was built with a compact sublithographic oolitic limestone from the region, although different types of limestones were introduced through restorations. Within this context, the main objective of this case study is the detection and identification of stone anomalies.

22.4.1.1 Methodologies

Fig. 22.1 illustrates the different NDT techniques used in this study to inspect the condition state of the main facade of the Monastery of Batalha.

Table 22.2 Review of combinations of non-destructive techniques with ground-penetrating radar (GPR).

Typology	Combined technique	Application	Contribution GPR	Contribution technique	References
Geophysics	Ultrasonic pulse velocity (UPV)	Estimation of thickness and quality of walls in stone building facades	Definition of discontinuities and changes in materials	Material information related to mechanical properties (quality)	Yalçıner et al. (2019)
Geophysics	Ultrasonic pulse velocity (UPV)	Detection of delamination, cracking and cavities in walls/facade and columns	Determination of crack axis and filling materials	Material information related to mechanical properties	Yalçıner et al. (2019)
Geophysics	Electrical resistivity tomography (ERT)	Detection of cracks, fractures, and voids in masonry	High-resolution detection	Low-resolution detection	De Donno et al. (2017)
Geophysics	Electrical resistivity tomography (ERT)	Monitoring of subsidence and settlement phenomenon	Microscale effects	Increased-depth information	Capozzoli et al. (2020)
Geophysics	Electrical resistivity tomography (ERT)	Study of the structural integrity of wall foundation, floors, and soil system (zonification, cavities, subsidence, and so on)	Characterization of the stratigraphy of the subsoil, revelation of the presence of natural and/or anthropic buried structures	Information about wet or fine-grained material. Increased-depth information	Evangelista et al. (2017)
Geophysics	Electrical resistivity tomography (ERT), electromagnetic induction (EMI)	Buried building detection and structure evaluation	Evaluation of the potential archaeological risk: precise detection of underground utilities	Evaluation of the potential archaeological risk: location of masonry pieces that could be destroyed during the excavation phase	Cozzolino et al. (2020)

Geophysics	Seismic tomography	Analysis of columns for load distribution evaluation prior restoration	Existence of unexpected anomalies in homogeneous materials	Definition of the inner geometry of the columns	Perez-Gracia et al. (2019)
Geophysics	Seismic tomography	Detection of delamination, cracking and cavities in walls/facade and columns	Accurate preliminary assessment of the internal structure	Information about material quality	Pérez-Gracia et al. (2013)
Geospatial	Seismic tomography	Detection of hidden targets or structural elements in stone walls	Detection of the arrangement of materials	Detection of damaged areas	Santos-Assuncao et al. (2016)
Geophysics	Microwave tomography, electromagnetic induction (EMI)	Detection and localization of rebar in concrete components	Determination of the rebar position	Estimation of rebar depth under cover layers	Zhou et al. (2018)
Geophysics	Microwave tomography, electromagnetic induction (EMI)	Estimation of thickness in casted concrete with embedded rebars	Determination of the rebar position	Estimation of rebar diameter and cover thickness	Sossa et al. (2019) and Zhou et al. (2018)
Geophysics	Microwave tomography	Detection of delamination, cracking and cavities in walls/facade and columns	Position of crack patterns, cavities and changes in material	Estimation of depth of cracks, voids and inner stone layers	Catapano et al. (2018)
Geospatial	Terrestrial laser scanning	Detection of cracks, fractures, and voids in limestone walls	Information about the inner structure	Information about the geometry of the external structure	Villarino et al. (2014) and Solla et al. (2020)

(*Continued*)

Table 22.2 (Continued)

Typology	Combined technique	Application	Contribution GPR	Contribution technique	References
Geospatial	Terrestrial laser scanning	Estimation of thickness in limestone walls	Information about changes in material and their depth	Information about the geometry of the external structure	Villarino et al. (2014)
Geospatial	Photogrammetry	Layer characterization of floor mosaics	Inner structure and presence of defects (internal cracking, moisture)	High-detail surface image, digital surface model	Caldeira et al. (2019)
Geospatial	Photogrammetry	Characterization of masonry arches for structural assessment	Inner ring stone thickness of the arch	Main external dimensions	Solla et al. (2012)
Geospatial	Infrared thermography (IRT)	Material degradation of masonry bridges	Knowledge of the bridge subsurface structure	Detection of anomalies related to material degradation close to the surface	Biscarini et al. (2020)
Mechanical	Gravimetric measurements	Detection of moisture in tuff bricks wall	Levels of dielectric permittivity and electrical conductivity	Water content to correlate with the GPR measurements	Agliata et al. (2018)
Mechanical	Drilling and penetration resistance, accelerometers	Inspection of timber floor/ceiling and location of beams	Location and length of the beams, distinction between timber and metal beams	Damage inspection at the ends of the beams	Fontul et al. (2018)

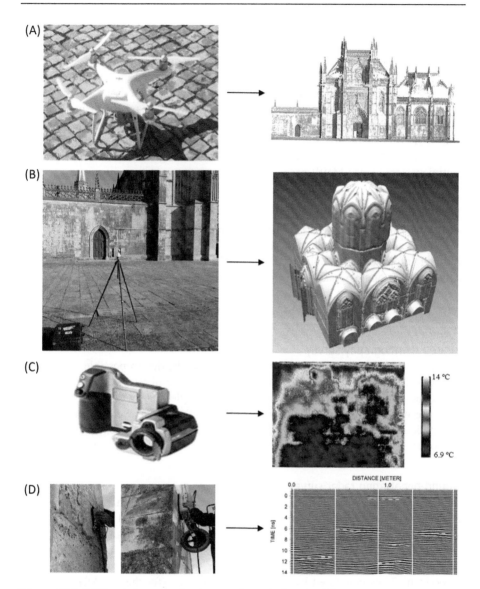

Figure 22.1 NDT techniques used to inspect the main facade of the Monastery of Batalha: (A) UAS and photogrammetry, (B) TLS, (C) IRT, and (D) GPR. *GPR*, Ground-penetrating radar; *IRT*, infrared thermography; *NDT*, non-destructive technique; *TLS*, terrestrial laser scanning; *UAS*, unmanned aircraft system.
Source: Adapted from Solla, M., Gonçalves, L. M. S., Gonçalves, G., Francisco, C., Puente, I., Providência, P., Gaspar, F., & Rodrigues, H. (2020). A building information modeling approach to integrate geomatic data for the documentation and preservation of cultural heritage. *Remote Sensing. 12*(24), 4028, https://doi.org/10.3390/rs12244028.

Unmanned aircraft system survey and orthophoto generation

This survey aimed for the automatic classification of visible pathologies of the facade based on RGB (red, green, blue) imaging acquired with a camera mounted on a multirotor UAS (unmanned aircraft system). A Phantom 4 Pro multirotor drone was used. An object-based image analysis was applied, with the K-nearest neighbor (KNN) classifier, in the eCognition Developer software. Image segmentation was performed using the multiresolution segmentation algorithm using the following setup: scale 65, shape 0.1, and compactness 0.4. Moreover, the orthophotos of the facade and roof of the monastery were generated using a structure from motion multiview stereo (SfM-MVS)-based approach.

Light Detection and Ranging survey and 3D model generation

Light Detection and Ranging (LiDAR) scanning was used to produce a highly accurate 3D model of the monastery. This 3D data will be then used to create the Building Information Modeling (BIM) model. To capture the main facade of the monastery, 12 and 46 scans were performed outdoors and indoors, respectively, generating point clouds with a spatial resolution of 7 mm. The TLS (Terrestrial Laser Scanner) used was a tripod-mounted Faro Focus 3D X330.

Infrared thermography survey

IRT was used to detect pathologies in the main facade, due to possible changes in the thermal diffusion of the wall or due to the presence of water/moisture. A thermographic Flir T335 camera was used, capturing single images as parallel as possible to the wall. A passive technique was used, and the images were taken when the solar radiation was not direct on the facade wall, avoiding heterogeneous heating of the stones.

Ground-penetrating radar survey

The GPR technique was used to identify different building materials and to characterize the internal structure of the facade wall. A ground-coupled system was used, consisting of a Proex control unit and three different antennas with central frequencies of 500 and 2300 MHz. The setting parameters used were trace-intervals of 2 cm (for the 500 MHz antenna) and 1 cm (for the 2300 MHz antenna) and time windows of 75 ns composed by 512 samples per trace (for the 500 MHz antenna) and 15 ns composed by 520 samples per trace (for the 2300 MHz antenna).

The GPR signals received were filtered with the ReflexW software, using the following processing sequence: time-zero correction, dewow, gain function, background removal, band-pass (Butterworth), and migration (Kirchhoff). To transform the travel time-distance axis into a depth axis in the radargrams, the radar-wave velocity was previously calibrated for two different types of stones found in the main facade, the original one and a more recent one from later reconstructions. The estimated velocities were 13.4 and 11.4 cm/ns for the original stones and the recent ones, respectively.

22.4.1.2 Results and main findings

Fig. 22.2 presents the classification of anomalies on the main facade of the monastery obtained by the KNN classifier, with an overall accuracy of 76.5% and a kappa coefficient of 0.71. For the anomalies identified, patina 1 and biogenic crust 1 are the severest levels of pathology (having darker tones) while patina 2 and biogenic crust 2 are still in an initial stage (lighter tones).

Fig. 22.3 shows the thermal images obtained for three wall spans (t1–t3 in Fig. 22.2). Lower temperatures are observed at the areas closest to the ground, most probably due to the presence of moisture (absorbed from the ground and infiltrated in the facade wall). Differences in the temperatures can be also associated with the different stones (materials and thicknesses) and pathologies observed as they imply different energy emissions. Lower temperatures (purple to blue tones) are thus associated with the presence of patina and thinner stones from recent restoration interventions.

Fig. 22.4 includes the GPR data produced in the areas t1–t3 of the main facade, showing the different ashlars identified in depth (highlighted as dashed orange lines). Moreover, a cavity (wall tomb) was detected in the internal skin of the facade at Zone 2.1 (highlighted into a red ellipse) in the form of a complex pattern of reflections at the end of the profile (from 1.4 to 2.2 m in Fig. 22.4B). It was realized that the ashlars used in recent restoration interventions were ∼ 10 cm thick, whereas the original stones were ∼ 50–60 cm thick. The thickness variation

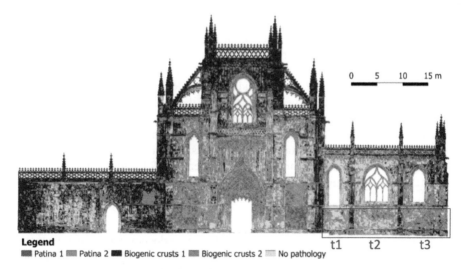

Figure 22.2 Orthoimage of the main facade of the monastery showing the automatic classification of pathologies obtained from the OBIA. *OBIA*, Object-based image analysis.
Source: Adapted from Solla, M., Gonçalves, L. M. S., Gonçalves, G., Francisco, C., Puente, I., Providência, P., Gaspar, F., & Rodrigues, H. (2020). A building information modeling approach to integrate geomatic data for the documentation and preservation of cultural heritage. *Remote Sensing. 12*(24), 4028, https://doi.org/10.1016/10.3390/rs12244028.

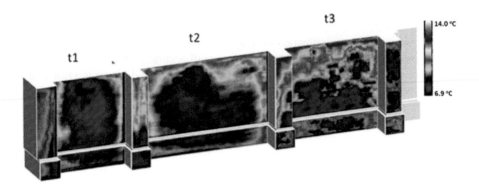

Figure 22.3 Infrared thermography in the areas t1–t3 of the main facade of the monastery. *Source*: Adapted from Solla, M., Gonçalves, L. M. S., Gonçalves, G., Francisco, C., Puente, I., Providência, P., Gaspar, F., & Rodrigues, H. (2020). A building information modeling approach to integrate geomatic data for the documentation and preservation of cultural heritage. *Remote Sensing*. *12*(24), 4028, https://doi.org/10.3390/rs12244028.

observed for a singular type of material is certainly caused by defects such as alveolization or by the unevenness of the inner face of the ashlars.

Furthermore, all the data produced was integrated into a BIM model. The BIM model was created with the Autodesk Revit software, using the 3D model obtained from the point clouds produced by the LiDAR and UAS survey. An object-oriented approach was assumed, which involved the creation of an initial wall object (Fig. 22.5). Next, the orthoimages obtained with the UAS, the IRT images, and the radargrams were inserted into the model, all images previously saved in JPEG format. The images providing exterior information were overlaid upon the model as wall and column claddings in a 3D plane (Fig. 22.5), with the dimensions and precise location defined by the point clouds. The GPR radargrams were positioned using field marks. To include the radargrams into the model, with their actual direction and dimensions, a fictitious wall object perpendicular to the plane of the facade was created for each radargram (Fig. 22.5) and then cladded with the radar image (Fig. 22.5). Moreover, metadata information was also added in the BIM model, such as the GPR system used and frequency antennas, IRT camera used, and date and time of data collection.

22.4.1.3   Final remarks

The use of different NDT techniques allowed for better documentation and diagnosis of the monastery. The UAS images allowed to automatically identify several pathologies in the stonework of the facade. The LiDAR enabled to generate a 3D model of the structure, which was then used to create the BIM model. The IRT was able to detect different materials, such as areas with patina, different ashlar thickness coming from different restoration works, and moist areas. Finally, the GPR was useful to estimate ashlar thickness, as well as to detect inner defects such as possible cracks or voids. Nevertheless, technical limitations were experienced during GPR data

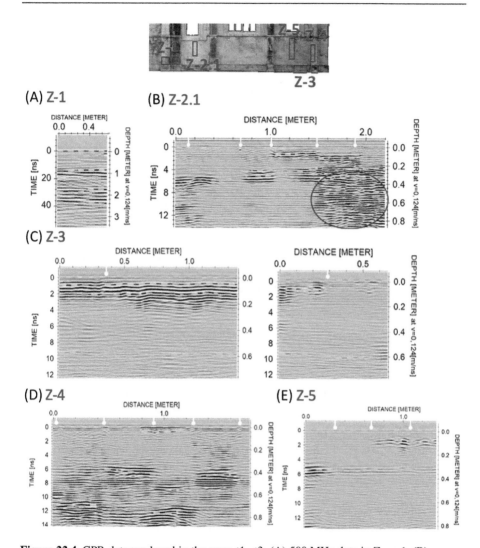

Figure 22.4 GPR data produced in the areas t1−t3: (A) 500 MHz data in Zone 1, (B) 2300 MHz data in Zone 2.1, (C) 2300 MHz data in Zone 3, (D) 2300 MHz data in Zone 4, and (E) 2300 MHz data in Zone 5. *GPR*, Ground-penetrating radar.
Source: Adapted from Solla, M., Gonçalves, L. M. S., Gonçalves, G., Francisco, C., Puente, I., Providência, P., Gaspar, F., & Rodrigues, H. (2020). A building information modeling approach to integrate geomatic data for the documentation and preservation of cultural heritage. *Remote Sensing. 12*(24), 4028, https://doi.org/10.3390/rs12244028.

acquisition to access the higher stonework of the facade. In this case, a ladder was used, but with the added difficulty of maneuvering heavy antennas.

The BIM allows for the integration of all the information produced by the different NDT techniques in a unique platform, thus facilitating the collaboration of

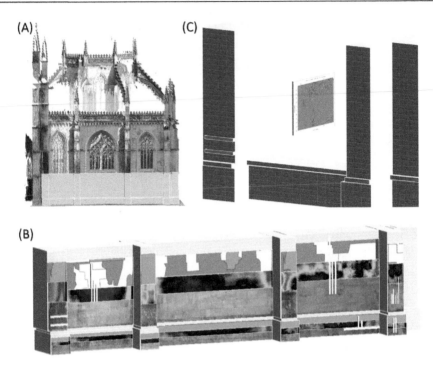

Figure 22.5 Data integration into the BIM model: (A) 3D model of the facade showing the wall object created, (B) resulted images integrated into the BIM model and fictitious wall-objects created for the radargrams, and (C) visualization of a radargram (cladded to a fictitious wall-object). *BIM*, Building Information Modeling.
Source: Adapted from Solla, M., Gonçalves, L. M. S., Gonçalves, G., Francisco, C., Puente, I., Providência, P., Gaspar, F., & Rodrigues, H. (2020). A building information modeling approach to integrate geomatic data for the documentation and preservation of cultural heritage. *Remote Sensing. 12*(24), 4028, https://doi.org/10.3390/rs12244028.

multidisciplinary teams to provide a better understanding of its real condition and preservation. However, heritage buildings are often composed of "nonstandard" elements, not defined by the existing libraries of building elements in the Revit platform. It is therefore necessary to create new architectural elements, or new component families, which is a time-consuming task.

22.4.2 Theater of Sao Carlos (Lisbon, Portugal)

The São Carlos National Theater in Lisbon (National Theater of Saint Charles) is one of Europe's most famous opera houses, dating from 1793 (TNSC, 2022), and beautifully preserved and is part of "The European Route of Historic Theatres" (Perspectiv, 2022). The main architect was José da Costa e Silva. This neoclassical building of Italian inspiration was listed as public interest property in 1928 and became a national monument in 1996. It is known that in 1888 an adjoining building was integrated for dressing rooms, rehearsal rooms, sewing, and wardrobe. In 1890, an iron curtain was placed between the stage and the audience. In 1908, it underwent interior remodeling by the architect Ventura Terra (noble hall, and so on) and outside, a shed was added and the arcades were closed by iron and glass doors. In 1935, it closed due to degradation. Major intervention works were carried out in 1938/40, followed by several other interventions in the subsequent decades.

The main purpose of this case study is to assess the structural health and safety conditions of the Noble Hall timber floor in the São Carlos National Theater (Fontul et al., 2018). Accurate information about the construction system of the Noble Hall floor is not available.

22.4.2.1 Methodologies

In a multidisciplinary approach, an extensive field work was carried out, involving: (1) assessment of the conservation state of timber beam ends, using drilling equipment (Fig. 22.6A); (2) GPR for identification of location and type of floor beams (Fig. 22.6B); and (3) dynamic characterization of the floor (Fig. 22.6C). These activities were further complemented with the development and calibration of a numerical model (Fig. 22.6C) of the floor structure, enabling the estimation of acceptable service loads taking into account acceptable deformation values and the estimated strength of the structural materials.

Drilling resistance

The state of conservation of timber beam ends embedded in masonry exterior walls was verified by assessing their resistance to drilling with the IML RESI B 300 equipment. This equipment drills a 3 mm diameter hole in the timber, allowing obtaining a profile of the drilling resistance variation along the element cross section (Machado et al., 2015; Tannert et al., 2010).

The procedure used consisted in obtaining a resistance profile away from the beams end area (90-degree drilling), a location normally not affected by biological deterioration, followed by obtaining the profile at the delivery area (45-degree inclined drilling) (Fig. 22.6A). The test points were decided based on the location of the beams, estimated through the use of georadar, referred later.

Ground-penetrating radar survey

The survey included the location and characterization of beams in the various areas of the floor by using two GPR systems equipped with two different ground- or air-coupled antennas (Fig. 22.6B). A ground-coupled pulsed systems manufactured by Malå Geoscience was used, composed of a Proex control unit and a bistatic antenna having a central frequency of 2.3 GHz, with the antennas in contact with the surface. Data acquisitions were carried out by distance with a trace-interval of 0.01 m or by time with a trace-interval of 0.01 second. An air-coupled Sir-20 pulsed system manufactured by GSSI Inc. was also used, consisting of an acquisition unit with a bistatic antenna having a central frequency of 1.8 GHz. Data acquisition was carried out by distance with trace-intervals of 0.01 m (corresponding to 100 scans/m).

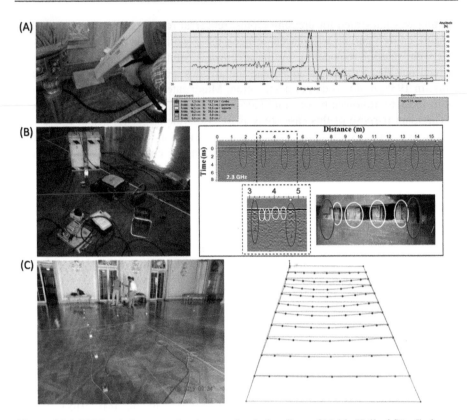

Figure 22.6 NDT techniques used to inspect the timber floor of Noble Hall of São Carlos Theater: (A) drilling resistance, (B) GPR, and (C) dynamic tests. *GPR*, Ground-penetrating radar; *NDT*, non-destructive technique.
Source: Adapted from Fontul, S., Solla, M., Cruz, H., Machado, J. S., Pajewski, L. (2018). Ground penetrating radar investigations in the Noble Hall of São Carlos Theater in Lisbon, Portugal. *Surveys in Geophysics*. *39*(6), 1125−1147, Available from: http://www.wkap.nl/journalhome.htm/0169-3298, https://doi.org/10.1007/s10712-018-9477-z and Cruz, H., Candeias, P., Machado J. S., Campos Costa, A., Fontul, S. (2017). Avaliação da segurança de pavimento do Salão Nobre do TNSC − uma abordagem multidisciplinar (in Portuguese). CREPAT 2017 - Congresso da Reabilitação do Património, 29−30 June Aveiro, Portugal.

Dynamic test

A dynamic test was performed on the pavement to characterize its dynamic response based on excitation produced by ambient noise. High-sensitivity piezo-electric accelerometers were used, and a data acquisition system composed of accelerometer conditioning modules, an NI SCXI-1600 USB acquisition board and a laptop computer. The sensors were fixed using easily removable adhesives.

Ten campaigns were carried out, employing 15 or 16 accelerometers, 4 of which were fixed (Fig. 22.6C). In each campaign, acceleration time series were measured with a duration of 20 minutes and a sampling frequency of 250 Hz to ensure the quality of the results in terms of spectral resolution (Cruz et al., 2017).

22.4.2.2 Results and main findings

Some aspects of the Noble Hall and the testing alignments are presented in Fig. 22.7.

The drilling resistance method was used to determine the deterioration level of the beams and their bearing capacity, given the fact that the decorative wood flooring would not allow direct access to timber members. This evaluation was necessary to assess the conservation state of the end of the beams embedded in masonry exterior walls (with windows in Fig. 22.7). The presence of roof leaks or rainwater combined with the porosity of masonry walls turns these areas into high hazard areas of exposure of timber to damp conditions.

The results obtained made it possible to assume that the wooden floor beams are sound or without signs of significant degradation by fungi or subterranean termites. The differences in the drilling resistance profile, shown in Fig. 22.8, are due to the difference between the entry angle of the needle and the beam face (90 degrees in the case of the test outside the delivery zone and 45 degrees in the delivery zone).

The GPR tests allowed establishing the geometry of the structure, type, and location of the beams in various areas of the floor. Since metal is an ideal reflector for electromagnetic waves, the identification of metallic beams is easier than in the case of timber beams (see Fig. 22.9). The validation of the georadar readings was done by comparison with the visual inspection of the cut sampling areas. The 2.3 GHz ground-coupled system has provided information useful to distinguish between timber and metal beams and understand their configuration (Fig. 22.9). This allows for setting up the location to place the accelerometers used for dynamic

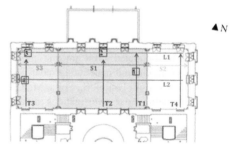

Figure 22.7 Aspect (at the left) and map of Noble Hall with the location of GPR profiles (at the right): two longitudinal (in blue L1 and L2) and four transverse (in red T1 to T4). *GPR*, Ground-penetrating radar.
Source: Adapted from Fontul, S., Solla, M., Cruz, H., Machado, J. S., Pajewski, L. (2018). Ground penetrating radar investigations in the Noble Hall of São Carlos Theater in Lisbon, Portugal. *Surveys in Geophysics. 39*(6), 1125–1147, Available from: http://www.wkap.nl/journalhome.htm/0169-3298, https://doi.org/10.1007/s10712-018-9477-z.

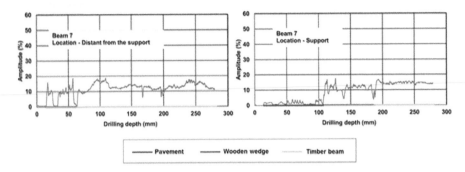

Figure 22.8 Example of results of drilling resistance: outside the support (at the left); next to the support (at the right).
Source: Adapted from Fontul, S., Solla, M., Cruz, H., Machado, J. S., Pajewski, L. (2018). Ground penetrating radar investigations in the Noble Hall of São Carlos Theater in Lisbon, Portugal. *Surveys in Geophysics*. *39*(6), 1125–1147, Available from: http://www.wkap.nl/journalhome.htm/0169-3298, https://doi.org/10.1007/s10712-018-9477-z.

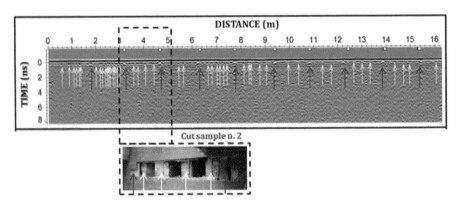

Figure 22.9 Timber (white arrows) and metal (red arrows) beams location detected with 2.3 GHz (ground-coupled) GPR antennas along L1 and photo of the cut sample that validate the results. *GPR*, Ground-penetrating radar.
Source: Adapted from Fontul, S., Solla, M., Cruz, H., Machado, J. S., Pajewski, L. (2018). Ground penetrating radar investigations in the Noble Hall of São Carlos Theater in Lisbon, Portugal. *Surveys in Geophysics*. *39*(6), 1125–1147, Available from: http://www.wkap.nl/journalhome.htm/0169-3298, https://doi.org/10.1007/s10712-018-9477-z.

test (Fig. 22.6C) and to develop the finite element numerical model used for the evaluation of the floor structural health.

Additionally, the GPR tests allowed for the detection of the lower ceiling structure. Fig. 22.10 shows a radargram obtained with 2.3 GHz antenna in the transversal direction T1 alignment through the middle of the floor, which has revealed the existence of a centered vaulted skeleton.

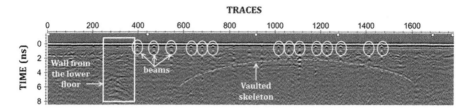

Figure 22.10 GPR results along T1 alignment showing the existence of a vaulted skeleton beneath the Noble Hall floor. *GPR*, Ground-penetrating radar.
Source: Adapted from Fontul, S., Solla, M., Cruz, H., Machado, J. S., Pajewski, L. (2018). Ground penetrating radar investigations in the Noble Hall of São Carlos Theater in Lisbon, Portugal. *Surveys in Geophysics*. *39*(6), 1125–1147, Available from: http://www.wkap.nl/journalhome.htm/0169-3298, https://doi.org/10.1007/s10712-018-9477-z.

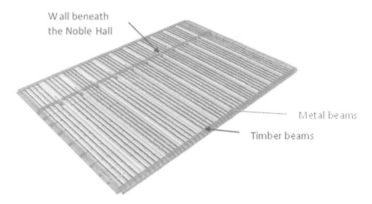

Figure 22.11 Finite element numerical model of the floor structure, set up based on GPR results and used for dynamic analysis. *GPR*, Ground-penetrating radar.
Source: Adapted from Cruz, H., Candeias, P., Machado J. S., Campos Costa, A., Fontul, S. (2017). Avaliação da segurança de pavimento do Salão Nobre do TNSC − uma abordagem multidisciplinar (in Portuguese). CREPAT 2017 - Congresso da Reabilitação do Património, 29−30 June Aveiro, Portugal.

Dynamic tests were performed to identify the structural characteristics and vibrational modes of the floor. The measurements were made with accelerometers placed on the floor surface. For this, it was essential to know the accurate location of the beams, their material, and extension. Fig. 22.6C shows the placement of the accelerometers, for seismic measurements, precisely where beam locations were identified with GPR (Cruz et al., 2017).

The application of stress waves and core drilling supported the predictions made about timber beams density (used for self-weigh load input) and modulus of elasticity. It was possible to set up a numerical model for further structural analysis based on the geometric information regarding the structure of the floor. A finite element model was developed (see Fig. 22.11) using the SAP2000 software for automatic calculation of structures, as a way to evaluate the behavior of the floor of the Great Hall.

The maximum loads and deflections for the different loading condition, including the permanent loads and the overload of use, were estimated. It was observed that the bearing capacity of the floor is conditioned by the bearing capacity of the timber beams and that the bending stress is more critical than the shear stress.

22.4.2.3   Final remarks

The study of the Noble Hall floor structure is an example of the advantages of a multidisciplinary approach in assessing heritage buildings, namely, the safety of timber structures, particularly when the degree of knowledge about the structure is low and the aim is to ensure that this activity has the least possible impact on existing structures.

Complementary NDT tests can provide integrated information not only on the geometry of the structure but also in the material properties and on the structure response to dynamic loads. Numerical modeling is possible in this way, based on the realistic evaluation of the heritage condition. The numerical model developed can be used as a reference for future evaluations, since any degradation of the pavement structure will manifest itself by reducing its vibration frequencies and/or changing modal configurations.

22.4.3   Cathedral of Mallorca (Balearic Islands, Spain)

The Cathedral of Mallorca is placed near the south Roman wall of the old city, just on a cliff above the Mediterranean Sea. Its construction began in 1230 and finished in 1587. The actual building is about 110 m long and 33 m wide, being the height 44 m. Although the cathedral is a Gothic building, other architectural styles due to many improvements, restorations, and enlargements highlight its importance. The ground on which part of the cathedral is built is unnatural fill terrain dating back to Roman times when anthropogenic fill was used to level the ground inside the wall that surrounded the city. This fact and the location of the building in the coastal cliff are causes of several damages most likely due to the building settlement. Therefore the assessment of the building requires a complete study including the analysis of structural damaged members and the study of the ground.

22.4.3.1   Methodologies

Detailed assessment of each part of the cathedral requires the use of more than one survey method. In some cases, those methods provide complementary data, being in other cases supplementary data. GPR and seismic tomography were used to assess structural parts of the building: columns and walls (Pérez-Gracia et al., 2013). Resistivity imaging passive seismic and GPR were used, combined with boreholes, to study the ground outside and under the cathedral (Pérez-Gracia et al., 2009).

Vibration measurements were used to determine the dynamic behavior of the structure (Caselles et al., 2015).

Vibration measurements

In the case of simple structures, the identification of modes of vibration is easily approached by comparing the modes measured at concrete points of the building. In these cases, longitudinal, transversal, and torsional modes can be identified by comparing the results from different vibration measurements. However, in more complex, asymmetrical, and big structures the procedure becomes more complex, appearing vibration modes that can be a combination of the three movements. To obtain accurate results, it is a mandatory correct selection of measurement points, covering as much as possible the different zones of the structure.

In the case of the Cathedral of Mallorca, the dimensions of the building point to low natural frequencies. Therefore measurements must be done during long time periods using low-frequency sensors. Time intervals for measurements were between 120 and 200 seconds. On the other hand, due to the complexity of the building with different superposed structures (as towers, vaults, and so on) requires punctual measurements of their natural frequencies separately. Therefore the accelerometers were placed at each one of the arches of the main nave, in two positions and also on the floor and on the top of the tower of the church. The seismometer was a Lennartz LE-3D/20s.

Refraction microtremor arrays (ReMi)

The ground structure under the cathedral could explain part of the damage of the building. Therefore its assessment requires the analysis of the shallow geology. ReMi survey was applied to characterize the soil, combined with GPR and other NDT surveys. The method consisted of the measurement of background noise to determine the mechanical shear wave velocity. This parameter is used to define and characterize physical changes in the materials.

The ambient noise was measured at different points of the cathedral floor using a multichannel seismograph Daqulink II (with 24 channels), and 24 geophones (with a resonant frequency of 4.5 Hz) placed on metallic plates glued to the floor tiles. The distribution of the geophones were in lines and in L configurations, covering in different measurements all the inside floor and part of the surrounding surface outside the cathedral (Pérez-Gracia et al., 2009).

Resistivity imaging

Measurements of the ground resistivity were used as complimentary information about the properties and characteristics of the shallowest part of the soil under the cathedral, and also as supplementary data to the GPR surveys. The objective was to determine changes in the ground based on the electrical characteristics of each material and each zone.

Measurements were carried out with and Ohmmapper TR1, a device that measures the electrical properties of soils without using invasive electrodes. The methodology consists of a transmitter and received joined by a coaxial cable. Both are moved along the surface during the data acquisition process, and the receiver

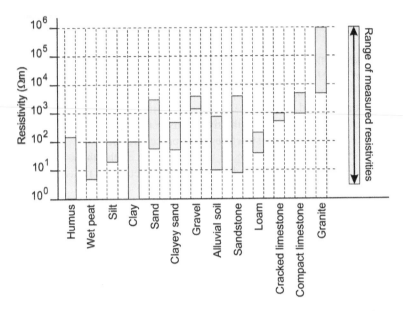

Figure 22.12 Resistivity of common soil materials compared with the range of values measured in the study.

measures a voltage that is affected by the resistivity of the media. This resistivity is obtained at each point considering the geometric factor of the device. As a result, continuous sections of resistivity are obtained and displayed as resistivity sections of the soil. It operates in a range between 3 and 100,000 Wm. The range of the resistivity of different and common soil materials is between 1 and 10^6 Wm, being highly dependent not only on the materials and the grain size but on the soil water content. Fig. 22.12 presents the range for several common soil materials compared with the range of measurements.

Seismic tomography

The inner structure of several columns was surveyed with seismic tomography. Data was acquired with high-frequency and sensitivity accelerometers. The tomography of the columns was based on the transmitted waves, picking only the first arrival to the sensors. The main advantages of this methodology are that it is easy to identify the wave and the signal is not obscured due to later arrivals that in most cases are due to multiple interactions in inner targets or heterogeneities. The wave source was a normalized and instrumented hammer, acquiring 63500 samples per second.

The tomographic images are determined based on the wave velocity and amplitude and are divided into pixels defined by the coverage of the wave trajectories. The dimension of each one of the pixels must be slightly higher than that resolution. Signals were recorded for frequencies higher than 10 kHz, and the resolution is 1/10 the period in the case of first arrivals. The characteristics of the pixel are determined by the wave trajectories crossing it. Therefore a pixel only can be defined if at least one wave trajectory crosses the pixel.

An important part of the survey is the preparation of the accelerometers on the surface of the structure because it is necessary to perfectly couple the sensor to the surface of the medium, but without damaging the building.

Ground-penetrating radar

The last survey applied to the study of the building was GPR. The method was applied in different parts of the structure: the church floor, the pavement outside the building, the walls, and the columns. 200 MHz and 400 MHz center frequency antennas were used in the study of floors and ground, while 900 MHz and 1.5 GHz center frequency antennas were applied in the assessment of columns and walls.

22.4.3.2 Results and main findings

Data acquisition planning

The assessment was focused in the analysis of possible causes of future damage due to changes in the ground. Additionally, several parts of the structure were also studied to know the actual state and the inner structure. The survey of the ground was designed considering separated zones outside and inside the building, characterized by boreholes. Fig. 22.13 shows the different zones and the position of the different surveys.

The assessment of structural members was focused on the study of several columns and walls, including the dynamic behavior analysis using vibration measurement points in the roof of the building. Fig. 22.14 shows the location of each one of the measurements. Three of the columns were detailed analyzed combining seismic tomography and GPR.

Combined analysis of the floor

The passive seismic survey data was processed after selecting the time windows with higher stationary noise. Hence, applying the fast Fourier transform to obtain the spectra, the quotient between the vertical and horizontal components was obtained and represented in front of the period. In each time window, average values and standard deviation are determined. Several characteristic results are shown in Fig. 22.15. The images allow for defining the values of the main period at each measurement point. Comparing all the data, the general result allows the identification of three zones that probably present significant changes in the ground. It is a piece of valuable information for planning and preparing further surveys with other geophysical methodologies. Fig. 22.16 presents this previous zonation of the ground. Those important changes in the seismic response most likely correspond to changes in the soil that could affect the structure.

The results from the passive seismic were compared to boreholes and to the ReMi survey. The study was focused inside the cathedral, where significant differences were observed in the passive seismic survey. This study pretends to determine the seismic shear wave velocity, analyzing their distribution in the medium. The soil mechanical properties are associated to these velocities. Several works

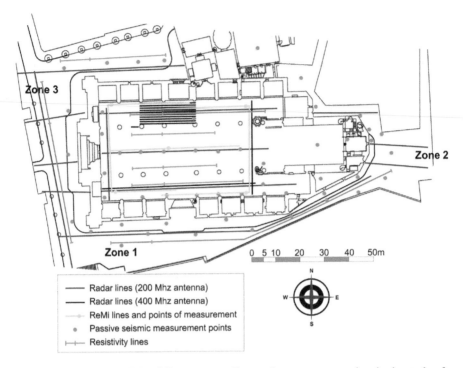

Figure 22.13 Location of the different survey lines and measurement points in the study of ground under the building.
Source: Adapted from Pérez-Gracia, V., Caselles, J. O., Clapes, J., Osorio, R., Martínez, G., Canas, J. A. (2009). Integrated near-surface geophysical survey of the Cathedral of Mallorca. *Journal of Archaeological Science. 36*(7), 1289–1299, Available from: http://www.elsevier.com/inca/publications/store/6/2/2/8/5/4/index.htt, https://doi.org/10.1016/j.jas.2009.03.001.

demonstrate the empirical relation between the shear wave velocity and the resistance to penetration in SPT standard tests (Akin et al., 2011). In general, higher s-wave velocities correspond to hardest soils. According the NEHRP (National Earthquake Hazard Reduction Program), in hard rock, the velocities are higher than 1500 m/s, decreasing to less than 180 m/s in the case of low compacted soils. The results obtained evidenced great differences in the soil compaction under the cathedral (Fig. 22.17).

Values of shear wave velocity between 800 and 1200 m/s correspond to limestone. This material appears at about 12 m deep in the North zone, and it is found at about 30 m depth in the South-West zone. Limestone was not detected in the East zone. This part is close to the clift, and the low velocity indicates probably anthropogenic filling materials. In the other zones of the cathedral, a zone with a wave velocity of about 600 m/s also appears, corresponding to conglomerates.

GPR survey of the floor allowed to study with higher detail the shallowest zone, arriving until 10 m. The resistivity imaging provides details of about 4 m depth.

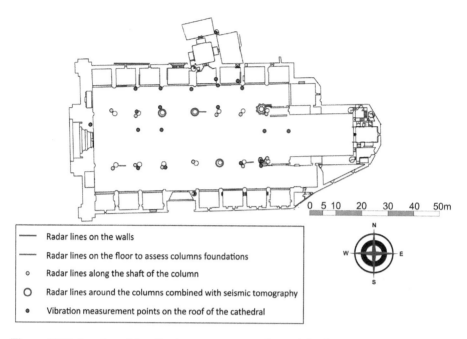

Figure 22.14 Location of the vibration measurement points and the GPR and seismic tomography applied to the study of structural members of the cathedral. *GPR*, Ground-penetrating radar.

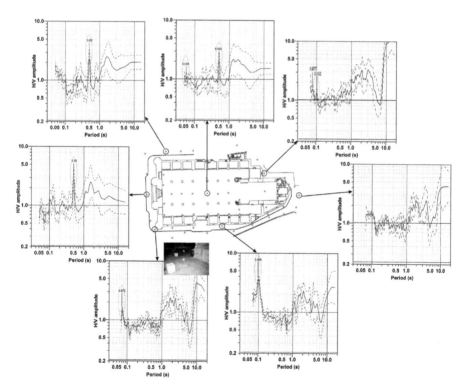

Figure 22.15 Representative results of the passive seismic survey, showing the difference in the main period.

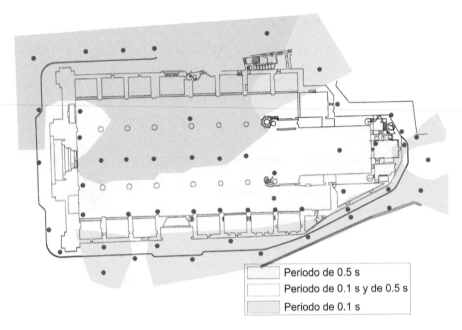

Figure 22.16 Zonation defined with the passive seismic.

Comparing data from GPR and resistivity to the seismic information and borehole (Pérez-Gracia et al., 2009), it is possible to deduce that the soil is composed by an anthropogenic filling that arrives to a depth higher than 40 m in the apse (East). The thickness of this layer of low compacted material is variable. In the main nave of the cathedral and close to the main entrance, under these poor materials, conglomerates exist. Fig. 22.18 presents an example of the data obtained outside the cathedral, along a radar line in the South zone and also along a resistivity survey line. Both are compared to the geology inferred from boreholes.

Combined analysis of structural members

Columns, walls, and foundations were analyzed with GPR. Additionally, the columns were also assessed with seismic tomography.

There were two objectives of this second study: defining the internal structure of walls, columns, and buttresses and detecting possible damaged zones (mainly the existence of cracks inside the materials). All this information together with the ground data was needed to determine an accurate model for a dynamic evaluation of the building (Pérez-Gracia et al., 2013).

GPR B-scans in walls, obtained with 900 MHz and 1 GHz antennas, indicated clearly the thickness of the structure and inside embedded targets, revealing that walls were built by three zones: two of them forming a skin of two layers of ashlars, the central one being a filling of heterogeneous material (Fig. 22.19). Radar images seem to point also to an irregular surface of the ashlars in the contact with filling. This specific pattern was also observed in the study of other structures

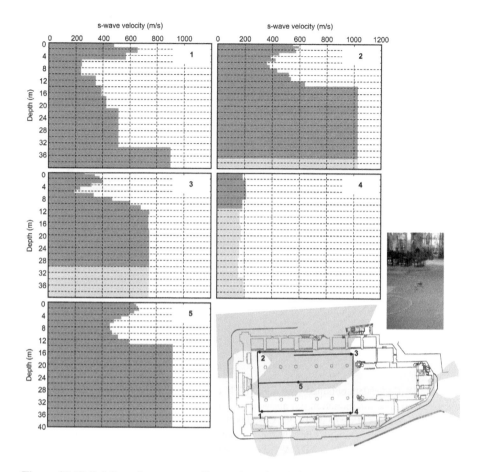

Figure 22.17 ReMi results corresponding to the points 1, 2, 3, 4, and 5.
Source: Adapted from Pérez-Gracia, V., Caselles, J. O., Clapes, J., Osorio, R., Martínez, G., Canas, J. A. (2009). Integrated near-surface geophysical survey of the Cathedral of Mallorca. *Journal of Archaeological Science. 36*(7), 1289−1299, Available from: http://www.elsevier.com/inca/publications/store/6/2/2/8/5/4/index.htt, https://doi.org/10.1016/j.jas.2009.03.001.

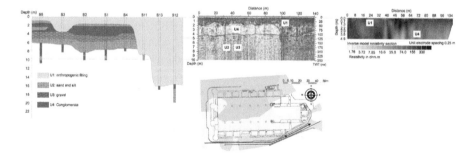

Figure 22.18 Geology defined using boreholes compared to GPR B-scan and pseudosections of resistivity imaging. *GPR*, Ground-penetrating radar.
Source: Adapted from Pérez-Gracia, V., Caselles, J. O., Clapes, J., Osorio, R., Martínez, G., Canas, J. A. (2009). Integrated near-surface geophysical survey of the Cathedral of Mallorca. *Journal of Archaeological Science. 36*(7), 1289−1299, Available from: http://www.elsevier.com/inca/publications/store/6/2/2/8/5/4/index.htt, https://doi.org/10.1016/j.jas.2009.03.001.

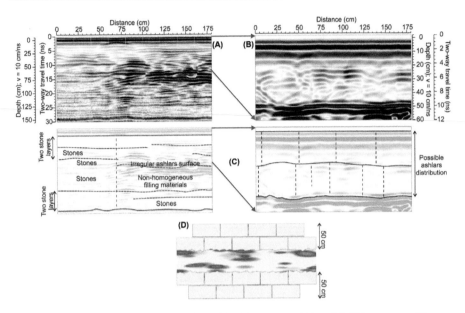

Figure 22.19 Radar data obtained with a 900 MHz antenna (A) and 1 GHz antenna (B) on a structural wall. The interpretation (C) allowed to estimate the structure of those walls (D). *Source*: Adapted from Pérez-Gracia, V., Caselles, J. O., Clapés, J., Martinez, G., Osorio, R. (2013). Non-destructive analysis in cultural heritage buildings: Evaluating the Mallorca cathedral supporting structures. *NDT & E International. 59*, 40–47, https://doi.org/10.1016/j. ndteint.2013.04.014.

(Pérez-Gracia et al., 2011). Those stones, with a polished external surface and a polished contact between other stones present a rough surface in the contact with filling, to assure a better contact.

Columns were assessed with seismic tomography and GPR, revealing the inner structure that consisted of files of stones rotated 90 degrees one from to the other (Fig. 22.20). Both radar data and seismic tomography detected the joints between the ashlars. Possible cracks were also detected with GPR, observed as anomalies near the joint. Seismic tomography defines those areas with possible damage as low velocity. Higher velocity values are indicators of good-quality stones, while lower velocities denote possible damage as cracking or lower stone quality.

Fig. 22.21 presents an example of detection of column foundations with GPR, showing a good matching to the observation in later excavations.

Vibration measurements for structural modal identification

The last analysis was focused to determining the dynamic characteristics of the building. The great dimensions of the building, its structural characteristics, and the difficult accessibility force the measurements in three phases: a first measurement in different points of the structure, a modeling of the structure dynamics, and a second measurement in the most conflictive points. Vibration measurement points were placed on the roof, over the arches and vaults (Fig. 22.22).

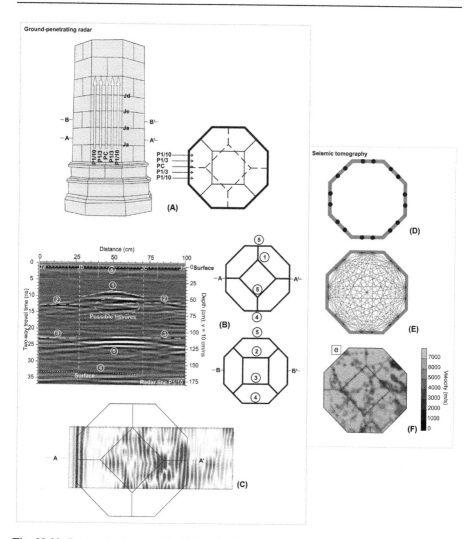

Fig. 22.20 Study of columns with GPR and seismic tomography. (A) Position of GPR radar lines. (B) B-scan with anomalies associated to the different stones and to possible cracking. (C) C-scan. (D) Position of geophones on the column. (E) Ideal seismic wave trajectories inside the column. (E) Tomography indicating the existence of low quality stones zones, most likely due to cracking, corresponding to low velocities. *GPR*, Ground-penetrating radar. *Source*: Adapted from Pérez-Gracia, V., Caselles, J. O., Clapés, J., Martinez, G., Osorio, R. (2013). Non-destructive analysis in cultural heritage buildings: Evaluating the Mallorca cathedral supporting structures. *NDT & E International. 59*, 40−47, https://doi.org/10.1016/j. ndteint.2013.04.014.

Results highlighted that the frequency for the first five modes of vibration at the different points was quite similar. However, the shape of the spectra at the different points differs greatly. Additionally, the analysis of the particle motion representation,

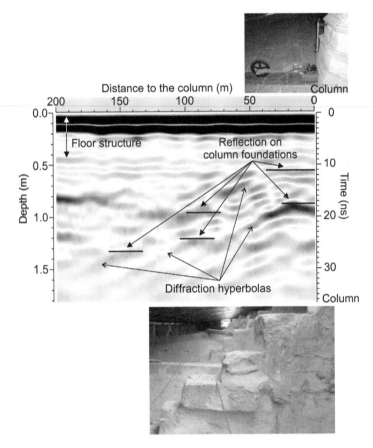

Figure 22.21 Example of detection of column foundations with GPR. *GPR*, Ground-penetrating radar.
Source: Adapted from Pérez-Gracia, V., Caselles, J. O., Clapes, J., Osorio, R., Martínez, G., Canas, J. A. (2009). Integrated near-surface geophysical survey of the Cathedral of Mallorca. *Journal of Archaeological Science. 36*(7), 1289−1299, Available from: http://www.elsevier.com/inca/publications/store/6/2/2/8/5/4/index.htt, https://doi.org/10.1016/j.jas.2009.03.001.

which represents the amplitude and shape of the movement of the measurement point for different phases, denotes important changes between measurement points. An example of the results is presented in Fig. 22.23.

Data indicates different modal frequencies. However, the most conclusive information is obtained by comparing the particle motion. For example, when comparing measurements along the line E-W, the first mode appears transversal in the apse (measurement point 18), but it becomes longitudinal in point 1, placed on the facade. The results conclude that the motion of the facade is purely longitudinal, being a very rigid structure. However, in the zone of the apse, the E-W component is coupled with the N-S component, indicating compression of the arches and

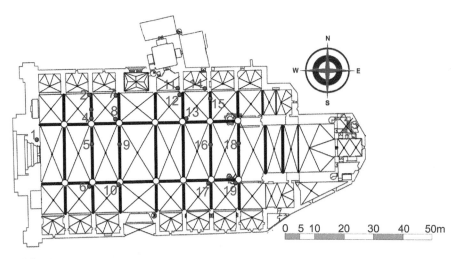

n● Vibration measurement points

Figure 22.22 Vibration measurement points, located on the roof structures.
Source: Adapted from Caselles, O., Martínez, G., Clapés, J., Roca, P., Pérez-Gracia, M. D. L. V. (2015). Application of particle motion technique to structural modal identification of heritage buildings. *International Journal of Architectural Heritage. 9*(3), 310–323, Available from: http://www.tandf.co.uk/journals/titles/15583058.asp, https://doi.org/10.1080/15583058.2013.784824.

vaults. Other walls present lower stiffness than the facade, and in some parts appear most likely rotation and flexion. The representation of the particle motion also allows to differentiate between zones with higher vibration levels and zones with low vibration. Asymmetries and distortion in the results could point to possible cracking in the structures or to the existence of added structures, such as the bell tower in this cathedral.

22.4.3.3   Final remarks

The assessment of the cathedral with different NDT techniques allowed to define significant parameters for modeling the structural behavior and for planning restoration and future conservation. The study of ground with passive seismic combined with GPR and resistivity imaging denoted an important change of the ground materials that could be the cause of cracking. This combination of techniques has shown its great utility and reliability in studies for the maintenance of large cultural heritage buildings, and the analysis of soil under and around the building provides zoning maps that can facilitate maintenance tasks. Passive seismic also provides information about the quality of the materials. In the assessment of the structure, GPR allows detection of moisture, cracking, and also the existence of buried targets. It is a useful technology, combined with seismic tomography in the most complex cases, to determine the constructive techniques, the shape and size of the different constructive elements, such as ashlars, filling mortar, or reinforcements.

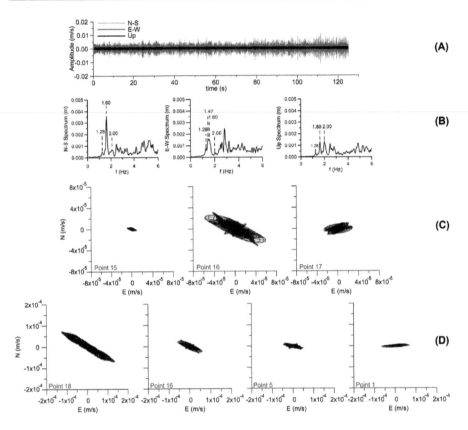

Figure 22.23 Vibration measurement results. (A) Time signal in point 16 for the three components. (B) Spectra of the signal recorded in point 16. (C) Particle vibration motion for the first mode of point 16 compared to the results in points aligned in N-S direction. (D) Particle vibration motion for the first mode of point 16 compared to the results in points aligned in E-W direction, being point 1 placed on the facade of the building. *GPR*, Ground-penetrating radar.
Source: Adapted from Caselles, O., Martínez, G., Clapés, J., Roca, P., Pérez-Gracia, M. D. L. V. (2015). Application of particle motion technique to structural modal identification of heritage buildings. *International Journal of Architectural Heritage. 9*(3), 310–323, Available from: http://www.tandf.co.uk/journals/titles/15583058.asp, https://doi.org/10.1080/15583058.2013.784824.

The GPR has also shown its ability to analyze the foundation of the load-bearing elements of the structure. However, the greatest limitation that has existed has been access to many areas of the cathedral. For example, in the study of the columns, it has only been possible to analyze the parts closest to the ground. The same has happened with the load-bearing walls. The combination with seismic tomography has proven to be very practical, since it has made it possible to determine very accurately the type of column that was studied, being able to determine the position and size of the central ashlars.

Finally, the dynamic analysis provides a more complete and comprehensive knowledge of the structure, because it is possible to determine the behavior for dynamic loads. However, vibration analysis could be more complete in the case of complementing the experimental tests with computer models.

22.4.4 Gothic basilicas (Barcelona, Spain): Santa Maria del Mar and Santa Maria del Pi

Medieval Barcelona was a relevant political and commercial city. Urbanism was developed and important buildings were built, especially in the Gothic period. Many old churches were redone in a new, more monumental style. It is the case with the cathedral or the basilica of Santa Maria del Pi. This last church was built over the remains of a 10th-century church previously demolished. In other cases, new basilicas were built, such as Santa Maria del Mar, which was built most likely over Roman remains.

Santa Maria del Mar basilica was built between 1329 and 1383, supported by the parishioners of the zone with work or financing the construction. After the construction, it suffered different damage and restorations as a consequence of a fire in 1397, an earthquake in 1428, and a later fire in 1936. The actual structure is a typical Mediterranean Gothic church, characterized by the columns' slenderness and the opening spaces between the three naves which are composed by four sections, being the height of the roof in the lateral naves similar to the distance in the central nave, and the distance between columns (15 m) higher than in other churches of similar characteristics; this particular structure produces the impression of a single large nave. The naves are covered with ribbed vaults, with artistic keystones at each vault. On the main facade, there are two octagonal side towers, and two buttresses on both sides of the central rose window. On the side facades, there are buttresses to support the loads of the arches.

Santa Maria del Pi was built between 1319 and 1391 with a Gothic style, over the remains of a Romanesque church placed outside the walls of the old city. The tower bell was built later, in 1460. Both the earthquake of 1228 and the War of Succession of 1714 caused significant damage to the building, some damage fixed in 1717. In 1863, the first restoration was carried out, which focused on the roofs of the chapels, on the doorway, and on the rear facade. With this first restoration, the building now has a Gothic main facade and a later Modernist one. There were two other important restorations, one in 1915 and the other in 1957. The latter, very important, was carried out because the building had suffered a great fire during the civil war of 1936–40. The basilica has a single nave, 45-meter long and 16.5-meter wide, consisting of seven rectangular sections covered with ribbed vaults. The maximum height in the nave is 27 m. The Gothic facade has a central rose window that was completely rebuilt in 1940. One of the side facades contains the only Romanesque vestige of the temple: a door from the initial church. The buttresses that support loads of the arches are located on these lateral facades. There are also buttresses on the rear facade.

The study of both basilicas was focused on the analysis of the structural characteristics and also on the evaluation of the ground, to prevent further damage. The structural assessment of those types of buildings requires knowledge of the structures typology, existence of restorations, possible damage, and soil properties. In many cases, these aspects are not documented, and a thorough prior inspection is required to obtain as much information as possible. The NDT surveys are useful methodologies to determine aspects such as changes in materials, damage due to humidity, and the existence of hidden restorations.

22.4.4.1 Methodologies

Santa Maria del Mar was assessed using GPR, seismic tomography, and passive seismic (Vega Perez-Gracia et al., 2019). The GPR evaluation was focused on two aspects: the analysis of the soil under the building and the assessment of walls, columns, and vaults. Seismic tomography was applied to columns.

The study of Santa Maria del Pi was dedicated to the evaluation of soil properties by means of passive seismic, and the study of the basilica vaults using GPR.

Ground-penetrating radar

Vaults were studied in both basilicas with a GPR with a 400 MHz antenna. Data were acquired along the lateral naves, over the chapels, and on the roof of the church, covering the whole surface. Radar lines across the different sectors of the main naves, along and across the vaults (see Fig. 22.24). Walls and columns were analyzed in Santa Maria del mar with a 900 MHz antenna and the floor of this church with 200 and 400 MHz antennas.

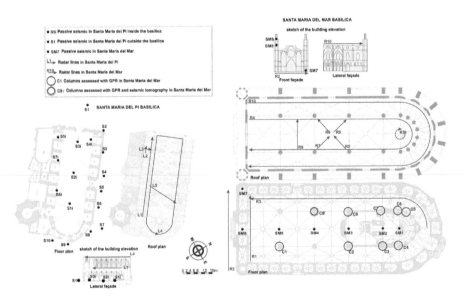

Figure 22.24 Position of radar lines and seismic measurement in both basilicas.

Seismic tomography

Seismic tomography was applied to one of the columns. The study required a seismometer, several sensors (accelerometers) that were placed around the column, and an instrumented hammer. The column was selected after a previous visual inspection due to the damages observed in the structure. A section of the column was selected, and 32 hammer-impact points were marked around the column, to obtain sufficient resolution.

Passive seismic

Passive seismic in Santa Maria del Pi consisted of data acquisition in 16 points, being 6 points inside the building, and 10 of those points around the basilica. Data (ambient noise) was acquired with a seismograph (see Fig. 22.24).

Passive seismic was also applied in Santa Maria del Mar to assess structures. Therefore the measurement points were placed at each one of the arches of the basilica and also on the two towers of the main facade (Fig. 22.24).

22.4.4.2 Results and main findings

Ground assessment

One of the causes of damage in historical buildings is the soil under the structure. Urban soils have been leveled on many occasions using anthropogenic fills. This type of material is usually less compacted than natural soil, so if the loads change due to building restorations or modifications to their structure, differential settlements may occur. Therefore cracking affects parts of the structure. In other cases, the humidity in the soil can affect the walls, foundations, and floors of the building. Water content in soils could be caused because of rainy periods but also because of breaks or leaks in pipes. The characteristics of soils could help drainage, thus preventing moisture from affecting the structures, but sometimes they can retain water near the buildings, so that moisture damage appears. In that case, efflorescence, damp spots, deterioration of paints, detachments, or falling ornamental elements are usually the visible effects of damage that is internal to the walls and floors. Hence, the assessment of ground under and around the structure could be a helpful tool to understand observed damage in visual inspections and can also facilitate decision-making in building restorations.

Different methodologies are useful in the study of soils. In these two examples, two different methods have been applied in each basilica.

GPR data was successful to determine the shallowest zone of the soil, arriving at 3 m deep with a 200 MHz antenna. In the zone where Santa Maria del Mar was built, a layer of about 1 m thick exists, composed of anthropogenic materials. A great number of anomalies most likely associated with ancient constructive vestiges are detected in this layer. It is consistent with the most recent archaeological excavations that detected significant Roman remains. The anomalies were observed in radar lines inside and outside the building. At higher depths, the random backscattering indicates the existence of gravels or other irregular materials (Fig. 22.25).

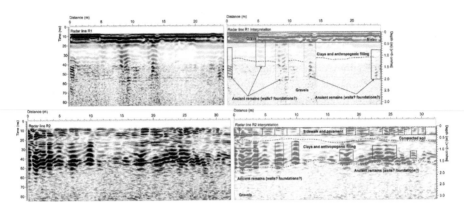

Figure 22.25 Two examples of the radar images of the ground in the zone of the basilica of Santa Maria del Mar.
Source: Adapted from Perez-Gracia, V., Santos-Assunçao, S., Caselles, O., Clapes, J., Sossa, V. (2019). Combining ground penetrating radar and seismic surveys in the assessment of cultural heritage buildings: The study of roofs, columns, and ground of the gothic church Santa Maria del Mar, in Barcelona (Spain). *Structural Control and Health Monitoring. 26*(4), e2327, https://doi.org/10.1002/stc.2327.

Passive seismic allowed to determine of different types of soil, being possible to separate between zones of superficial fills with little compaction and zones of more compacted soil. The representation of the quotient between the horizontal to vertical components produces complex spectra that are associated with a complex soil, possibly formed by a large number of layers of thin material (see Fig. 22.26). Moreover, the analysis of periods separates two main groups of measurement points: those with smaller periods, near 0.1 second, and those with higher periods (near 1 second). The detection of the periods near 0.1 second is connected to poorly compacted soils near the surface. Therefore it is possible to define a map of types of soils depending on those periods (see Fig. 22.26), being most likely anthropogenic filling the poorly compacted materials (Caselles et al., 2007). It is noticeable that most of the structure is built over the filling.

Both GPR and passive seismic provide information about the soil but referred to different aspects. GPR offers detailed information about the shallowest part of the soil. It is possible to distinguish between the most superficial materials, allowing the detection of buried targets under the building. However, the maximum depth in this study is less than 3 m, and, although it is possible to differentiate between layers, there is no information about the mechanical or dynamic properties of the materials. In turn, the data obtained by passive seismic contain information on deeper materials, although they do not allow the detail that is observed in radar images. This method is suitable to distinguish between zones characterized by different mechanical properties. Hence, these two techniques provide complementary information, and their application in the same study could expand the data obtained, allowing a better interpretation of the results and the development of more accurate models.

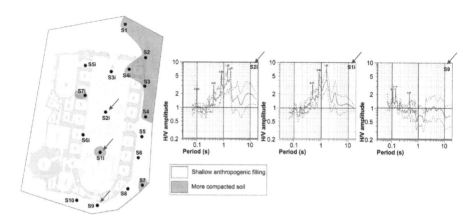

Figure 22.26 The analysis of H/V ratios indicates quite complex soil. The study of the periods allows us to separate between two zones. The arrows indicate the points where the examples of spectra are associated.

Assessment of arches and vaults

The extremely large loads that introduce the arches and vaults in gothic churches is surprising, especially in the case of the Basilica of Santa Maria del Mar, since the columns are very slender and are farther apart than in other Gothic constructions. To solve the problem that could be caused by loads on the columns, the structures supported by the columns were lightened as much as possible. The construction method, which was observed during the study with GPR, consisted of introducing hollow ceramic elements into the structural mass of the arcades and ceilings. The study in this church focused, therefore, on the analysis of the structures to determine the way in which they were built, detecting embedded elements, and differentiating between materials.

However, the study of the roof of Santa Maria del Pi focused on the analysis of moisture damage, which was believed to have been detected in previous visual inspections.

In both cases, the GPR was used as the most feasible method due to accessibility, study time, and information that it could provide. In the assessment of cultural heritage, both the detection of types of structures and the assessment of damage due to moisture are usual objectives.

Structures in Santa Maria del Mar

Radar data was acquired along lines on the roof (see Fig. 22.24). The roof structure of this church is supported by vaults and arches that transmit the loads to the columns that, in turn, transmit them to the foundations. It has been estimated that each of the columns of the basilica support about 500 tons, reaching a stress at its base that can reach 3.3 MPa (Vega Perez-Gracia et al., 2019). The thickness of the roof is higher near the top of the columns, and smaller just over the keystone of the vaults (about 50 cm), which are built with 20 cm thick stones. Between the tiles of the roof and the ashlars of the vaults, there is filling material that consists of a

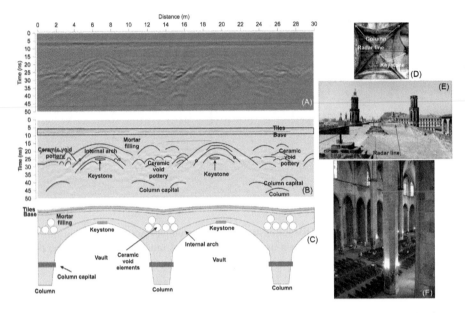

Figure 22.27 (A) Example of radar data obtained in the roof of the basilica. (B) Radar data interpretation marking the anomalies. (C) Possible model defined from the B-scan. (D) Vault and the corresponding radar line (on the roof) crossing the vault. (E) Roof of the basilica. (F) Image of the slender columns.

mortar made up of lime, small stones, pieces of ceramics, and construction remains. The study of the filling in cultural heritage masonry vaults points to the existence of lightened fillings in the typical Mediterranean Gothic (Casquero, 2015) to reduce possible damage in columns.

Results confirmed the existence of the lightened filling (Fig. 22.27). In the B-scans, it is possible to interpret the tiles and the base of the roof, and a homogeneous mortar. However, in the thicker areas (corresponding to the base of the vaults), important hyperbolic anomalies indicate the existence of large elements placed in a more or less orderly manner. The existence of some hyperbolas below the first ones, whose shape corresponds to a signal propagation velocity somewhat higher, seems to show that the elements in which the reflections occur are hollow. The result is consistent with the hypothesis of a filler mortar lightened by means of hollow ceramic jars. This is an interesting and skillful technique to lighten the loads that the thickness of the material would generate in the columns. In this way, it is possible to generate tall and large spaces through slender columns, without undermining the safety of the construction.

Moisture in the vaults in Santa Maria del Pi

The great fire during the Spanish Civil War forces an intense restoration in 1957. To reinforce some structures, metallic elements were used on the roofs. The fall of small stones in 2005 caused part of the building to be protected with nets. Subsequently, visual inspections determined the existence of more damage, especially on the roof of the building, where the effects of the corrosion of the metallic elements were seen at some points. The shape of the roof is designed to conduct rainwater to certain points, from where it drains. However, the corrosion of the metallic elements seemed to indicate that the drainage system was damaged at some points and the water was leaking, producing humidity in the construction materials. The GPR assessment was focused on the detection of the zones with higher water content, to limit the problem and be able to act mainly in the damaged ones. Radar lines covered the perimeter of the roof (see Fig. 22.24) because in the perimeter the drainages for rainwater exist.

B-scan images demonstrate the existence of a roof structure similar to the one observed in Santa Maria del Mar: the keystones and the vaults are defined clearly, being also detected the use of lightened filling on the vaults of the structure. In some parts, the wave amplitude increases significantly, denoting the existence of dampness (see Fig. 22.28).

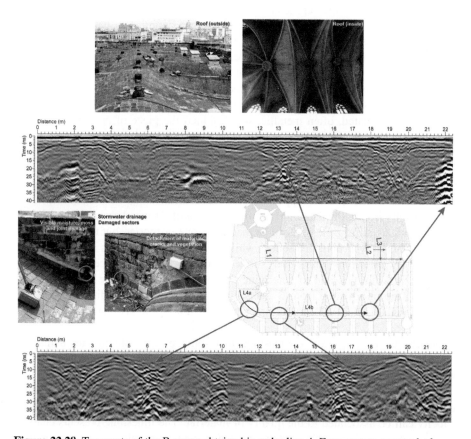

Figure 22.28 Two parts of the B-scans obtained in radar line 4. Four sectors are marked because of the great increase of the amplitude, showing possible humidity due to filtration in the stormwater drainage of the building.

The drainage structure is holes in the base of the vaults that move the water accumulated in the roof to the street. However, in some parts, filtration affects the structure of the vaults.

Therefore the GPR inspection was useful enough to determine those sectors that might be deeply analyzed for further restorations. The changes in the wave amplitude denote changes in the materials due to the existence of water content.

Assessment of columns

The columns of Santa Maria del Mar were also studied with GPR and seismic tomography. The objective was to determine the type of construction and possible damaged zones. Some columns presented visible damages and even the existence of metallic materials, most likely from past restorations after the fire during the Spanish Civil War.

The load at each one of the columns is about 500 Tm and is built with good-quality limestone obtained in the mountains surrounding Barcelona. The eight columns from the base to the capital are 17 m, they are separated by 15 m from one to the other. The constructive methodology of Gothic columns is based in massive stones connected to the others. However, due to the cost of stones, in many cases, mortar filled the inner part of the structure. In general, there were three methods to build columns (Fig. 22.29): the first one was with rotated solid pieces of stone placed in rows that were rotated 90 degrees or another angle, one with respect to the other; the second one was with solid pieces of stone with a central hole in which an anchor was located, often metallic, which joined the piece of one row with that of the next row; and the third type of construction was using solid pieces of stone placed in such a way that there was a central hole that was filled with mortar.

The GPR and seismic assessment of columns were focused on the analysis of the constructive technique, and also to determine the possible existence of damage because some of the columns appeared with visible deterioration on the surface.

The results from GPR data indicate the existence of a rotated structure in the column, around a central zone that can be made of stone or mortar, but that in alternate rows presents a hyperbolic reflection, while in the other rows, it is a characteristic anomaly of reflections on a flat surface parallel to the surface (Fig. 22.30). However, it is not possible to determine the exact geometry of the inner structure only using the information from B-scans. It is also not possible to distinguish materials, although similar wave travel times point to homogeneous materials in all the column.

Notwithstanding, B-scans demonstrate that some of the columns are damaged, mainly in the most superficial zone. The existence of metallic linkage in some joints is also visible due to the strong reflection produced in the targets (Fig. 22.30).

The seismic tomography allowed to clarify the inner structure of the columns, revealing the shape and the position of each element. In addition, the seismic wave velocity can be associated to the quality of material. The limestone presents characteristics velocities from 3000 to 6000 m/s, being the higher velocities corresponding to better stone quality. The measured velocities in the case of the columns ashlars are between 3000 and 3500 m/s, indicating, most likely, a medium quality of the

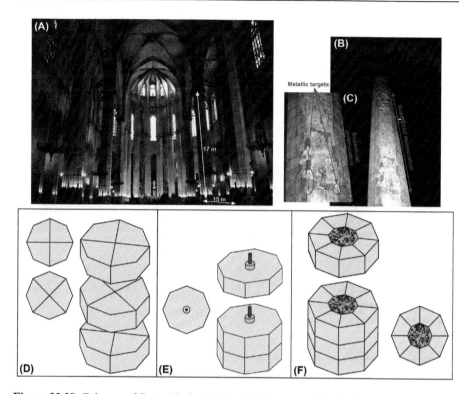

Figure 22.29 Columns of Santa Maria del Mar. (A) Photograph inside the church. (B) The column that was assessed with seismic tomography; the damage is visible on the surface. (C) Details of the damaged column: corroded metallic pieces, and detachment of the external part of the ashlars. (D) Constructive method consisting on rotating the stones. (E) Constructive method based on the linkage of stones by means of a central connection. (F) Columns constructed with mortar in the center of the stones.

stones. The mortar between the stones is also clearly determined, being the characteristic velocity of about 500 m/s (Fig. 22.30).

Some areas with velocities between 1500 and 2200 m/s indicate the damaged sectors in the column (Perez-Gracia et al., 2019). Damage affects mainly the most superficial sectors and also the contact between stones inside the column.

22.4.4.3 Final remarks

The assessment of these two Gothic basilicas was focused on the different problems previously observed in visual inspections. The study of soil was only applied in the case of one basilica. In this case, passive seismic pointed to a quite complex soil that could be cause of differential settlement of the structure due to partial subsidence of the ground, which can lead to cracks and even to the fall of construction materials.

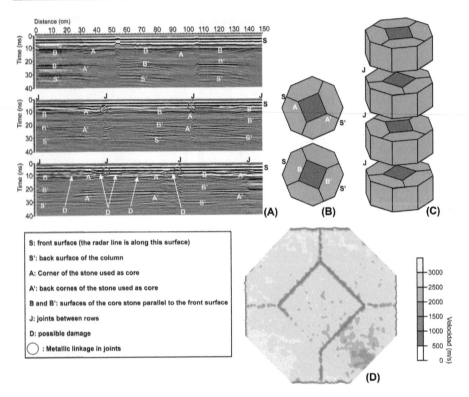

Figure 22.30 (A) Three B-scans obtained in three different columns. The third image corresponds to the most damaged column. (B) Position of the stones, with the targets associated to each B-scan anomaly. (C) Rotation of the rows in the same column. (D) Seismic tomography.

Source: Adapted from Perez-Gracia, V., Santos-Assunçao, S., Caselles, O., Clapes, J., Sossa, V. (2019). Combining ground penetrating radar and seismic surveys in the assessment of cultural heritage buildings: The study of roofs, columns, and ground of the gothic church Santa Maria del Mar, in Barcelona (Spain). *Structural Control and Health Monitoring. 26*(4), e2327, https://doi.org/10.1002/stc.2327.

In both cases, GPR was used to study the vaults and arches. These surveys could be carried out because it was possible to access the roof, from where the radar data was acquired. In both buildings, an interesting construction technique of the vaults was detected: the use of lightened filler between the roof pavement and the arches' ashlars. This lightened filler consisted of large hollow ceramic pots placed in the mortar. This technique allowed the construction of very slender columns separated from each other. Although the B-scans obtained in both basilicas are quite similar, the damage due to rainwater filtrations introduces significant changes in the radar data obtained in Santa Maria del Pi. The damaged zones appear in the radargrams associated with an important increase of the amplitude of the wave reflected on the pots and stones of the mortar.

Finally, in the study of columns of Santa Maria del Mar, the combination of GPR with seismic tomography demonstrates an interesting potential. The information provided by seismic tomography is complete: it is possible to define the size, the shape, and the position of the inner ashlars. Moreover, the wave velocity could indicate the quality of the materials and also the existence of damaged zones. However, data acquisition is time-consuming, and in many cases (as the case presented in this study), only a single section is obtained. The identification of the shape, size, and position of the ashlars is more difficult with radar data. However, combining a single section with a more complete study of the columns by means of GPR is a powerful solution to define the inner structures. In addition, GPR detects damages and also the existence of metallic targets used in past restorations.

22.5 Final remarks and future perspectives

The previous sections present a review of the GPR applications in cultural heritage assessment, combined with other technologies. Several case studies exemplify these combined applications in different structures. Nowadays, the use of combined methodologies is a successful approach to obtaining enough data about the structures, but it is also a process to save time during data acquisition. For example, the combination of seismic tomography and GPR data acquired in a common offset is a way to obtain the maximum information in the shortest time possible. Seismic tomography is a process that requires many measurement points and is time-consuming if it is necessary to analyze large structural members, such as columns. Hence, obtaining a single section in a column, combined with GPR data acquisition is the most advantageous way to assess those targets. In addition, the combination of both methodologies also provides supplementary information.

Despite the good results, there are some aspects that limit the application of this technology. The difficulty to reach high walls or columns forces the study to focus on the lower part of these structures in many cases. Even in the case of scaffolding use, data acquisition results are time-consuming, although the time of assembly of the scaffolding is not considered. In other cases, the difficulty is the access to narrow zones or to inspect ceilings when the access to the roof is not viable.

Future developments to solve these problems are focused on the development of robotic platforms, aerial or terrestrial, with different NDT sensors, as proposed for application in road inspection procedures (Wang et al., 2022) or wall-climbing robots with a built-in GPR, for reinforced concrete inspection (Garrido & Sattar, 2021). The adherence of the robot to the wall is achieved by means of neodymium magnets. The miniaturization of antennas (Howlader & Sattar, 2016; Howlader et al., 2016) is also a future development, to facilitate the transport of frequency equipment between 500 and 100 MHz by means of robots that can be attached to reinforced concrete walls. The combination of different sensors with GPR in the robotic platforms could also improve the assessment of structures, even artificial 3D vision cameras can be used (Roudari et al., 2019).

Autonomous robot platforms with GPR were also proposed for landmine detection (Bechtel et al., 2018). A discussion about the application of those technologies can be found in the work of Hoła and Schabowicz (2010), Perez-Gracia et al. (2022), and Solla et al. (2021). The use of GPR in drones has been also tested in experimental research. The application for soil moisture mapping (Wu et al., 2019), snow cover mapping (Vergnano et al., 2022), to map archaeological sites and indigenous settlements (Sonnemann et al., 2016), landmine detection (Šipoš et al., 2017), or in rescues (Chandra & Tanzi, 2018), has been performed to identify new fields of application. However, the importance of coordination, appropriate communication and exchange of information are issues that must be solved for a successful survey in buildings.

Another main challenge is the need to maximize the quality and accuracy of the results obtained and to minimize subjectivity in the interpretational process. Recently, deep learning techniques and their application to signal processing and automatic object detection became state of the art thanks to their detection speed and accuracy compared to conventional image processing techniques (Khudoyarov et al., 2020; Zhang et al., 2020). However, these advanced learning processing techniques have not yet been properly explored with GPR, mainly due to the limitations in data availability. To overcome this limitation, synthetic data is used to support training. Another accurate progress must be based on improving the detection of multiple objects and the management of complex scenarios (Travassos et al., 2021).

Finally, the survey results could be implemented in BIM. Several works developed the integration of GPR data into BIM. One example is the detection of rebar that could be accurately detected and introduced in building models (Xiang, Ou, et al., 2021; Xiang, Rashidi, et al., 2021) or the application in road assessment (D'Amico et al., 2022). In the case of cultural heritage and more complex structures, the integration of GPR data with the map of damage and orthophotos could be time-consuming but provides complete information of the building to make decisions about conservation (Solla et al., 2020). However, this integration of different data in building models could be used not only for the structure conservation management but also for cultural heritage diffusion mapping historical buildings (Tucci et al., 2019). Therefore the proposal of a huge cultural heritage site represented in a BIM database could be a vast documentation for rehabilitation, diffusion, and preventive conservation (Mahmoud, 2018).

Acknowledgments

This project has received funding from the Xunta de Galicia through the project ENDITí (ED431F 2021/08). Mercedes Solla acknowledges the grant RYC2019−026604−I funded by MCIN/AEI/10.13039/501100011033 and by "ESF Investing in your future." Work produced with the support of a 2022 Leonardo Grant for Researchers and Cultural Creators, BBVA Foundation. The BBVA Foundation takes no responsibility for the opinions, statements, and contents of this project, which are entirely the responsibility of its authors.

References

Agliata, R., Bogaard, T. A., Greco, R., Mollo, L., Slob, E. C., & Steele-Dunne, S. C. (2018). Noninvasive estimation of moisture content in tuff bricks by GPR. *Construction and Building Materials*, *160*, 698−706. Available from https://doi.org/10.1016/j.conbuildmat.2017.11.103.

Akin, M. K., Kramer, S. L., & Topal, T. (2011). Empirical correlations of shear wave velocity (Vs) and penetration resistance (SPT-N) for different soils in an earthquake-prone area (Erbaa-Turkey). *Engineering Geology*, *119*(1−2), 1−17. Available from https://doi.org/10.1016/j.enggeo.2011.01.007.

Perspectiv. (2022). *Association of Historic Theatres in Europe*. http://www.perspectiv-online.org/pages/en/european-route.php?lang = EN.

Bechtel, T., Pochanin, G., Truskavetsky, S., Dimitri, M., Ruban, V., Orlenko, O., Byndych, T., Sherstyuk, A., Viatkin, K., Crawford, F., Falorni, P. Bulletti. A. Capineri, L. (2018). Terrain analysis in Eastern Ukraine and the design of a robotic platform carrying GPR sensors for landmine detection. *17th International Conference on Ground Penetrating Radar*, GPR 2018, Institute of Electrical and Electronics Engineers Inc. United States. http://ieeexplore.ieee.org/xpl/mostRecentIssue.jsp?punumber = 8410205.

Biscarini, C., Catapano, I., Cavalagli, N., Ludeno, G., Pepe, F. A., & Ubertini, F. (2020). UAV photogrammetry, infrared thermography and GPR for enhancing structural and material degradation evaluation of the Roman masonry bridge of Ponte Lucano in Italy. *NDT and E International*, *115*. Available from https://doi.org/10.1016/j.ndteint.2020.102287, https://www.journals.elsevier.com/ndt-and-e-international.

Caldeira, B., Oliveira, R. J., Teixidó, T., Borges, J. F., Henriques, R., Carneiro, A., & Peña, J. A. (2019). Studying the construction of floor mosaics in the Roman Villa of Pisões (Portugal) using noninvasive methods: High-resolution 3D GPR and photogrammetry. *Remote Sensing*, *11*(16). Available from https://doi.org/10.3390/rs11161882, https://res.mdpi.com/d_attachment/remotesensing/remotesensing-11-01882/article_deploy/remotesensing-11-01882.pdf.

Calia, A., Lettieri, M., Leucci, G., Matera, L., Persico, R., & Sileo, M. (2013). The mosaic of the crypt of St. Nicholas in Bari (Italy): Integrated GPR and laboratory diagnostic study. *Journal of Archaeological Science*, *40*(12), 4162−4169. Available from https://doi.org/10.1016/j.jas.2013.06.005, http://www.elsevier.com/inca/publications/store/6/2/2/8/5/4/index.htt.

Capozzoli, L., De Martino, G., Polemio, M., & Rizzo, E. (2020). Geophysical Techniques for Monitoring Settlement Phenomena Occurring in Reinforced Concrete Buildings. *Surveys in Geophysics*, *41*(3), 575−604. Available from https://doi.org/10.1007/s10712-019-09554-8.

Caselles, O., Canas, J.A., Clapés, J., & Osorio, R. (2007). *Estudio constructivo-estructural de la iglesia de Santa María del Pi*.

Caselles, O., Martínez, G., Clapés, J., Roca, P., & Pérez-Gracia, M. D. L. V. (2015). Application of particle motion technique to structural modal identification of heritage buildings. *International Journal of Architectural Heritage*, *9*(3), 310−323. Available from https://doi.org/10.1080/15583058.2013.784824, http://www.tandf.co.uk/journals/titles/15583058.asp.

Casquero, A. R. (2015). Caracterización estructural de los rellenos situados en el trasdós de bóvedas de edificios históricos. *Doctoral dissertation, (in spanish)*, *348*.

Cataldo, R., De Donno, A., De Nunzio, G., Leucci, G., Nuzzo, L., & Siviero, S. (2005). Integrated methods for analysis of deterioration of cultural heritage: The Crypt of \Cattedrale di Otranto\. *Journal of Cultural Heritage*, *6*(1), 29−38. Available from https://doi.org/10.1016/j.culher.2004.05.004, http://www.elsevier.com.

Catapano, I., Ludeno, G., Soldovieri, F., Tosti, F., & Padeletti, G. (2018). Structural assessment via ground penetrating radar at the Consoli Palace of Gubbio (Italy. *Remote Sensing*, *10*(2), 45. Available from https://doi.org/10.3390/rs10010045.

Chandra, M., & Tanzi, T. J. (2018). Drone-borne GPR design: Propagation issues. *Comptes Rendus Physique.*, *19*(1−2), 72−84. Available from https://doi.org/10.1016/j.crhy.2018.01.002, https://comptes-rendus.academie-sciences.fr/physique.

Cintra, D. C. B., Manhaes, P. M. B., Fernandes, F. M. C. P., Roehl, D. M., Araruna Júnior, J. T., & Sánchez Filho, E. S. (2020). Evaluation of the GPR (1.2 GHz) technique in the characterization of masonry shells of the Theatro Municipal do Rio de Janeiro. *Revista IBRACON de Estruturas e Materiais*, *13*(2), 274−297. Available from https://doi.org/10.1590/s1983-41952020000200006.

Cozzolino, M., Gentile, V., Mauriello, P., & Peditrou, A. (2020). Non-destructive techniques for building evaluation in urban areas: The case study of the Redesigning Project of Eleftheria Square (Nicosia, Cyprus. *Applied Sciences*, *10*(12), 4296. Available from https://doi.org/10.3390/app10124296, https://www.mdpi.com/2076-3417/10/12/4296.

Cruz, H., Candeias, P., Machado J.S., Campos Costa, A., Fontul, S. (2017). *Avaliação da segurança de pavimento do Salão Nobre do TNSC − uma abordagem multidisciplinar* (in Portuguese). CREPAT 2017 - Congresso da Reabilitação do Património, 29−30 June Aveiro, Portugal.

Dabas, M., Camerlynck, C., & Camps, P. F. I. (2000). Simultaneous use of electrostatic quadrupole and GPR in urban context: Investigation of the basement of the Cathedral of Girona (Catalunya, Spain. *Geophysics*, *65*(2), 526−532. Available from https://doi.org/10.1190/1.1444747.

D'Amico, F., Bianchini Ciampoli, L., Di Benedetto, A., Bertolini, L., & Napolitano, A. (2022). integrating non-destructive surveys into a preliminary BIM-oriented digital model for possible future application in road pavements management. *Infrastructures*, *7*(1), 10. Available from https://doi.org/10.3390/infrastructures7010010.

Danese, M., Sileo, M., & Masini, N. (2018). Geophysical methods and spatial information for the analysis of decaying frescoes. *Surveys in Geophysics*, *39*(6), 1149−1166. Available from https://doi.org/10.1007/s10712-018-9484-0, http://www.wkap.nl/journalhome.htm/0169-3298.

De Donno, G., Di Giambattista, L., & Orlando, L. (2017). High-resolution investigation of masonry samples through GPR and electrical resistivity tomography. *Construction and Building Materials*, *154*, 1234−1249. Available from https://doi.org/10.1016/j.conbuildmat.2017.06.112.

Diz-Mellado, E., Mascort-Albea, E. J., Romero-Hernández, R., Galán-Marín, C., Rivera-Gómez, C., Ruiz-Jaramillo, J., & Jaramillo-Morilla, A. (2021). Non-destructive testing and finite element method integrated procedure for heritage diagnosis: The Seville Cathedral case study. *Journal of Building Engineering*, *37*, 102134. Available from https://doi.org/10.1016/j.jobe.2020.102134.

Evangelista, L., de Silva, F., d'Onofrio, A., Di Fiore, V., Silvestri, F., di Santolo, A. S., Cavuoto, G., Punzo, M., & Tarallo, D. (2017). Application of ERT and GPR geophysical testing to the subsoil characterization of cultural heritage sites in Napoli (Italy). *Measurement*, *104*, 326−335. Available from https://doi.org/10.1016/j.measurement.2016.07.042.

Fontul, S., Solla, M., Cruz, H., Machado, J. S., & Pajewski, L. (2018). Ground penetrating radar investigations in the Noble Hall of São Carlos Theater in Lisbon, Portugal. *Surveys in Geophysics*, *39*(6), 1125−1147. Available from https://doi.org/10.1007/s10712-018-9477-z, http://www.wkap.nl/journalhome.htm/0169-3298.

Garrido, G. G., & Sattar, T. P. (2021). An autonomous wall-climbing robot for inspection of reinforced concrete structures: SIRCAUR. *Journal of Artificial Intelligence and Technology*, *1*(3), 188−196.

Garrido, I., Solla, M., Lagüela, S., & Fernández, N. (2020). IRT and GPR techniques for moisture detection and characterisation in buildings. *Sensors*, *20*(22), 6421. Available from https://doi.org/10.3390/s20226421.

Garrido, I., Solla, M., Lagüela, S., Rasol, M., & Rasulo, Ao (2022). Review of infrared thermography and ground-penetrating radar applications for building assessment. *Advances in Civil Engineering*, *2022*, 1−20. Available from https://doi.org/10.1155/2022/5229911.

González-Drigo, R., Pérez-Gracia, V., Di Capua, D., & Pujades, L. G. (2008). GPR survey applied to Modernista buildings in Barcelona: The cultural heritage of the College of Industrial Engineering. *Journal of Cultural Heritage*, *9*(2), 196−202. Available from https://doi.org/10.1016/j.culher.2007.10.006, http://www.elsevier.com.

Guadagnuolo, M., Faella, G., Donadio, A., & Ferri, L. (2014). Integrated evaluation of the Church of S. Nicola di Mira: Conservation versus safety. *NDT & E International*, *68*, 53−65. Available from https://doi.org/10.1016/j.ndteint.2014.08.002.

Howlader, M. O. F., & Sattar, T. P. (2016). Miniaturization of dipole antenna for low frequency ground penetrating radar. *Progress In Electromagnetics Research C*, *61*, 161−170. Available from https://doi.org/10.2528/PIERC15103004, http://www.jpier.org/PIERC/pierc61/17.15103004.pdf.

Howlader, M. O. F., Sattar, T. P., & Dudley, S. (2016). Development of a wall climbing robotic ground penetrating radar system for inspection of vertical concrete structures. *International Journal of Mechanical and Mechatronics Engineering*, *10*(8), 1382−1388.

Hoła, J., & Schabowicz, K. (2010). State-of-the-art non-destructive methods for diagnostic testing of building structures - anticipated development trends. *Archives of Civil and Mechanical Engineering*, *10*(3), 5−18. Available from https://doi.org/10.1016/s1644-9665(12)60133-2, http://www.acme.pwr.wroc.pl/repository/282/online.pdf.

Işık, N., Halifeoğlu, F. M., & İpek, S. (2022). Detecting the ground-dependent structural damages in a historic mosque by employing GPR. *Journal of Applied Geophysics*, *199*. Available from https://doi.org/10.1016/j.jappgeo.2022.104606, http://www.elsevier.com/inca/publications/store/5/0/3/3/3/3/.

Khudoyarov, S., Kim, N., & Lee, J. J. (2020). Three-dimensional convolutional neural network−based underground object classification using 3D GPR data. *Structural Health Monitoring*, *19*(6), 1884−1893. Available from https://doi.org/10.1177/1475921720902700.

Leucci, G., De Giorgi, L., Ditaranto, I., Miccoli, I., & Scardozzi, G. (2021). Ground-penetrating radar prospections in Lecce Cathedral: New data about the crypt and the structures under the church. *Remote Sensing*, *13*(9), 1692. Available from https://doi.org/10.3390/rs13091692.

Machado, J. S., Riggio, M., & D'Ayala, D. (2015). Assessment of structural timber members by non- and semi-destructive methods. *Construction and Building Materials*, *101*(Part 2), 1155−1278. Available from https://doi.org/10.1016/j.conbuildmat.2015.11.032, http://doi.org/10.1016/j.conbuildmat.2015.11.032.

Mahmoud, Y. S. (2018). Revitalizing cultural heritage through building information modeling (BIM). *International Journal of Multidisciplinary Studies in Architecture and Cultural Heritage*, *1*(2), 34−52.

Martínez-Garrido, M. I., Fort, R., Gómez-Heras, M., Valles-Iriso, J., & Varas-Muriel, M. J. (2018). A comprehensive study for moisture control in cultural heritage using non-destructive techniques. *Journal of Applied Geophysics*, *155*, 36−52. Available from https://doi.org/10.1016/j.jappgeo.2018.03.008, http://www.elsevier.com/inca/publications/store/5/0/3/3/3/3/.

Novo, A., Lorenzo, H., Rial, F. I., & Solla, M. (2010). Three-dimensional ground-penetrating radar strategies over an indoor archaeological site: Convent of Santo Domingo (Lugo, Spain). *Archaeological Prospection*, *17*(4), 213−222. Available from https://doi.org/10.1002/arp.386.

Núñez-Nieto, X., Solla, M., Novo, A., & Lorenzo, H. (2014). Three-dimensional ground-penetrating radar methodologies for the characterization and volumetric reconstruction of underground tunneling. *Construction and Building Materials*, *71*, 551−560. Available from https://doi.org/10.1016/j.conbuildmat.2014.08.083.

Perez-Gracia, V., Santos-Assunçao, S., Caselles, O., Clapes, J., & Sossa, V. (2019). Combining ground penetrating radar and seismic surveys in the assessment of cultural heritage buildings: The study of roofs, columns, and ground of the gothic church Santa Maria del Mar, in Barcelona (Spain. *Structural Control and Health Monitoring*, *26*(4), e2327. Available from https://doi.org/10.1002/stc.2327.

Perez-Gracia, V., Solla, M., & Fontul, S. (2022). Analysis of the GPR signal for moisture detection: application to heritage buildings. *International Journal of Architectural Heritage*. Available from https://doi.org/10.1080/15583058.2022.2139652, http://www.tandf.co.uk/journals/titles/15583058.asp.

Pérez-Gracia, V., Di Capua, D., Caselles, O., Rial, F., Lorenzo, H., González-Drigo, R., & Armesto, J. (2011). Characterization of a Romanesque bridge in Galicia (Spain. *International Journal of Architectural Heritage*, *5*(3), 251−263. Available from https://doi.org/10.1080/15583050903560249, http://www.tandf.co.uk/journals/titles/15583058.asp.

Pérez-Gracia, V., Caselles, J. O., Clapés, J., Martinez, G., & Osorio, R. (2013). Non-destructive analysis in cultural heritage buildings: Evaluating the Mallorca cathedral supporting structures. *NDT & E International*, *59*, 40−47. Available from https://doi.org/10.1016/j.ndteint.2013.04.014.

Pérez-Gracia, V., Caselles, J. O., Clapes, J., Osorio, R., Martínez, G., & Canas, J. A. (2009). Integrated near-surface geophysical survey of the Cathedral of Mallorca. *Journal of Archaeological Science*, *36*(7), 1289−1299. Available from https://doi.org/10.1016/j.jas.2009.03.001, http://www.elsevier.com/inca/publications/store/6/2/2/8/5/4/index.htt.

Roudari, S.S., Okore-Hanson, T., Hamoush, S., Yi, S., & Megri, A. (2019). *Robotic non-destructive evaluation of RC structures using 3D vision camera, IE, and GPR*. American Society for non-destructive Testing. ASNT Research Symposium Annual Conference, Las Vegas.

Rucka, M., Wojtczak, E., & Zielińska, M. (2020). Interpolation methods in GPR tomographic imaging of linear and volume anomalies for cultural heritage diagnostics. *Measurement*, *154*, 107494. Available from https://doi.org/10.1016/j.measurement.2020.107494.

Santos-Assunçao, S., Perez-Gracia, V., Caselles, O., Clapes, J., & Salinas, V. (2014). Assessment of complex masonry structures with GPR compared to other non-destructive testing studies. *Remote Sensing*, *6*(9), 8220−8237. Available from https://doi.org/10.3390/rs6098220, http://www.mdpi.com/2072-4292/6/9/8220/pdf.

Santos-Assuncao, S., Perez-Gracia, V., Salinas, V., Caselles, O., Gonzalez-Drigo, R., Pujades, L. G., & Lantada, N. (2016). GPR backscattering intensity analysis applied to detect paleochannels and infilled streams for seismic nanozonation in urban environments. *IEEE Journal of Selected Topics in Applied Earth Observations and Remote Sensing*, *9*(1), 167−177. Available from https://doi.org/10.1109/JSTARS.2015.2466235, http://ieeexplore.ieee.org/xpl/RecentIssue.jsp?punumber = 4609443.

Šipoš, D., Planinšič, P., & Gleich, D. (2017). Slovenia on drone ground penetrating radar for landmine detection. *2017 1st International Conference on Landmine: Detection, Clearance and Legislations*, LDCL 2017. Institute of Electrical and Electronics Engineers Inc. https://doi.org/10.1109/LDCL.2017.7976931, 9781509058235.

Solla, M., Caamaño, J. C., Riveiro, B., & Arias, P. (2012). A novel methodology for the structural assessment of stone arches based on geometric data by integration of photogrammetry and ground-penetrating radar. *Engineering Structures, 35*, 296−306. Available from https://doi.org/10.1016/j.engstruct.2011.11.004.

Solla, M., Gonçalves, L. M. S., Gonçalves, G., Francisco, C., Puente, I., Providência, P., Gaspar, F., & Rodrigues, H. (2020). A building information modeling approach to integrate geomatic data for the documentation and preservation of cultural heritage. *Remote Sensing, 12*(24), 4028. Available from https://doi.org/10.3390/rs12244028.

Solla, M., Lagüela, S., Fernández, N., & Garrido, I. (2019). Assessing rebar corrosión through the combination of non-destructive GPR and IRT methodologies. *Remote Sensing, 11.*

Solla, M., López-Leira, J. M., Alonso-Díaz, A., & Rodríguez, J. L. (2022). Ground-penetrating radar and geotechnical analyses to investigate the foundation settlements of an Indiana House in NW Spain. *International Journal of Architectural Heritage.* Available from https://doi.org/10.1080/15583058.2022.2138631, http://www.tandf.co.uk/journals/titles/15583058.asp.

Solla, M., Pérez-Gracia, V., & Fontul, Sa (2021). A review of GPR application on transport infrastructures: Troubleshooting and best practices. *Remote Sensing, 13*(4), 672. Available from https://doi.org/10.3390/rs13040672.

Sonnemann, T. F., Hung, J. U., & Hofman, C. L. (2016). Mapping indigenous settlement topography in the caribbean using drones. *Remote Sensing, 8*(10). Available from https://doi.org/10.3390/rs8100791, http://www.mdpi.com/journal/remotesensing.

Sossa, V., Pérez-Gracia, V., González-Drigo, R., & Mezgeen, A. (2019). Lab non-destructive test to analyze the effect of corrosion on ground penetrating radar scans. *Remote Sensing, 11*(23).

Tannert, T., Dietsch, P., Köhler, J., & Verlag, S. (2010). *Drill resistance*. COST E 55 report (pp. 72−74). Unpublished content.

TNSC. (2022). *Teatro Nacional de São Carlos*. http://tnsc.pt/.

Travassos, X. L., Avila, S. L., & Ida, N. (2021). Artificial neural networks and machine learning techniques applied to ground penetrating radar: A review. *Applied Computing and Informatics, 17*(2), 296−308.

Tucci, G., Betti, M., Conti, A., Corongiu, M., Fiorini, L., Matta, C., Kovačević, C., Borri, C., & Hollberg, C. (2019). BIM for museums: an integrated approach from the building to the collections. *International Archives of the Photogrammetry, Remote Sensing & Spatial Information Sciences, XLII-2/W11*, 1089−1096. Available from https://doi.org/10.5194/isprs-archives-XLII-2-W11-1089-2019.

Vergnano, A., Franco, D., & Godio, A. (2022). Drone-borne ground-penetrating radar for snow cover mapping. *Remote Sensing, 14*(7), 1763. Available from https://doi.org/10.3390/rs14071763.

Villarino, A., Riveiro, B., Gonzalez-Aguilera, D., & Sánchez-Aparicio, L. J. (2014). The integration of geotechnologies in the evaluation of a wine cellar structure through the finite element method. *Remote Sensing, 6*(11), 11107−11126. Available from https://doi.org/10.3390/rs61111107, http://www.mdpi.com/2072-4292/6/11/11107/pdf.

Wang, S., Sui, X., Leng, Z., Jiang, J., & Lu, G. (2022). Asphalt pavement density measurement using non-destructive testing methods: current practices, challenges, and future vision. *Construction and Building Materials*, *344*, 128154. Available from https://doi. org/10.1016/j.conbuildmat.2022.128154.

Wu, K., Rodriguez, G. A., Zajc, M., Jacquemin, E., Clément, M., De Coster, A., & Lambot, S. (2019). A new drone-borne GPR for soil moisture mapping. *Remote Sensing of Environment*, *235*, 111456. Available from https://doi.org/10.1016/j.rse.2019.111456.

Xiang, Z., Ou, G., & Rashidi, A. (2021). Automated translation of rebar information from GPR data into as-built BIM: A deep learning-based approach. Computer vision and pattern recognition. *ASCE International Conference on Computing in Civil Engineering*. Unpublished content. https://doi.org/10.48550/arXiv.2110.15448, 2021.

Xiang, Z., Rashidi, A., & Ou, G. (2021). Automated framework to translate rebar spatial information from GPR into BIM. *Journal of Construction Engineering and Management*, *147* (10). Available from https://doi.org/10.1061/(ASCE)CO.1943-7862.0002141.

Yalçıner, C. Ç., Büyüksaraç, A., & Kurban, Y. C. (2019). non-destructive damage analysis in Kariye (Chora) Museum as a cultural heritage building. *Journal of Applied Geophysics, 171*.

Zhang, J., Yang, X., Li, W., Zhang, S., & Jia, Y. (2020). Automatic detection of moisture damages in asphalt pavements from GPR data with deep CNN and IRS method. *Automation in Construction*, *113*, 103119. Available from https://doi.org/10.1016/j. autcon.2020.103119.

Zhou, F., Chen, Z., Liu, H., Cui, J., Spencer, B., & Fang, G. (2018). Simultaneous estimation of rebar diameter and cover thickness by a GPR-EMI dual sensor. *Sensors, 18*(9), 2969. Available from https://doi.org/10.3390/s18092969.

Index

Note: Page numbers followed by "*f*" and "*t*" refer to figures and tables, respectively.

Printed and bound by CPI Group (UK) Ltd, Croydon, CR0 4YY

03/10/2024

01040420-0010